de Gruyter Expositions in Mathematics 1

The Analytical and Topological Theory of Semigroups

Trends and Developments

Editors

Karl Heinrich Hofmann
Jimmie D. Lawson
John S. Pym

Walter de Gruyter · Berlin · New York 1990

Editors

Karl Heinrich Hofmann
Fachbereich Mathematik
Technische Hochschule Darmstadt
Schlossgartenstr. 7
D-6100 Darmstadt, FRG

Jimmie D. Lawson
Department of Mathematics
Louisiana State University
Baton Rouge, LA 70803
USA

John S. Pym
Department of Pure Mathematics
University of Sheffield
Sheffield, S10 2TN
England

1980 Mathematics Subject Classification (1985 Revision): Primary: 22-02; 22A15; 22A20; 22A25; 22A26; 22E05; 22E15; 22E30; 22E45; 22E60; 43-02; 43A01; 43A07; 43A35; 43A60; 43A65. Secondary: 06B30; 11Axx; 14L99; 17B05; 20M20; 49E15; 60-02.

∞ Printed on acid-free paper which falls within the guidelines of the ANSI to ensure permanence and durability.

Library of Congress Cataloging-in-Publication Data

The Analytical and topological theory of semigroups : trends and developments / editors, Karl Heinrich Hofmann, Jimmie D. Lawson, John S. Pym
p. cm. – (De Gruyter expositions in mathematics ; 1)
Based on a conference held Jan. 30–Feb. 4 1989, hosted by the Mathematisches Forschungsinstitut Oberwolfach.
Includes bibliographical references.
ISBN 3-11-012489-0 (Berlin : cloth : alk. paper), – ISBN 0-89925-698-8 (N.Y. : cloth : alk. paper)
1. Topological semigroups - Congresses. I. Hofmann, Karl Heinrich. II. Lawson, Jimmie D. III. Pym, J.S. (John Sydney). 1938- . IV. Series.
QA387.A46 1990
512'.2–dc20 90-3845

Deutsche Bibliothek Cataloging-in-Publication Data

The **analytical and topological theory of semigroups** : trends and developments / ed.: Karl Heinrich Hofmann ... – Berlin ; New York : de Gruyter, 1990
(De Gruyter expositions in mathematics ; 1)
ISBN 3-11-012489-0
NE: Hofmann, Karl Heinrich [Hrsg.]; GT

Typeset by the authors using TEX. TEX conversion: Danny Lee Lewis, Berlin. Printing: Ratzlow-Druck, Berlin. Binding: Dieter Mikolai, Berlin. Cover design: Thomas Bonnie, Hamburg.

Preface

The best mathematics is the most mixed-up mathematics, those disciplines in which analysis, algebra and topology all play a vital role.
A. D. Wallace

ALEXANDER DONIPHAN WALLACE, one of the nestors of the topological theory of semigroups, wrote these lines in a letter on topological semigroups:[1] "It is a most difficult subject", he said. "It is unintuitive in the sense that its theorems are *not* generalizations of theorems about groups. The standard methods will not immediately work." Two decades of subsequent development have added to WALLACE's remarks a weight far beyond what he might have imagined, and the evolution of mathematics as a whole has naturally caused the emphasis in semigroup theory to shift. In the fifties, when WALLACE was a dominant figure of the early activity in topological semigroups, the accent was on algebraic topology and the topology of continua[2]. In the early sixties, MOSTERT brought the theory of compact transformation groups to bear on the subject[3].

In the present collection, however, the reader will not find any article devoted to the general theory of topological semigroups. To our knowledge, the last survey which contained a substantial amount about this field was published in the mid-seventies[4]. Interest has moved, both to theories with more structure than joint continuity of multiplication, and to others with less.

Theories with enriched structure represent the more recent development. At the 1981 Conference on the Algebraic and Topological Theory of Semigroups[5] HOFMANN and LAWSON reported on a theory which had just attained an early stage

---

1 Semigroup Forum **7** (1974), pp. 20 ff.

2 Wallace, A. D., *On the structure of topological semigroups*, Bull. Amer. Math. Soc. **61** (1955), 95–112

3 Hofmann, K. H., and P. S. Mostert, *Applications of transformation groups to problems in topological semigroups*, in: Proc. Conf. Transformation Groups, P. S. Mostert, Ed., Springer-Verlag Berlin etc. 1968

4 Hofmann, K. H., *Topological semigroups: History, theory, applications*, Jahresber. Dt. Math. Ver. **78** (1976), 9–59

5 Hofmann, K. H., H. Jürgensen, and H. J. Weinert, Eds., "Recent Developments in the Algebraic, Analytical, and Topological Theory of Semigroups", Springer Lecture Notes in Math. **998** (1983)

of maturity: the Lie theory of subsemigroups of Lie groups. This line of analysis in semigroup theory emerged at the right time to fit into a general trend of high activity in geometry - meaning primarily differential geometry and algebraic geometry. The current advanced state of this theory is documented in this book[6]. The sophisticated blend of algebraic geometry and semigroups was pioneered by PÚTCHA and RENNER. The cross-connections between the Lie theory of semigroups and other fields are just emerging: examples are systems and geometric control[7], and the unitary representation theory of Lie groups[8].

Developments in the direction of greater generality have often been influenced by other branches of analysis. The inspiration for the investigation of semigroups with only separately continuous multiplications came from DE LEEUW and GLICKSBERG[9]. A little later, J. L. TAYLOR provided further motivation for this study by showing that the spectra of measure algebras on locally compact groups – crucial objects in harmonic analysis – were natural compact semitopological semigroups[10]. The close relationship with analysis at this time is documented in WILLIAMSON's survey article of 1967[11]. Although important algebraic aspects of the basic structure theory of compact topological semigroups carry over to the semitopological case, there are wide differences – compare the books of HOFMANN and MOSTERT and of RUPPERT[12]. A dramatic example was provided by the construction by BROWN and MORAN (following vital progress by WEST) of singly generated compact separately continuous semigroups with extremely complex subsemigroups of idempotents. Their construction relied heavily on harmonic analysis; only recently a derivation has been found in the spirit of pure topological algebra[13]. A vital tool in developing a theory of compact semitopological semigroups largely free of functional analytic methods has been the Ellis-Lawson Theorem which usually provides a substantial set of points at which joint continuity holds[12,13].

From an abstract point of view, it appears natural to try to progress from multiplications continuous in each variable separately to multiplications continuous in just one. It was shown by RUPPERT in the early seventies that, for compact right topological semigroups, the structure theory of the minimal ideal still holds; but

6 This volume: Chapters Hofmann, Hilgert, Lawson, Kupka, Renner

7 This volume: Chapter Kupka

8 This volume: Chapter Hilgert

9 de Leeuw, K. and I. Glicksberg, *Applications of almost periodic compactifications*, Acta Math. **105** (1961), 63–97

10 Taylor, J. L., "Measure algebras", CBMS Regional Conference Series **16**, Amer. Math. Soc., Providence, 1972

11 Williamson, J. H., *Harmonic analysis on semigroups*, J. London Math. Soc. **41** (1967), 1–41

12 Hofmann, K. H., and P. S. Mostert, "Elements of Compact Semigroups", Merrill, Ohio, 1966; Ruppert, W., "Compact Semitopological Semigroups: an Intrinsic Theory", Springer Lecture Notes in Mathematics **1079**, 1984

13 This volume: Chapter Ruppert

there is a much older body of theory, for such semigroups had already arisen in connection with topological dynamics[14]. The emphasis on compactness in the theory is justified by the fact that topological groups and semigroups have a range of significant compactifications which are right topological semigroups[15]. There are applications to fields other than transformation groups. Strikingly successful has been HINDMAN's use of the semigroup $\beta\mathbb{N}$ in combinatorial number theory. The most elegant example here is the recent proof of van der Waerden's Theorem on the existence of arithmetic progressions in partitions of the integers, in which the original detailed and complicated combinatorial methods have been replaced by general and direct structural arguments[16].

The close relationship with functional and harmonic analysis continues. Several other contributions to this volume provide further evidence[17].

From January 30 through February 4, 1989, the Mathematisches Forschungsinstitut Oberwolfach hosted a conference on the analytical and topological theory of semigroups. This book is one of its many fruits. The organizers of the conference invited various authors to review the status in their special field. The presentations, discussions and interactions at the conference resulted in a final updating and polishing of the surveys prepared for the occasion. The resulting articles are collected here. We have deliberately aimed for something different from a volume of proceedings. Indeed many more specialized contributions communicated at the conference are being published elsewhere. We hoped that we could inspire sixteen authors to bring forth contributions which would make up a book, as uniform in its goal and purpose as one could conceivably expect, a source of information and reference on the analytical and topological theory of semigroups. The authors have responded admirably to our plea and we now have cause to express our thanks to all of them.

In limiting the coverage of a mathematical area one has to draw certain lines. Most are of a purely organizational nature. Semigroups of operators were not the subject of the conference and they are not represented in this book. A recent source which sprang from another conference in Oberwolfach makes up for their omission here[18].

One other thing we observe with considerable regret. It is not longer possible, as it still was in 1981, to unite mathematicians working in the algebraic theory of semigroups on the one hand with those working in the analytical and topological theory on the other. Both fields have become too vast to be covered at one

---

14 Ellis, R., "Lectures on Topological Dynamics", Benjamin, New York, 1969

15 Berglund, J. F., H. D. Junghenn and P. Milnes, "Analysis on Semigroups. Function Spaces. Compactifications. Representations", Wiley, New York, 1989

16 This volume: Chapter Hindman

17 This volume: Chapters Baker, Berg, Heyer, Lau

18 Special Issue on Semigroups and Differential Operators, Davies, E. B., J. A. Goldstein, and R. Nagel, Eds., Semigroup Forum **38** (1989) 127–266

conference in Oberwolfach. At Wallace's time topological and algebraic semigroup theory were still dwelling comfortably in one house. Perhaps we pay the price today for the enormous development of semigroup theory that Wallace has predicted. He had foreseen the forces that would accelerate its development. "Semigroups appear more naturally in a physical universe than in a geometric one ... [They] might be regarded as exemplars of irreversible actions... The great majority of phenomena are irreversible, ergo one must have semigroups[1]."

There is no question in my mind but that topological semigroups must, by sheer necessity, play a dominant role in describing reality.

A. D. Wallace

Acknowledgements

First and foremost, editors and authors thank the Mathematisches Forschungsinstitut Oberwolfach for its logistical and financial support of the meeting and for the hospitality the participants experienced during the conference in the unimitable fashion due to this institution alone.

The authors of this volume have to thank numerous funding agencies and institutions:

Christian Berg's work was completed while the author was visiting URA 750, Global Analysis at the University of Nancy 1, France, invited by the Centre National de la Recherche Scientifique.

Neil Hindman gratefully acknowledges support received from the National Science Foundation (USA) via grant DMS-8520873.

A portion of Karl H. Hofmann's research was done while he held an Academy Fellowship of the Volkswagen Foundation for which he is grateful. Most of this survey was written while he enjoyed the renewed hospitality of Tulane University.

Anthony To-Ming Lau's research is supported by an NSERC grant.

Jimmie D. Lawson gratefully acknowledges the support from the National Science Foundation during the period that this research was done.

Michael W. Mislove's work was partially supported by the Office of Naval Research and by the Louisiana State Board of Regents.

Wolfgang A. F. Ruppert gratefully acknowledges support received from the Alexander von Humboldt-Stiftung during part of the period while he was working on the subject covered by his survey.

Many authors cooperated in preparing a source file in TEX. In typesetting the contributions of authors without access to TEX the editors had the valuable assistance of Ms. Meredith Mickel at Tulane university and Ms. Edith Seitz at the Technische Hochschule Darmstadt; surely many unnamed TEXnologists helped individual authors with their TEXscript. The final uniform format of the book was created at the Technische Hochschule Darmstadt. In this work the editors had the indefatigable active support of W. A. F. Ruppert who redrew most of the diagrams and figures in the text and lend them his expertise with TEX. We especially thank him for his participation in the editing of this work.

May 1990

Karl H. Hofmann, Darmstadt
Jimmie D. Lawson, Baton Rouge
John S. Pym, Sheffield

Table of Contents

Part I. Lie theory and algebraic geometry

Part II. The compact case

Part III. Functional analysis on semigroups

Part IV. Applications: Systems theory, number theory, probability, topology

Part I

Lie theory and algebraic geometry

Lie groups and semigroups

Karl H. Hofmann

Introduction

The contours of a *Lie theory of semigroups* have become clear and distinct in the last years. It deals, in essence, with the theory of subsemigroups of Lie groups – more specifically, it deals with the question of how semigroups get there in the first place, with their applications, and with their structure theory. The question of embedding semigroups with enough topological or differentiable structure into Lie groups is treated by LAWSON [13]; HILGERT looks at applications in analysis [4], and we shall be concerned with the status of the structure theory of subsemigroups of Lie groups.

It would be preposterous to speak about *all* subsemigroups of Lie groups. One does not consider *all* subgroups of Lie groups in Lie group theory either. Rather one concentrates on subgroups which have a Lie algebra and are determined by it. In the Lie group $\mathbb{R}$ this leaves not many subgroups; in $\mathbb{R}^n$ it covers all vector subspaces. In the case of subsemigroups of Lie groups the situation is similar. We shall focus our attention on subsemigroups which are determined by their one-parameter subsemigroups. We will allow a little leeway; the specifics remain to be seen. At any rate, in $\mathbb{R}$, the semigroups we find in this way are the two half-lines, and in $\mathbb{R}^n$ we have to take all convex closed cones into consideration. These provide a good preliminary model of our objects. However, certain other types of subsemigroups of Lie groups are quite relevant even though they have no analog in group theory. In a connected Lie group, there are no proper open subgroups, but there are plenty of interesting open subsemigroups, notably those containing the identity in their closure. In $\mathbb{R}^n$ this includes all open half-space semigroups (and their intersections, the open convex cones). A theory of subsemigroups of Lie groups must also deal with open subsemigroups.

The basic feature of Lie group theory is that Lie groups and their subgroups are almost completely described by their infinitesimal structure at the origin, and this infinitesimal structure is encoded into the Lie algebra. Thus a large portion of Lie group theory is Lie algebra theory. Local Lie group theory is a historical episode and a pedagogical device providing preliminary information prior to the more sophisticated global theory. Every local Lie group is the identity neighborhood of a global one. (We stick with finite dimensional Lie groups.) Hence Lie group theory has actually two parts: The infinitesimal theory and the global theory. By contrast, the Lie theory of semigroups has to deal with a local theory for good mathematical reasons; there do exist local semigroups in Lie groups which cannot be identity neighborhoods of a subsemigroup of a Lie group. If we could show that every local semigroup is an identity neighborhood of *some* topological semigroup – be it subsemigroup of a Lie group or not – then this would be a sophisticated result opening up new avenues leading distinctly beyond Lie group theory. Our present knowledge has not carried us that far. We shall deal in our survey of the Lie theory of semigroups with the parts we have mentioned:

1. The infinitesimal theory.
2. The local theory.
3. The global theory.

The *history* of topological semigroup has many roots. It is perhaps remarkable that SOPHUS LIE himself addressed transformation *semigroups* under the name of "continuous groups". One can even recognize convex cones in LIE's writings. (See [8], [10].) In functional analysis a large branch of research covers the theory of strongly continuous one-parameter semigroups of operators on Banach spaces going back to E. HILLE in the late forties; this is not our concern here, because the Lie theory of semigroups presented in this book is a multiparameter theory. The study of compact semigroups was initiated by A. D. WALLACE in the fifties. This theory gave much momentum to the investigation of topological semigroups and later entered functional analysis in the context of almost periodic and weakly almost periodic functions. The impact of compact semigroups is still quite noticeable today (see the chapters by HINDMAN, MISLOVE, PYM, RUPPERT and TROALLIC in this volume). If a compact semigroup is contained in a group, it is itself a group. The theory presented here and in the chapters by HILGERT [4] and LAWSON [13] is, therefore, quite different since it is largely, but not exclusively, concerned with subsemigroups of Lie groups. With an eye on the applications one could say that the theory of compact semigroups is applied in studies of the ergodic behavior of semigroups while the Lie theory of semigroups studies the nature of a semigroup around the origin and how its infinitesimal properties influence large scale behavior.

It had not been noticed by workers in the topological and analytical theory of semigroups that certain basic aspects of the Lie theory of semigroups were recognized and developed by LOEWNER whose work originates from the study of functions of one complex variable and merges into the investigation of functions with generalized monotonicity properties and from there into transformation semigroups. Only now, after L. Bers made LOEWNER's work accessible in a volume of collected papers [14], it is possible to follow LOEWNER's access to a Lie theory of semigroups. Remarkably, the idea of what we call a *Lie wedge* and a portion of the basic aspects of an infinitesimal theory of semigroups were clearly formulated by LOEWNER in 1948 in an address to the American Mathematical Society. They were published 1950 in the Bulletin of the AMS in an article entitled *Some classes of functions defined by difference or differential inequalities* (see [14], pp. 149,

136). Proofs of the results formulated in this paper on basic Lie semigroup theory were not given and their future publication was promised; but they never emerged. From 1955 on, LOEWNER published papers on applications of what we would call the Lie theory of semigroups on topics like the semigroup of totally positive matrices, transformation semigroups, semigroups of holomorphic self-maps of domains, transformation semigroups invariant under the groups of isometries of Euclidean and non-Euclidean geometries, partial orders defined by linear semigroups, semigroups of monotone transformations of higer order on a real interval [14]. He summarized a good deal of his research in an address to the American Mathematical Society in 1962; this overview was published in the Bulletin of the AMS in 1964 under the title *On semigroups in analysis and geometry* (see [14], p. 437–451). It appears that LOEWNER and his work had no immediate followers.

However, in the dissertation of L. J. M. ROTHKRANTZ "Transformatiehalfgroepen van nietcompacte hermitesche symmetrische Ruimten" at Amsterdam in 1980 the issue of semigroups of holomorphic self-maps of domains was resumed and amplified, and new vistas towards the relation of semigroups and the geometry of symmetric spaces were opened. Research on transformation semigroups of manifolds with geometric structure is one of the current applications of Lie semigroup theory. One type of geometric structure is the assignment of a Lorentzian metric; the investigation of such manifolds leads to questions of global causality which have been recently attacked with semigroup methods [6].

Considerable momentum in the field of semigroups in Lie groups and of convex cones in Lie algebras is due to investigation of invariant cones in Lie algebras initiated in a seminal paper by E. B. VINBERG on *Invariant cones and orderings in Lie groups* in 1980 [16]. The theory of invariant cones in Lie algebras has become a fairly rich theory as we shall see below. Substantial contributions are due to G. I. OL'SHANKSKIĬ, S. PANEITZ, S. KUMARESAN and A. RANJAN, HILGERT and HOFMANN, SPINDLER. For detailed references we refer to [7].

In a natural way, the Lie theory of semigroups emerged in systems and geometric control theory in the mid-seventies in the work of GAUTHIER, HIRSCHORN, JURDEVIC, KUPKA, SALLET, SUSSMANN. The reader will find more information on this aspect in KUPKA's chapter on *Semigroups in systems theory* in this volume. The systematic development of the Lie theory of semigroups [7] shows that its link to systems theory is a two-way street: Significant basic results in the Lie theory of semigroups are established with methods from geometric control theory.

The connection between the Lie theory of semigroups and representation theory of Lie groups, notably the analytic extension of representation was pioneered by G. I. OL'SHANSKIĬ; this relationship continues to be of vital interest as is exemplified by the oscillator semigroup whose significance was discovered recently by HOWE. More details about this link are found in the chapter by HILGERT on *Applications of Lie semigroups in analysis* in this volume.

The first attempt at describing a basic Lie theory of semigroups was published by HOFMANN and LAWSON in the proceedings of a Conference at Oberwolfach in 1981

[11]. Today research on the foundations and the applications of the Lie theory of semigroups is in full flux. For a comprehensive discussion of the Lie theory of semigroups at an approximately up-to-date level and for the historical attributions of individual results the reader is referred to [7]. That book is a good source of the theory and its status up to about 1988; however, the state of our present knowledge reaches already beyond what is presented there. We shall indicate in this survey certain aspects of the latest status of the theory and the problems which we still face today. We cannot attempt a presentation of its full genealogy and all of its ramifications.

1. The infinitesimal theory

If S is a subsemigroup of a Lie group G with exponential function $\exp\colon \mathfrak{g} \to G$ then the set $\mathbf{L}(S) \subseteq \mathfrak{g}$ of all vectors in the Lie algebra $\mathfrak{g}$ of G which are tangent (better: subtangent) to the set $\exp^{-1} S$ at 0 is a *convex cone* W. We also call it a wedge, because in general it has an edge $W \cap -W$, that is a maximal vector subspace. If S is closed, then each vector $x \in \mathbf{L}(S)$ gives a one-parameter subsemigroup $t \mapsto \exp t{\cdot}x\colon \mathbb{R}^+ \to S$ of S, and all one-parameter subsemigroups of S are so obtained. (There is the purely geometric question how a *subtangent vector x of a set M at the origin* of a euclidean vector space should be defined; it suffices to say that it is of the form $x = \lim r_n{\cdot}x_n$ with $x_n \to 0$ in M and $r_n > 0$.) We also have to know that a *wedge* in a euclidean space is an additively and topologically closed convex set containing the origin.

Proposition 1.1. *The tangent wedge* $\mathbf{L}(S)$ *of a subsemigroup* S *of a Lie group is a wedge* W *satisfying*

$$e^{\operatorname{ad} X} W \subseteq W \quad \textit{for all} \quad X \in W \cap -W. \quad \square \tag{1}$$

The wedge $\mathbf{L}(S)$ is the *"infinitesimal"* object associated with S. This leads us to the following definition:

Definition 1.2. A *Lie wedge* is a wedge W in a finite dimensional Lie algebra $\mathfrak{g}$ satisfying (1).

Lie wedges

The "infinitesimal" theory of semigroups in Lie groups therefore is the theory of Lie wedges in Lie algebras. Every *pointed cone* ($W \cap -W = \{0\}$) is trivially a Lie wedge. Every Lie subalgebra is a Lie wedge; more generally, the edge of a Lie wedge is a Lie algebra, but this is not sufficient for a wedge to be a Lie wedge as examples show. The linear span $W-W$ of a Lie wedge need not be a Lie algebra.

Convex cones and wedges have a rich geometric structure. In particular, with every point $x \in W$ one can associate the set $\mathbf{L}(W-x)$ of all vectors subtangent to W at x. This set is a convex cone, and its edge T_x is the *tangent space to* W *at* x. The following description of a Lie wedge uses only geometry and Lie brackets:

Proposition 1.3. *A wedge* W *in a Lie algebra is a Lie wedge if and only if*

$$[x, W\cap -W] \subseteq T_x \quad \textit{for all} \quad x \in W. \quad \Box \tag{2}$$

In fact, if W has interior points, it suffices to test equation (2) for all those x in which T_x is a hyperplane; such points are called C^1*-points*. If W does not have interior points we can still speak of C^1-points because W always has interior in $W-W$ (with the caveat that we only consider finite dimensional vector spaces!); so a point of W is a C^1-point if it is a C^1-point in $W-W$.

Corollary 1.4. *A wedge* W *in a Lie algebra is a Lie wedge if and only if it satisfies*

$$[x, W\cap -W] \subseteq T_x \quad \textit{for all} \quad x \in C^1(W). \quad \Box \tag{3}$$

(For additional information see [7], Theorem II.1.12.)

We know numerous examples of Lie wedges in prototypical Lie algebras such as $\mathrm{sl}(2,\mathbb{R})$. (See [7], [16].) However, we do not even have a start of a reasonably general theory or Lie wedges or its classification. Thus one of the major projects remaining is vaguely expressed as follows:

Problem P.1. Formulate feasible principles of classifications of Lie wedges in finite dimensional Lie algebras and develop a general structure theory. $\Box$

One must keep in mind that *every pointed cone in any Lie algebra is a Lie wedge*! If W is a Lie wedge in the Lie algebra $\mathfrak{g}$ of a, say, simply connected Lie group G, then the analytic subgroup H with Lie algebra $\mathfrak{h} = W\cap -W$ need not be closed as simple examples show [5]. If, however, H is closed then $\mathfrak{g}/\mathfrak{h}$ may be identified with the tangent space of the homogeneous manifold G/H at H; it contains the pointed cone $W/(W\cap -W)$ which is invariant under the action of H on $\mathfrak{g}/h$ induced by the adjoint action. There are indications that this is the starting point of a general theory. One of the best understood class of Lie wedges is the class of wedges which below we shall call Ol'shanskiĭ wedges, but even in their case, a complete theory is still missing.

Lie semialgebras

In a Lie algebra $\mathfrak{g}$ one finds convex symmetric open neighborhoods B of 0 such that the CAMPBELL–HAUSDORFF series $x*y = x+y+\frac{1}{2}\cdot[x,y]+H_3(x,y)+\cdots$ converges for $x, y \in B$ and defines a local multiplication $*: B\times B \to \mathfrak{g}$. Such neighborhoods B are called *Campbell-Hausdorff-neighborhoods* or CH-*neighborhoods* for short. If $\mathfrak{g}$ is the Lie algebra of a Lie group, we have $\exp(x*y) = \exp x \exp y$ for $x, y \in B$. The function $\lambda_x^*: B \to \mathfrak{g}$ given by $\lambda_x^*(y) = x * y$ is locally a diffeomorphism; its derivative $d\lambda_x^*(0): \mathfrak{g} \to \mathfrak{g}$ is a linear operator given explicitly through a power series in $\operatorname{ad} x$. Indeed if we write $g(z) = z(1 - e^{-z})^{-1}$ then $g(\operatorname{ad} x): \mathfrak{g} \to \mathfrak{g}$ is a linear operator defined for all x close enough to 0, depending analytically on x, and agreeing with $d\lambda_x^*$ wherever both are defined. The operators $g(\operatorname{ad} x)$ play an important role in Lie theory in general and in the Lie theory of semigroups in particular. More about this can be found in [7].

If $\mathfrak{h}$ is a Lie subalgebra, then $(B \cap \mathfrak{h}) * (B \cap \mathfrak{h}) \subseteq \mathfrak{h}$. For a Lie wedge W, in general, $(B\cap W)*(B\cap W)$ is not contained in W. Numerous examples are given in [7]. It is therefore a special class of Lie wedges for which this is the case; semigroups with such tangent objects are locally ruled by one-pameter semigroups just as Lie groups are.

Definition 1.5. (i) A *Lie semialgebra* W in a Lie algebra $\mathfrak{g}$ is a wedge for which there is a CH-neighborhood B such that

$$(W \cap B) * (W \cap B) \subseteq W.$$

(ii) A wedge W in Lie algebra $\mathfrak{g}$ is called *invariant* provided that

$$e^{\operatorname{ad} x} W = W \quad \text{for all} \quad x \in \mathfrak{g}.$$

We note that, just in the case of Lie wedges, the idea of a Lie semialgebra generalizes the concept of a Lie subalgebra. But in contrast with Lie wedges, a powerful general theory is available for Lie semialgebras. In fact there is hope for a complete classification of Lie semialgebras which are not Lie algebras provided we accept our state of knowledge of invariant wedges as satisfactory. A vector space is an invariant wedge if and only if it is an ideal. We shall see below that there is a profound and extensive (not yet entirely complete) classification theory available for invariant wedges. On the surface, very little connection seems to exist between Lie semialgebras and invariant wedges. However, a deeper inspection of Lie semialgebras shows that they have a very strong tendency to be invariant. What this means needs to be specified, notably since Lie semialgebras generalize Lie subalgebras and invariant wedges generalize ideals; it would be absurd to claim that subalgebras had a tendency to be ideals.

The first thing one observes for both Lie semialgebras and invariant wedges is their characterisation in terms of the Lie bracket and the geometry of convex cones.

Proposition 1.6. *For a wedge W in a Lie algebra the following conditions are equivalent:*
(1) *W is a Lie a semialgebra.*
(2) *$[x, T_x] \subseteq T_x$ for all $x \in W$.*
(3) *$[x, T_x] \subseteq T_x$ for all $x \in C^1(W)$.*
(4) *$g(\operatorname{ad} x)T_x = T_x$ for all sufficiently small $x \in C^1(W)$.*
(5) *$g(\operatorname{ad} x)\mathbf{L}(W - x) = \mathbf{L}(W - x)$ for all sufficiently small $x \in W$.*
Also, the following statements are equivalent:
(i) *W is an invariant wedge.*
(ii) *$[x, \mathfrak{g}] \subseteq T_x$ for all $x \in W$.*
(iii) *$[x, \mathfrak{g}] \subseteq T_x$ for all $x \in C^1(W)$.* □

For details see [7], Theorems II.1.14 and II.2.14.

Even though we know little about Lie wedges in general, at the present stage of our knowledge, one class of Lie wedges stands out as particularly interesting in the applications. Indeed suppose that the Lie algebra $\mathfrak{g}$ is the direct sum $\mathfrak{k} \oplus \mathfrak{p}$ of a subalgebra $\mathfrak{k}$ and a vector subspace $\mathfrak{p}$ containing a pointed cone C such that $e^{\operatorname{ad} \mathfrak{k}} C \subseteq C$. Then $W = \mathfrak{k} \oplus C$ is a Lie wedge. If, in addition $[\mathfrak{p}, \mathfrak{p},] \subseteq \mathfrak{k}$ and $\operatorname{ad} c$ has real spectrum for every $c \in C$, then W has the following additional property:

$$(5_C) \qquad g(\operatorname{ad} x)\mathbf{L}(W - x) = \mathbf{L}(W - x) \text{ for all } x \in C.$$

This invariance condition of the Lie wedge W is weaker than condition (5) above which characterizes Lie semialgebras.

Definition 1.7. A wedge W in a Lie algebra $\mathfrak{g}$ is called an *Ol'shanskiĭ wedge* if it is of the form $\mathfrak{k} \oplus C$ with a Lie subalgebra $\mathfrak{k}$ and a pointed wedge C such that $e^{\operatorname{ad} \mathfrak{k}} C \subseteq C$ and condition (5_C) is satisfied.

We shall return to this class of Lie wedges in Proposition 3.9 and the discussion which follows that proposition.

With the preceding results we recognize very easily that we have a hierarchy:

The condition $[x, W \cap -W] \subseteq T_x$, for all $x \in W$, means Lie wedge,
$[x, T_x] \subseteq T_x$, for all $x \in W$, means Lie semialgebra,
$[x, \mathfrak{g}] \subseteq T_x$, for all $x \in W$, means invariant wedge.

Since $W \cap -W \subseteq T_x \subseteq \mathfrak{g}$ is true for all $x \in W$, it is clear that every invariant wedge is a Lie semialgebra and that every semialgebra is a Lie wedge. None of the reverse implications is true in general. Example 1.8 below will show, in particular, that a Lie semialgebra need not be invariant.

Proposition 1.6 was a translation of analytic conditions into a blend of algebraic and geometric conditions. We have an essentially complete structure theory on Lie semialgebras which we describe now. The original Definition 1.5 of a Lie semialgebra did not allow us to conclude that the intersection of an arbitrary family of Lie semialgebras is again a Lie semialgebra. However with the aid of Proposition 1.6 it is not hard to verify that given an arbitrary Campbell–Hausdorff neighborhood B in a Lie algebra $\mathfrak{g}$, a wedge is a Lie semialgebra if and only if $(W \cap B)*(W \cap B) \subseteq W$. Notice that now the size of the neighborhood longer depends on W as it certainly did in the original definition. It is therefore true that *the intersection of any family of Lie semialgebras is a Lie semialgebra.* It is now a reasonable viewpoint to aim for a description of Lie semialgebras as intersections of certain basic Lie semialgebras. We present some examples:

Example 1.8. (a) If $\mathfrak{h}$ is a *hyperplane subalgebra* of a Lie algebra $\mathfrak{g}$ (that is, a hyperplane which is also a subalgebra) then the two closed half-spaces bounded by $\mathfrak{h}$ are Lie semialgebras, called *half-space semialgebras.* Special cases:

(aa) The hyperplane subalgebras of $\mathrm{sl}(2, \mathbb{R})$ are the planes tangent to the zero cone of the Cartan-Killing form B of sl(2) which is given by $B(X, Y) = 8{\cdot}\mathrm{tr}\, XY$. In particular, for $X = \begin{pmatrix} a & b \\ c & -a \end{pmatrix}$ we have $\mathrm{tr}\, X^2 = 2(a^2 + bc) = -2 \det X$; thus the zero cone in question is simply the set of all $X \in \mathrm{sl}(2)$ with $\det X = 0$. The plane tangent to this cone at X is the set of solutions of the equation $2ax + cy + bz = 0$.

(ab) If $\mathfrak{h}$ is any hyperplane in $\mathfrak{g}$ containing the commutator algebra $\mathfrak{g}'$ then $\mathfrak{h}$ is a half-space semialgebra.

(ac) If $\mathfrak{g}$ is the two-dimensional non-abelian Lie algebra then every one-dimensional vector subspace is a hyperplane subalgebra (exactly one of which is an ideal). The last example has a generalization which yields a simple but instructive class of examples:

(ad) Let $\mathfrak{g}$ denote the class of all Lie algebras $\mathfrak{g}$ on which there is is a linear form ω such that $[x, y] = \omega(x){\cdot}y - \omega(y){\cdot}x$. Any such algebra is isomorphic to an algebra of matrices

$$\begin{pmatrix} x_0{\cdot}E_n & v \\ 0 & 0 \end{pmatrix}, \quad v = \begin{pmatrix} x_1 \\ \vdots \\ x_n \end{pmatrix}, \quad (x_0, \ldots, x_n) \in \mathbb{R}^{n+1},$$

where $n = 1, \ldots$ and where E_n denotes the $n \times n$-unit matrix. This class of Lie algebras is called *almost abelian* since it contains all abelian ones and has the properties that all vector subspaces are subalgebras and all wedges are Lie semialgebras.

(b) All intersections of half-spaces semialgebras are Lie semialgebras. These are called *intersection semialgebras* (A. EGGERT).

(c) All invariant wedges in a Lie algebra are Lie semialgebras. (See Proposition 1.6.)

(d) If W_{inv} is an invariant wedge and W_{int} is an intersection semialgebra then $W_{\text{inv}} \cap W_{\text{int}}$ is a Lie semialgebra.

The most general type of Lie semialgebra we have seen in these examples is that of 1.8.d. To EGGERT one owes the crowning result of Lie semialgebra theory which is based onto much previous work collected in [7]. It allows an essentially complete characterization of Lie semialgebras and deserves to be called the *Classification Theorem for Lie Semialgebras.*

Theorem 1.9. (EGGERT) *Every Lie semialgebra W is of the form $W_{\text{inv}} \cap W_{\text{int}}$, where W_{inv} is an invariant wedge and W_{int} is an intersection semialgebra.* □

If we accept the highly developed classification theory of invariant wedges which we shall describe below as available background, obviously, there remains the task to describe the intersection algebras. Let us describe some of the background theory.

First we introduce the concept of the characterstic function of a Lie semialgebra. If $x \in C^1(W)$ then by Proposition 1.6(3) the operator $\operatorname{ad} x$ induces on the one-dimensional space $\mathfrak{g}/T_x$ a linear operator which is multiplication by a scalar $\lambda(x)$. The function $\lambda\colon C^1(W) \to \mathbb{R}$ is called the *characteristic function* of W. Clearly $\lambda(x) = 0$ if and only if $[x, \mathfrak{g}] \subseteq T_x$. Consequently, a Lie semialgebra is invariant if an only if its characterstic function is identically 0. The characteristic function is continuous (see [7], II.2.34). It was first introduced and applied by HILGERT and HOFMANN, but the crucial result is due to LAWSON.

Proposition 1.10. (J. D. LAWSON) *If W is a generating semialgebra in $\mathfrak{g}$ with characteristic function λ, then $\lambda(x) \neq 0$ implies $[T_x, T_x] \subseteq T_x$.* □

These results emphasize the significance of hyperplane subalgebras. We now have complete information on these. There is a unique largest ideal $\Delta = \Delta(\mathfrak{g})$ which is contained in all hyperplane subalgebras, and there are two further characteristic ideals Δ_a and Δ_s such that $\mathfrak{g}/\Delta$ is the direct sum of its radical Δ_s/Δ a unique Levi complement $\Delta_a/\Delta \cong \mathrm{sl}(2, \mathbb{R})^m$ for some number m, and the radical $\mathfrak{r}$ is a metabelian algebra (i.e., its commutator algebra is abelian) of a special structure rather precisely known. In particular, it is the direct sum of its center and an ideal whose morphisms onto the two dimensional nonabelian algebra separate the points. (See [9].) Specifically, $\mathfrak{r}$ is the semidirect sum of a Cartan subalgebra $\mathfrak{c}$ and a direct sum of characteristic ideals $\mathfrak{m}_\alpha$ where α ranges through a finite set B of non-zero linear forms on $\mathfrak{r}$ such that $[x,m] = \alpha(x)\cdot m$ for $(x,m) \in \mathfrak{r} \times \mathfrak{m}_\alpha$. These linear forms α are called *base roots* of $\mathfrak{g}$. The center $\mathfrak{z}$ of $\mathfrak{r}$ is contained in every Cartan algebra.

The ideals Δ_s and Δ_a can be easily found: Firstly, in order to find Δ_s consider $\mathfrak{g}/\mathrm{rad}\,\mathfrak{g}$ and let J be the sum of all ideals of this quotient not isomorphic to $\mathrm{sl}(2,\mathbb{R})$. Then the inverse image of J in $\mathfrak{g}$ is Δ_s. The algorithm to find Δ_a is more complicated, but can be explicitly described (see [9]). Since Δ_a always contains $\mathfrak{g}''$, the problem is readily reduced to metabelian Lie algebras; however, these still occur in an amazing variety, while those of the type $\mathfrak{g}/\Delta_a$ are of a fairly special form.

If we accept these results, we can easily classify the hyperplane subalgebras in a Lie algebra $\mathfrak{g}$. Each hyperplane subalgebra $\mathfrak{h}$ contains Δ. Thus for the classification we may just as well assume that $\mathfrak{g}$ is *reduced* in the sense that $\Delta = \{0\}$.

Proposition 1.12. (Hofmann [9]) *Any reduced Lie algebra $\mathfrak{g}$ is of the form*

$$\mathfrak{g} = (\mathfrak{c} \oplus \mathfrak{m}) \oplus \mathfrak{s}, \quad \mathfrak{m} = \bigoplus_{\alpha \in B} \mathfrak{m}_\alpha, \quad \mathfrak{s} \cong \mathrm{sl}(2)^m.$$

Here the $\mathfrak{m}_\alpha$ are characteristic ideals such that $[x,m] = \alpha(x)\cdot m$ for $x \in \mathfrak{g}$, $m \in \mathfrak{m}_\alpha$ and where $\mathfrak{c}$ is the (abelian) Cartan algebra of the radical $\mathfrak{c}\oplus\mathfrak{m}$ of $\mathfrak{g}$. The hyperplane subalgebras $\mathfrak{h}$ fall into the following classes:

(a) *The $\mathfrak{h}$ containing $\mathfrak{m} \oplus \mathfrak{s}$. These $\mathfrak{h}$ are the ones which are ideals.*

(b_α) *For each base root $\alpha \in B$, the $\mathfrak{h}$ of the form*

$$\mathfrak{h} = \gamma(\mathfrak{c}) \oplus \sum_{\substack{\beta \in B \\ \beta \neq \alpha}} \mathfrak{m}_\beta \oplus \mathfrak{h}_\alpha \oplus \mathfrak{s}$$

with a hyperplane $\mathfrak{h}_\alpha$ in $\mathfrak{m}_\alpha$ and an inner automorphism γ. All of these are conjugate under Aut($\mathfrak{g}$). *The largest ideal of $\mathfrak{g}$ contained in $\mathfrak{h}$ is*

$$(\gamma(\mathfrak{c}) \cap \ker \alpha) \oplus (\mathfrak{m} \cap \mathfrak{h}) \oplus \mathfrak{s}.$$

(c) *The $\mathfrak{h}$ of the form $(\mathfrak{c} \oplus \mathfrak{m}) + x^\perp$ where $x^\perp$ denotes the annihilator of x in $\mathfrak{s}$ with respect to the Cartan-Killing form B on $\mathfrak{s}$ and where x is an element in one of the simple summands of $\mathfrak{s}$ with $B(x) = 0$. These $\mathfrak{h}$ are conjugate under* Aut($\mathfrak{g}$). *The maximal ideal contained in $\mathfrak{h}$ is*

$$(\mathfrak{c} \oplus \mathfrak{m}) \oplus \mathfrak{s}_x,$$

where $\mathfrak{s}_x$ is the sum of all simple summands not containing x. □

Let us use this result to associate with a general Lie algebra a number of canonically defined concepts.

Definition 1.13. (i) Let $\mathfrak{g}$ be a Lie algebra and $\mathfrak{h}$ a hyperplane subalgebra. We say that $\mathfrak{h}$ *is of abelian type* if $\mathfrak{h}/\Delta$ is of type (a) in the preceding Proposition 1.12 and *of semisimple type* if $\mathfrak{h}/\Delta$ is of type (c) in the preceding theorem. A linear functional α of $\mathfrak{g}$ is said to be a *base root* if it vanishes on Δ_s and induces on $\mathfrak{g}/\Delta_s$ a base root in the sense spelled out in the discussion preceding Proposition 1.12. The set of all base roots is again denoted B. For each $\alpha \in B$ we say that $\mathfrak{h}$ is of type α if $\mathfrak{h}/\Delta$ is of type (b_α) in Proposition 1.12.

We denote with $\mathcal{H}_{\mathrm{ab}}$, $\mathcal{H}_{\mathrm{ss}}$, and $\mathcal{H}_\alpha$ the set of hyperplane subalgebras of abelian type, semisimple type, and of type α, respectively.

(ii) Let W be a generating semialgebra of $\mathfrak{g}$. We let C^1_{ab} denote the set of all C^1-points such that $T_x \in \mathcal{H}_{\mathrm{ab}}$, and we define C^1_{ss} and C^1_α in the same vein. We note that $L_x(W)$ for $x \in C^1(W)$ is the half-space bounded by T_x and containing W. We define

$$W_{\mathrm{ab}} = \bigcap_{x\in C^1_{\mathrm{ab}}} L_x(W), \qquad W_{\mathrm{ss}} = \bigcap_{x\in C^1_{\mathrm{ss}}} L_x(W), \qquad W_\alpha = \bigcap_{x\in C^1_\alpha} L_x(W),$$

(iii) Lie algebras of the type $\mathfrak{c} \oplus \mathfrak{m}$ in Proposition 1.12 are called *special metabelian algebras*. They are exactly the solvable reduced Lie algebras.

The preceding definitions and Proposition 1.12 allow a complete description of all intersection semialgebras W as intersections of unique intersection semialgebras canonically associated with W (see [9]). Specifically, if W is an intersection semialgebra, it is a canonical intersection of semialgebras $W_{\mathrm{ab}} \cap W_{\mathrm{ss}} \cap \bigcap_{\alpha\in B} W_\alpha$.

Cartan subalgebras and root decomposition are instrumental in the entire theory. If $\mathfrak{g}$ contains a Lie semialgebra W with interior points such that $W \cap -W$ does not contain a nontrivial ideal of $\mathfrak{g}$, then all Cartan subalgebras are abelian, and for each x int the interior of W, the eigenvalues of $\operatorname{ad} x$ are either real or purely imaginary. Since regular elements are dense, we have Cartan subalgebras intersecting the interior of W, and associated roots are either real or imaginary. The former are responsible for the intersection Lie semialgebras, the latter for the invariant wedge W_{inv}. We shall see in the next section that in the classification of invariant wedges, Cartan algebras and root decompositions are required even for the formulation of the principal results.

The preceding remarks and EGGERT's Classification Theorem 1.9 yield the *Prime Decomposition Theorem for Semialgebras*.

Theorem 1.14. (i) *Any generating Lie semialgebra W in a Lie algebra $\mathfrak{g}$ can be written as*

$$W = W_{\text{ab}} \cap W_{\text{ss}} \cap \bigcap_{\alpha \in B} W_\alpha \cap W_{\text{inv}}$$

with an invariant wedge W_{inv}.

(ii) *If the largest ideal contained in $W \cap -W$ is $\{0\}$, then all Cartan algebras of $\mathfrak{g}$ are abelian, and if $\mathfrak{h}$ is a Cartan algebra meeting the interior of W, then $\mathfrak{g} = \mathfrak{g}_r + \mathfrak{g}_i$ with subalgebras satisfying $\mathfrak{g}_r \cap \mathfrak{g}_i = \mathfrak{h}$ with root decompositions $\mathfrak{g}_r = \mathfrak{h} \oplus \mathfrak{g}_r^+$ and $\mathfrak{g}_i = \mathfrak{h} \oplus \mathfrak{g}_i^+$ such that $\mathfrak{g}_r^+$ is the sum of root spaces for real roots and $\mathfrak{g}_i^+$ is the sum of root spaces for purely imaginary roots. Moreover, $W_r = \mathfrak{g}_r \cap W$ is an intersection algebra in the special metabelian Lie algebra $\mathfrak{g}_r$ and $W_i = \mathfrak{g}_i \cap W$ is an invariant wedge in $\mathfrak{g}_i$. The wedge $W_r \oplus \mathfrak{g}_i^+$ is the intersection algebra*

$$W_{\text{ab}} \cap W_{\text{ss}} \cap \bigcap_{\alpha \in B} W_\alpha$$

in $\mathfrak{g}$ and $W_i \oplus g_r^+$ is the invariant algebra W_{inv} in $\mathfrak{g}$. □

The algebra $\mathfrak{g}_i$ is of a type more precisely described in the section on invariant wedges below.

Notice that W_{ab} contains the commutator algebra and is invariant.

In [7] a Lie semialgebra is called *trivial* if it contains the commutator algebra. (It is known, e. g., that a generating Lie semialgebra must be trivial if $\mathfrak{g}$ is nilpotent or the underlying real Lie algebra of a complex Lie algebra. See [7], Chapter II, Section 7.) Thus a Lie semialgebra W is trivial if and only if $W = W_{\text{ab}}$. The "prime factor" W_{ss} may be invariant, too, although this is likely to be rather the exception.

Under special conditions imposed on either the Lie algebra $\mathfrak{g}$, or the geometry of the Lie semialgebra W, or the position of W in relation to characteristic subalgebras such as the nilradical, a considerable amount of information is available which may be found in [7]. In particular, intersection semialgebras "have corners" if one of the prime factors "has corners" or if there are several prime factors. Sufficiently "round" semialgebras are either invariant or are contained in almost abelian Lie algebras (see Example 1.8(ad)). Details are described in [7].

As a typical example of a result which yields rather sharp information in special circumstances we cite another result of A. Eggert's [2]. Here the description of the structure of the wedge is done by building up from smaller components rather than by intersecting "prime factors".

Proposition 1.15. (A. Eggert) *If $\mathfrak{g}$ is a Lie algebra which is isomorphic to* $\mathrm{sl}(2,\mathbb{R})^m \times \mathfrak{j}$ *with a Lie algebra $\mathfrak{j}$ in which every hyperplane algebra is an ideal,* *then every generating Lie semialgebra is isomorphic to* $W_1 \times \cdots \times W_m \times V$ *with a generating Lie semialgebra* W_j *in* $\mathrm{sl}(2, \mathbb{R})$ *and an invariant wedge* V *in* $\mathfrak{j}$. □

Invariant wedges

If W is an invariant wedge in a Lie algebra $\mathfrak{g}$, then $W - W$ is an ideal in which W is generating, and $W \cap -W$ is an ideal $\mathfrak{j}$ such that $W/\mathfrak{j}$ is a pointed cone in $\mathfrak{g}/\mathfrak{j}$. It is therefore no substantial loss of generality when we restrict our attention to pointed generating invariant cones W in a Lie algebra $\mathfrak{g}$.

The first observation is that Lie algebras supporting such cones are relatively special. The key for the understanding of invariant wedges in a Lie algebra are the Cartan subalgebras. We say that a Cartan algebra $\mathfrak{h}$ is *compactly embedded* if $e^{\mathrm{ad}\,\mathfrak{h}}$ is dense in a torus subgroup of the automorphism group $\mathrm{Aut}\,\mathfrak{g}$ of $\mathfrak{g}$. Every Lie algebra $\mathfrak{g}$ with a generating pointed invariant cone W has a compactly embedded Cartan algebra $\mathfrak{h}$, and all such are conjugate and intersect the interior of W. The intersection $\mathfrak{h} \cap W$ determines W uniquely, thus the compactly embedded Cartan algebras $\mathfrak{h}$ provide a tool for the possible classification of invariant cones. It is therefore of little surprise that the compactly embedded Cartan algebras are at the focus of the theory of invariant cones in Lie algebras.

Much of the structure of Lie algebras supporting invariant pointed cones with inner points is described in [7] and in more recent work of K. H. Spindler [21], [22]. We collect some of this information in the following theorem:

Theorem 1.16. *Suppose that* $\mathfrak{g}$ *is a Lie algebra containing some pointed generating invariant cone. Then* $\mathfrak{g}$ *contains a compactly embedded Cartan algebra* $\mathfrak{h}$*, and all of these are conjugate. All Cartan subalgebras are abelian. There is a unique maximal compactly embedded subalgebra* $\mathfrak{k}$ *and a unique Levi complement* $\mathfrak{s} \supseteq \mathfrak{k}'$ *such that* $\mathfrak{h} \subseteq \mathfrak{k}$ *and* $\mathfrak{h} = (\mathfrak{h} \cap \mathfrak{s}) \oplus (\mathfrak{h} \cap \mathfrak{r})$ *with the radical* $\mathfrak{r}$*. The center* $\mathfrak{z}$ *of* $\mathfrak{k}$ *is non-zero. The nilradical* $\mathfrak{n}$ *is metabelian and* $\mathfrak{n} \cap \mathfrak{h}$ *is the center of* $\mathfrak{g}$*. The distinguished Levi complement* $\mathfrak{s}$ *satisfies* $[\mathfrak{s}, \mathfrak{h}] \subseteq \mathfrak{s}$ *so that* $\mathfrak{s} + \mathfrak{h}$ *is a reductive algebra. Further,* $\mathfrak{g}$ *decomposes uniquely into a direct sum of* $\mathfrak{k}$*-submodules* $\mathfrak{k} \oplus \mathfrak{p}_s \oplus \mathfrak{p}_n$ *such that* $\mathfrak{s} + \mathfrak{h} = \mathfrak{k} \oplus \mathfrak{p}_s$ *is a Cartan decomposition and* $\mathfrak{p}_n \subseteq \mathfrak{n}$ *is a symplectic* $\mathfrak{s} + \mathfrak{h}$*-module. One has* $[\mathfrak{p}_s, \mathfrak{p}_s] \subseteq \mathfrak{k}$ *and* $[\mathfrak{p}_n, \mathfrak{p}_n] \subseteq \mathfrak{n} \cap \mathfrak{h}$. □

The Cartan subalgebra $\mathfrak{h}$ is just a copy of $\mathbb{R}^n$ where n is the rank of $\mathfrak{g}$. But it does carry geometrical structure inherited from the Lie algebra. Firstly, $\mathfrak{h}$ is the Cartan algebra of $\mathfrak{k}$ and as such supports a Weyl group $\mathcal{W}$ generated by the reflections at the zero-hyperplanes of the roots $\omega \in \Omega_k^+$. (Alternatively, $\mathcal{W}$ may be defined as the factor group of the normalizer of T modulo its centralizer in the group of all inner automorphisms of $\mathfrak{g}$. See [7].) Secondly, if $x \in \mathfrak{g}^\omega$ is any root vector, then $\tau_x = (\mathrm{ad}\,x)^2|\mathfrak{h}$ is a rank one endomorphism of $\mathfrak{h}$ such that $\tau_x(h) = \omega(h)\cdot[Ix, x]$. We denote the set of all of these rank one operators τ_x of $\mathfrak{h}$ with root vectors x for roots $\omega \in \Omega_p^+$ by $\mathcal{C}$. Their kernels range through the finite set $\{\omega^{-1}(0) \mid \omega \in \Omega\}$. The image $\mathrm{im}(\tau_x) = \mathbb{R}\cdot[Ix, x]$, as $x \neq 0$ ranges through $\mathfrak{g}^\omega$ depends only on ω for $\omega \in \Omega_k^+ \cup \Omega_{ps}^+$. In solvable algebras where one has enormous leeway in constructing algebras for almost any preassigned root system (see [15]) this is no longer true. We notice that $\mathcal{W}$ and $\mathcal{C}$ are linked through the requirement that $\mathcal{C}$ is invariant under conjugation by the elements of $\mathcal{W}$. The preceding results gave us information on the structure of a Lie algebra $\mathfrak{g}$ which supports invariant pointed generating cones. So far we have not talked about the cones W themselves except that we noted that the intersection $W \cap \mathfrak{h}$ with a fixed compactly embedded Cartan algebra determined W uniquely. This is the key for a classification. If we know which pointed generating cones C of $\mathfrak{h}$ are the traces of invariant pointed cones W then we know the possible W. In this sense the following theorem due to Hilgert and Hofmann gives a classification of the invariant cones W in $\mathfrak{g}$. (See [7], Chapter III, [21] and [22].)

Theorem 1.18. *Let $\mathfrak{g}$ be a Lie algebra supporting invariant pointed generating cones and suppose that $\mathfrak{h}$ is a compactly embedded Cartan subalgebra. Then for a pointed generating cone C in $\mathfrak{h}$ the following statements are equivalent:*
(1) *$C = W \cap \mathfrak{h}$ for a unique invariant pointed generating cone W of $\mathfrak{g}$.*
(2) *C is invariant under the Weyl group $\mathcal{W}$ and the set $\mathcal{C}$ of rank one endomorphisms of $\mathfrak{h}$.* □

Experience shows that this criterion is very manageable in concrete situations. The sources given above discuss the low dimensional examples such as $\mathrm{sl}(2,\mathbb{R})$, the solvable 4-dimensional oscillator algebra (which has analogs in every higher even dimension), $\mathrm{so}(2,1)$, $\mathrm{gl}(2,\mathbb{R})$. We observe further that Theorem 1.18 reduces the classification of invariant pointed generating cones in Lie algebras to a geometric problem in $\mathbb{R}^n$:

Problem P.2. Suppose that $\mathfrak{h}$ is a finite dimensional real Hilbert space with two orthogonal decompositions

$$\mathfrak{h} = \mathfrak{h}_r \oplus \mathfrak{h}_s = \mathfrak{z} \oplus \mathfrak{h}', \quad \mathfrak{h}_r \subseteq \mathfrak{z}, \quad \mathfrak{h}' \subseteq \mathfrak{h}_s$$

and equipped with a finite set of hyperplanes $\mathcal{H} = \mathcal{H}_k \,\dot\cup\, \mathcal{H}_{ps} \,\dot\cup\, \mathcal{H}_{pn}$. Let $\mathcal{C} = \mathcal{C}_{ps} \,\dot\cup\, \mathcal{C}_{pn}$ denote a set of rank one endomorphisms whose kernels are the hyperplanes in $\mathcal{H}_p = \mathcal{H}_{ps} \cup \mathcal{H}_{pn}$. Let $\mathcal{W}$ be a Coxeter group generated by reflections in the hyperplanes of $\mathcal{H}_k$ and assume that all of these contain $\mathfrak{z}$, whence $\mathfrak{z}$ is fixed and $\mathcal{W}$ operates on $\mathfrak{h}'$. Suppose that $\mathcal{C}$ is invariant under conjugation by elements of $\mathcal{W}$ and that

$$\mathrm{im}\,\tau \subseteq \begin{cases} \mathfrak{h}_s & \text{for } \tau \in \mathcal{C}_{ps}, \\ \mathfrak{h}_r & \text{for } \tau \in \mathcal{C}_{pn}. \end{cases}$$

Determine all pointed generating cones which are invariant under both $\mathcal{W}$ and $\mathcal{C}$. □

The spirit of this final step in the classification of invariant cones in a Lie algebra corresponds to the classification of complex simple Lie algebras, which is reduced to a geometric problem in euclidean spaces with root systems. In this case the Dynkin diagrams provide a further reduction to combinatorics and graph theory; we have not reached this stage here, and since the examples exhibit whole continua of invariant cones there is some doubt that such a reduction is possible.

For tables of simple Lie algebras supporting invariant cones see [15], pp. 137ff. Invariant cones occur in the following simple Lie algebras: $\mathrm{sl}(2,\mathbb{R})$, $\mathrm{su}(p,q)$ with $1 < q \leq p$, $\mathrm{su}(p,1)$ with $1 < p$, $\mathrm{so}(p,2)$ with $2 \leq p$, p odd, $\mathrm{so}(2,1)$, $\mathrm{sp}(n,\mathbb{R})$ with $3 \leq n$, $\mathrm{so}^*(2n)$ with $4 \leq n$, $e_{6(2)}$, $e_{7(-25)}$.

As far as solvable algebras are concerned, invariant wedges are completely classified, too. (See [7], [15], pp. 142ff.) The root system can be completely arbitrary in this case. In fact, NEEB has given the following construction of solvable Lie algebras satisfying the conclusions of Theorem 1.17:

Theorem 1.19. (NEEB) *Let $\mathfrak{h} = \mathfrak{z} \oplus \mathfrak{h}_1$ denote the direct sum of two arbitrary vector spaces and let Ω^+ denote a finite set of linear functionals on $\mathfrak{h}$. Suppose that there is an $h \in \mathfrak{h}$ with $\omega(h) > 0$ for all $\omega \in \Omega^+$ and that $\mathfrak{z} = \bigcap_{\omega\in\Omega^+} \ker\omega$.*

For each $\omega \in \Omega^+$ let $\mathfrak{g}^\omega$ denote a real vector space with a complex structure I and equipped with a nonzero bilinear symmetric map $Q^\omega\colon \mathfrak{g}^\omega \times \mathfrak{g}^\omega \to \mathfrak{z}$ with $Q^\omega(x, Ix) = 0$ for all nonzero elements x in $\mathfrak{g}^\omega$. Then

$$\mathfrak{g} = \mathfrak{h} \oplus \bigoplus_{\omega\in\Omega^+} \mathfrak{g}^\omega,$$

when endowed with the bracket

$$[h + \sum_{\omega\in\Omega^+} x_\omega,\, h' + \sum_{\omega\in\Omega^+} x'_\omega] = -\sum_{\omega\in\Omega^+} \big(Q^\omega(Ix_\omega, x'_\omega) + \omega(h)\cdot Ix'_\omega - \omega(h')\cdot Ix_\omega\big),$$

is a Lie algebra satisfying the conclusions of Theorem 1.17 *with Ω^+ as set of positive roots with respect to the Cartan algebra $\mathfrak{h}$.*

If $\mathcal{C}$ is the set of all rank one operators $\mathfrak{h} \mapsto \omega(h)\cdot Q^\omega(x_\omega, x_\omega)$ of $\mathfrak{h}$ with $\omega \in \Omega^+$ and $x_\omega \in \mathfrak{g}^\omega$, then every convex pointed generating cone C in $\mathfrak{h}$ satisfying $\mathcal{C}C \subseteq C$ is the trace $W \cap \mathfrak{h}$ of an invariant pointed generating cone W in $\mathfrak{g}$, and all pointed generating invariant cones in $\mathfrak{g}$ are so obtained. □

This theorem clarifies the solvable situation completely. SPINDLER has given a universal construction of general Lie algebras which can support invariant pointed generating cones in [22]

2. The local theory

Let $\mathfrak{g}$ again denote a finite dimensional Lie algebra and B a CH-neighborhood (see the first paragraph of the preceding section). A *local semigroup with respect to* B is a subset $S \subseteq B$ such that $0 \in S$ and $(S * S) \cap B \subseteq S$. We write $\mathbf{L}(S)$ for the set of subtangent vectors of S at 0. We know that $\mathbf{L}(S)$ is a Lie wedge. If G is a Lie group with Lie algebra $\mathfrak{g}$ and exp: $\mathfrak{g} \to G$ the exponential function, then $\exp S$ will be a local subsemigroup of G with respect to $U = \exp B$ provided B is so small that $\exp|B$ maps B bijectively onto its image U. The basic fact to record in this context is the local version of Lie's (Third) Fundamental Theorem which was established in a series of articles by HILGERT, HOFMANN and LAWSON. (For details see [7], Chapter IV.)

Theorem 2.1. (The Fundamental Theorem, Local Version) *For a finite dimensional real Lie algebra* $\mathfrak{g}$ *and a wedge* W *in* $\mathfrak{g}$, *the following statements are equivalent:*

(1) W *is a Lie wedge.*

(2) *There is a CH-neighborhood* B *and a local semigroup with respect to* B *such that* $W = \mathbf{L}(S)$. □

We notice that for a given W, *both* B and S depend on W. If W is a pointed cone and thus a Lie wedge, the widening of W will generally necessitate a reduction of B in size. The construction of S is not a trivial matter, notably if W has an edge. It is important to point out that S is not determined canonically; if W is pointed there are short cuts to an easy way of getting a suitable S (see [7]).

Quite in contrast with the corresponding situation for Lie groups, where a subalgebra $\mathfrak{h}$ of the Lie algebra $\mathfrak{g}$ of a Lie group G will always generate an analytic subgroup H with $\mathbf{L}(H) = \mathfrak{h}$, it is by no means true that every Lie wedge W in $\mathfrak{g}$ is the subtangent wedge $\mathbf{L}(S)$ of a subsemigroup of G even if G is simply connected. Examples are easy to come by, e. g., in the Heisenberg group (see [7], Examples VI.1.6.). Thus Theorem 2.1 is the best possible if we wish to realize a given Lie wedge in a Lie algebra as the tangent object of a semigroup in a Lie group – such a realization is possible only locally. The local theory leading to the local version of Lie's Fundamental Theorem for Lie wedges and semigroups has a distinct control theoretical flavor and the details are rather involved. We have remarked that there are examples of Lie wedges W in each compact not semisimple Lie algebra $\mathfrak{g}$ of rank at least three such that in any connected Lie group G with Lie algebra $\mathfrak{g}$ the analytic subgroup H with Lie algebra $W \cap -W$ is not closed. (See [5].)

With all the information we have in the local theory, unresolved questions remain.

Problem P.3. Suppose that W is a Lie wedge in the Lie algebra $\mathfrak{g}$ of a simply connected Lie group G containing a closed subgroup H with $\mathbf{L}(H) = W \cap -W$.

Is there an open neighborhood U of H in G and a subset $S \subseteq U$ with $\mathbf{L}(S) = W$ and $SS \cap U \subseteq S$? □

It is not unreasonable to expect that the methods developed by LAWSON for dealing with manifolds with invariant wedge fields or a variation of the them of rerouting (see [7]) might yield a solution of this problem.

3. The global theory

One aspect of the global theory is the general theory of subsemigroups of Lie groups and the extent to which they are determined by the set of one-parameter subsemigroups in their closures. Another aspect, however, is the question of the life of Lie-like semigroups outside Lie groups. The latter is probably one of the most unsettling aspects of the Lie theory of semigroups since our knowledge is still so fragmentary. We know these semigroups to exist; some of them are cancellative and defined on analytic manifolds with analytic semigroup multiplication. (See [13] and [7], Chapter VII.)

Lie's Fundamental Theorem – global version

Lie's Fundamental Theorem in its global and final version would have to be concerned with the question whether every Lie wedge is the tangent wedge at the identity of some topological semigroup. It would have to read something like this:

Conjecture 3.1. (The Fundamental Theorem, global version) *Suppose that G is a simply connected Lie group, H a closed analytic subgroup, and W is a Lie wedge in the Lie algebra $\mathfrak{g}$ whose edge $W \cap -W$ is the Lie algebra $\mathfrak{h}$ of H. Then one finds a topological semigroup S with group of units H_S, open neighborhoods N of H_S in S and U of H in G, and a homeomorphism $f: N \to S_U\ S_U \subseteq G$ such that the following conditions are satisfied:*

(i) $S_U S_U \cap U \subseteq S_U$.
(ii) $s,\ t,\ st \in U$ *implies* $f(st) = f(s)f(t)$,
(iii) f *maps* H_S *isomorphically onto* H,
(iv) $\mathbf{L}(S_U) = W$. □

Moreover, if this conjecture is correct, is there a morphism $F: S \to G$ which extends f?

In other words, every Lie wedge is the tangent wedge of a semigroup if its edge generates a closed analytic subgroup.

How much of the Fundamental Theorem do we know to be true? W. WEISS has given two constructions which show that Conjecture 3.1 is true if W is pointed. He shows

Theorem 3.2. ([24]) *If W is a pointed cone in a the Lie algebra $\mathfrak{g}$ of a Lie group G, then there are arbitrarily small neighborhoods B of 0 in $\mathfrak{g}$ for which there is a topological semigroup S and a continuous homomorphism $f\colon S \to G$ and for which* $\exp|B\colon B \to U$ *is a diffeomorphism of B onto an open neighborhood of 1 in G such that*

(i) *$f|f^{-1}(U)\colon f^{-1}(U) \to S_B$ is a homeomorphism onto a local semigroup S_B with respect to U,*

(ii) $\mathbf{L}(S_B) = W$. □

WEISS has two constructions for Theorem 3.2. The first is a category theoretical method yielding a free semigroup subject to given suitable local data. His second construction utilizes a combination of analysis, geometry and topological semigroup theory. He considers absolutely continuous curves γ starting from 1 in G and having their tangent vectors $\gamma'(t)$ in the translate $d\lambda_{\gamma(t)}(1)(W)$ of the wedge W, where $\lambda_g x = gx$ defines left translation in G. The choice of an invariant form which is positive with respect to the cone field allows him to consider a set of such curves which are suitably parametrized, and this set turns out to be a locally compact topological semigroup with respect to the operation of the concatenation of trajectories. The end point evaluation defines a morphism from this semigroup into G, and inside the kernel congruence of this homorphism, a suitable congruence can be identified such that the factor semigroup is the desired semigroup S. It is this technique which promises the best results towards Conjecture 3.1. We note that a proof of Conjecture 3.1 in its full generality would settle Problem P.2. in the affirmative. (See also the J. D. Lawson's chapter in this book.)

The globality of Lie wedges

When is a Lie wedge W in the Lie algebra $\mathfrak{g}$ of a Lie group G the subtangent wedge of a subsemigroup of G? This is by no means always the case, even for pointed cones, for which Conjecture 3.1 is proven to be true. The question is answered in principle by the following theorem (see Theorem VI.5.1 of [7]; the version given there has some extra frills).

Theorem 3.3. *Suppose that the Lie algebra $\mathfrak{g}$ of a Lie group G is generated as a Lie algebra by a Lie wedge W in $\mathfrak{g}$ and suppose that there is a closed analytic subgroup whose Lie algebra is the edge of W. Then there is a semigroup S in G with $\mathbf{L}(S) = W$ if and only if there is a smooth function $f\colon G \to \mathbb{R}$ which is linked with W in the following fashion:*

(i) *For all vectors $X \in d\lambda_g(\mathbf{1})(W)$ the number $df(g)(X)$ is nonnegative.*

(ii) *For all vectors $X \in W$ with $-X \neq W$ the number $df(\mathbf{1})(X)$ is positive.*

The existence of a smooth function $f: G \to \mathbb{R}$ satisfying (i) *and $df(\mathbf{1}) \neq 0$ is sufficient for the conclusion that the semigroup generated by* $\exp W$ *is proper (i.e., is different from G).* □

An enriched version of the preceding theorem is given by NEEB in [15], Theorem I.3.16. For the last statement of the theorem see [15], Proposition I.3.28. NEEB recently succeeded in improving one implication in Theorem 3.3 in a remarkable fashion. He shows:

Theorem 3.4. (NEEB) *Suppose that the Lie algebra $\mathfrak{g}$ of a Lie group G is generated as a Lie algebra by a Lie wedge W in $\mathfrak{g}$ and suppose that there is a closed analytic subgroup whose Lie algebra is the edge of W. Then there is a semigroup S in G with $\mathbf{L}(S) = W$ if and only if there is a smooth function $f: G \to \mathbb{R}$ such that for all vectors $X \in d\lambda_g(\mathbf{1})(W)$ with $-X \notin d\lambda_g(\mathbf{1})(W)$ the number $df(g)(X)$ is positive.* □

We say that a Lie wedge W in the Lie algebra of a Lie group G is *global* if there is a subsemigroup S in G with $\mathbf{L}(S) = W$.

The following consequence turned out to be very useful in many applications.

Proposition 3.5. *Suppose that $W_1 \subseteq W_2$ are Lie wedges in the Lie algebra of a Lie group G such that the analytic subgroup generated by the edge of W_1 is closed and that*

$$W_1 \setminus -W_1 \subseteq W_2 \setminus -W_2,$$

equivalently, that

$$W_1 \cap H(W_2) \subseteq H(W_1).$$

Then W_1 is global if W_2 is global. □

There are examples which illustrate the importance of the hypothesis that the subgroup generated by the edge of W_1 be closed (see [15], Example I.3.18.).

K.-H. NEEB has given a great variety of globality results in [15]. The role of the group of units of a semigroup in the context of globality is illustrated in the following result (see [15], Proposition I.3.20):

Proposition 3.6. (NEEB) *If G is a connected Lie group and W a generating Lie wedge in $\mathfrak{g}$, and if S denotes the semigroup generated by* $\exp W$ *in G, then the following statements are equivalent:*

(1) *W is global in G.*

(2) $\mathbf{L}(\overline{S} \cap \overline{S}^{-1}) \subseteq W \cap -W$.

(3) *The analytic subgroup generated by* $\exp(W \cap -W)$ *is closed in* G *and the relation* $\lim hs_n = \mathbf{1}$ *for elements* h, $s_n \in S$ *implies that* h *is a unit in* S. □

NEEB has shown that the functions f that occurred in Theorem 3.3 exist in abundant supply for global Lie wedges. In fact, he proves the following result ([15], Theorem II.2.5):

Theorem 3.7. (NEEB) *If* W *is a global generating Lie wedge in the Lie algebra* $\mathfrak{g}$ *of a connected Lie group* G, *then the set of linear forms* $df(\mathbf{1})$ *on* $\mathfrak{g}$ *as* f *ranges through all smooth functions on* G *satisfying the conditions* (i) *and* (ii) *of* Theorem 3.3 *is the algebraic interior of* W^*, *the wedge of linear forms which are nonnegative on* W. □

(The algebraic interior of a wedge V is the interior of V with respect to the linear span $V - V$.) The last conclusion of the theorem implies that there are enough functions f of the type described in Theorem 3.7 such that W is the intersection of the closed half-spaces containing W and bounded by the hyperplanes $df(\mathbf{1})^{-1}(0)$. NEEB has in fact established a duality theory between cones of smooth positive functions on G and closed infinitesimally generated subsemigroups S of G [17].

OL'SHANSKIĬ was the first to observe that there are even invariant cones which are not global in simple simply connected Lie groups, and for simple Lie groups he determined necessary and sufficient criteria for an invarinant wedge to be global (see [20] and [16]).

NEEB has in fact found some new results concerning globality in simple algebras. It is probably still premature to say that the question of the globality of invariant cones has reached a conclusive form. It was shown independely by GICHEV [3] and NEEB [15] that a pointed invariant generating cone in the Lie algebra of a simply connected solvable group is always global. In conjunction with Theorem 1.19 above this information gives us a clear picture of the globality problem in simply connected solvable Lie groups.

For Lie groups in general the formulation of necessary and sufficient conditions for an invariant pointed generating wedge to be global is, therefore, still an open problem. However, NEEB recently established the following result:

Theorem 3.8. *Every pointed generating invariant cone in the Lie algebra of an arbitrary simply connected Lie group contains an invariant pointed generating and global cone.* □

Globality of Ol'shanskiĭ wedges

There is one very interesting special class of Lie wedges which for which the globality issue is of great interest in the applications. Special but typical instances of these were discovered by OLSHANSKIĬ. Indeed in [7] the following result is proved (see Theorem V.4.57):

Proposition 3.9. *Let G be a connected Lie group with a Lie algebra $\mathfrak{g}$ which is a direct sum of a Lie algebra $\mathfrak{k}$ and a vector space $\mathfrak{p}$ satisfying $[\mathfrak{k}, \mathfrak{p}] \subseteq \mathfrak{p}$ and $[\mathfrak{p}, \mathfrak{p}] \subseteq \mathfrak{k}$. Suppose that C is a pointed cone in $\mathfrak{p}$ such that $e^{\operatorname{ad} \mathfrak{k}} C \subseteq C$ and that each* $\operatorname{ad} c$ *with $c \in C$ has a real spectrum. Suppose that the subgroup H generated by* $\exp \mathfrak{k}$ *and the set $S = (\exp C)H$ are closed in G. Assume further that H meets the set* $\exp -C \exp C$ *only in $\{\mathbf{1}\}$ and that 0 is the only central element of $\mathfrak{g}$ contained $C - C$ mapped onto $\mathbf{1}$ by* $\exp$. *Then S is a semigroup with $\mathbf{L}(S) = C \oplus \mathfrak{k}$ and S is generated by* $\exp(C \oplus \mathfrak{k})$. *In addition, the function $(X, h) \mapsto (\exp X)h : C \times H \to S$ is a homeomorphism.* □

In particular, the Ol'shanskiĭ wedge $\mathfrak{k}+C$ is global in G. We know of no example in which the closedness of S is not automatic in the presence of the other hypotheses. A typical example is the group $G = \mathrm{Sl}(2, \mathbb{C})$ with $\mathfrak{k} = \mathrm{sl}(2, \mathbb{R})$, $H = \mathrm{Sl}(2, \mathbb{R})$ and $\mathfrak{p} = i \cdot \mathrm{sl}(2, \mathbb{R})$ with $C = i \cdot \mathcal{K}$ where $\mathcal{K}$ is one half of the standard double cone in $\mathrm{sl}(2, \mathbb{R})$. Then $S = (\exp i \cdot \mathcal{K}) \, \mathrm{Sl}(2, \mathbb{R})$ is a subsemigroup of G homeomorphic to $\mathcal{K} \times \mathbb{R}^2 \times S^1$ with $\mathbf{L}(S) = i \cdot \mathcal{K} \oplus \mathrm{sl}(2, \mathbb{R})$. This Ol'shanskiĭ wedge is not a Lie semialgebra. The universal covering semigroup of S is homeomorphic to $\mathcal{K} \times \mathbb{R}^3$ and is locally isomorphic to S, but is not (algebraically) isomorphic to any subsemigroup of a Lie group. For more examples fitting the situation described in Theorem 3.8 see [7], pp. 438ff. The following conjecture sounds reasonable:

Conjecture 3.10. Every generating Ol'shanskiĭ wedge $W = \mathfrak{k} \oplus C$ with $C \subseteq \mathfrak{p}$ and $[\mathfrak{p}, \mathfrak{p}] \subseteq \mathfrak{k}$ in the Lie algebra $\mathfrak{g} = \mathfrak{k} \oplus \mathfrak{p}$ of a simply connected Lie group is global. □

A partial confirmation of this conjecture was proved by DÖRR [1]. He calls an invariant generating wedge W in a Lie algebra *adapted* if for a Levi decomposition $\mathfrak{r} \oplus \mathfrak{s}$ of the Lie algebra one has $W \cap \mathfrak{s} = p_{\mathfrak{s}}(W)$ where $p_{\mathfrak{s}}$ is the projection onto $\mathfrak{s}$ along $\mathfrak{r}$.

Proposition 3.11. (DÖRR) *Suppose that $\mathfrak{g} = \mathfrak{k} \oplus \mathfrak{p}$ with $\mathfrak{p} = i \cdot \mathfrak{k}$ and that G is the simply connected real Lie group with Lie algebra $\mathfrak{g}$, the complexification of $\mathfrak{k}$. Let $C = i \cdot W$ with an invariant pointed generating cone W which is adapted in $\mathfrak{k}$. Then the Ol'shanskiĭ wedge $\mathfrak{k} \oplus C$ is global in G.* □

Problem P.4. Prove or disprove Conjecture 3.10. □

The interior of infinitesimally generated semigroups

One very important aspect of subsemigroups in Lie groups is that they usually have interior points. "Usually" means that this is true whenever they have a subtangent wedge $W = \mathbf{L}(S)$ which generates the Lie algebra $\mathfrak{g}$ of the group as a Lie algebra. In the context of a Lie theory of semigroups, this is not a vast restriction. The Lie theory of semigroups can only capture those features which somehow are coded in $\mathbf{L}(S)$. Thus given S, and assuming that $\mathbf{L}(S)$ is defined (see Section 1 of Chapter V of [7]), we may always take the Lie algebra $\mathfrak{h}$ generated in $\mathfrak{g}$ by $\mathbf{L}(S)$. There is a unique analytic subgroup H generated by $\exp \mathfrak{h}$ and on H there is a unique topology and analytic structure making it into a Lie group H_L. Information on the subsemigroup $S \cap H_L$ is the very best we can expect a Lie theory to yield, and $\mathbf{L}(S \cap H_L) = \mathbf{L}(S)$ generates $\mathbf{L}(H_L) = \mathfrak{h}$.

The best results we now have on the interior of a subsemigroup of a Lie group in fact show that the boundary of such a semigroup cannot be very ragged. From each boundary point there is even an analytic curve directly into the interior.

Theorem 3.12. (Hofmann and Ruppert) *Let S be a subsemigroup of a Lie group and suppose that there is a subset $E \subseteq \mathfrak{g}$ generating the Lie algebra $\mathfrak{g}$ of G such that* $\exp \mathbb{R}^+ \cdot E \subseteq S$. *Then there are elements $X_1, \ldots, X_n \in \mathbb{R}^+ \cdot E$ such that the analytic curve $\phi \colon \mathbb{R}^+ \to G$ given by $\phi(t) = \exp t \cdot X_n \cdots \exp t \cdot X_1$ satisfies $\phi(0) = \mathbf{1}$ and $\phi(\mathbb{R}^+ \setminus \{0\}) \subseteq \operatorname{int} S$.*

Moreover, for any $X \in \mathbf{L}(S)$ there is an analytic curve $\gamma \colon \mathbb{R}^+ \to G$ such that $\gamma(0) = \mathbf{1}$, $\dot\gamma(0) = X$, and $\gamma(\mathbb{R}^+ \setminus \{0\}) \subseteq \operatorname{int} S$. □

(For a proof see [12].) It is a simple fact to establish that once a semigroup S in a topological group has inner points and $\mathbf{1} \in \overline{\operatorname{int} S}$, then the interior of the semigroup is a dense ideal. In a Lie group G with a subsemigroup S satisfying the conditions of the preceding theorem, the path components of $\overline{S}$ are closed and open in $\overline{S}$ and are connected by smooth arcs.

Divisible semigroups

A subsemigroup S of a group G is called *divisible* if for each $s \in S$ and each natural number $n \in \mathbb{N}$ there is an element $x \in S$ with $x^n = s$. By a result of Hofmann and Lawson (see [7], Chapter V, Section 6, pp. 459ff.) we know

Proposition 3.13. *A closed subsemigroup S of a connected Lie group G is divisible if and only* $\exp X = S$ *where $W = \mathbf{L}(S)$ is the tangent wedge of S.* □

There is a very reasonable conjecture:

Conjecture 3.14. The Lie wedge W of a closed divisible subsemigroup S of a connected Lie group is a Lie semialgebra. □

There is also a concept of local divisibility of a semigroup, and it can be shown that the Lie wedge of a locally divisible semigroup is a Lie semialgebra (see [7], Remark V.6.1).

In full generality, Conjecture 3.14 seems to be curiously hard to prove. In [7] it was shown that the conjecture is true in special cases which include all S with singleton groups $S \cap S^{-1}$ of units.

Problem P.5. Prove Conjecture 3.14 or else refute it by constructing a counterexample.

References

[1] Dörr, N., On Ol'shanskiĭ's semigroup, submitted, 1989

[2] Eggert, A., Zur Klassifikation von Semialgebren, Thesis, TH Darmstadt 1988

[3] Gichev, Y. M., Invariant orderings in solvable Lie groups, Siberian Math. J. 30 (1989), 57–69

[4] Hilgert, J., Applications of Lie semigroups in analysis, this Volume

[5] Hilgert, J., and K. H. Hofmann, On Sophus Lie's Fundamental Theorem, J. Funct. Analysis 67 (1986), 209–216

[6] —, On the causal structure of homogeneous manifolds, Preprint, Technische Hochschule Darmstadt, 1989

[7] Hilgert, J., K. H. Hofmann, and J. D. Lawson, "Lie groups, convex cones, and semigroups," Oxford University Press, Oxford 1989

[8] Hofmann, K. H., Semigroups in the 19th century? in: Theory of Semigroups, Conf. Theor. Appl. Semigroups Greifswald, 1984, Proceedings, Math. Gesellschaft d. DDR, Berlin 1985

[9] —, A memo on hyperplane subalgebras, Geom. Dedicata, to appear

[10] Hofmann, K. H., and J. D. Lawson, On Lie's Fundamental Theorems I, Indag. Math. 45 (1983), 453-466

[11] —, Foundations of Lie semigroups, in: Lecture Notes in Math. 998 (1983), 128–201

[12] Hofmann, K. H., and W. A. F. Ruppert, On the interior of subsemigroups of Lie groups, Trans. Amer. Math. Soc., to appear

[13] Lawson, J., Embedding semigroups into Lie groups, this Volume

[14] Loewner, Ch., "Collected Papers", Lipman Bers, Ed., Birkhäuser Boston, Basel, 1988

[15] Neeb, K.-H., "Globalität von Lie-Keilen", Diplom Thesis, Technische Hochschule Darmstadt, 1988

[16] —, More is true on SL(2), Preprint Darmstadt, 1989

[17] —, The duality between subsemigroups of Lie groups and monotone functions, Trans. Amer. Math. Soc., to appear

[18] —, Conal orders on homogeneous spaces with complete Riemannian metrics, submitted

[19] —, Globality in semisimple Lie groups, in: Dissertation, Technische Hochschule Darmstadt 1989

[20] Ol'shanskiĭ, G. I., Invariant orderings on simple Lie groups. The solution to E. B. VINBERG'S problem, Funct. Anal. and Appl. 18 (1984), 28–42

[21] Spindler, K., "Invariante Kegel in Liealgebren", Mitteilungen Math. Sem. Giessen 188 (1988)

[22] —, Some remarks on Levi complements and roots in Lie algebras with cone potential, Preprint, Conf. Anal. Topol. Theory of Semigroups, Oberwolfach 1989

[23] Vinberg, E. B., Invariant cones and orderings in Lie groups, Funct. Anal. and Appl. 14 (1980), 1–13

[24] Weiss, W., Local Lie-semigroups and open embeddings into global topological semigroups, Preprint, TH Darmstadt, 1989

Applications of Lie semigroups in analysis

Joachim Hilgert

This survey is an attempt to point out how the Lie theory of semigroups occurs in more classical parts of analysis. This means that I want to describe situations where semigroups and/or their tangent objects have actually been used to solve problems which *did not arise* in the semigroup context. The diversity of those applications made it necessary that I restricted myself to presenting the semigroup tools together with a short indication how they are applied rather than the application in full detail. To compensate for this I have included a fairly long list of references.

Analytic continuation of unitary representations

The abstract setting for this section will be the following. Let G be a Lie group and $\pi: G \to U(\mathcal{H})$ a unitary representation, i.e., a group homomorphism into the group of unitary operators $U(\mathcal{H})$ of a Hilbert space $\mathcal{H}$ such that the map $G \times \mathcal{H} \to \mathcal{H}$ defined by $(g, f) \mapsto \pi(g)f$ is continuous. From π we want to construct

$$\widetilde{\pi}: \Gamma \to C(\mathcal{H}),$$

where
- $C(\mathcal{H})$ is the semigroup of contractions, i.e., norm decreasing maps $\mathcal{H} \to \mathcal{H}$
- Γ is a complex manifold and $\widetilde{\pi}$ is holomorphic as a vector valued map
- G is the Shilov boundary of Γ and $\widetilde{\pi}$ is an analytic continuation of π
- Γ is a semigroup and $\widetilde{\pi}$ is a representation of Γ.

1. The metaplectic and the oscillator semigroup

Before we go into more theoretical aspects of the analytic continuation we describe a few relevant examples. Our first example has been studied by various physicists in the context of nuclear models. The objects of interest for these physicists were a class of integral operators on the Bargman-Fock space $\mathcal{F}_n$ of entire functions on $\mathbb{C}^n$ with the L^2-norm given by the measure $d\mu(\zeta) = \pi^{-n} e^{\overline{\zeta}^t \zeta} d\zeta$. The key observation was that these integral operators formed a semigroup, so the determination of a matrix semigroup for which the integral operators form a representation would be useful in replacing calculations with operators by simple matrix calculations. At least up to constants one could do that for a subsemigroup of the complex symplectic group. In [17] KRAMER, MOSHINSKY and SELIGMAN showed that it is possible to extend the projective representation of $\mathrm{Sp}(1,\mathbb{R})$ on the Bargmann-Fock-space defined by the uniqueness of the canonical commutator relations [2] to a subsemigroup with interior in $\mathrm{Sp}(1,\mathbb{C})$ and applied this representation to the nuclear cluster model (see also [16]). The corresponding analytic extension for the symplectic groups of arbitrary dimension were described in [5]. Later BRUNET [3] proved that the projective representation can be "integrated" to a contractive representation of a double covering semigroup of the aforementioned complex semigroup. The Shilov boundary of this covering semigroup is the metaplectic group and the representation restricts to the metaplectic representation. We give the precise definitions below.

We call a function on $\mathbb{C}^n$ of *Gaussian type* if it is of the form $\zeta \mapsto e^{-\frac{1}{2}\zeta^t A\zeta}$ where A is a symmetric complex matrix. Note that such functions are holomorphic on all of $\mathbb{C}^n$. Further we call a function of Gaussian type on $\mathbb{C}^n$ a *Gaussian function* if it belongs to the Bargmann-Fock Hilbert space $\mathcal{F}_n$. Let

$$X = \begin{pmatrix} A & B \\ B^t & D \end{pmatrix}$$

be an element of Ω_{2n}, the Siegel domain of complex symmetric $2n \times 2n$-matrices such that $X^*X < \mathbf{1}$. Then we set

$$K_X(\zeta,\overline{\omega}) = e^{-\frac{1}{2}(\zeta^t A\zeta + 2\zeta^t B\overline{\omega} + \overline{\omega}^t D\overline{\omega})} = e^{-\frac{1}{2}v^t X v},$$

where $v^t = (\zeta^t, \overline{\omega}^t)$. The corresponding kernel operator

$$f \mapsto \left(\zeta \mapsto \int_{C^n} K_X(\zeta,\omega) f(\omega) d\mu(\omega)\right)$$

will also denoted by K_X, i.e., we have

$$K_X f(\zeta) = \int_{C^n} e^{-\frac{1}{2}(\zeta^t A\zeta + 2\zeta^t B\overline{\omega} + \overline{\omega}^t D\overline{\omega})} f(\omega) d\mu(\omega).$$

The convergence of the integral can be shown using an isometry between $L^2(\mathbb{R}^n)$ and $\mathcal{F}_n$. The semigroup property follows from the multiplication law below ([5], 3.6).

Let $X, Y \in \Omega_{2n}$ and

$$X = \begin{pmatrix} A & B \\ B^t & D \end{pmatrix}, Y = \begin{pmatrix} \widetilde{A} & \widetilde{B} \\ \widetilde{B}^t & \widetilde{D} \end{pmatrix},$$

then we have

$$K_X \circ K_Y = \frac{1}{\left(\det(\mathbf{1} - \widetilde{A}D)\right)^{\frac{1}{2}}} K_Z \colon \mathcal{F}_n \to \mathcal{F}_n$$

where

$$Z = \begin{pmatrix} A + \left(B(\mathbf{1} - \widetilde{A}D)^{-1}\widetilde{A}B^t\right)^s & -B(\mathbf{1} - \widetilde{A}D)^{-1}\widetilde{B} \\ -\widetilde{B}^t(\mathbf{1} - \widetilde{A}D)^{-1}B^t & \widetilde{D} + \left(\widetilde{B}^t D(\mathbf{1} - \widetilde{A}D)^{-1}\widetilde{B}\right)^s \end{pmatrix}.$$

Here $C^s = \frac{1}{2}(C + C^t)$.

In [2], BARGMANN gives a realization of the projective representation of the symplectic group coming from the Stone-von Neumann Theorem via kernel operators on $\mathcal{F}_n$. He does not use $\mathrm{Sp}(n, \mathbb{R})$ but the isomorphic group $G = \mathrm{U}(n,n) \cap \mathrm{Sp}(n, \mathbb{C})$. Note that G is the set of all complex $2n \times 2n$-matrices of the form

$$g = \begin{pmatrix} A & B \\ \overline{B} & \overline{A} \end{pmatrix},$$

where A and B are $n \times n$-block matrices, which satisfy

$$\begin{aligned} AA^* - BB^* &= \mathbf{1} \\ A^t B &= B^t A \end{aligned}$$

or, equivalently,

$$\begin{aligned} A^* A - B^t \overline{B} &= \mathbf{1} \\ A^t \overline{B} &= B^* A . \end{aligned}$$

From this it follows that A is invertible and that the matrices $\overline{B}A^{-1}$ and $-A^{-1}B$ are symmetric. It is shown in [2], §3 that the projective representation of G on $\mathcal{F}_n$ is given by $g \mapsto F_g(\zeta, \overline{\omega})$, where

$$F_g(\zeta, \overline{\omega}) = e^{\frac{1}{2}(\zeta^t \overline{B} A^{-1} \zeta + \zeta^t (A^{-1})^t \overline{\omega} + \overline{\omega}^t A^{-1} \zeta - \overline{\omega}^t A^{-1} B \overline{\omega})}.$$

This means that F_g is a kernel operator of Gaussian type with matrix

$$X_g = -\begin{pmatrix} \overline{B}A^{-1} & (A^{-1})^t \\ A^{-1} & -A^{-1}B \end{pmatrix}.$$

In [5] BRUNET and KRAMER formally extend these kernels by simply replacing $\overline{B}$ by an arbitrary C and then find conditions in which the resulting kernels yield decent operators. The result is that it makes sense to write the above formula for a subsemigroup of contractions in the complex symplectic group, where the hermitean form which is being contracted is not a positive definite one. More precisely, if V is a complex vector space and $B: V \times V \to \mathbb{C}$ a nondegenerate hermitian form, we call

$$S_B = \{g \in \mathrm{Gl}(V): B(gv, gv) \leq B(v, v) \ \ \forall v \in V\}$$

the semigroup of *B-contractions*. Its tangent wedge $\mathbf{L}(S_B) = \{x \in \mathrm{gl}(V): e^{\mathbb{R}^+ x} \subseteq S_B\}$ is then given by

$$\mathbf{L}(S_B) = \{x \in \mathrm{gl}(V): B(xv, v) + B(v, xv) \leq 0\}. \tag{1.1}$$

(See [11].) Note that the interior S_B^o of S_B is given by the above formula with $\leq$ replaced by $<$. Now we can describe our subsemigroup of $\mathrm{Sp}(n, \mathbb{C})$ and its relation to the Gaussian kernel operators (see [9]).

Lemma 1.1. *Let $B: \mathbb{C}^n \times \mathbb{C}^n \to \mathbb{C}$ be the hermitian form given by the matrix*

$$L = \begin{pmatrix} -\mathbf{1} & 0 \\ 0 & \mathbf{1} \end{pmatrix}$$

and S_B the semigroup of B-contractions. Then

$$\phi(\begin{pmatrix} A & B \\ B^t & D \end{pmatrix}) = \begin{pmatrix} -(B^t)^{-1} & -(B^t)^{-1}D \\ A(B^t)^{-1} & -B + A(B^t)^{-1}D \end{pmatrix}$$

defines a map $\phi: \mathcal{D}_\Omega \to S_B^o$ where $\mathcal{D}_\Omega = \{X \in \Omega_{2n}: \det(B) \neq 0\}$. The map ϕ is invertible with inverse $\psi: S_B^o \to \mathcal{D}_\Omega$ given by

$$\psi(\begin{pmatrix} A & B \\ C & D \end{pmatrix}) = -\begin{pmatrix} CA^{-1} & (A^t)^{-1} \\ A^{-1} & -A^{-1}B \end{pmatrix}. \qquad \square$$

Proposition 1.2. (i) *The set $S_\Omega^\sharp = \{(cK_X) \in GK_C: X \in \mathcal{D}_\Omega\}$ is a subsemigroup of GK_C of Gaussian kernel operators and the map $\phi: \mathcal{D}_\Omega \to S_B^o$ induces a semigroup homomorphism $\phi: S_\Omega^\sharp \to S_B^o$.*

(ii) *The set*

$$S_\Omega = \{(cK_X) \in S_\Omega^\sharp: c^2 = \det(-B)\},$$

where

$$X = \begin{pmatrix} A & B \\ B^t & D \end{pmatrix},$$

is a subsemigroup of $S_\Omega^\sharp$ *and the semigroup homomorphism* $\phi: S_\Omega \to S_B^o$ *is a double covering.* □

The proposition above shows that the double covering of S_B^o together with its canonical map into S_Ω can be viewed as an example of our abstract setting as stated at the beginning of this section. Following BRUNET we call this semigroup the *metaplectic semigroup.*

Another example of a semigroup extension, which arises in a similar way, and in fact turns out to be essentially the same, can be found in HOWE's paper [14]. His goal, however, is purely mathematical, namely the proof of certain estimates for symbols of pseudo-differential operators. On the other hand the actual application of the resulting semigroups is also of a technical nature. Here, too one considers certain integral operators, this time on $L^2(\mathbb{R}^n)$, and shows that they can be viewed as a representation of a semigroup. The algebraic structure of this semigroup, or rather a simple extension of this semigroup, is then what is used in the proof of the estimates. Again we give the precise definitions.

We call a function on $\mathbb{R}^n$ a function of *Gaussian type* if it is of the form $\xi \mapsto e^{-\frac{1}{2}\xi^t A \xi}$ where A is a symmetric complex matrix. It is integrable if the real part of A is positive definite. We call a function of Gaussian type a *Gaussian function* if it is integrable or, equivalently, if the real part of A is positive definite, i.e., if A belongs the *generalized Siegel upper halfplane* $\mathcal{S}_n$.

Let $\mathcal{S}_{2n}$ be the Siegel upper halfplane of complex symmetric $2n \times 2n$-matrices with positive definite real part and let

$$X = \begin{pmatrix} A & B \\ B^t & D \end{pmatrix}$$

be an element of $\mathcal{S}_{2n}$. Then we set

$$K_X(\xi, \eta) = e^{-\frac{1}{2}(\xi^t A \xi + 2\xi^t B \eta + \eta^t D \eta)} = e^{-\frac{1}{2} v^t X v},$$

where $v^t = (\xi^t, \eta^t)$. Again the corresponding kernel operator

$$f \mapsto (\xi \mapsto \int_{\mathbb{R}^n} K_X(\xi, \eta) f(\eta) d\eta)$$

will also be denoted by K_X, i.e., we have

$$K_X f(\xi) = \int_{\mathbb{R}^n} e^{-\frac{1}{2}(\xi^t A \xi + 2\xi^t B \eta + \eta^t D \eta)} f(\eta) d\eta.$$

As in the previous case one derives the semigroup property from the multiplication law of the integral operators (see [14], 3.2.2).

Let $X, Y \in \mathcal{S}_{2n}$

$$X = \begin{pmatrix} A & B \\ B^t & D \end{pmatrix}, Y = \begin{pmatrix} \widetilde{A} & \widetilde{B} \\ \widetilde{B}^t & \widetilde{D} \end{pmatrix},$$

then we have

$$K_X \circ K_Y = \frac{(2\pi)^n}{\det(D + \widetilde{A})^{\frac{1}{2}}} K_Z \colon L^2(\mathbb{R}^n) \to L^2(\mathbb{R}^n)$$

where

$$Z = \begin{pmatrix} A - B(D + \widetilde{A})^{-1} B^t & -B(D + \widetilde{A})^{-1} \widetilde{B} \\ -\widetilde{B}^t (D + \widetilde{A})^{-1} B^t & \widetilde{D} - \widetilde{B}^t (D + \widetilde{A})^{-1} \widetilde{B} \end{pmatrix}.$$

As in in the case of the metaplectic semigroup it is possible to find a smaller semigroup by restricting the scalars to certain square roots depending on the kernel. More precisely, let

$$\mathcal{D}_o = \{X = \begin{pmatrix} A & B \\ B^t & D \end{pmatrix} \in \mathcal{S}_{2n} \colon \det B \neq 0\}.$$

Then the sets $S_o^\sharp = \{(cK_X) \in GK_{\mathbb{R}} \colon X \in \mathcal{D}_o\}$ and

$$S_{tw} = \{(cK_X) \in S_o^\sharp \colon c^2 = \det(-\frac{B}{2\pi})\},$$

where

$$X = \begin{pmatrix} A & B \\ B^t & D \end{pmatrix},$$

are subsemigroups of the semigroup of Gaussian kernel operators $(GK_{\mathbb{R}}, \circ)$. Yet Howe does not use this semigroup directly, but instead uses the *Weyl-transform*, a partial Fourier transform, and in this way replaces the Gaussian functions, viewed as integral operators, by Gaussian functions together with a *twisted convolution*.

The *Weyl transform* maps Schwartz functions on $\mathbb{R}^{2n}$ to kernel operators on $L^2(\mathbb{R}^n)$ via

$$\rho(F) = K_{\rho(F)}$$

where

$$K_{\rho(F)}(\xi, \eta) = \int_{\mathbb{R}^n} F(\xi - \eta, \tau) e^{\pi i (\xi + \eta)^t \tau} d\tau.$$

Proposition 1.3. ([14], 13.2) *Let* $v^t = (\xi^t, \eta^t)$ *and*

$$X = \begin{pmatrix} A & B \\ B^t & D \end{pmatrix} \in \mathcal{S}_{2n}.$$

Then

$$K_{\rho(F_X)} = \frac{(2\pi)^{\frac{n}{2}}}{(\det D)^{\frac{1}{2}}} K_{\widetilde{X}}$$

with

$$\widetilde{X} = \begin{pmatrix} A - (B - i\pi)D^{-1}(B^t - i\pi) & -A + (B - i\pi)D^{-1}(B^t + i\pi) \\ -A + (B + i\pi)D^{-1}(B^t - i\pi) & A - (B + i\pi)D^{-1}(B^t + i\pi) \end{pmatrix} \in \mathcal{S}_{2n}. \qquad \square$$

We denote the map $X \mapsto \widetilde{X}$ by $\widetilde{\rho}\colon \mathcal{S}_{2n} \to \mathcal{S}_{2n}$.

Proposition 1.4. ([14], §7 and [13]) *Let* $S(\mathbb{R}^{2n})$ *be the space of Schwartz functions on* $\mathbb{R}^{2n}$ *then*

$$\rho\colon (S(\mathbb{R}^{2n}), *_{tw}) \to (S(\mathbb{R}^{2n}), \circ)$$

is an involutive algebra isomorphism, where $*_{tw}$ *denotes twisted convolution, i.e.,*

$$F_1 *_{tw} F_2(v) = \int_{\mathbb{R}^{2n}} F_1(w) F_2(v - w) e^{-\pi i w^t J v}\, dw$$

with

$$J = \begin{pmatrix} 0 & \mathbf{1} \\ -\mathbf{1} & 0 \end{pmatrix}$$

and $\circ$ *the composition of integral operators on* $L^2(\mathbb{R}^n)$. $\square$

This now shows that the Weyl transform yields a canonical isomorphism

$$\rho\colon (GK_{\mathbb{R}}, *_{tw}) \to (GK_{\mathbb{R}}, \circ).$$

For the semigroup of Gaussian functions with twisted convolution we can give a subsemigroup via squareroots and a double covering onto a semigroup of contractions in the complex symplectic group. Unfortunately the construction is not as straightforward as in the case of the metaplectic semigroup. We have to consider the *operator Cayley transform* defined by

$$c_{op}(x) = (x + \mathbf{1})(x - \mathbf{1})^{-1}$$

whenever the inverse of $x - \mathbf{1}$ exists. We note that $(x+\mathbf{1})(x-\mathbf{1})^{-1} - \mathbf{1} = (x+\mathbf{1})(x-\mathbf{1})^{-1} - (x-\mathbf{1})(x-\mathbf{1})^{-1} = 2(x-\mathbf{1})^{-1}$ so that we can apply the Cayley transform twice.

Remark 1.5. Set $D_c = \{x \in \mathrm{gl}(V) : \det(x - \mathbf{1}) \neq 0\}$.
(i) $c_{op}^2 : D_c \to D_c$ is the identity.
(ii) $S_B^o \subseteq D_c$.
(iii) $c_{op} : \mathbf{L}(S_B) \cap D_c \to S_B \cap D_c$ is a bijection. □

Now we consider the hermitean form $B_{\mathbb{R}}$ on $\mathbb{C}^n$ given by the matrix

$$iJ = i\begin{pmatrix} 0 & \mathbf{1} \\ -\mathbf{1} & 0 \end{pmatrix}.$$

The subsemigroup of $\mathrm{Sp}(n, \mathbb{C})$ consisting of all elements which are contractions w.r.t. $B_{\mathbb{R}}$ will be denoted by $S_{B_{\mathbb{R}}}$. Note that it follows from (1.1) that the edge of $\mathbf{L}(S_{B_{\mathbb{R}}})$ is $\mathrm{sp}(n, \mathbb{R})$. In fact we have

$$B_{\mathbb{R}}(Xv, v) + B_{\mathbb{R}}(v, Xv) = 2\,\mathrm{Re}\big(B_{\mathbb{R}}(v, Xv)\big) = 2\,\mathrm{Re}(iv^* JXv).$$

Lemma 1.6. *The map $\beta : \mathrm{Mat}(2n, \mathbb{C}) \to \mathrm{Mat}(2n, \mathbb{C})$ defined by $\beta(X) = -\frac{i}{\pi}JX$ induces a linear isomorphism $\beta : \mathcal{S}_{2n} \to \mathrm{int}\,\mathbf{L}(S_{B_{\mathbb{R}}})$ which maps the set $\mathcal{D}_{tw} = \{X \in \mathcal{S}_{2n} : \det(X + i\pi J) \neq 0\}$ onto D_c (see* Remark 1.5*).* □

Now we are ready to describe HOWE's semigroup.

Proposition 1.7. ([14], §12) *The set $S_{tw}^\sharp = \{(cK_X) \in GK_{\mathbb{R}} : X \in \mathcal{D}_{tw}\}$ is a subsemigroup of $(GK_{\mathbb{R}}, *_{tw})$ and the map $(c, X) \mapsto c_{op}(-\frac{i}{\pi}JX)$ induces a semigroup homomorphism $S_{tw}^\sharp \to S_{B_{\mathbb{R}}}^o$. Moreover the set*

$$S_{tw} = \{(cK_X) \in S_{tw}^\sharp : c^2 = \frac{\det(X + i\pi J)}{(2\pi)^{2n}}\}$$

is a subsemigroup of $S_{tw}^\sharp$ and the semigroup homomorphism $c_{op} \circ \beta : S_{tw} \to S_{B_{\mathbb{R}}}^o$ is a double covering. □

HOWE calls the semigroup S_{tw} the *oscillator semigroup* and gives various references to work which comes close to defining this semigroup. A remarkable feature of this semigroup is that it contains many classical operators. The Shilov boundary of the the oscillator semigroup is the double covering of the real symplectic group, i.e., the metaplectic group, but it is not trivial to show that the semigroup representation is an analytic continuation of the metaplectic representation. One way of doing this is to use the fact that the oscillator semigroup is isomorphic to the metaplectic semigroup. The isomorphism is given via the isometry $U: L^2(\mathbb{R}^n) \to \mathcal{F}_n$ given by (see [1, 2])

$$Uf(\zeta) = \int_{\mathbb{R}^n} U(\zeta, \xi) f(\xi) d\xi,$$

where

$$U(\zeta, \xi) = \pi^{-\frac{n}{4}} e^{-\frac{1}{2}(\zeta^2+\xi^2)+\sqrt{2}\zeta^t \xi}.$$

This isometry leads to a map $U_{n,n}$ from $(GK_{\mathbb{R}}, \circ)$ to $(GK_C, \circ)$ which, on the level of matrices is given by $\alpha: X \mapsto c_{op}(X)^{-1}$, a bijection from $\mathcal{S}_m$ to Ω_m with inverse $Y \mapsto -c_{op}(Y)$. It turns out that the maps $\alpha: \mathcal{S}_{2n} \to \Omega_{2n}$ and $\widetilde{\rho}: \mathcal{S}_{2n} \to \mathcal{S}_{2n}$ induce bijections $\alpha: \mathcal{D}_o \to \mathcal{D}_\Omega$ and $\widetilde{\rho}: \mathcal{D}_{tw} \to \mathcal{D}_o$, respectively.

Note that the map $\theta: S_o \to S_o$ given by

$$\theta(cK_X) = (2\pi)^{\frac{n}{2}} cK_{2\pi X}$$

is an automorphism. If we now denote by $\widetilde{\theta}$ the multiplication by 2π we can describe the interplay of the metaplectic and the oscillator semigroup by the following commutative diagram with bijective vertical maps.

$$\begin{array}{ccccccccc}
(GK_{\mathbb{R}}, *_{tw}) & \supseteq & S_{tw} & \longrightarrow & \mathcal{D}_{tw} & \xrightarrow{\beta} & \operatorname{int}\mathbf{L}(S_{B_{\mathbb{R}}}) & \xrightarrow{c_{op}} & S^o_{B_{\mathbb{R}}} \\
\downarrow \rho & & \downarrow \rho & & \downarrow \widetilde{\rho} & & & & \\
(GK_{\mathbb{R}}, \circ) & \supseteq & S_o & \longrightarrow & \mathcal{D}_o & & & & \\
\uparrow \theta & & \uparrow \theta & & \uparrow \widetilde{\theta} & & & & \downarrow c_{geo} \\
(GK_{\mathbb{R}}, \circ) & \supseteq & S_o & \longrightarrow & \mathcal{D}_o & & & & \\
\downarrow U_{n,n} & & \downarrow U_{n,n} & & \downarrow \alpha & & & & \\
(GK_C, \circ) & \supseteq & S_\Omega & \longrightarrow & \mathcal{D}_\Omega & & \xrightarrow{\phi} & & S^o_B.
\end{array}$$

The only map that has not been described before is c_{geo}, a *geometric Cayley transform*, i.e., an inner automorphism of Sp(n, $\mathbb{C}$) given by $c_{geo}(g) = h_o g h_o^{-1}$ with

$$h_o = \frac{1}{\sqrt{2}} \begin{pmatrix} 1 & i \\ 1 & -i \end{pmatrix}.$$

Added in proof. The Fock realization of the oscillator semigroup has been studied independently by Folland (cf. G. B. Folland, Harmonic analysis in Phase space, Ann. Math. Studies **122**, Princeton Univ. Press,1989).

2. Hardy spaces and realization of the holomorphic discrete series

Let G be a semisimple Lie group with finite center. A central object of the harmonic analysis for G is the decomposition of the (left) regular representation of G

on $L^2(G)$ into a direct integral of irreducible representations. The representations which occur in the regular representation come in finitely many series. These series are in one-to-one correspondence with the conjugacy classes of *Cartan subgroups* of G. In [8] GELFAND and GINDIKIN put forth a program to realize these representations on spaces of holomorphic functions on certain complex domains associated to the conjugacy classes of Cartan subgroups in a natural way. Their approach is as follows.

Suppose that G is contained in a complex Lie group G_C with Lie algebra $\mathfrak{g}_C$, the complexification of $\mathfrak{g} = \mathbf{L}(G)$. Further let σ be the involutive automorphism of G_C whose set of fixed points is G. Now let $H^1, \ldots, H^k$ be a set of representatives for the conjugacy classes of Cartan subgroups of G. These Cartan subgroups have complexifications H_C^j which are defined by $\mathbf{L}(H^j) = \mathbf{L}(H_C^j) \cap \mathfrak{g}$. Note that for each H^j we find a complementary group H_j in H_C^j defined by

$$H_j = \{g \in H_C^j \colon \sigma(g) = g^{-1}\}.$$

It turns out that $\bigcup_{j=1}^k GH_jG = G_C$. In order to obtain a *disjoint* union one has to replace H_j by certain "*Weyl chambers*". One can show [40] that these "Weyl chambers" are of the form $\exp ic_j$, where c_j is an open convex cone in the Lie algebra $\mathfrak{h}^j$ of H^j. The sets $G(\exp ic_j)G$ are open in G_C. They are the domains mentioned above. The space $\mathcal{H}(c_j)$ of holomorphic maps on the domain $G(\exp ic_j)G$ consists of *all* holomorphic maps $f\colon G(\exp ic_j)G \to \mathbb{C}$ which satisfy

$$\sup_{g_1 \in G, h \in \exp ic_j} \int_G |f(g_1^{-1} h g_1 g)|^2 dg < \infty. \tag{2.1}$$

GELFAND's and GINDIKIN's paper contains no proofs, and it took nine years until there appeared a paper, [40], providing proofs for some of their claims. STANTON, its author, showed that in the case that H_j is a *compact* Cartan subgroup and c_j a certain Weyl chamber, the space above is indeed non-empty. In order to show this he proved that it is possible to analytically continue the representations $\pi_\lambda\colon G \to U(\mathcal{H}_\lambda)$ of the holomorphic discrete series from G to $G(\exp ic_j)G$ and then calculated that all the matrix units $g \mapsto (\pi(g)v, w)_{\mathcal{H}_\lambda}$ for $v, w \in \mathcal{H}_\lambda$ are contained in $\mathcal{H}(c_j)$. Neither GELFAND and GINDIKIN nor STANTON noticed that the domain to which they analytically continued the holomorphic discrete series was a *subsemigroup* of G_C. The one who *did* notice that was G.I. OL'SHANSKIĬ. As early as 1982, he worked out a version of the Gelfand-Gindikin program which was far more complete than that of STANTON (see [31]). Unfortunately, his work only appeared in a Russian proceedings volume and therefore was virtually inaccessible. Let us describe OL'SHANSKIĬ's work (note that he uses the right regular representation). The first observation is that the domain $G(\exp ic_j)G$ can be written as $G(\exp iW^o)$ where W is an invariant (under inner automorphisms) convex cone in $\mathfrak{g}$ with interior W^o. But it is known from earlier work of OL'SHANSKIĬ's [28] that this set is an open subsemigroup of G_C. We denote it by $\Gamma^o(W)$ and its closure by $\Gamma(W)$. Now we can replace the definition of $\mathcal{H}(c_j)$ given above by saying, a holomorphic mapping $f: \Gamma^o(W) \mapsto \mathbb{C}$ belongs to $H = \mathcal{H}(W)$ if

$$\|f\|_H = \sup_{\gamma \in \Gamma^o(W)} \int_G |f(\gamma g)|^2 dg < \infty. \tag{2.2}$$

Note that in this formulation we can understand the definition of $\mathcal{H}(W)$ not only as an *analogue* of the classical Hardy spaces [41], but even as a generalization. In fact if we set $G = \mathbb{R}$ and $W = \mathbb{R}^+$ then G_C is $\mathbb{C}$ and $\Gamma^o(W)$ the upper half plane. Thus the inequality above is just the defining inequality of the classical Hardy space. We note at this point that one of the strong points of OL'SHANSKIĬ's treatment of the Gelfand-Gindikin program is its flexibility.

It is easy to see that one has a representation of $\Gamma(W)$ on $\mathcal{H}(W)$. In fact, for $f \in \mathcal{H}(W)$ and $\gamma \in \Gamma(W)$ we define a function $\mathcal{T}(\gamma)f$ on $\Gamma^o(W)$ by the formula

$$\mathcal{T}(\gamma)f(\gamma_1) = f(\gamma_1\gamma).$$

Theorem 2.1. *The following statements hold under the present circumstances:*
(i) *$\mathcal{H}(W)$ is a Hilbert space w.r.t. the norm $\|\cdot\|_H$.*
(ii) *There exists an isometry $I: H \to L^2(G)$ such that for an arbitrary function $f \in H$ and an arbitrary sequence $\gamma_1, \gamma_2, \ldots$ in $\Gamma^o(W)$ which converges to* **1**, *the sequence $\{\gamma_j \cdot f\}$ converges to If w.r.t. the metric of $L^2(G)$.*
(iii) *I commutes with right translations from G, i.e., $I\mathcal{T}(g) = R(g)I$.*
(iv) *$\mathcal{T}(\cdot)$ is a holomorphic representation of the semigroup $\Gamma^o(W)$ on $\mathcal{H}(W)$.* □

Up to this point it is not yet clear whether all these statements are trivial, i.e., whether $\mathcal{H}(W)$ is non-empty. But OL'SHANSKIĬ also computes the image of I. We have to introduce some more notation in order to describe the result. Let π be an arbitrary unitary representation of the group G on any Hilbert space $\mathcal{H}$. To each $X \in i\mathfrak{g}$ one can associate the operator $\pi(X)$ on $\mathcal{H}$ which is determined by the condition

$$\pi(\exp itX) = \mathrm{Exp} it\pi(X), \quad \forall t \in \mathbb{R}.$$

Note that here Exp is a formal expression which really means that $i\pi(X)$ is the infinitesimal generator of the unitary one parameter group $t \mapsto \pi(\exp itX)$. We say that $\pi(X) \leq 0$ if the spectrum of the operator $\pi(X)$ is contained in the halfline $(-\infty, 0]$. Finally we say that the representation π is *W-admissible* if $\pi(X) \leq 0$ for all $X \in iW$.

Theorem 2.2. *$I(H)$ is the biggest $R(G)$-invariant subspace of L such that the corresponding unitary representation is W-admissible.* □

OL'SHANSKIĬ also gave a characterization of those representations which are W-admissible with respect to some invariant cone W. They are exactly the highest weight representations (see [15], [28]). If W is the minimal invariant cone in $\mathfrak{g}$ then the W-admissible representations which occur in the regular representation are exactly the representations from the holomorphic discrete series. Analogous results are possible also for certain non Riemannian symmetric spaces (see the work of OLAFSSON and ØRSTED [24], [25] as well as [12]).

3. The analytic continuation procedure

We now give a more detailed description of OL'SHANSKIĬ's analytic continuation procedure. If G is a Lie group and S is a subsemigroup of G then the tangent object $\mathbf{L}(S)$ (according to [11]) is a Lie wedge in $\mathfrak{g} = \mathbf{L}(G)$ which means that it is a closed convex cone which is invariant under the inner automorphisms of the form $e^{\operatorname{ad} x}$ with x contained in the edge $\mathbf{L}(S) \cap -\mathbf{L}(S)$ of $\mathbf{L}(S)$. If now G is a real Lie group inside a complexification G_C and S is a subsemigroup of G_C containing G then $\mathbf{L}(S)$ is of the form $\mathfrak{g} + iW$ where W is a wedge in $\mathfrak{g}$ which is invariant under all inner automorphisms of $\mathfrak{g}$. OL'SHANSKIĬ showed in [28] that for simple $\mathfrak{g}$ the semigroup generated by $\mathbf{L}(S)$ is just the product $(\exp iW)G$ where this product is even a *direct* product of topological spaces. The same result is true for solvable groups.

In this section we consider the following situation. Let G_C be a complex Lie group with Lie algebra $\mathfrak{g}_C$ and $\mathfrak{g}$ be a real form of $\mathfrak{g}_C$. We assume that G, the analytic subgroup of G_C with Lie algebra $\mathfrak{g}$, is closed. Let W be a proper generating invariant cone in $\mathfrak{g}$ such that the set $\Gamma(W) = (\exp iW)G$ is a closed subsemigroup of G_C. Moreover we assume that the map $G \times W \to \Gamma(W)$, defined by $(g, x) \to (\exp ix)g$ is a homeomorphism and even a diffeomorphism when restricted to $G \times W^o$. Finally we assume that there exists a (real) automorphism σ of G_C whose differential is the complex conjugation of $\mathfrak{g}_C$ with respect to the real form $\mathfrak{g}$.

Let $\pi: G \to U(\mathcal{H})$ be a unitary strongly continuous representation and $\mathcal{H}^\infty$ the set of C^∞-vectors for π. Then for any $x \in \mathfrak{g}$ the mapping $\phi: t \mapsto \pi(\exp tx)$ is a strongly continuous unitary one-parameter group in $\mathcal{H}$ so that by Stone's Theorem there is a unique, densely defined, skew adjoint, infinitesimal generator $-iH_x$ for ϕ. On $\mathcal{H}^\infty$ this generator is given by $d\pi(x)$ where $d\pi: \mathfrak{g} \to L(\mathcal{H}^\infty)$ is the Lie algebra representation associated with π. This infinitesimal representation extends to the universal enveloping algebra $\mathcal{U}(\mathfrak{g}_C)$ of $\mathfrak{g}_C$ and we set

$$C(\pi) = \{y \in \mathfrak{g}: (id\pi(y)\xi \mid \xi) \leq 0 \ \forall \xi \in \mathcal{H}^\infty\} \tag{3.1}$$

where $(\mid)$ is the inner product on $\mathcal{H}$. In other words $C(\pi)$ is the set of elements of $\mathfrak{g}$ for which $id\pi(x)$ is *negative* (see [6], Section 4.1). The elements of $C(\pi)$ will be called *negative elements.*

Proposition 3.1. *$C(\pi)$ is a closed convex cone in $\mathfrak{g}$ which is invariant under* $\mathrm{Ad}\, G$. □

For our invariant cone W in $\mathfrak{g}$ we denote the set of all unitary representations $\pi: G \to U(\mathcal{H})$ for which W consists of negative elements by $\mathcal{A}(W)$.

Definition 3.2. Let S be a semigroup with unit and $^\sharp: S \to S$ a bijection such that $s^{\sharp\sharp} = s$ and $(s_1 s_2)^\sharp = s_2^\sharp s_1^\sharp$ for all $s, s_1, s_2 \in S$. Then $^\sharp$ is called an *involution* on S and the pair $(S,^\sharp)$ is called a *semigroup with involution.*

If $\mathcal{H}$ is a (complex) Hilbert space and $C(\mathcal{H}) = \{T \in B(\mathcal{H}): ||T|| \leq 1\}$ is the set of all *contractions* on $\mathcal{H}$ then $(C(\mathcal{H}),^*)$ is a semigroup with involution, where T^* is the adjoint of T w.r.t. the inner product on $\mathcal{H}$. We provide $\Gamma(W)$ with an involution setting $g^\sharp = \sigma(g)^{-1}$ for all $g \in G_C$. The only thing we have to do in order to verify that $^\sharp$ is an involution of $\Gamma(W)$ is to show that it leaves $\Gamma(W)$ invariant. But this follows from $\big((\exp x)g\big)^\sharp = \sigma\big((\exp x)g\big)^{-1} = \big((\exp -x)g\big)^{-1} = g^{-1}(\exp x) \in \Gamma(W)$.

Now suppose that $\pi \in \mathcal{A}(W)$. Then for any $x \in W$ the operator $id\pi(x)$ generates a selfadjoint contraction semigroup (see [6], Theorem 4.6) which we denote by $t \mapsto T_x(t)$.

Definition 3.3. Let $(S,^\sharp)$ be a topological semigroup with involution, then a semigroup homomorphism $\rho: S \to C(\mathcal{H})$ is called a *contractive representation of* $(S,^\sharp)$ if it preserves the involution and is continuous w.r.t. the weak operator topology on $C(\mathcal{H})$. A contractive representation is called *irreducible* if there is no closed non-trivial subspace of $\mathcal{H}$ which is invariant under $\rho(S)$.

A contractive semigroup representation $\rho: \Gamma(W) \to C(\mathcal{H})$ is called *holomorphic* if the map $\rho \mid_{\Gamma(W)^o}$ is holomorphic (for the definition of holomorphy in Hilbert-(Banach) spaces see [42], Section 4.4).

We want to construct a holomorphic representation of $\Gamma(W)$ starting from a representation $\pi \in \mathcal{A}(W)$. The definition of the analytic continuation is straightforward. In fact, for $\gamma = (\exp tx)g \in \Gamma(W)$ we define $\tau(\gamma) = T_x(t)\pi(g)$. Then we get a strongly continuous map $\tau: \Gamma(W) \to C(\mathcal{H})$ which preserves the involutions. The problem now is to show that τ has all the desired properties. The main problem is to show the holomorphy.

First one notes that if $f: \Gamma(W) \to \mathbb{C}$ is continuous and $f \mid_{\Gamma(W)^o}$ is holomorphic such that f vanishes identically on G, then $f \equiv 0$. Then one has to use NELSON's theory of analytic vectors (see [42]).

Lemma 3.4. *There exists a neighborhood $\mathcal{U}$ of $\mathbf{1}$ in G_C and a holomorphic mapping $\widetilde{\tau}_\xi: \mathcal{U} \to \mathcal{H}$ for each $\xi \in \mathcal{H}'$, where $\mathcal{H}'$ is a dense subspace of $\mathcal{H}$ consisting of analytic vectors, such that*

$$\widetilde{\tau}_\xi(g) = \pi(g)\xi \qquad \text{for all } g \in G \cap \mathcal{U}. \qquad \square$$

Lemma 3.5. *The following assertions hold:*
(i) $\tau(\gamma)\xi = \widetilde{\tau}_\xi(\gamma)$ *for all* $\gamma \in \Gamma(W) \cap \mathcal{U}, \xi \in \mathcal{H}^\omega$.
(ii) $\gamma \mapsto \tau(\gamma)\xi$ *is holomorphic for all* $\xi \in \mathcal{H}$ *and* $\gamma \in \Gamma(W)^o \cap \mathcal{U}$.
(iii) *The map* $\gamma \mapsto \tau(\gamma)\xi$ *is real analytic on* $\Gamma(W)^o$ *for all* $\xi \in \mathcal{H}^\omega$, *where* $\mathcal{H}^\omega$ *is the set of analytic vectors in* $\mathcal{H}$ *for* π.
(iv) *The map* $\gamma \mapsto \tau(\gamma)$ *is holomorphic on* $\Gamma(W)^o$. □

Theorem 3.6. *Let $\pi \in \mathcal{A}(W)$ then π extends uniquely to a holomorphic representation τ of $\Gamma(W)$.*

Proof. We know that the map $\gamma \mapsto \tau(\gamma)$ is holomorphic on $\Gamma(W)^o$ and extends π. Moreover we know $\tau(\gamma^\sharp) = \tau(\gamma)^*$. If now $g \in G$ is fixed then $\gamma \mapsto \tau(\gamma g)$ and $\gamma \mapsto \tau(\gamma)\pi(g)$ are holomorphic on $\Gamma(W)^o$ and coincide on G. Thus we have

$$\tau(\gamma g) = \tau(\gamma)\pi(g) = \tau(\gamma)\tau(g).$$

Now we fix $\gamma_o \in \Gamma(W)$ and consider $\gamma \mapsto \tau(\gamma_o\gamma)$ as well as $\gamma \mapsto \tau(\gamma_o)\tau(\gamma)$. Again we know that both maps are holomorphic on $\Gamma(W)^o$ $\big($since $\Gamma(W)^o$ is a semigroup ideal in $\Gamma(W)\big)$ and by the above they agree on G. This shows that τ is a semigroup homomorphism. The uniqueness of the extension is clear. □

It is also possible to prove a converse of Theorem 3.6. Note first that the restriction to G of a holomorphic representation is always a unitary representation. In fact we have:

Remark 3.7. Let $(S,^\sharp)$ be a semigroup with involution and $\rho: S \to C(\mathcal{H})$ a contractive representation. Set $H = \{s \in S: s^\sharp s = s^\sharp s = \mathbf{1}\}$ then H is a subgroup of S and the restriction of ρ to H is a strongly continuous unitary representation. □

Theorem 3.8. *Let τ be a holomorphic representation of $\Gamma(W)$ then any $x \in W$ is a negative element for $\pi = \tau\,|_G$, so that $\pi \in \mathcal{A}(W)$.*

Proof. If $x \in W$ we consider the map

$$z \mapsto \tau(\exp zix) = \tau\big(\exp(i \operatorname{Re} zx) \exp(-\operatorname{Im} zx)\big)$$

for $\operatorname{Re} z \geq 0$ and the hypothesis says that it is holomorphic for $\operatorname{Re} z > 0$. On $\operatorname{Re} z = 0$ this map is just $it \mapsto \tau(\exp -tx) = \pi(\exp -tx)$. Since $\tau(\gamma^\sharp) = \tau(\gamma)^*$ we know that $\tau(\exp tix)$ is self adjoint for $t \geq 0$. If $d\tau(ix)$ is the infinitesimal generator of $\tau(\exp tix)$ then we see that $d\tau(ix) = id\pi(x)$ and the claim follows. □

Analytic extensions of semigroup representations

OL'SHANSKIĬ's original interest in semigroups did not come from the Gelfand-Gindikin program. It rather occurred to him that he could use semigroups in the study of representations of infinite dimensional analogues of certain classical groups. A key ingredient in his "semigroup method" is again an analytic continuation procedure – this time from a representation of a local semigroup to a representation of a related Lie group. This procedure is due to two physicists, LÜSCHER and MACK [22], and can be viewed as an extension of the Hille-Phillips Theorem for one parameter semigroups.

4. Representations of infinite dimensional classical groups

The purpose of OL'SHANSKIĬ's work (see [26], [27], [32]) is to determine those unitary representations of the infinite dimensional groups $\mathrm{SO}_0(p,\infty)$, $\mathrm{U}(p,\infty)$, $\mathrm{Sp}(p,\infty)$ for $p = 0, 1, 2, \ldots$ which are admissible in a sense to be specified below. In order to simplify notation we restrict ourselves to the case $\mathrm{U}(p,\infty)$. This group can be realized on the complex Hilbert space $l^2(\mathbb{N})$ as a subgroup of the set of invertible operators preserving the (non-definite) inner product $J(\xi,\xi) = \langle \xi,\xi\rangle = -|\xi_1|^2 - \cdots - |\xi_p|^2 + |\xi_{p+1}|^2 + \ldots$ We denote by $\mathcal{G}(n)$ the group of all such operators that preserve all but the first n elements of the canonical basis of $l^2(\mathbb{N})$. Then we define

$$\mathcal{G} = \mathrm{U}(p,\infty) = \bigcup_{n\in\mathbb{N}} \mathcal{G}(n).$$

Further we let $\mathcal{G}_n$ be the subgroup of $\mathcal{G}$ consisting of all operators which fix the first n basis vectors. If now π is a unitary representation which is continuous w.r.t. the inductive limit topology, i.e., when restricted to the $\mathcal{G}(n)$, and $\mathcal{H}$ the Hilbert space on which the $\pi(g)$ operate, then we denote by $\mathcal{H}_n$ the space of vectors in $\mathcal{H}$ which are fixed by the elements $\pi(g)$ with $g \in \mathcal{G}_n$. The representation π is called *admissible* if the space $\mathcal{H}_\infty = \bigcup_{n\in\mathbb{N}} \mathcal{H}_n$ is dense in $\mathcal{H}$. Ol'shanskii shows that this is equivalent to continuity w.r.t. the norm topology on $\mathcal{G}$.

If one views the elements of $\mathcal{G}$ as infinitely large matrices it is clear what it means to associate to $g \in \mathcal{G}(k)$ its *left upper corner* of size n. We denote this map by θ_n and note that its image is contained in the semigroup $\Gamma(n)$ of J-contractions, where J is the hermitean form from above but restricted to the span of the first n basis vectors. More is true:

Lemma 4.1. *For k large enough we have*
(i) $\theta_n\big(\mathcal{G}(k)\big) = \Gamma(n)$.
(ii) *If $g_1, g_2 \in \mathcal{G}(k)$, then $\theta(g_1) = \theta(g_2)$ if and only if $g_1 = ug_2v$ for some elements $u, v \in \mathcal{G}_n(k) = \mathcal{G}(k) \cap \mathcal{G}_n$.*
(iii) *The map $\theta_n: \mathcal{G}(k) \to \Gamma(n)$ is proper and open.* □

Next we assume that π is an admissible representation of $\mathcal{G}$ and n is so large that $\mathcal{H}_n$ is non-zero. Let $P_n: \mathcal{H} \to \mathcal{H}_n$ be the orthogonal projection and $C(\mathcal{H}_n)$ the semigroup of contractions on $\mathcal{H}_n$ w.r.t. the Hilbert inner product. The lemma above now guarantees the existence of a continuous map $\widetilde{\pi}_n: \Gamma(n) \to C(\mathcal{H}_n)$ defined by $\widetilde{\pi}_n\big(\theta(g)\big) = P_n\pi(g)|_{\mathcal{H}_n}$.

Theorem 4.2. *$\widetilde{\pi}_n: \Gamma(n) \to C(\mathcal{H}_n)$ is a holomorphic semigroup representation.* □

Note that we are in the following situation. $G = \mathrm{Gl}(n, \mathbb{C})$ is a Lie group with involution $\sigma(x) = -X^\sharp$, where $^\sharp$ denotes the adjoint w.r.t. the form J. Let $\mathfrak{g}_\pm = \{x \in \mathfrak{g}: d\sigma(x) = \pm x\}$ and $H = \{g \in G: \sigma(g) = g\}_o$. Further $C = \{x \in \mathfrak{g}: x = x^\sharp, \langle x\xi, \xi\rangle \leq 0\ \forall \xi \in \mathbb{C}^n = l^2(1, \ldots, n)\} \subset \mathfrak{g}_-$ is an $\mathrm{Ad}\, H$-invariant cone and $\Gamma(n)$ has $\mathfrak{g}_+ + C$ as tangent cone.

Let $\mathfrak{g}^* = \mathfrak{g}_+ + i\mathfrak{g}_-$ and suppose that G^* denotes the simply connected Lie group with $\mathbf{L}(G^*) = \mathfrak{g}^*$. In our case we have $\mathfrak{g}^* = \mathfrak{u}(p, q) \oplus \mathfrak{u}(p, q)$ so that we may view $\mathfrak{g}$ as a part, the diagonal, of $\mathfrak{g}^*$.

Theorem 4.3. (LÜSCHER and MACK) *Let $\rho: \Gamma(n) \to C(\mathcal{H})$ be a strongly continuous involutive representation, then there exists a representation $\pi: G^* \to U(\mathcal{H})$ such that*

$$d\pi(ix) = id\rho(x)\ \ \forall x \in C$$

and

$$d\pi(x) = d\rho(x)\ \ \forall x \in \mathfrak{g}_+. \qquad \square$$

We ought to note here that the Lüscher–Mack Theorem is true for weaker hypotheses. In particular it would be sufficient to have a local semigroup instead of $\Gamma(n)$.

If one applies the Lüscher Mack theorem to all the $\widetilde{\pi}_n: \Gamma(n) \to C(\mathcal{H}_n)$ then one obtains representations $\pi_n^*: \mathcal{G}^*(n) \to U(\mathcal{H}_n)$ for large n. These representations may be collected to give one representation $\pi^*: \mathcal{G}^* \to U(\mathcal{H})$, where $\mathcal{G}^* = \bigcup_{n\in\mathbb{N}} \mathcal{G}_n^*$. It turns out that due to the holomorphy of the representations $\widetilde{\pi}_n$ one can describe π^* in terms of two sequences of natural numbers (roughly to be interpreted as highest weights for the subgroups $\mathrm{U}(\infty) \times \mathrm{U}(\infty)$ of $\mathcal{G}^*$). Finally one notes that with the embedding $\mathcal{G} \subseteq \mathcal{G}^*$ one has $\pi = \pi^*|_{\mathcal{G}}$ so that π can also be described in terms of the sequences mentioned above (see [32]).

Laplace transforms

It is well known that the Fourier transform as an integral operator on $\mathbb{R}$ has a group theoretical interpretation. The generalizations of the Fourier transform to Lie groups play an important role in (non-commutative) harmonic analysis. Whereas the Laplace transform, an integral operator on the semigroup $\mathbb{R}^+$, is also important in classical analysis, until recently there seems to have been no attempt to generalize it to a wider class of semigroups. We will describe below some work of MIZONY (see [23]) and FARAUT (together with the physicist VIANI, see [7]) which can be viewed as first steps in this direction.

5. Semigroup theoretical interpretation of special functions

Let G/H be a symmetric space and $G = KAN$ the Iwasawa decomposition of G. Consider the boundary K/M of the Riemannian symmetric space G/K, where M is the centralizer of A in K. Note that $K/M = G/P$ with $P = MAN$ the corresponding minimal parabolic subgroup. This shows that G operates on K/M. Now let $\mathcal{O}$ be open in K/M and set

$$S(\mathcal{O}) = \{g \in G : g^{-1} \cdot (\mathcal{O}) \subseteq \mathcal{O}\}.$$

It is clear that $S(\mathcal{O})$ is a semigroup. If $\mathcal{O}$ is an H-orbit we can consider the image measure $d\mu$ on $\mathcal{O}$ of the Haar measure dh on H. The *Poisson kernel* $P_{\mathcal{O}}: S(\mathcal{O}) \times \mathcal{O} \to \mathbb{R}^+$ is defined by

$$P_{\mathcal{O}}(g, \eta) = \frac{d(g^{-1})^* \mu(\eta)}{d\mu(\eta)}$$

for $g \in S(\mathcal{O})$ and $\eta \in \mathcal{O}$.

Proposition 5.1. (a) $P_{\mathcal{O}}(g_1 g_2, \eta) = P_{\mathcal{O}}(g_1, \eta) P_{\mathcal{O}}(g_2, g_1^{-1} \cdot \eta)$
(b) $P_{\mathcal{O}}(h, \eta) = 1$ *for all* $h \in H, \eta \in \mathcal{O}$. □

This proposition shows that for any $\alpha \in \mathbb{C}$

$$\pi_\alpha(g) f(\eta) = P_{\mathcal{O}}(g, \eta)^\alpha f(g^{-1} \cdot \eta) \tag{5.1}$$

defines a representation of $S(\mathcal{O})$ on the space of functions on $\mathcal{O}$. If the function $\pi_\alpha(g)1 \in L^1(\mathcal{O}, d\mu)$ then we set

$$\Phi(\alpha, g) = \int_{\mathcal{O}} (\pi_\alpha(g)1)(\eta) d\mu(\eta). \tag{5.2}$$

Remark 5.2. Let $\pi_\alpha(g)1 \in L^1(\mathcal{O}, d\mu)$ for all $g \in S(\mathcal{O})^o$, then
(a) $\Phi(\alpha, g) = \Phi(\alpha, h_1 g h_2)$ for all $h_1, h_2 \in H$.
(b) $\Phi(\alpha, g_1)\Phi(\alpha, g_2) = \int_H \Phi(\alpha, g_1 h g_2) dh$. □

Now we assume that the open semigroup $S(\mathcal{O})^o$ admits a Cartan decomposition

$$S(\mathcal{O})^o = H S_A^o H \tag{5.3}$$

where $S_A^o \subseteq A$ and the Haar measure of G on $S(\mathcal{O})^o$ decomposes as $dg = dh d\nu(s) dh$. Then the second formula from Remark 5.2 with $s_1, s_2 \in S_A^o$ can be written as

$$\Phi(\alpha, s_1)\Phi(\alpha, s_2) = \int_{S_A^o} K(s_1, s_2, s)\Phi(\alpha, s) d\nu(s), \tag{5.4}$$

for some kernel function K. Now we can define a *convolution product* and a *spherical Laplace transform* by

$$f *_{\mathcal{O}} g(s) = \int_{S_A^o} \int_{S_A^o} K(s_1, s_2, s) f(s_1) g(s_2) d\nu(s_1) d\nu(s_2) \tag{5.5}$$

for $f, g \in C_c^\infty(S_A^o)$ and

$$\mathcal{L}_{\mathcal{O}}(f)(\alpha) = \int_{S_A^o} f(s)\Phi(\alpha, s) d\nu(s) \tag{5.6}$$

for all α with $(\pi_\alpha(s)1) \in L^1(\mathcal{O}, d\nu)$ for all $s \in S(\mathcal{O})^o$. We obtain

Proposition 5.3. *Let $f, g \in C_c^\infty(S_A^o)$ then*

$$\mathcal{L}_{\mathcal{O}}(f *_{\mathcal{O}} g)(\alpha) = \mathcal{L}_{\mathcal{O}}(f)(\alpha)\mathcal{L}_{\mathcal{O}}(g)(\alpha). \quad \square$$

With this purely formal machinery at hand we can indicate Mizony's results. They are of a fairly technical nature, so we give just one theorem omitting some details. For the full story we refer to [23].

Theorem 5.4. *Let $G = \mathrm{SO}_0(1,n)$ and $H = \mathrm{SO}_0(1,n-1)$ then S_A^o can be identified with the positive reals such that*

$$d\nu(t) = \frac{1}{(\sinh t)^{n-1}} dt.$$

The function $\Phi(\alpha,t)$ can be defined for $\alpha = \rho - i\lambda$ where $\rho = \frac{n-1}{2}$ and $\mathrm{Im}\lambda > \rho - 1$. The $\Phi(\alpha,\cdot)$ are exactly the Jacobi functions of the second kind. □

Finally we note that Mizony also gives an inversion formula for the spherical Laplace transform in the case $G/H = \mathrm{SO}_0(1,n)/\mathrm{SO}_0(1,n-1)$.

6. Ordered spaces and partial diagonalization of integral operators

Let $X = G/H$ be a symmetric space with an invariant ordering $\leq$, i.e., $x \leq y$ implies $g{\cdot}x \leq g{\cdot}y$ for all $g \in G$. We assume that the intervals $[y,x] = \{z \in X : y \leq z \leq x\}$ are compact. A kernel $K(x,y)$ on X is called a *Volterra kernel* if it is continuous on $\Gamma = \{(x,y) : y \leq x\}$ and vanishes outside Γ. We denote the set of all Volterra kernels by $V(X)$. The product of two Volterra kernels is given by

$$K_1 \sharp K_2(x,y) = \int_{[y,x]} K_1(x,z) K_2(z,y) dz. \tag{6.1}$$

A Volterra kernel is called *invariant* if $K(g{\cdot}x, g{\cdot}y) = K(x,y)$ for all $x,y \in X$. The set of invariant Volterra kernels is denoted by $V(X)^\natural$. Note that $V(X)$ as well as $V(X)^\natural$ form an algebra under $\sharp$.

Let x_o be a base point of X fixed by H. Then we can identify an invariant Volterra kernel with a function $f : G \to \mathbb{C}$ which is continuous on $S = \{g \in G : g{\cdot}x_o \geq x_o\}$, vanishes outside S and is bi-invariant under H. Using this one finds

Proposition 6.1. $V(X)^\natural$ *is commutative.* □

Now we suppose that S is of the type $S(\mathcal{O})$ satisfying the various hypotheses from the previous paragraph. Then the f_K coming from invariant Volterra kernels can be viewed as functions on S_A^o.

Proposition 6.2. *Let K_1, K_2 be Volterra kernels and f_{K_1} and f_{K_2} the associated functions on S_A^o. Suppose that f_{K_1} and f_{K_2} lie in the domain of the spherical Laplace transform, then*

$$\mathcal{L}_{\mathcal{O}}(f_{K_1}) \mathcal{L}_{\mathcal{O}}(f_{K_2}) = \mathcal{L}_{\mathcal{O}}(f_{K_1 \sharp K_2}).$$ □

Now suppose that G acts on $\mathbb{R}^k$ and that $\mathbb{R}^k$ is ordered by a proper generating cone. Consider the integral equation

$$A(u,v) = B(u,v) + \int_{\mathbb{R}^k} N(u,u')A(u',v)du' \tag{6.2}$$

where A, B and N are Volterra kernels with respect to the ordered space $\mathbb{R}^k$. We make the following assumptions:

1) The equation (6.2) is G-invariant, where G acts simultaneously on the two variables.
2) For all u and v under consideration (this may be less than $\mathbb{R}^k$) the interval $[v,u]$ is contained in a union of G-orbits $\bigcup_{p\in P} G\cdot u_p$ such that the isotropy group of u_p is H for all $p \in P$.
3) As a measure space we can write

$$\bigcup_{p\in P} G\cdot u_p = P \times G/H$$

with a finite measure dp on P.

4) The orderings of $\bigcup_{p\in P} G\cdot u_p$ and G/H are compatible, i.e., if $u = (p,x) \geq (q,y) = v$ then $x \geq y$.

Then we can write

$$\begin{aligned}\int_{\mathbb{R}^k} N(u,u')A(u',v)du' &= \int_P \int_{G/H} N((p,x),(p',x'))A((p',x'),(q,y))dx'dp' \\ &= \int_P \int_{[y,x]} N((p,x),(p',x'))A((p',x'),(q,y))dx'dp' \\ &= \int_P N_{p,p'} \sharp A_{p',q}(x,y)dp'\end{aligned}$$

where $N_{p,p'}(x,x') = N((p,x),(p',x'))$ and $A_{p',q}(x',y) = A((p',x'),(q,y))$. Note that $N_{p,p'}$ and $A_{p',q}$ are G-invariant kernels. They are Volterra type kernels because of the compatibility (the continuity condition has to be weakened a little since $N_{p',p}$ and $A_{p',q}$ may jump to zero within Γ). Thus we finally may rewrite (6.2) as

$$\mathcal{L}_{\mathcal{O}}(f_{A_{p,q}}) = \mathcal{L}_{\mathcal{O}}(f_{B_{p,q}}) + \int_P \mathcal{L}_{\mathcal{O}}(f_{N_{p,p'}})\mathcal{L}_{\mathcal{O}}(f_{A_{p',q}})dp' \tag{6.3}$$

which is the anounced partial diagonalization. For a concrete example involving $G/H = \mathrm{SO}_0(1,n)/\mathrm{SO}_0(1,n-1)$ we refer to [7].

Differential equations and causality

Semigroups can be used in the study of the causal structure of homogeneous manifolds. Viewing causal (or timelike) paths as solutions of certain differential inclusions this can certainly be viewed as a part of analysis. On the other hand the methods involved are more geometrical than analytical in nature and moreover quite similar to the methods which come from the interplay between Lie semigroups and geometric control theory. Thus for this topic we only refer to the original papers [19, 20], [10], [18] and KUPKA's article in these proceedings.

7. Stability of causal differential equations

In this section we describe a theorem of PANEITZ on the asymptotic behaviour of certain differential equations. It has been used in the context of quantization for curved space-times (see [35], [36], [38] and [39]). We include this theorem because the infinite dimensional analogue of the invariant cone in $\mathrm{sp}(n, \mathbb{R})$ plays a decisive role in it and note in passing that also the invariant cones in $\mathrm{su}(2, 2)$ show up in the further development of PANEITZ' theory as Killing vectors for positive conserved (Noether) quantities (see [38]).

Let $\mathcal{H}$ be a *real* Hilbert space with inner product $(\cdot, \cdot)$. For a given complex structure J on $\mathcal{H}$, i.e $J = J^t$ and $J^2 = I$, we set

$$\omega_J(v, w) = (v, Jw)$$

and

$$\langle v, w\rangle_J = (v, w) + i\omega_J(v, w)$$

so that $(\mathcal{H}, \langle\cdot, \cdot\rangle_J)$ is a *complex* Hilbert space with respect to the complex structure J. Note that ω_J is a symplectic form on $\mathcal{H}$. We set

$$\mathrm{Sp}(\mathcal{H}) = \{g \in \mathrm{Gl}(\mathcal{H})\colon g^t Jg = J\}$$

and

$$\mathrm{sp}(\mathcal{H}) = \{X \in \mathrm{B}(\mathcal{H})\colon X^t J + JX = 0\}.$$

Consider the equation

$$\frac{d}{dt}v(t) = A(t)v(t)$$

where $A(t) \in \mathrm{Sp}(\mathcal{H})$ and $v(t) \in \mathcal{H}$, and the corresponding operator equation

$$\frac{d}{dt}S(t) = A(t)S(t). \tag{7.1}$$

The following theorem of Paneitz gives sufficient conditions for a unique solution of (7.1) to exist and show a decent asymptotic behaviour.

Theorem 7.1. *Let* $A: \mathbb{R} \to C = \{X \in \mathrm{sp}(\mathcal{H}): \omega_J(Xv, v) \geq 0 \ \forall v \in \mathcal{H}\}$ *be a strongly continuous and norm bounded map such that*

$$\int_{-\infty}^{\infty} ||A(t)||dt < 2.$$

Then (7.1) *has a unique solution* $S: [-\infty, \infty] \to \mathrm{Sp}(\mathcal{H})$ *with* $S(-\infty) = I$ *and* $(S(t) + I)$ *is invertible for all* t. □

Note that C is invariant under the adjoint action of $\mathrm{Sp}(\mathcal{H})$. The operator $S = S(\infty)$ is interpreted as a scattering operator. The proof of Theorem 7.1 shows that it is the Cayley transform of an element $Y \in C$. It is important for the applications to physics to know whether S can be viewed as a unitary operator with respect to some $\langle \, , \, \rangle_{J'}$, i.e., whether S commutes with some complex structure J' in the $\mathrm{Sp}(\mathcal{H})$-orbit of J. In view of the ambiguity of the various quantization procedures one also wants to know the degree of uniqueness of such a J'.

Theorem 7.2. *Let* A, S, Y *be as above and assume that*

$$\int_{-\infty}^{\infty} A(t)dt \in C^o = \{X \in \mathrm{sp}(\mathcal{H}): \omega_J(Xv, v) \geq k||v||^2 \ \forall v \in \mathcal{H} \text{ for some } k > 0\}.$$

Then $Y \in C^o$ *and* S *commutes with a unique* J' *in the* $\mathrm{Sp}(\mathcal{H})$*-orbit of* J. □

The last statement of Theorem 7.2 is related to the fact that each element of C^o is conjugate under $\mathrm{Sp}(\mathcal{H})$ to a skew hermitean operator.

References

[1] Bargmann, V., On a Hilbert space of analytic functions and an associated integral transform, Part I, Comm. Pure. Appl. Math. 14, 187– 214 (1961)

[2] —, Group representations on Hilbert spaces of analytic functions, in Analytic Methods in Mathematical Physics, Gilbert and Newton, eds. Gordon and Breach, New York 1968

[3] Brunet, M., The metaplectic semigroup and related topics, Reports on Math. Phys.22, 149–170 (1985)

[4] Brunet, M., and P. Kramer, Semigroups of length increasing transformations, Reports on Math. Phys. 15 (1979), 287–304

[5] —, Complex extension of the representation of the symplectic group associated with the canonical commutation relations, Reports on Math. Phys.17, 205–215 (1980)

[6] Davies, E. B., One-parameter semigroups, Acad. Press, London 1980

[7] Faraut, J., and G. Viano, Volterra algebra and the Bethe-Salpeter equation, J. Math. Phys. 27 (1986), 840–846

[8] Gelfand, I. M., and S. Gindikin, Complex manifolds whose skeletons are semisimple real Lie groups and the analytic discrete series of representations, Funct. Anal. and Appl. 7 (1977), 19–27

[9] Hilgert, J., A note on Howe's oscillator semigroup, Ann. Inst. Fourrier 39 (1989), 1–25

[10] Hilgert, J. and K. H. Hofmann, On the causal structure of homogeneous manifolds, Preprint, 1989

[11] Hilgert, J., K. H. Hofmann and J. D. Lawson, Lie Groups, Convex Cones and Semigroups, Oxford University Press, 1989

[12] Hilgert, J., G. 'Olafsson and B. Ørsted, Hardy spaces on affine symmetric spaces, Preprint, 1989

[13] Howe, R., Quantum mechanics and partial differential equations, J. Funct. Anal. 38, 188–254 (1980)

[14] —, The oscillator semigroup, in: "The mathematical heritage of Hermann Weyl", Proc. Symp. Pure Math. 48 (ed. R.O. Wells), AMS, Propvidence 1988

[15] Kashiwara, M., and M. Vergne, On the Shale Segal Weil representation and harmonic polynomials, Inventiones Math. 44 (1978), 1–47

[16] Kramer, P., Composite particles and symplectic (semi-)groups, in Group Theoretical Methods in Physics, P. Kramer and A. Rieckers eds., LNP 79, Springer, Berlin 1978

[17] Kramer, P., M. Moshinsky and T. H. Seligman, Complex extensions of canonical transformations and quantum mechanics, in Group theory and its applications III, E. Loeble ed., Acad. Press, New York 1975

[18] Lawson, J. D., Ordered manifolds, invariant cone fields and semigroups, Preprint, 1988

[19] Levichev, A., Sufficient conditions for the nonexistence of closed causal curves in homogeneous space-times, Izvestia Phys. 10 (1985), 118–119

[20] —, Some methods in the investigation of the causal structure of homogeneous Lorentzian manifolds, Preprint (Russian), 1986

[21] —, Left invariant orders on special affine groups, Sib. Math. J. 28 (1987), 152–156 (Russian)

[22] Lüscher, M., and G. Mack, Global conformal invariance in quantum field theory, Comm. Math. Phys. 41, 203–234 (1975)

[23] Mizony, M., Semi-groupes de Lie et fonctions de Jacobi de deuxieme espèce, Thèse d'état, Université de Lyon I, 1987

[24] Olafsson, G. and B. Ørsted, The holomorphic discrete series for affine symmetric spaces I, J.Funct. Anal. 81 (1988), 126–159

[25] —, The holomorphic discrete series of an affine symmetric space and representations with reproducing kernels, Preprint, Odense, 1988

[26] Ol'shanskiĭ, G. I., Unitary representations of the infinite dimensional classical groups U(p,∞), SO_0(p,∞), Sp(p,∞) and the corresponding motion groups, Funct. Anal. and Appl. 12 (1978), 32–44

[27] —, Construction of unitary representations of the infinite dimensional classical groups, Sov. Math. Dokl. 21 (1980), 66–70

[28] —, Invariant cones in Lie algebras, Lie semigroups and the holomorphic discrete series, Funct. Anal. Appl. 15 (1981), 275–285

[29] —, Convex cones in symmetric Lie algebras, Lie semigroups, and invariant causal (order) structures on pseudo-Riemannian symmetric spaces, Sov. Math. Dokl. 26 (1982), 97–101

[30] —, Invariant orderings on simple Lie groups, the solution to E. B. Vinberg's problem, Funct. Anal. and Appl. 16 (1982), 311–313

[31] —, Complex Lie semigroups, Hardy spaces and the Gelfand-Gindikin program, in Problems in group theory and homological algebra, Jaroslavl 1982 (Russian)

[32] —, Infinite dimensional classical groups of finite R-rank: Description of representations and asymptotic theory, Funct. Anal. Appl. 18 (1982), 28–42

[33] —, Unitary repesentations of the infinite symmetric group: a semigroup approach, in: Representations of Lie groups and Lie algebras, Akad. Kiado, Budapest (1985)

[34] Paneitz, S., Invariant convex cones and causality in semisimple Lie algebras and groups, J. Funct. Anal. 43 (1981), 313–359

[35] —, Unitarization of symplectics and stability for causal differential equations in Hilbert space, J. Funct. Anal.41 (1981), 315–326

[36] —, Essential unitarization of symplectics and applications to field quantization, J. Funct. Anal.48 (1982), 310–359

[37] —, Determination of invariant convex cones in simple Lie algebras, Arkiv för Mat. 21 (1984), 217–228

[38] —, Global solutions of the hyperbolic Yang-Mills fields and their sharp asymptotics, In Proceedings of Symposia in Pure Math. 45, Part 2, F. Browder ed., 1986

[39] Segal, I. E., Nonlinear wave equations, in Non-linear Partial Differential Operators and Quantization procedures, S.I. Anderson and H. Doebner editors, LNM 1037 (1983), 115–141

[40] Stanton, R. J., Analytic extension of the holomorphic discrete series, Amer. J. Math. 108 (1986), 1411–1424

[41] Stein, E., and G. Weiss, Introduction to harmonic analysis on euclidean spaces, Princeton University Press, Princeton 1967

[42] Warner, G., Harmonic analysis on semi-simple Lie groups I, Springer, New York 1972

Embedding semigroups into Lie groups

Jimmie D. Lawson

A commutative semigroup can be embedded in a group if and only if it is cancellative. The group is obtained as the equivalence classes of the relation $\cong$ on the set of ordered pairs defined by $(a, b) \cong (c, d)$ if $ad = bc$. In this fashion one obtains the integers from the positive integers, the rationals from the integers, and the field of quotients of an integral domain (where slight adjustments to accomodate 0 must be included in the latter cases).

The problem of algebraically embedding a non-commutative semigroup into a group (actually of embedding a non-commutative ring into a division ring) was investigated as early as 1931 by Ore [25], who showed that a semigroup could be embedded into a group of left quotients if it was cancellative and right reversible. Dubreil [8] noted that the converse also held. Cancellativity remains a necessary condition for group embeddability in general; Malcev in [23] showed that this condition was not sufficient. Necessary and sufficient conditions are known, but they are rather complicated to state and verify. The general problem has given rise to a significant body of literature. We refer the reader to Chapter 12 of [7] for the principal results in this area. The question of algebraic embeddability is a specter looming in the background even in the topological setting.

The earliest consideration of the problem of embedding a topological semigroup into a topological group appears to be the work of Peck [26], Tamari [30], and Gelbaum, Kalisch, and Olmsted [9] in the early fifties. This was followed up by the work of Schieferdecker [29] and Rothman [27]. These authors concentrated on the commutative case, although Schieferdecker also considered right reversible semigroups. A careful study of the case that a right reversible cancellative topological semigroup embeds as an open subsemigroup in a topological group of left quotients was carried out by McKilligan in [24]. Weinert [31] considered the more general problem of forming a semigroup T of right quotients with respect to a subsemigroup Σ of S and finding a topology on T making it a topological semigroup into which S embedded. An attempt to give general necessary and sufficient conditions for the embedding of a topological semigroup into a topological group may be found in the work of Christoph [6], but the conditions are quite technical and complex.

The question of embeddability of a semigroup into a group in the topological setting is a natural enough question in and of itself, but its study has been given additional impetus from other considerations. In the study of compact abelian semigroups by investigators in the field of compact semigroups, semigroups which were locally compact cones arose in a natural way ([2], [19]). The question of embedding these cones into topological vector spaces was then an obvious one and was taken up by Brown and Friedberg [2], Keimel [20], and Lawson and Madison [22] (the latter paper gave an unrestricted positive answer). Later, in an attempt to establish matrix representations for a special class of compact semigroups, Brown and Friedberg [3] established a group embedding theorem into Lie groups for a special class of right reversible semigroups. In the study of differentiable semigroups (see [10]), questions of embeddability and local embeddability into Lie groups arise in a natural way in attempting to determine the differentiable and algebraic structure. Graham was able to establish embeddability into a Lie group in a neighborhood of 1 in the finite dimensional case, and his results were extended in [12, Section VII.2].

Another line of investigation that led to the question of group embeddability had its historical roots in Hilbert's fifth problem. Hilbert's fifth problem, to show that a locally Euclidean topological group is a Lie group, evidently has a semigroup version: is a locally Euclidean semigroup an analytic semigroup, i.e., does it admit the structure of an analytic manifold so that multiplication is analytic and the underlying topology agrees with the original topology? The conjecture in this generality is too broad, however, since the existence of too many commuting idempotents is incompatible with differentiability. Indeed, the real line with the operation $xy = \min\{x, y\}$ is a canonical counterexample to this conjecture. One sure way to remove this annoyance is to require cancellation, an axiom sufficiently strong to deny the existence of any non-identity idempotents. The amended conjecture now becomes: is a cancellative semigroup on a Euclidean manifold an analytic semigroup? The answer in this case is yes, was first proved by R. Houston [16] in his dissertation, and first appeared in print in [4]. An alternate proof was given by Hofmann and Weiss, and it is their approach as modified in [12] that we sketch in Section 4. The proof in either case involves questions of (local) group embeddings. Of course, there are analytic semigroups (even linear ones) which are not cancellative – the matrix semigroup of $n \times n$ matrices under multiplication, for example – so that this result is not a characterization.

During the last decade a Lie theory for subsemigroups of Lie groups has emerged. The major components of this theory may be found in the monograph [12]. As part of a comprehensive theory, one would like an abstract theory of such semigroups (independent of the group in which they occur). From another direction (multiparameter) semigroups of operators which are "Lie-like" have arisen in certain settings in analysis and mathematical physics (e.g., Howe's oscillator semigroups [17]; see [11] for additional references). Several natural questions arise. How is the class of topological semigroups which are embeddable (or locally embeddable) in a Lie group characterized? How comprehensive is this class of semigroups among the cancellable semigroups? What is the relationship with the class of differentiable semigroups? Can the group in which the semigroup was embedded be recaptured from the semigroup alone? Such questions motivate a careful study of appropriate conditions and techniques for embedding a topological semigroup into a Lie group. Our principal goal in this paper is to survey what is known about this last problem.

1. The free group on a topological semigroup

There is an inclusion functor from the category of groups and homomorphisms to the category of semigroups and homomorphisms. The left adjoint to this inclusion functor assigns to a semigroup S a group $G(S)$ and a homomorphism $\gamma: S \to G(S)$ (we also write γ as γ_S if we wish to show its dependence on S) with the property that given any group H and any homomorphism $\alpha: S \to H$, there exists a unique homomorphism $\beta: G(S) \to H$ such that $\beta \circ \gamma = \alpha$.

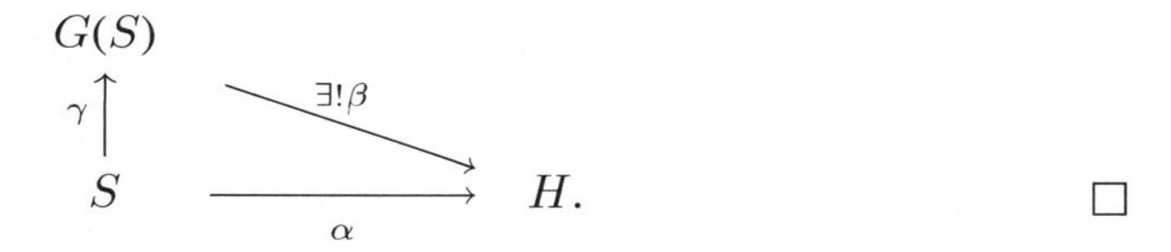

□

The pair $(G(S), \gamma)$ is called the *free group on* S. It is unique up to an isomorphism which commutes with the homomorphisms from S.

Alternately $G(S)$ can be constructed by first forming the free group $F(S)$ with alphabet S and dividing out the smallest congruence $\cong$ so that the natural inclusion $\gamma: S \to G(S) = F(S)/\cong$ is a homomorphism. We refer to Chapter 12 of [7] for a detailed treatment of $G(S)$. It follows directly from this construction that the semigroup generated by $\gamma(S) \cup \gamma(S)^{-1}$ is $G(S)$. It also follows from the universal property of the preceding paragraph that the semigroup S is algebraically embeddable in a group if and only if $\gamma: S \to G(S)$ is injective.

In general $G(S)$ is a quite complicated object to compute. For this reason in both the topological and algebraic setting the attention has often been focused on situations where $G(S)$ is a group of quotients. In the case S is commutative and cancellative, then $G(S)$ is just the standard group of quotients, and in the case that S is cancellative and right reversible ($Sa \cap Sb \neq \emptyset$ for all $a, b \in S$), then $G(S)$ is the group of left quotients of the form $a^{-1}b$ for $a, b \in S$ (see Chapter 12 of [7]).

We adapt the construction of the free group on S to the topological setting. Let S be a topological semigroup (we assume that topological semigroups are **always** Hausdorff). We consider all topologies on $G(S)$ which make $G(S)$ into a not necessarily Hausdorff topological group and for which the mapping $\gamma: S \to G(S)$ is continuous. The indiscrete topology is one such. It is straightforward to verify that the join of all these topologies in the lattice of topologies on $G(S)$ is again a topology making $G(S)$ a topological group for which $\gamma: S \to G(S)$ is continuous. This topology is called the *quotient topology* on $G(S)$.

Proposition 1.1. *Let S be a topological semigroup and let $(G(S), \gamma)$ be the free group on S. The quotient topology of $G(S)$ is characterized by the properties*
(i) *$\gamma: S \to G(S)$ is continuous,*
(ii) *$G(S)$ is a topological group, and*
(iii) *given any (not necessarily Hausdorff) topological group H and given any continuous homomorphism $\alpha: S \to H$, then there exists a unique continuous homomorphism $\beta: G(S) \to H$ such that $\beta \circ \gamma = \alpha$.*

Proof. Properties (i) and (ii) follow from the definition of the quotient topology on $G(S)$ given in the paragraph preceding the proposition. Given any (not necessarily Hausdorff) topological group H and any continuous homomorphism $\alpha: S \to H$, then there exists a (unique) homomorphism $\beta: G(S) \to H$ such that $\beta \circ \gamma = \alpha$, since $G(S)$ is the free group on S. The topology on $G(S)$ consisting of all inverse images under β of open sets in H makes $G(S)$ into a topological group, and γ is continuous with respect to this topology. Hence the quotient topology on $G(S)$ is finer than this topology. It follows that β is continuous.

Conversely suppose that $G(S)$ is equipped with a topology satisfying the three conditions of the proposition. Since the quotient topology on $G(S)$ is the finest satisfying (i) and (ii), it follows that the quotient topology is finer than the given topology. The mapping $\gamma: S \to G(S)$, where $G(S)$ is equipped with the quotient topology, is continuous and hence there exists a continuous $\beta: G(S) \to G(S)$ such that $\beta \circ \gamma = \gamma$ by property (iii). It follows (from uniqueness considerations) that β must be the identity on $G(S)$ and hence the given topology is finer than the quotient topology. They thus agree. □

Proposition 1.1 leads to the following definition.

Definition 1.2. Let S be a topological semigroup, let G be a group equipped with a topology, and let $\gamma: S \to G$ be a homomorphism. If the pair (G, γ) satisfies the three properties of Proposition 1.1, then the pair is called the *free topological group on S*.

Remark 1.3. Proposition 1.1 guarantees the existence of the free topological group on S, and shows that algebraically the underlying group is the free group on S. It is straightforward to verify that it is unique up to a topological isomorphism that commutes with the homomorphisms from S.

There appears to be no systematic study of the free topological group on S in the literature. We derive some of its elementary properties.

Corollary 1.4. *The topological semigroup S embeds (both topologically and algebraically) into a topological group if and only if $\gamma: S \to G(S)$ is a topological embedding.*

Proof. If γ is a topological embedding, then S embeds into the topological group $G(S)$. Conversely if the topological semigroup S embeds in any topological group H, then a direct argument using property (iii) of Proposition 1.1 yields that γ is an embedding. □

In light of these results there are several natural questions that arise.

Q1. Under what conditions is γ an embedding (or a local homeomorphism)?

Q2. Under what conditions is $G(S)$ Hausdorff?

Q3. Under what conditions does S (or some non-empty open subset of S) map onto an open subset of $G(S)$?

Q4. Under what conditions is $G(S)$ a Lie group?

Q5. Under what conditions can $G(S)$ be explicitly described or identified?

The next results show that the free topological group on a semigroup is the same as the free topological group on any of its ideals. It will sometimes be convenient in computing a free topological group to pass to an ideal which exhibits more desirable behavior than the original semigroup.

Lemma 1.5. *Let I be an ideal of a semigroup S and let $f: I \to H$ be a homomorphism into a group. Then there exists a unique homomorphic extension $F: S \to H$ of f given by $s \mapsto \big(f(y)\big)^{-1} f(ys)$ for any $y \in I$.*

Proof. We first verify that F is a homomorphism. Let $s, t \in S$. We have

$$\begin{aligned} F(t)f(y) &= f(y)^{-1}f(yt)f(y) = f(y)^{-1}f(yty) \\ &= f(y)^{-1}f(y)f(ty) = f(ty). \end{aligned}$$

Thus $F(s)F(t)f(y) = f(y)^{-1}f(ys)f(ty)$. On the other hand

$$F(st)f(y) = f(y)^{-1}f(yst)f(y) = f(y)^{-1}f(ysty) = f(y)^{-1}f(ys)f(ty).$$

By cancellation in H, we conclude $F(s)F(t) = F(st)$.

Suppose now that $E: S \to H$ is another homomorphic extension of f. Then

$$f(y)E(s) = E(y)E(s) = E(ys) = f(ys)$$

and hence $E(s) = f(y)^{-1}f(ys) = F(s)$. □

Proposition 1.6. *Let S be a topological semigroup, and let I be an ideal in S. Let $\gamma_S: S \to G(S)$ be the free topological group on S. Then $\gamma_I: I \to G(S)$ is the free topological group on I, where γ_I is the restriction of γ_S to I. Conversely if $\gamma_I: I \to G(I)$ is the free topological group on I, then the unique homomorphic extension $\gamma_S: S \to G(I)$ is the free topological group on S.*

Proof. We show $(G(S), \gamma_I)$ is the free topological group on I. Let $f: I \to H$ be a continuous homomorphism into a topological group H. By Lemma 1.5 the mapping F given by $F(s) = f(y)^{-1} f(ys)$, where $y \in I$, is a homomorphic extension of f; it is clearly continuous. Hence there exists $\beta: G(S) \to H$ such that $\beta \circ \gamma_S = F$. Then $\beta \circ \gamma_I = f$. Now if $\delta: G(S) \to H$ satisfies $\delta \circ \gamma_I = f$, then it must be the case that $\delta \circ \gamma_S = F$ since the extension of f to S is unique. Thus $\delta = \beta$.

Let now $(G(I), \gamma_I)$ be the free topological group on I. Again by Lemma 1.5 there exists an unique homomorphic extension $\gamma_S: S \to G(I)$. As in the previous paragraph this extension is continuous. Let $F: S \to H$ be a continuous homomorphism to a topological group H. Then there exists an unique continuous homomorphism $\beta: G(I) \to H$ such that $\beta \circ \gamma_I$ equals the restriction f of F to I. Then $\beta \circ \gamma_S$ is an extension of f, and hence must equal F by uniqueness of extension. □

2. Open embeddings

We shall be particularly concerned in this paper with the case that a semigroup S embeds as an open subsemigroup (or at least as a semigroup with interior) in the free topological group on S. Just as cancellativity is a necessary condition for a semigroup to algebraically embed in a group, there is an important necessary topological condition for an open embedding.

Definition 2.1. A topological semigroup S satisfies condition (INT) if

(INT) given $a, b \in S$ and V open containing b, there exist open sets U containing a and N containing ab (resp. ba) such that $x \in U$ implies $N \subseteq xV$ (resp. $N \subseteq Vx$).

Some version of condition (INT) appears wherever authors have studied open embeddings into groups (see "continuity of inversion" in [26], Property F in [27], Theorem 4.1 in [24], and Lemma 4 in [4], from which we take our formulation). The next proposition shows why.

Proposition 2.2. *Let S be a topological semigroup, G a topological group, and $\alpha: S \to G$ a continuous homomorphism which is a local homeomorphism. Then S satisfies condition* (INT).

Proof. Let $a, b \in S$. Pick an open set N_1 containing ab such that α restricted to N_1 is a homeomorphism onto an open subset of G. Pick open sets U_1 containing a and V_1 containing b such that $U_1 V_1 \subseteq N_1$.

Let V be an open set containing b. We may assume without loss of generality that V is chosen small enough so that $V \subseteq V_1$. Note that since α is a local homeomorphism, $\alpha(V)$ is an open set containing $\alpha(b) = \alpha(a)^{-1}\alpha(ab)$. Choose open sets P containing $\alpha(a)$ and Q containing $\alpha(ab)$ such that $P^{-1}Q \subseteq \alpha(V)$. Pick an open set U with $a \in U \subseteq U_1$ such that $\alpha(U) \subseteq P$ and an open set N with $ab \in N \subseteq N_1$ such that $\alpha(N) \subseteq Q$. If $z \in N$ and $x \in U$, then $\alpha(x)^{-1}\alpha(z) \in \alpha(V)$, so $\alpha(x)^{-1}\alpha(z) = \alpha(y)$ for some $y \in V$. Then

$$\alpha(z) = \alpha(x)\alpha(x)^{-1}\alpha(z) = \alpha(x)\alpha(y) = \alpha(xy).$$

Since $xy \in U_1 V_1 \subseteq N_1$ and α is injective on N_1, it follows that $z = xy$. Thus $N \subseteq xV$ for all $x \in U$. The other direction follows similarly. □

Remark 2.3. Condition (INT) implies that the left and right translation mappings on S are open mappings (see Lemma VII.1.15 of [12]).

The condition (INT) is entirely satisfactory in the case that S is a cancellative right (or left) reversible semigroup.

Proposition 2.4. *Let S be a cancellative right reversible topological semigroup. Then S satisfies condition* (INT) *if and only if the group $G(S)$ of left quotients admits a (unique) topology making it a Hausdorff topological group such that S embeds as an open subsemigroup. In this case $G(S)$ endowed with this topology is the free topological group on S.*

Proof. Proposition 2.2 gives the implication in one direction. The implication in the other direction follows from Proposition VII.1.28 of [12] or results of McKilligan [24]. That $G(S)$ is the free topological group on S follows from the fact that it is the free group on S (as remarked earlier) and the fact that the topology given $G(S)$ is the quotient topology from of the mapping $(s,t) \mapsto s^{-1}t: S \times S \to G(S)$. Thus any homomorphism from S to a topological group G extends continuously to $G(S)$. Related results may be found in [31]. □

There are some standard hypotheses that force condition (INT) to hold.

Proposition 2.5. *Let S be a cancellative topological semigroup. Then S satisfies condition* (INT)

(i) *if S is locally Euclidean at each of its points, or*

(ii) *if S is a differentiable manifold without boundary (modeled on a Banach space) and if each of the left and right translation mappings is a diffeomorphism from S into S.*

Proof. Part (i) is a consequence of the Invariance of Domain Theorem (see Proposition VII.1.24 of [12] or Lemma 4 of [4]). The differentiable case follows from the Parameterized Mapping Theorem for differentiable mappings (see part (iii) of Proposition VII.2.11 of [12]). □

One of the basic facts about topological semigroups with identity is that the identity tends to sit on the boundary, where the appropriate notion of boundary depends on the context (see, for example, Section B.7 of [14]). Thus in studying the embeddability of semigroups into groups, it is desirable to consider more general semigroups than those on manifolds without boundary. Now if S is a subsemigroup of a topological group and if the interior of S is non-empty, then the interior forms an ideal. Hence our strategy for proving general embedding theorems is to first find some ideal that embeds as an open subsemigroup, and then extend this embedding to the whole semigroup. We formalize this approach.

Definition 2.6. A subset A of a topological space X is called an *admissible subset* of X if the interior of A in X is dense in A. If X is understood we sometimes refer to A simply as an *admissible set.* A T_3-space X is called a *Euclidean manifold with generalized boundary* if there exists an n such that each point of X possesses a neighborhood which is homeomorphic to some admissible set in $\mathbb{R}^n$.

Proposition 2.7. *Let S be a cancellative topological semigroup, let I be an ideal of S, and let $f: I \to G$ be an embedding of I into a topological group G such that $f(I)$ is an open set in G. If there exists $y \in I$ such that the mapping $\lambda_y: S \to I$ defined by $\lambda_y(s) = ys$ is a homeomorphism onto an admissible subset of I, then the embedding f extends* (*uniquely*) *to an embedding $F: S \to G$ such that $F(S)$ is an admissible subset of G.*

Proof. Since f is a homeomorphism onto the open set $f(I)$ it carries the admissible set yS to an admissible subset of G, and this set is carried to another admissible subset by left translation by $f(y)^{-1}$. It then follows from Lemma 1.5 that $F(S)$ is an admissible subset of G. Since F is the composition of three homeomorphisms, it also is a homeomorphism onto its image. □

Remark 2.8. If a cancellative topological semigroup S homeomorphically embeds into a topological group, then the left and right translations λ_y and ρ_y respectively must be homeomorphisms from S into S for each $y \in S$ (since in the group one can multiply by y^{-1}).

3. The local quotient construction of Brown–Houston

Throughout this section and the next S denotes a cancellative topological semigroup that satisfies property (INT) of Definition 2.1.

In this and the coming sections we show how to associate with S a number of related structures that play an important role in proving group embedding theorems about S. We have seen in Section 1 how to construct for S the free topological group $G(S)$ on S. However, this construction is not explicit enough to be helpful in proving very many properties about $G(S)$. It would be much more useful to be able to represent $G(S)$ as left or right quotients of elements of S, but of course this is not possible in general. Such a representation is possible in a neighborhood of the identity, however, via the Brown–Houston construction of local quotients [4], which we review in this section. We associate with S in a canonical way a local group and a local action of this local group on S.

We first recall the notion of a local group (the definitions of a local group found in the literature vary slightly, but all amount essentially to the same thing).

Definition 3.1. A *local group* is a system (G, e, θ, m) consisting of a Hausdorff topological space G, an element $e \in G$, a function $\theta\colon \operatorname{dom}\theta \to G$, and a function m (partial multiplication) defined on a subset $\operatorname{dom}(m) \subseteq G \times G$ into G. Furthermore, there must exist an open set Γ containing e such that $\Gamma \subseteq \operatorname{dom}\theta$ and $\Gamma \times \Gamma \subseteq \operatorname{dom}(m)$ and such that the following conditions are also satisfied:

i) the restriction of m to $\Gamma \times \Gamma$ is continuous;

ii) if $a, b, c, m(a,b), m(b,c) \in \Gamma$, then $m(m(a,b),c)$ and $m(a, m(b,c))$ are defined and equal;

iii) if $a \in \Gamma$, then $m(a,e) = a = m(e,a)$;

iv) θ restricted to Γ is continuous and $m(a,\theta(a)) = e = m(\theta(a),a)$ for $a, \theta(a) \in G$.

A local group is a *locally Euclidean local group* if e has an open neighborhood homeomorphic in the relative topology to Euclidean n-space for some n. A local group G is a *local Lie group* provided G is an analytic manifold, m restricted to $\Gamma \times \Gamma$ is analytic, and θ restricted to Γ is analytic.

The local group is generally denoted more compactly by G. The element e is the identity element. We usually write ab (or some other multiplicative notation) for $m(a,b)$ and a^{-1} for $\theta(a)$.

Definition 3.2. Let G and H be local groups. We say that G and H are *locally isomorphic* if there exist an open set U containing e_G, an open set V containing e_H and a homeomorphism $h\colon U \to V$ such that for $a, b \in U$, we have $ab \in U$ iff

$h(a)h(b) \in V$ and in this case $h(ab) = h(a)h(b)$. The function h is called a *local isomorphism*.

One of the principal results of [18] is the local version of the solution of Hilbert's fifth problem; to wit, a locally Euclidean local group is locally isomorphic to a local Lie group (Theorem 107). In this case there is a corresponding finite dimensional real Lie algebra and a locally defined exponential mapping. One may take one of the Lie groups corresponding to this Lie algebra and show that the original local Lie group is locally isomorphic to this group (since the Campbell-Hausdorff multiplication on a neighborhood of 0 in the Lie algebra is locally isomorphic both to that of the local Lie group and the Lie group via the corresponding exponential mappings).

Definition 3.3. If A, B are subsets of S, then we say *A right reverses in B* if given $s, t \in A$, then $sB \cap tB \neq \emptyset$, i.e., there exist $b, c \in B$ such that $sb = tc$.

Our goal is to build a local group of quotients whose members are of the form $a^{-1}b$ for suitable $a, b \in S$. (Left quotients will be more convenient for our purposes than right quotients.) The approach taken is a standard one for such constructions: we obtain the local group of quotients as a set of equivalence classes of an appropriate equivalence relation $\equiv$ on a suitable subset of $S \times S$. We then let $a \setminus b$ denote the equivalence class of (a, b). Since $(a \setminus b)(b \setminus c) = a \setminus c$, we define a partial operation $\cdot$ on pairs by $(a, b) \cdot (b, c) = (a, c)$. It will then be necessary that $\equiv$ is a congruence with respect to this operation.

Let $a, b, c, d \in S$ and let $x, y \in V$, where V is some non-empty subset of S. If S were a subsemigroup of a group and if $ax = by$ and $cx = dy$, then $a^{-1}b = xy^{-1} = c^{-1}d$. These observations motivate the following definition:

Definition 3.4. For $a, b, c, d \in S$ and a non-empty subset V of S, we define

$$S \times_V S = \{(a, b) \in S \times S \colon ax = by \quad \text{for some} \quad x, y \in V\}$$

and the *left quotient relation* on $S \times_V S$ by

$$(a, b) \overset{\text{def}}{\equiv} (c, d) \qquad \text{if} \qquad ax = by, \ cx = dy \quad \text{where} \quad x, y \in V.$$

Note that the relation $\equiv$ depends on V.

The next lemma is a basic calculational tool.

Lemma 3.5. (Malcev's Condition) *Suppose $ax = cy$, $bx = dy$, and $au = cv$ for $a, b, c, d, u, v, x, y \in S$. If $yS \cap vS \neq \emptyset$, then $bu = dv$.*

Proof. There exist $s, t \in S$ such that $ys = vt$. Then $axs = cys = cvt = aut$, so by cancellation $xs = ut$. We now obtain $but = bxs = dys = dvt$. Again by cancellation we obtain $bu = dv$. □

In [23] Malcev showed that any subsemigroup of a group must satisfy Lemma 3.5 (without the intersection requirement), but that there existed cancellative semigroups which did not. Hence not every cancellative semigroup is embeddable in a group. Lemma 3.5 is crucial in establishing congruence properties of the relation $\equiv$.

Theorem 3.6. *Suppose $V \subseteq S$ is a non-empty open set such that V right reverses in V' and V' right reverses in S for some V'. Then $\equiv$ is an equivalence relation on $S \times_V S$. Denote the equivalence class of (a, b) by $a \setminus b$. Then $Q(S, V) \stackrel{\text{def}}{=} S \times_V S / \equiv$ equipped with the quotient topology induced by $S \times_V S$ is a local group with respect to the operation $a \setminus b * c \setminus d = p \setminus r$ if there exist $p, q, r \in S$ such that $(a, b) \equiv (p, q)$ and $(c, d) \equiv (q, r)$. The diagonal is an equivalence class for $\equiv$ and is the identity of $Q(S, V)$; the inverse of $a \setminus b$ is $b \setminus a$.*

Proof. See Theorem VII.3.1 of [12]. □

Open sets of the desired type V of Theorem 3.6 exist in S in abundance according to the following lemma, a special case of Corollary VII.1.17 of [12] or Corollary 4.2 of [4].

Lemma 3.7. *Let $b \in S$. Then there exist open sets V, V' such that $b \in V$, V right reverses in V' and V' right reverses in S.* □

We wish to define a local right action of the local group $Q(S, V)$ on S. We first make precise the notion of a local right action.

Definition 3.8. Let X be a space and G a local group. A *local right action* of G on X is a continuous function $(x, g) \mapsto xg$ defined on an open subset of $X \times G$ containing $X \times \{e\}$ with values in X such that

(i) for each $x \in X$, $xe = x$;

(ii) there exists a neighborhood Ω of $X \times \{e\} \times \{e\}$ in $X \times G \times G$ such that for all $(x, g, g') \in \Omega$, we have that gg', $(xg)g'$, and $x(gg')$ are defined and $(xg)g' = x(gg')$.

The local right action is *locally simply transitive* if given $x \in X$, there exists an open set N_x in G containing e such that the function $g \mapsto xg \colon N_x \to X$ is defined on all of N_x and is a homeomorphism onto an open set containing x.

Note that since translation by an element of S is an open mapping, then ρ_s, right translation by s, has range the open left ideal Ss. We consider a more general notion of a right translation.

Definition 3.9. A *partial right translation* is an injective open mapping ρ with domain and range non-empty open left ideals of S satisfying $(xy)\rho = x(y\rho)$ for all $y \in \operatorname{dom}\rho$ and $x \in S$. Note that we compose partial right translations on the right. It is an elementary exercise (see, for example, [7]) that, if ρ is a partial right translation, then so is ρ^{-1}, and that, if ρ and μ are partial right translations which compose (range$(\mu) \cap$ dom$(\rho) \neq \emptyset$), then $\mu\rho$ is a partial right translation with appropriately curtailed domain.

We want to define a local right action of the local group $Q(S, V)$ of Theorem 3.6 on S in such a way that each member of $Q(S, V)$ acts as a partial right translation on S. To this end we define $\rho(a, b)$ for $(a, b) \in S \times_V S$ by $x\rho(a, b) = y$ if there exist $u, v \in V$ such that $au = bv$ and $xu = yv$, i.e., if $(a, b) \equiv (x, y)$.

Theorem 3.10. *For each $(a, b) \in S \times_V S$, the function $\rho(a, b)$ is well-defined, is a partial right translation, and is equal to $\rho_u(\rho_v)^{-1}$ if $au = bv$. The function $a \setminus b \mapsto \rho(a, b)$ from $Q(S, V)$ to the partial semigroup of partial right translations under the operation of composition (where defined) is a (well-defined) monomorphism onto its image. Furthermore, the partial function $\big(s, \beta(a, b)\big) \to s\rho(a, b)$ is a locally simply transitive local right action of $Q(S, V)$ on S.*

Proof. See Theorem VII.3.6 of [12]. See Lemma 7 of [4] for a related result. □

Suppose now that G is a topological group that is locally isomorphic to $Q(S, V)$. Then we can use the local isomorphism to define a locally transitive right local action of G on S in the obvious way.

Corollary 3.11. *Let G be a topological group which is locally isomorphic to $Q(S, V)$. Then the local isomorphism defines a locally simply transitive local right action of G on S. There exists $\mathcal{D} \subseteq S \times \mathcal{N}(e)$ (where $\mathcal{N}(e)$ denotes the set of open subsets containing e_G) such that the following conditions are satisfied:*

(i) *$g \mapsto ag\colon N \to aN$ is a homeomorphism onto the open set aN for all $(a, N) \in \mathcal{D}$;*

(ii) *$ae = a$;*

(iii) *Let $a \in S$. If we set $\mathcal{D}_a = \{W \in \mathcal{N}(e)\colon (a, W) \in \mathcal{D}\}$, then $\mathcal{D}_a$ is a basis of open neighborhoods of e such that $V \subseteq W$, $V \in \mathcal{N}(e)$, and $W \in \mathcal{D}_a$ imply $V \in \mathcal{D}_a$.* □

4. The double sheaf and analytic structures

In this section we present the double sheaf construction of Hofmann and Weiss [15]. Given a cancellative semigroup S and a group G locally isomorphic to $Q(S, V)$, the double sheaf is a semigroup with homomorphic projections onto both G and S making it a sheaf over both (indeed it is a covering space of S). An analytic structure on G induces one on the double sheaf which in turn induces one on S.

Every locally simply transitive local right action of a topological group G on a topological space X gives rise to a topology on $X \times G$ finer than the product topology such that the projection into X is a covering projection, the projection into G is a local homeomorphism (thus yielding a sheaf), and the mapping $\big((x, h), g\big) \mapsto (xg, hg)$ is again a locally simply transitive local right action. In the case that X is a semigroup on which G acts by partial right translations, then one may induce on $X \times G$ a continuous semigroup multiplication such that the projections are homomorphisms.

Let S be a cancellative semigroup which satisfies condition (INT), and let G be a Hausdorff topological group which is locally isomorphic to $Q(S, V)$. Then there is a locally simply transitive local right action of G on S. We define $\mathcal{D}$ as in Corollary 3.11. For every $(a, N) \in \mathcal{D}$ and $h \in G$, let

$$B(a, N, h) = \{(ag, hg) : g \in N\} \subseteq S \times G.$$

Lemma 4.1. *The set $\mathcal{B} = \{B(a, N, h) : (a, N) \in \mathcal{D}, h \in G\}$ is a base for a Hausdorff topology τ on $S \times G$ finer than the product topology. The coordinatewise multiplication is continuous with respect to this topology.*

Proof. See Lemma VII.3.8 of [12] or Lemma 2.1 of [15]. □

Definition 4.2. The topological semigroup $\Sigma = (S \times G, \tau)$ is called the *double sheaf* of S and G. The projections onto the factors are denoted $\sigma_S : \Sigma \to S$ and $\psi_S : \Sigma \to G$.

Proposition 4.3. *The mappings σ_S and ψ_S are continuous surjective homomorphisms, the mapping $\psi_S : \Sigma \to G$ is a local homeomorphism onto G, and the mapping $\sigma_S : \Sigma \to S$ is a covering projection.*

Proof. See Theorem VII.3.11 of [12]. □

We turn now to considerations of analyticity. Suppose that the group G is a Lie group. In a standard way one can use the local homeomorphism $\psi_S : \Sigma \to G$ to induce a unique analytic structure on Σ such that ψ is an analytic mapping. Then the covering projection σ_S is used to induce an analytic structure on S. One verifies directly that the multiplication mappings on Σ and S are then analytic with respect to these induced structures. (See [15] or Theorem VII.3.13 of [12] for details.)

Theorem 4.4. *Suppose that G is a Lie group. Then there exist unique analytic structures on Σ and S such that ψ_S and σ_S are local diffeomorphisms. With respect to these analytic structures, multiplication is analytic on Σ and S.* □

Remark 4.5. The proof of 4.4 does not depend on the fact that G is a real Lie group and works equally well in the case that G is a complex Lie group. In this case we obtain complex analytic structures on S and Σ and the multiplication in this case is complex analytic.

Corollary 4.6. *Let S be a cancellative topological semigroup on a Euclidean manifold. Then S admits an analytic structure for which the multiplication is analytic.*

Proof. By Proposition 2.5, S satisfies condition (INT). Since $Q(S, V)$ is locally homeomorphic to S (since its local action on S is locally simply transitive by Theorem 3.10), the local group $Q(S, V)$ is locally Euclidean. Hence by Jacoby's result [18] it is a local Lie group, and thus is locally isomorphic to a global Lie group G, which has a (unique) analytic structure. Then (as in Corollary 3.11) there is a corresponding local right action of G on S, and we apply Theorem 4.4. □

Remark 4.7. We observe that in the infinite dimensional case for a cancellative semigroup S satisfying (INT) what one needs to induce an analytic structure on S by the same argument is that there exists a Lie group locally isomorphic to $Q(S, V)$.

Suppose now that S is a cancellative semigroup on a manifold modeled on a Banach space and the manifold is equipped with a C^k-differentiable structure where $1 \leq k \leq \infty$, or $k = \omega$ in the case that the differentiable structure is an analytic structure, such that multiplication is a C^k mapping and left and right translations are diffeomorphisms onto open subsets of S. By Proposition 2.5, S satisfies condition (INT). Let G denote the local group $Q(S, V)$. The proof of Theorem VII.2.15 in [12] adapts to this setting to show that given $s^2 \in S$ and an appropriately small neighborhood N of e in G, then the C^k-structure induced on N by the homeomorphism of part (i) of Corollary 3.11 (where s^2N is given the relative C^k-structure from S) makes N into a C^k local group. (Indeed the mapping $(s^2g, s^2h) \mapsto s^2gh$ can be rewritten as $(s^2g, s^2h) \mapsto \big(\rho(s)^{-1}(s^2g)\big)\big(\lambda(s)^{-1}(s^2h)\big)$, where $\lambda(s)$ ($\rho(s)$) is left (right) translation by s.) By the uniqueness of C^k-structures for differential local groups (see e.g. Theorem VII.2.30 of [12]), this C^k-structure on N must agree locally with the C^k-structure on G generated by the unique analytic structure on G. Thus the homeomorphism of part (i) of Corollary 3.11 is a C^k-diffeomorphism between some neighborhood N of e with the restricted C^k-structure generated by the analytic structure on G and s^2N with the relative C^k-structure from S. It follows from the way the analytic structure on S given in Theorem 4.4 was constructed that this homeomorphism is an analytic diffeomorphism between s^2N and N in that case also. It now follows readily by composing with $\lambda(s)^{-1}$ that the C^r-structure on S is that induced by the analytic structure which we previously constructed. We thus obtain a strengthened version of Corollary 4.6.

Proposition 4.8. *Let S be a cancellative topological semigroup on a manifold. For $1 \le k \le \omega$, S admits a unique C^k-structure consistent with the topology for which the multiplication is a C^k-mapping and translations are C^k-diffeomorphisms onto open subsets.* □

We consider now the extent to which the construction of Σ is functorial. Of course in this case one needs first of all to select the group G in a functorial way. If S is locally connected, then $Q(S,V)$ will be locally connected (since it is locally homeomorphic to S), and hence if G is a group locally isomorphic to $Q(S,V)$, it will be locally connected. Thus the identity component will be open in G and hence also locally isomorphic to $Q(S,V)$. Then the universal covering group of G, which we denote $\widetilde{G}(S)$ when it exists, will be uniquely determined in terms of S. We let $\Sigma(S)$ denote the Σ corresponding to the choice $G = \widetilde{G}(S)$.

Proposition 4.9. *Let $f: S \to T$ be a continuous homomorphism of locally connected cancellative semigroups satisfying* (INT), *and suppose that there exist universal covering groups $\widetilde{G}(S)$ and $\widetilde{G}(T)$ such that $Q(S,V)$ (resp. $Q(T,W)$) is locally isomorphic to $\widetilde{G}(S)$ (resp. $\widetilde{G}(T)$). Then there exist unique continuous homomorphisms $\widetilde{G}(f): \widetilde{G}(S) \to \widetilde{G}(T)$ and $\Sigma(f): \Sigma(S) \to \Sigma(T)$ such that the following diagram commutes:*

$$\begin{array}{ccc}
\widetilde{G}(S) & \xrightarrow{\widetilde{G}(f)} & \widetilde{G}(T) \\
\psi \uparrow & & \uparrow \psi \\
\Sigma(S) & \xrightarrow{\Sigma(f)} & \Sigma(T) \\
\sigma \downarrow & & \downarrow \sigma \\
S & \xrightarrow{f} & T.
\end{array}$$

If $\widetilde{G}(S)$ and $\widetilde{G}(T)$ are Lie groups, then f is analytic with respect to the induced analytic structures on S and T.

Proof. The homomorphism $\widetilde{G}(f)$ is defined in such a way that given $s \in S$, there exists N open in $\widetilde{G}(S)$ such that $f(sg) = f(s)\widetilde{G}(f)(g)$ for $g \in N$. The mapping $\widetilde{G}(f)$ is the unique extension of this local mapping to a global homomorphism on $\widetilde{G}(S)$ (this extension is possible since $\widetilde{G}(S)$ is a universal covering group). The mapping $\Sigma(f)$ is easily defined from f and $\widetilde{G}(f)$. See Proposition VII.3.16 of [12] or Proposition 2.5 of [15] for additional details. □

5. Connected semigroup coverings

In this section let S be a cancellative topological semigroup which satisfies (INT), *is connected, locally connected and has a countable base, and, in addition, has an associated universal covering group $\widetilde{G}(S)$ which is locally isomorphic to $Q(S, V)$. Then $\Sigma = (S \times \widetilde{G}(S), \tau)$ again denotes the double sheaf from the previous section.*

Let $\kappa(x)$ denote the connected component of a point $x \in \Sigma$. Since the product of connected sets is connected, the components give rise to a congruence defined by $x\kappa y$ if and only if $\kappa(x) = \kappa(y)$. The factor semigroup Σ/κ is denoted $\Xi = \Xi(\Sigma)$ and $\kappa\colon \Sigma \to \Xi$ is the natural homomorphism. We come now to a crucial lemma.

Lemma 5.1. *The semigroup Ξ consisting of the connected components of Σ is a group. Therefore there exists a uniquely determined component C of Σ which is a subsemigroup.*

Proof. Let $\lambda, \mu \in \Xi$. It must be shown that there exist components $\nu, \xi \in \Xi$ such that $\lambda\nu = \mu = \xi\lambda$. Choose $(s, g) \in \Sigma$ with $\kappa(s, g) = \lambda$. Since S is connected and locally connected, it is a standard theorem of covering spaces that each component of Σ maps onto S. In particular, $s^2 \in \sigma_S(\mu)$, i.e., there exists an element $h \in \Sigma$ such that $\kappa(s^2, h) = \mu$. We then define $\nu = \kappa(s, g^{-1}h)$ and $\xi = (s, hg^{-1})$. Then

$$\begin{aligned}\lambda\nu = \kappa(s, g)\kappa(s, g^{-1}h) = \kappa(s^2, gg^{-1}h) = \\ = \kappa(s^2, hgg^{-1}) = \kappa(s, hg^{-1})\kappa(s, g) = \xi\lambda.\end{aligned}$$

Since equations can be solved on both sides, Ξ is a group. Then the identity element C of Ξ satisfies $C^2 \subseteq C$ and it is the only component with this property. □

Definition 5.2. The unique connected component of $\Sigma(S)$ which is a subsemigroup is denoted by $\widehat{S}$. The restrictions of σ_S and ψ_S to $\widehat{S}$ are again denoted by σ_S and ψ_S and we depend on context to indicate which one is being considered.

There now results a connected version of Proposition 4.9.

Theorem 5.3. (i) *There exist a connected semigroup $\widehat{S}$ and continuous homomorphisms $\sigma_S\colon \widehat{S} \to S$ and $\psi_S\colon \widehat{S} \to \widetilde{G}(S)$, where σ_S is also a covering projection and ψ_S is a local homeomorphism onto an open subset of G. If $\widetilde{G}(S)$ is a Lie group, then there exist unique analytic structures on S and $\widehat{S}$ such that ψ_S and σ_S are local analytic isomorphisms.*

(ii) *Let $f\colon S \to T$ be a continuous homomorphism, where S and T both satisfy the hypotheses of part* (i). *Then there exists a continuous semigroup homomorphism $\widehat{f}\colon \widehat{S} \to \widehat{T}$ such that the following diagram commutes:*

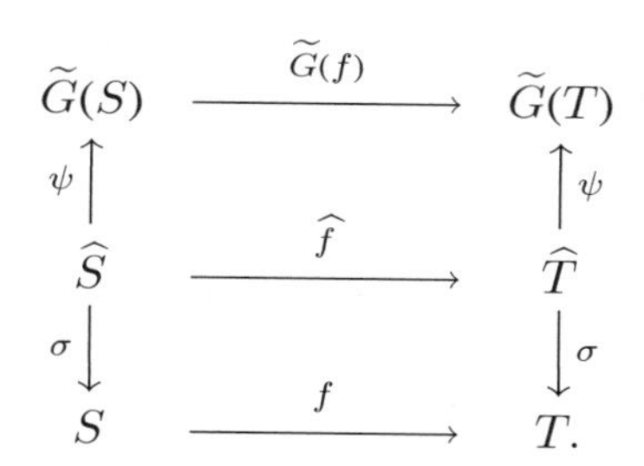

$$\begin{array}{ccc} \widetilde{G}(S) & \xrightarrow{\widetilde{G}(f)} & \widetilde{G}(T) \\ \psi\uparrow & & \uparrow\psi \\ \widehat{S} & \xrightarrow{\widehat{f}} & \widehat{T} \\ \sigma\downarrow & & \downarrow\sigma \\ S & \xrightarrow{f} & T. \end{array}$$

If $\widetilde{G}(S)$ is a Lie group, then $\widehat{f}$ is an analytic mapping.

Proof. The proof follows essentially from the restriction of the results in Proposition 4.9. See Theorem VII.3.19 of [12] or Theorem 3.3 of [15]. □

It is not clear a priori whether $\widehat{S}$ is a proper subset of $\Sigma(S)$. This is shown to be the case by a standard covering space calculation involving fibers and weights.

Lemma 5.5. *Let X be a connected, locally connected space with a countable base and let $p\colon \widehat{X} \to X$ be a covering projection, where $\widehat{X}$ is connected. Then $\widehat{X}$ has a countable base.* □

We have previously defined a local right action of $\widetilde{G}(S)$ on Σ. We now define a (global) left action by $g \cdot (s, h) = (s, gh)$. It is immediate from the definition of the topology on Σ that $(s, h) \mapsto (s, gh)$ is a homeomorphism that carries each σ-fiber onto itself. It follows that each $g \in \widetilde{G}(S)$ acts as a deck transformation on

Σ. We consider the *stabilizer group* G_S of $\widehat{S}$ consisting of all $h \in \widetilde{G}(S)$ such that $h \cdot \widehat{S} \subseteq \widehat{S}$. Since $\big(h,(s,g)\big) \mapsto (s,hg)$ defines a left action of $\widetilde{G}(S)$ on Σ, it follows that G_S is a subgroup. Since G_S acts transitively on the fibers of the connected covering space $\sigma_S\colon \widehat{S} \to S$, it follows by standard results of covering spaces that it is the group of deck transfomations for this covering space. Furthermore, since for $(s,g) \in \widehat{S}$, the collection $\{B(s,N,hg)\colon h \in G_S\}$ is a collection of pairwise disjoint open sets in $\widehat{S}$ and $\widehat{S}$ has a countable base by the preceding lemma, it follows that G_S is countable.

Proposition 5.6. *Define* $\phi\colon \widetilde{G}(S) \to \Xi$ *by* $\phi(g) = g \cdot \widehat{S}$. *Then* ϕ *is a surjective homomorphism with kernel the countable central subgroup* G_S.

Proof. One checks directly that ϕ is a surjective homomorphism with kernel G_S. Hence G_S is a normal subgroup. We remarked earlier that it is countable as a consequence of Lemma 5.5. Since for $h \in G_S$, $\{ghg^{-1}\colon g \in \widetilde{G}(S)\}$ is a connected subset of the countable completely regular space G_S, it follows that this set is a singleton, i.e., G_S is central. (See Proposition VII.3.21 of [12] or Proposition 3.5 of [15] for additional details.) □

At this point one might conjecture that the central subgroup G_S of $\widetilde{G}(S)$ is closed. This conjecture appears plausible, but remains unproved in general. However, it is known to be valid for several important cases.

Proposition 5.7. *We have a commuting diagram*

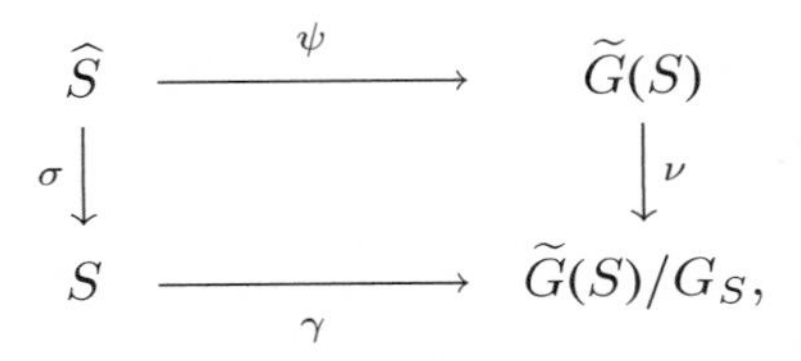

The homomorphism γ_S *is a local homeomorphism onto an open subsemigroup of the topological group* $\widetilde{G}(S)/G_S$ *in the case that* G_S *is discrete (equivalently closed) in* $\widetilde{G}(S)$. *This happens if*

(i) S *is simply connected, in which case* $S = \widehat{S}$ *and* G_S *is trivial, or*
(ii) $\widetilde{G}(S)$ *is a Lie group such that its Lie algebra has trivial center, or*
(iii) *there exists a homomorphism from* S *to some group* H *which is one-to-one on some non-empty open subset of* S.

Proof. If σ_S identifies two points of $\widehat{S}$, then they are in the same fiber. Hence their $\widetilde{G}(S)$-coordinates differ by a left translate of some member of G_S, i.e., they are in the same coset of G_S. It follows that there is induced a homomorphism $\gamma_S: S \to \widetilde{G}(S)/G_S$ making the diagram commute. If G_S is discrete, then it is closed, and hence $\widetilde{G}(S)/G_S$, is a topological group. Since the mappings $\widetilde{G}(S) \to \widetilde{G}(S)/G_S$, $\psi_S: \widehat{S} \to \widetilde{G}(S)$ and $\sigma_S: \widehat{S} \to S$ are all local homeomorphisms, it follows that γ_S is also, and hence that its image is open.

If S is simply connected, then σ_S is a homeomorphism. Hence $G_S = \{e\}$, and $\gamma_S = \psi_S$. If $\widetilde{G}(S)$ is a Lie group such that its Lie algebra has no center, then any central subgroup of $\widetilde{G}(S)$ must be discrete (since the adjoint representation will be a local homeomorphism whose kernel is the center). Hence in this case G_S will be discrete.

See Proposition VII.3.25 of [12] for the proof of the last part. □

6. The free topological group on *S* revisited

We are now ready to apply the machinery of the preceding sections to the problem of identifying the free topological group for the type of semigroups we have been considering.

We need at this point some additional facts about local right actions.

Definition 6.1. Let X be a space and suppose we are given a local right action of a local group G on X. For $x \in X$, we define the *G-orbit* of x, denoted Orb(x) to be the set of all points $y \in X$ such that there exist $g_1, \ldots, g_n \in G$ and $x_0, \ldots, x_n \in X$ with $x_0 = x$, $x_i = x_{i-1}g_i$ for $1 \leq i \leq n$ (in particular, $x_{i-1}g_i$ is defined), and $x_n = y$. We abbreviate the last conditions by writing $y = xg_1 \cdots g_n$.

Lemma 6.2. *Suppose that X is a space and that there is given a locally simply transitive local right action of a local group G on S. Then for $x \in X$, the orbit of x is both open and closed and hence contains the component of x.*

Proof. The proof is straightforward. See Lemma VII.3.23 of [12]. □

Theorem 6.3. *Let S be a connected, locally connected cancellative topological semigroup with a countable base satisfying* (INT) *and suppose there exists a universal covering group $\widetilde{G}(S)$ locally isomorphic to $Q\,(S, V)$. Then $\gamma_S: S \to \widetilde{G}(S)/G_S$, where the latter is equipped with the quotient topology induced by the natural homomorphism from $\widetilde{G}(S)$, is the free topological group on S. Hence we may identify $G(S)$ with $\widetilde{G}(S)/G_S$.*

Proof. Let $\theta\colon S \to H$ be a homomorphism into a group H. We wish to construct a homomorphism $\theta'\colon \widetilde{G}(S)/G_S \to H$ so that $\theta' \circ \gamma_S = \theta$. This will establish that $\gamma_S\colon S \to \widetilde{G}(S)/G_S$ is the free group on S since it shows that the desired universal property is satisfied. Note that the uniqueness of θ' follows from the fact that $\gamma_S(S)$ generates $G(S)$.

In the manner of Proposition 4.9 and Remark 4.10, one defines a local homomorphism $\Delta\colon \widetilde{G}(S) \to H$ such that $\theta(s)\Delta(g) = \theta(sg)$ for small g. Then Δ extends to a global homomorphism. Since for small g the mapping Δ is given by $g \mapsto \theta(s)^{-1}\theta(sg)$, it follows that Δ is continuous if θ is (and if H is a topological group; actually one only needs separate continuity of multiplication on H). Observe that θ preserves the local action if we use the homomorphism Δ to define a right action of $\widetilde{G}(S)$ on H.

If we consider additionally the right action of multiplication of $\widetilde{G}(S)$ on itself, then there is induced a local right action of $\widetilde{G}(S)$ on $S\times\widetilde{G}(S)$ which sends $\big((s,h),g\big)$ to (sg, hg). The topology τ on $\Sigma = S\times\widetilde{G}(S)$ is defined precisely in such a way so that this local right action is locally simply transitive.

Suppose $g \in G_S$. Then by definition of G_S for $(s,h) \in \widehat{S}$, we have $(s, gh) \in \widehat{S}$. Since $\widehat{S}$ is a component in Σ, it follows from Lemma 6.2 that there exist $g_1, \dots, g_n \in \widetilde{G}(S)$ such that $(s, gh) = (sg_1\cdots g_n, hg_1\cdots g_n)$. By Proposition 5.6 the group G_S is central, so $hg = gh = hg_1\cdots g_n$. By cancellation $g = g_1\cdots g_n$. Thus $\Delta(g) = \Delta(g_1)\cdots\Delta(g_n)$. We have also that $s = sg_1\cdots g_n$ and hence $\theta(s) = \theta(sg_1\cdots g_n)$. We have seen in an earlier paragraph that $\theta(sg') = \theta(s)\Delta(g')$ for $g' \in G$ such that sg' is defined, and by induction $\theta(sg_1\cdots g_n) = \theta(s)\Delta(g_1)\cdots\Delta(g_n)$. Again by cancellation of $\theta(s)$,

$$e_H = \Delta(g_1)\cdots\Delta(g_n) = \Delta(g).$$

Since g was an arbitrary element of G_C, if follows that the kernel of Δ contains G_S. Hence there exists a mapping $\theta'\colon \widetilde{G(S)}/G_S \to H$ such that $\theta' \circ \nu = \Delta$, where ν is the natural homomorphism from $\widetilde{G(S)}$ to $\widetilde{G(S)}/G_S$.

To complete the proof, we must show that $\theta' \circ \gamma_S = \theta$. This follows from a standard diagram chasing argument. □

We consider an important special case.

Theorem 6.4. *Let S be a cancellative topological semigroup on a paracompact connected Euclidean manifold.*

(i) *If S is algebraically embeddable in a group, then $G(S)$ is a Lie group and the mapping $\gamma\colon S \to G(S)$ is an open embedding. In particular, S is topologically isomorphic to an open subsemigroup of a Lie group.*

(ii) *If S is simply connected, then $G(S)$ is a simply connected Lie group, and $\gamma\colon S \to G(S)$ is a local homeomorphism.*

Proof. A paracompact connected Euclidean manifold has a countable base. By Proposition 2.5, S satisfies condition (INT). As in Corollary 4.6, there exists a Lie group G locally isomorphic to $Q(S, V)$, and we might as well take it to be simply connected, and hence $\widetilde{G}(S)$. If S algebraically embeds in a group, then this embedding must factor through the free group $G(S)$ on S. It follows that γ is a monomorphism in this case. Both cases now follow from Proposition 5.7 and Theorem 6.3. □

By means of Propositions 1.6 and 2.7 the preceding results can be generalized to semigroups with well-behaved ideals. The next theorem is illustrative of this procedure.

Theorem 6.5. *Let S be a topological semigroup with a countable base on a connected Euclidean manifold with generalized boundary and suppose that S can be algebraically embedded in a group. Then $G(S)$, the free group on S, admits the structure of a Lie group. If there exists a in the interior of S such that $s \mapsto as: S \to S$ is an embedding onto an admissible subset of S, then the homomorphism $\gamma: S \to G(S)$ is an embedding onto a subsemigroup which is an admissible subset of $G(S)$.*

Proof. Since S can be embedded in a group, it is cancellative. Consider the set I of interior points of S. It follows from the Invariance of Domain Theorem that I is an ideal of S. We can apply Theorem 6.4 to I and conclude that $\gamma_I: I \to G(I)$ is a topological isomorphism onto an open subsemigroup of the Lie group $G(I)$. One now applies Propositions 1.6 and 2.7. □

Suppose now that S is a topological semigroup that contains an open ideal I such that I is a cancellative semigroup, has a countable base, and is connected and locally Euclidean. Suppose that $\delta: S \to G$ is a continuous homomorphism into a connected Lie group G which is a homeomorphism on some open set $U \subseteq I$ to an open subset $\delta(U)$ in G. We are interested in the relationship between the pair (G, δ) and the free topological group $(G(S), \gamma)$ on S. We argue along the lines of the proof of Theorem 6.4 that $G(I)$ is a Lie group and that $\gamma_I: I \to G(I)$ is a local homeomorphism (see, in particular, (iii) of Proposition 5.7). By Proposition 1.6 we may identify $G(S)$ with $G(I)$ and take γ_S to be the unique extension of γ_I to S. By the universal property, there exists a continuous homomorphism $h: G(S) \to G$ such that $h \circ \gamma_S = \delta$. Since there exists an open subset U of I such that the restrictions of δ and γ to U are homeomorphisms onto open subsets of G and $G(S)$ respectively, it follows that the restriction of h is a homeomorphism from $\gamma(U)$ to $\delta(U)$. Thus h is a local homeomorphism, and hence must be a covering projection from $G(S)$ onto G, since the kernel of h must be a discrete central subgroup. The mapping γ_S is then a lifting of δ for this covering projection h.

Theorem 6.6. *Let S be a topological semigroup containing an open ideal I such that I is a cancellative semigroup, has a countable base, and is connected and locally Euclidean, and let $\delta: S \to G$ be a continuous homomorphism into a connected Lie group G which is a homeomorphism on some open set $U \subseteq I$ to an open subset $\delta(U)$ in G. Then there exists a largest subgroup A of the fundamental group $\pi(G)$ such that δ lifts to the covering group G_A of G corresponding to A. The free topological group on S can be taken to be G_A together with the lift of δ.*

Proof. By the remarks before the statement of the theorem the free topological group on S consists of a covering group of G and a lift of δ. By the universal property of the free topological group on S, it is the largest such. □

Corollary 6.7. *Let S be a connected subsemigroup of a connected Lie group G such that the interior of S in G is dense in S. Then $G(S)$ is the largest connected covering group of G in which S lifts.* □

In [28] Ruppert considers certain topological semigroups S which can be embedded into topological groups as generating subsets. He seeks conditions under which the isomorphism type of the semigroup determines uniquely the isomorphism type of the group and says in this case that S is *uniquely group embeddable.* In this case the unique topological group into which S embeds as a generating set must be $G(S)$ (see Corollary 1.4). In the case that S is a connected subsemigroup of a Lie group with dense interior, the previous corollary is useful in his computations. Among other things, Ruppert obtains the following result:

Theorem 6.8. *Let S be an open subsemigroup of a connected Lie group G and suppose that either G is solvable or S is invariant in G. Then S is uniquely group embeddable, and so $G = G(S)$. In the particular case that G is nilpotent, then S is both left and right reversible.* □

One also obtains the following corollary from Theorem 6.6.

Corollary 6.9. *Let S be a topological semigroup containing an open ideal I such that I is a cancellative semigroup, has a countable base, and is connected and locally Euclidean, and let $\delta: S \to G$ be a continuous homomorphism into a simply connected Lie group G which is a homeomorphism on some open set $U \subseteq I$ to an open subset $\delta(U)$ in G. Then (G, δ) is the free topological group on S.* □

Example 6.10. If $S = \mathrm{Sl}(2, \mathbb{R})^+$, the semigroup of all matrices in $\mathrm{Sl}(2, \mathbb{R})$ with non-negative entries (cf. Proposition V.4.19 of [12]), then the inclusion $i: S \to \mathrm{Sl}(2, \mathbb{R})$ is not the inclusion into the free group on S; the lifting $j: S \to \widetilde{\mathrm{Sl}}(2, \mathbb{R})$ however is (by Corollary 6.9).

Example 6.11. Let C be one half of the invariant double cone in $\mathrm{sl}(2,\mathbb{R})$ consisting of those points where the Cartan-Killing quadratic form takes on non-positive values. (See the latter part of Section V.4 of [12] for an extended discussion of this example.) We consider the wedge $W = \mathrm{sl}(2,\mathbb{R})+iC$ in $\mathrm{sl}(2,\mathbb{C}) = \mathrm{sl}(2,\mathbb{R})+i\,\mathrm{sl}(2,\mathbb{R})$. Then $S = \mathrm{Sl}(2,\mathbb{R})\cdot\exp(iC)$ is a subsemigroup of the simply connected complex Lie group $\mathrm{Sl}(2,\mathbb{C})$ with dense interior consisting of the product of $\mathrm{Sl}(2,\mathbb{R})$ and the exponential of the interior of iC in $i\,\mathrm{sl}(2,\mathbb{R})$. There is now a covering semigroup $\widehat{S}$ of S such that the group of units of $\widehat{S}$ is $\widetilde{\mathrm{Sl}}(2,\mathbb{R})$, the universal covering group of $\mathrm{Sl}(2,\mathbb{R})$ (see Example V.4.59 of [12]). By Corollary 6.9 the covering projection from $\widehat{S}$ to S followed by the inclusion into $\mathrm{Sl}(2,\mathbb{C})$ gives the free topological group on $\widehat{S}$. Since the mapping is not an embedding it follows that the cancellative semigroup $\widehat{S}$ does not even algebraically embed into a group.

The preceding results and examples give useful insights into the question of the extent to which a generating subsemigroup of a Lie group determines the Lie group in which it sits.

There are several fundamental questions that remain unanswered. We include two here and one at the conclusion of the next section.

Problem A. Given a cancellative topological semigroup S on a connected Euclidean manifold, does there exist a continuous homomorphism into a Lie group which is a local homeomorphism? This will be true if and only if the embedding into $G(S)$ is a local homeomorphism (an analog of Corollary 1.4). By Proposition 5.7 and Theorem 6.3, the problem is equivalent to the problem of whether the subgroup G_S of $\widetilde{G}(S)$ is always discrete (equivalently closed).

Problem B. Let S be a cancellative semigroup on a connected Euclidean manifold for which there exists a continuous homomorphism into a Lie group which is a local homeomorphism (we say that S is *locally embeddable* in a Lie group). Does there exist a cancellative semigroup T which is a covering semigroup of a subsemigroup of a Lie group such that S embeds in T?

7. Local semigroups

The constructions given earlier in the paper modify to the case of local semigroups. Since we only consider neighborhoods around an identity 1, these might be more appropriately called local monoids. However, we retain the traditional terminology of "local semigroup."

The main example that we have in mind of a local semigroup is that of a neighborhood of the identity in a topological semigroup (with the multiplication restricted to the neighborhood). It is necessary to restrict because one gets embeddings into a group locally much more generally than global embeddings. For example, a compact topological semigroup never embeds in a group unless it is already a group, but frequently locally embeds near the identity. Local semigroups also arise naturally by composing the positive time (partial) transformations of a family of local flows on a manifold. There are no general methods available for embedding these local semigroups into global ones. One can also consult Chapter IV of [12] for the theory (and additional examples) of local semigroups in Lie groups.

The local group into which a local semigroup is embedded is again constructed via the Brown–Houston machinery of building local quotients [4]. The results of this section may be found proved in Chapter VII.1 of [12] or in [21].

We begin with a formal definition of a local semigroup (which we always assume to be equipped with an identity). Again there are minor variations in the way that this concept is defined in the literature.

Definition 7.1. A *local semigroup* (*with identity*) is a system $(T, 1, m)$ such that T is a T_3-space, $1 \in T$, and $m\colon \mathrm{dom}(m) \to T$ is a function (partial multiplication) such that $\mathrm{dom}(m) \subseteq T \times T$. Furthermore, there must exist an open set Σ containing 1 such that $\Sigma \times \Sigma \subseteq \mathrm{dom}(m)$ and such that the following conditions are satisfied:

(i) the restriction of m to $\Sigma \times \Sigma$ is continuous;

(ii) if $a, b, c, m(a, b), m(b, c) \in \Sigma$, then $m(m(a, b), c)$ and $m(a, m(b, c))$ are defined and equal;

iii) if $a \in \Sigma$, then $m(a, 1) = a = m(1, a)$.

One customarily denotes $m(a, b)$ by ab and this will be our usual practice in the sequel. By a mild abuse of terminology we often refer to the local semigroup simply as T.

The local semigroup is a *cancellative local semigroup* (*with identity*) if Σ can also be chosen so that the multiplication is cancellative on Σ, i.e., so that $a, b, c \in \Sigma$ and $ab = ac$ or $ba = ca$ implies that $b = c$.

An open neighborhood N of the identity in a local semigroup is called a *standard neighborhood* if all quadruple products of elements of N are defined and associative, if N is contained in one of the neighborhoods Σ guaranteed in the definition of a local semigroup, and if the multiplication is cancellative on N^2. One establishes directly that a cancellative local semigroup contains a basis of standard neighborhoods at 1.

Given a cancellative local semigroup, we wish to embed it locally in a local group of quotients.

For V an open subset of S we again define the *left quotient relation* on $S \times S$ by

$$(a, b) \stackrel{\text{def}}{\equiv} (c, d) \quad \text{if} \quad ax = by, \; cx = dy \quad \text{where} \quad x, y \in V.$$

We wish to choose some open set U containing 1 and an appropriate open set $V \subseteq S$ such that $\equiv$ is a closed equivalence relation on $U \times U$ and the multiplication induced on the equivalence classes by $(a,b)(b,c) = (a,c)$ makes $Q(U,V) \stackrel{\text{def}}{=} U \times U/\equiv$ a local group, where $Q(U,V)$ is given the quotient topology from $U \times U$. However, the case that 1 has a Euclidean neighborhood is far too restrictive in this setting, and so we need to modify our earlier requirement about condition (INT) to one appropriate to this context.

Definition 7.2. Let T be a cancellative local semigroup and let S be a standard neighborhood. A point $b \in S$ is a *pseudo-interior point* if given $a \in S$ and an open set V containing b, there exist open sets U, N with $a \in U$, $ab \in N$ (resp. $ba \in N$) such that $x \in U$ implies $N \subseteq xV$ (resp. $N \subseteq Vx$). The *pseudo-interior* S° of S consists of those points which possess neighborhoods made up entirely of pseudo-interior points.

Theorem 7.3. *Let T be a cancellative local semigroup and let S be a standard neighborhood with a dense pseudo-interior. Then there exist an open neighborhood $U \subseteq S$ of 1 and an open set $V \subseteq S^\circ$ such that $Q(U,V)$ is a local group and the mapping $1^\triangle : U \to Q(U,V)$ given by $s \mapsto 1 \setminus s$ is a continuous monomorphism which restricted to $S^\circ \cap U$ is a topological embedding onto an open subset of $Q(U,V)$. If some point of $U \cap S^\circ$ has a Euclidean neighborhood, then $Q(U,V)$ is locally isomorphic to a Lie group.*

If U is identified with its image in $Q(U,V)$, then every member of $Q(U,V)$ has a representation of the form $s^{-1}t$ for $s,t \in U$.

If, additionally, each neighborhood of 1 contains a point p such that the mapping $x \mapsto xp : S \to Sp$ is a homeomorphism and $Sp \cap S \subseteq S^\circ$, then the mapping $s \mapsto \beta(s,1)$ is a topological embedding on all of U. □

Corollary 7.4. *Let S be a locally compact locally cancellative local semigroup with identity and suppose that S is homeomorphic to an admissible subset of $\mathbb{R}^n$. Then S is locally embeddable into a finite dimensional Lie group G such that the image of the embedding is an admissible subset of G. Alternately, S is locally topologically isomorphic to some local subsemigroup on an admissible subset of a finite dimensional real Lie algebra with the restricted Campbell-Hausdorff multiplication.* □

The local group of quotients that we have been constructing corresponds in the local case to the free topological group $G(S)$ that we constructed in the global case. We make this idea precise.

Definition 7.5. Let S and T be local semigroups with identities e and f resp. A *local homomorphism* is a continous function $\alpha : N \to T$, where N is some neighborhood of e, such that $\alpha(e) = f$ and for $x, y, xy \in N$, we have $\alpha(xy) = \alpha(x)\alpha(y)$.

Note that the restriction of a local homomorphism to any smaller neighborhood of e is again a local homomorphism. Thus if we wish to restrict down to smaller neighborhoods in the domain or codomain where certain properties hold, we may do so and simply restrict the local homomorphism.

Definition 7.6. Let S be a local semigroup. A *local group of quotients* for S is a mapping $i: U \to G$, where U is an open set containing e in S, G is a local group, and i is both a local homomorphism and a homeomorphism onto an admissible subset of G.

Remark 7.7. Let $i: U \to G$ be a local group of quotients for S. Pick a set B open in G contained in $i(U) \cap \Gamma$, where Γ is a standard neighborhood in G. (This is possible since $i(U)$ is admissible and contains the identity.) For $i(s) \in B$ we have that $i(s)^{-1}i(U)$ and $i(U)i(s)^{-1}$ are neighborhoods of the identity. Hence the elements in some neighborhood of the identity in G can be written as left quotients and right quotients of the embedded image of U. This justifies the terminology "local group of quotients."

Proposition 7.8. *Let S be a local semigroup. Then S admits a local group of quotients if and only if e has a standard neighborhood Σ such that*
(i) *the multiplication is cancellative on Σ,*
(ii) *Σ° is dense in Σ,*
(iii) *for every neighborhood U of e, there exists $p \in U$ such that $\Sigma p \cap \Sigma \subseteq \Sigma^\circ$ and the mapping $x \mapsto xp: \Sigma \to \Sigma p$ is a homeomorphism.* □

The next proposition yields the universal property of the local group of quotients.

Proposition 7.9. *Let S be a local semigroup, $i: U \to G$ a local group of quotients for S, and let α be a local homomorphism to a local group H. Then there exists a local homomorphism γ from G to H such that $\gamma \circ i = \alpha$ on some neighborhood of e. If δ is another local homomorphism from G to H such that $\delta \circ i = \alpha$ on some neighborhood of e, then δ and γ agree on some neighborhood of the identity $i(e)$.* □

The next corollary states that a local group of quotients is unique up to local isomorphism.

Corollary 7.10. *Let $i: U \to G$ and $j: V \to H$ be local groups of quotients for S. Then there exists a local topological isomorphism γ from G to H such that $\gamma \circ i = j$ on some neighborhood of e.* □

The next proposition gives the sense in which embedding in local quotient groups is functorial.

Proposition 7.11. *Let S and T be local semigroups with local groups of quotients $i_S: U_S \to Q(S)$ and $i_T: U_T \to Q(T)$ and let $\alpha: S \to T$ be a local homomorphism. Then there exists a local homomorphism $Q(\alpha): Q(S) \to Q(T)$ such that $Q(\alpha) \circ i_S = i_T \circ \alpha$ on some neighborhood of e_S. Any two such local homomorphisms agree on some neighborhood of $i_S(e_S)$. If $\alpha(V)$ has non-empty interior in T for each open set in S containing e, then $Q(\alpha)$ is a locally open mapping (i. e. its restriction to some neighborhood of the identity if open).* □

We recall some basic facts about local Lie groups. One may consult Chapter III of [1] for a statement of these results in the most general setting. According to Lie's Fundamental Theorems each local Lie group G has associated to it in a functorial way a completely normable Lie algebra $\mathbf{L}(G)$. Conversely, a completely normable Lie algebra gives rise to a local Lie group $(\mathbf{L}(G), *)$, namely the locally defined Campbell-Hausdorff multiplication on some neighborhood of 0. These operations are inverse operations in the sense that G is locally analytically isomorphic to $(\mathbf{L}(G), *)$ (via the local inverse of the exponential mapping), and conversely if L is a completely normable Lie algebra, then the Lie algebra of $(L, *)$ is naturally isomorphic (as a completely normable Lie algebra) to L. The local isomorphism from G to $(\mathbf{L}(G), *)$ is called a *canonical chart*. In this section we adapt the idea of a canonical chart to the local semigroup setting.

Definition 7.12. A *canonical embedding* for a local semigroup S is a mapping $i: U \to L$ which is a local group of quotients (see Definition 7.6) for which L is a completely normable Lie algebra equipped with the local group structure arising from the Campbell-Hausdorff multiplication.

The existence of a canonical embedding is closely tied to the existence of a local embedding into a local Lie group.

Proposition 7.13. *Let S be a local semigroup. Then a canonical embedding exists for S if and only if S has a local group of quotients which is a local Lie group.* □

The next proposition shows that associating a Lie algebra with a local semigroup by means of a canonical homomorphism is a functorial construction.

Proposition 7.14. *Let $i: U \to \mathcal{L}(S)$ and $j: V \to \mathcal{L}(T)$ be canonical embeddings for S and T respectively. If α is a local homomorphism from S to T, then there exists a unique continuous Lie algebra homomorphism $\mathcal{L}(\alpha): \mathcal{L}(S) \to \mathcal{L}(T)$ such that the diagram*

$$\begin{array}{ccc} S & \xrightarrow{\quad\alpha\quad} & T \\ {\scriptstyle i}\downarrow & & \downarrow{\scriptstyle j} \\ \mathcal{L}(S) & \xrightarrow[\mathcal{L}(\alpha)]{} & \mathcal{L}(T) \end{array}$$

is commutative, i.e., such that $\mathcal{L}(\alpha) \circ i = j \circ \alpha$ *on a neighborhood of* e *in* S. □

The essential uniqueness of a canonical embedding (if it exists) now follows.

Corollary 7.15. *Let* $i: U \to L_1$ *and* $j: V \to L_2$ *be canonical embeddings for a local semigroup* S. *Then there exists a unique isomorphism of completely normable Lie algebras* $\Gamma: L_1 \to L_2$ *such that* $\Gamma \circ i = j$. □

The preceding results are important for the study of local semigroups which have a local group of quotients which is a Lie group. These results show that without any essential loss of generality, one might as well assume that such local semigroups are local subsemigroups in a completely normable Lie algebra equipped with the Campbell-Hausdorff multiplication. One can then bring to bear the machinery of Lie groups and Lie algebras in order to develop their theory, see e.g. [13] or [12].

Problem C. Let W be a Lie wedge in a Lie algebra L. Does there exist a cancellative topological (analytic) semigroup S on a manifold with generalized boundary such that S is locally isomorphic to a local semigroup with tangent wedge W? Does the interior of such a semigroup embed locally in a Lie group? If S is assumed simply connected, to what extent is it unique?

References

[1] Bourbaki, N., "Groupes et algebrès de Lie", Chap. II, Hermann, Paris, 1971

[2] Brown, D. R., and M. Friedberg, Representation theorems for uniquely divisible semigroups, Duke Math. J. 35 (1968), 341–352

[3] —, Linear representations of certain compact semigroups, Trans. Am. Math. Soc. 160 (1971), 453–465

[4] Brown, D. R., and R. S. Houston, Cancellative semigroups on manifolds, Semigroup Forum 35 (1987), 279–302

[5] Carruth, J. H., J. A. Hildebrant, and R. J. Koch, "The Theory of Topological Semigroups", Vols. I and II, Marcel Dekker, New York, 1983 and 1986

[6] Christoph, Jr., F. T., Embedding topological semigroups in topological groups, Semigroup Forum 1 (1970), 224–231

[7] Clifford, A., and G. Preston, "The Algebraic Theory of Semigroups", Vols. I & II, Amer. Math. Soc., 1961 and 1967

[8] Dubreil, P., Sur les problèmes de l' immersion et la théorie des modules, C. R. Acad. Sci. Paris 216 (1943), 625–627

[9] Gelbaum, B., G. K. Kalisch and J. M. H. Olmsted, On the embedding of topological semigroups and integral domains, Proc. Amer. Math. Soc. 2 (1951), 807–821

[10] Graham, G. Differentiable semigroups, "Proceedings Conference on Semigroups in Oberwolfach 1981," Springer Lecture Note in Mathematics 998 (1983), Springer Verlag, 57–127

[11] Hilgert, J. A note on Howe's oscillator semigroup, 1988, preprint

[12] Hilgert, J., K. H. Hofmann, and J. D. Lawson, "Lie Groups, Convex Cones, and Semigroup", Oxford Press, 1989, xxxvii+645pp

[13] Hofmann, K. H., and J. D. Lawson, Foundations of Lie semigroups, "Proceedings Conference on Semigroups in Oberwolfach 1981", Springer Lecture Note in Mathematics 998 (1983), Springer Verlag, 128–201

[14] Hofmann, K. H., and P. S. Mostert, "Elements of Compact Semigroups", Merrill, Columbus, 1966

[15] Hofmann, K. H., and W. Weiss, Note on cancellative semigroups on manifolds, Semigroup Forum

[16] Houston, R., "Cancellative semigroups on manifolds", Ph.D. dissertation, University of Houston, 1973

[17] Howe, R., The oscillator semigroup, 1987, preprint

[18] Jacoby, R., Some theorems on the structure of locally compact local groups, Ann. Math. 66 (1957), 36–69

[19] Keimel, K., Eine Exponentialfunktion für kompakte abelsche Halbgruppen, Math. Zeit. 96 (1967), 7–25

[20] —, Lokal kompakte Kegelhalbgruppen and deren Einbettung in topologische Vektorräume, Math. Zeit. 99 (1967), 405–428

[21] Lawson, J. D., Embedding local semigroups into groups, Proc. N.Y. Acad. Sci., (to appear)

[22] Lawson J. D., and B. Madison, On congruences and cones, Math. Zeit. 120 (1971), 18–24

[23] Malcev, A., On the immersion of an algebraic ring into a field, Math. Ann. 113 (1937), 686–691

[24] McKilligan, S. A., Embedding topological semigroups in topological groups, Proc. Edinburgh Math. Soc. 17 (1970/71), 127–138

[25] Ore, O. Linear equations in noncommutative fields, Ann. of Math. 32 (1931), 463–477

[26] Peck, J. E. L., "The embedding of a topological semigroup in a topological group and its generalisations", Ph. D. dissertation, Yale University, 1950

[27] Rothman, N. J., Embedding of topological semigroups, Math. Ann. 139 (1960), 197–203

[28] Ruppert, W. A. F., On open subsemigroups of connected groups, Semigroup Forum 39 (1989), 347–362

[29] Schieferdecker, E., Einbettungssätze für topologische Halbgruppen, Math. Ann. 131 (1956), 372–384

[30] Tamari, D., Sur l'immersion d'un semi-groupe topologique dans un groupe topologique, Colloques Internationaux du Centre National de la Recherche Scientifique, Algèbre et Théorie des Nombres 24 (1950), 217–221

[31] Weinert, H. J., Semigroups of right quotients of topological semigroups, Trans. Am. Math. Soc. 147 (1970), 333–348

Algebraic varieties and semigroups

Lex E. Renner

Introduction

The purpose of this chapter is to survey the theory of linear algebraic semigroups. The reader will likely be familiar with other important categories of semigroups (semitopological, topological, compact, etc.). It is now safe to say that algebraic varieties with semigroup structure yield some of the most interesting and richly structured examples of topological semigroups. On the other hand, the axiomatic theory of algebraic semigroups is now in a fairly complete state, due to the efforts of M. S. Putcha and the author. Putcha has recently published a monograph [14] which makes the basic theory accessible to anyone with a modest background in algebraic groups and semigroup theory.

The first five sections of this chapter summarize the most important general structural properties of algebraic semigroups, focusing mainly on irreducible, regular monoids. Section six deals with the generalized Bruhat decomposition for reductive monoids. Here, we obtain a decomposition into double cosets of a certain solvable group indexed by the elements of a certain finite inverse monoid. Section seven deals with rational representations and conjugacy classes of semisimple elements. Here we describe an attractive generalization of the basic results about representations of reductive groups. Section eight contains a description of some classification theorems for regular, irreducible monoids. This part of the theory has not yet been completed, but the two basic types; reductive and solvable regular monoids, have been classified in detail. Section nine contains a brief sketch of several other recent developments related to algebraic monoids. First we describe how the basic theory of algebraic monoids can be recast in a purely combinatorial/semigroup theoretic way, thereby extending the theory to monoids with unit groups with BN-pairs. We also describe, in this section, how the theory of algebraic monoids fits into the theory of embeddings of spherical homogeneous spaces. Finally, we outline a procedure, due to De Concini and Procesi, which calculates the rational cohomology ring of certain complex algebraic varieties closely related to algebraic monoids.

1. Definitions and examples

Definition 1.1. Let k be an algebraically closed field. An *algebraic semigroup* S is an affine, algebraic variety together with an associative morphism $\mu: S \times S \to S$ of algebraic varieties. If S has an identity element $1 \in S$ for μ, we say S is an *algebraic monoid*. An algebraic variety is *irreducible* if it cannot be expressed as the union of two, proper, closed, non-empty subsets. By the Hilbert basis theorem, any algebraic variety is a union of a finite number of irreducible, closed subsets. Furthermore, there are only a finite number of maximal irreducible closed subsets and these are called *components*. If S is an algebraic monoid, there is a unique irreducible component $S^0 \subseteq S$ such that $1 \in S^0$. Furthermore, S^0 is an algebraic monoid. An algebraic monoid is *irreducible* if $S = S^0$. In such a case, $S = \overline{G}$ (Zariski closure), where $G \subseteq S$ is the unit group of S. This results from the fact that the unit group is open in S, while any nonempty, open set of an irreducible variety is dense.

Alternatively, algebraic semigroups may be described (in the dual category) as *bialgebras*. A bialgebra is a k-algebra A with coassociative morphism $\nabla: A \to A \otimes A$. For example, $S = \{(x, y) \in k^2 \mid x^2 = y^3\}$ is represented by $A = k[x, y]/(x^2 - y^3)$ with $\nabla(x) = x \otimes x$ and $\nabla(y) = y \otimes y$. This approach is sometimes useful technically for constructing monoids with desirable properties. One obtains S from A by Hilbert's Nullstellensatz as the set of maximal ideals of A.

A *morphism* φ of algebraic semigroups S and T is a morphism $\varphi: S \to T$ of algebraic varieties such that $\varphi(xy) = \varphi(x)\varphi(y)$ for $x, y \in S$. A morphism of algebraic monoids is a morphism of algebraic semigroups such that $\varphi(1_S) = 1_T$. For an interesting example, consider $\varphi: M_2(k) \to M_3(k)$ defined by

$$\varphi\begin{pmatrix} a & c \\ b & d \end{pmatrix} = \begin{pmatrix} a^2 & ab & b^2 \\ 2ac & ad + bc & 2bd \\ c^2 & cd & d^2 \end{pmatrix}.$$

Examples 1.2. (a) $S = M_n(k)$. Now S is linear as an algebraic variety matrix multiplication is defined by quadratic polynomial functions of the matrix coordinates. As is well known, S is a regular monoid. Notice that any Zariski closed subsemigroup of S is also an algebraic semigroup.

(b) *Finite Semigroups*. Let S be a finite semigroup. Then if $A = \mathrm{Hom}(S, k)$, define $\nabla: A \to A \otimes A \cong \mathrm{Hom}(S \times S, k)$ by the rule $\nabla(f)(s, t) = f(st)$. It is easy to see that (A, ∇) is a bialgebra and that the associated semigroup is isomorphic to S.

(c) *D-monoids*. Let H be a closed, connected subgroup of

$$D_n(k)^* = k^* \times \cdots \times k^*,$$

the set of invertible, diagonal, $n \times n$ matrices. Let $S = \overline{H} \subseteq D_n(k)$, the Zariski closure of H in $D_n(k)$, the set of diagonal matrices. Then S is a semilattice of groups. Furthermore, $\mathcal{U}(S)$ is isomorphic to the face lattice of a rational polytope [4]. Such an S can be described axiomatically as a *D-monoid*; namely an irreducible, algebraic monoid S such that $k[S]$ is spanned over k by $\{x \in k[S] \mid \nabla(x) = x \otimes x\}$.

(d) *Semidirect Products.* Let S and T be algebraic semigroups and suppose we have a morphism of varieties $\gamma: S \times T \to T$ such that
$\gamma(s, t_1 t_2) = \gamma(s, t_1)\gamma(s, t_2)$ for all $s \in S$ and for all $t_1, t_2 \in T$, and
$\gamma(s_1 s_2, t) = \gamma(s_1, t)\gamma(s_2, t)$ for all $s_1, s_2 \in S$ and $t \in T$. Write t^s for $\gamma(s, t)$. Then $S \times T$ is an algebraic semigroup with $(s_1, t_1)(s_2, t_2) = (s_1 s_2, t_1^{s_2} t_2)$.

(e) *Algebraic Rees Construction.* Let Γ and I be affine varieties, and suppose S is an algebraic semigroup. Suppose $P: \Gamma \times I \to S$ is a morphism of algebraic varieties. Let $S' = I \times S \times \Gamma$ and define $(x, s, y)(x', s', y') = (x, sP(y, x')s', y')$. Then S' is an algebraic semigroup with this multiplication.

(f) *Solvable Algebraic Monoids.* An irreducible monoid S is called *solvable* if $G(S)$ is a solvable algebraic group. By the Lie-Kolchin Theorem [11] and Theorem 2.1 below, any solvable monoid S is isomorphic to a closed submonoid of $T_n(k)$ (the upper triangular monoid) for some n. Using this fact, one can construct a universal morphism $\pi: S \to Z$, to a D-monoid, such that $\pi \mid \overline{T}: \overline{T} \to Z$ is an isomorphism, for any maximal torus $T \subseteq G = G(S)$. Furthermore, π induces a bijection on regular $\mathcal{J}$-classes.

(g) *Monoids from Representations.* Let G be a connected, algebraic group and let $\rho: G \to G\ell_n(k)$ be a rational representation. Define $M(\rho) = \overline{\rho(G)k^*}$ (Zariski closure in $M_n(k)$), where $k^* \subseteq G\ell_n(k)$ is identified with the nonsingular, central diagonal matrices. It is easy to see that $M(\rho)$ is an irreducible, algebraic monoid with unit group $\rho(G)k^*$.

(h) *Unit Groups.* A basic problem is to characterize the algebraic groups that can occur as unit groups of irreducible, algebraic monoids which are not groups. The answer, due independently to Waterhouse [30] and the author [20], asserts that $G = G(S) \subsetneq S$ if and only if $X(G)$, the character group of G, is nontrivial. Furthermore, if G occurs this way, then we can find an irreducible S with $G(S) = G$ and $0 \in S$. In particular, no semisimple group G is the unit group G of an irreducible monoid $S \neq G$.

2. General semigroup theoretic properties

Any algebraic semigroup S is *strongly π-regular* ($s\pi r$) (or group bound). Precisely, if $x \in S$, then $x^n \mathcal{H} e$ for some $e \in E(S)$. In fact, the integer n can be chosen independently of $x \in S$. Thus, there is always an abundance of idempotents, and in particular (for monoids), if $E(S) = \{1\}$, then S is an algebraic group.

If $e \in E(S)$ then eSe is an algebraic monoid with unit group H_e the $\mathcal{H}$-class of e. In particular, H_e is an algebraic group.

If $\mathcal{U}(S)$ denotes the set of regular J-classes of S, then $\mathcal{U}(S)$ is a finite poset.

One of the more fundamental results of the elementary theory is the following representability theorem.

Theorem 2.1. ([14], Theorem 3.15). *Let S be an algebraic semigroup. Then for some $n > 0$, there exists a morphism $\rho\colon S \to M_n(k)$, of algebraic semigroups, such that ρ is a closed embedding of algebraic varieties. If, further, S is a monoid, we may choose ρ so that $\rho(1) = I_n$.* □

The basic idea behind 2.1 is right translation of functions. First of all, we may assume S is a monoid (since $S^1 = S \cup \{1\}$ is an algebraic monoid, and $S \subseteq S^1$ is a closed subsemigroup). Define $\gamma\colon S \to \mathrm{End}(k[S])$ by the rule $\gamma_s(f)(t) = f(ts)$. It is easy to check that γ is a morphism of monoids. To obtain the morphism ρ, one chooses a finite dimensional subspace $V \subseteq k[S]$ such that
(i) $\gamma_s(V) \subseteq V$ for each $s \in S$, and
(ii) V generates $k[S]$ as a k-algebra.
It then turns out that $\rho\colon S \to \mathrm{End}(V)$, $\rho(s)(v) = \gamma_s(v)$, satisfies the conclusion of 2.1.

Concerning two-sided ideals, we have the following two basic results.

Proposition 2.2. ([14], Corollary 3.30) *If $I \subseteq S$ is a two-sided ideal, then the Rees quotient, $\overline{I}/I$ is a nil semigroup. In particular, $E(I) = E(\overline{I})$.* □

Proposition 2.3. ([21]) *Let S be an irreducible monoid. If $P \subseteq S$ is a prime ideal (so that $S \setminus P$ is multiplicatively closed), then P is closed in S. Furthermore, there exists a morphism $\chi\colon S \to k$ of algebraic monoids such that $\chi^{-1}(0) = P$.* □

Using 2.3, it is deduced in [20] that any irreducible *quasiaffine* algebraic monoid variety is actually affine. This result was inspired by the corresponding result of Chevalley for algebraic groups.

3. Irreducible algebraic monoids

As one would expect, irreducible algebraic semigroups enjoy a lot of finer structural properties. In this section we focus mainly on irreducible monoids. For more information about irreducible semigroups the reader should consult chapter 5 of Putcha's monograph [14].

Proposition 3.1. ([14], Theorem 5.10) *Let S be irreducible. Then $\mathcal{U}(S)$ is a finite lattice.* □

It is an interesting open problem to characterize these lattices, especially in the light of PUTCHA's theory in [18] (see 9.1 for a summary of that paper). This has been done for some interesting classes of reductive monoids in [19]. See Figures 1, 2 and 3.

Very early in the development, one learns that unit groups of irreducible monoids are "big". The next result records how this intuition is reflected in Green's relations.

Proposition 3.2. ([15], Theorem 13) *Let S be an irreducible monoid, $a, b \in S$, $e, f \in E(S)$. Let $G = G(S)$.*

(a) $\mathcal{J} = \mathcal{D}$.

(b) $a\mathcal{R}b$ *if and only if* $aG = bG$.

(c) $a\mathcal{L}b$ *if and only if* $Ga = Gb$.

(d) $a\mathcal{J}b$ *if and only if* $GaG = GbG$.

(e) $e\mathcal{J}f$ *if and only if* $geg^{-1} = f$ *for some* $g \in G$. □

The main point behind (b), (c) and (d) is the following basic fact from algebraic group theory: If $G \times X \to X$ is a regular action of an algebraic group on an algebraic variety, then any orbit is open in its closure. The proof of (e) is more subtle and the reader can find it in chapter six of [14].

For $e \in E(S)$ we define $C_G^r(e) = \{g \in G \mid ge = ege\}$, the *right centralizer of* e *in* G. Similarly, $C_G^\ell(e) = \{g \in G \mid eg = ege\}$ and $C_G(e) = C_G^r(e) \cap C_G^\ell(e)$.

Proposition 3.3. *Let S be an irreducible monoid and let $e \in E(S)$.*

(a) $eSe \subseteq \overline{C_G(e)}$.

(b) $eS \subseteq \overline{C_G^r(e)}$.

(c) $Se \subseteq \overline{C_G^\ell(e)}$.

(d) *The morphism $\varphi\colon C_G(e) \to H_e$, $\varphi(g) = ge$, of algebraic groups is surjective.* □

For proof see Theorem 6.16 of [14]. It follows readily from 3.3 that any idempotent of S is in the closure of some maximal torus of G.

The following result is the starting point for a description of the idempotent set in terms of the unit group and the lattice of regular $\mathcal{J}$-classes.

Proposition 3.4. ([19]) *Suppose $e, f \in E(S)$ and $e\mathcal{J}f$. Then $e\mathcal{R}f$ if and only if $C_G^r(e) = C_G^r(f)$. Similarly, $e\mathcal{L}f$ if and only if $C_G^\ell(e) = C_G^\ell(f)$.* □

The proof of 3.4 follows easily using the basic fact, for irreducible semigroups, that if $e\mathcal{J}f$, then there exist $e_1, e_2, f_1, f_2 \in E(S)$ such that $e\mathcal{R}e_1\mathcal{L}f_1\mathcal{R}f$ and $e\mathcal{L}e_2\mathcal{R}f_2\mathcal{L}f$.

Consistent with PUTCHA's philosophy to characterize semigroup theoretic properties group theoretically and vice versa, one obtains the following characterization of solvable monoids.

Proposition 3.5. ([14], Corollary 6.2) *Let S be an irreducible monoid with zero. The following are equivalent.*
(a) *$G(S)$ is solvable.*
(b) *S is a semilattice of Archimedean semigroups.* □

Recall that a semigroup S is *Archimedean* if for any $a, b \in S$, $a|b^i$ for some $i > 0$. Notice that the implication $(a) \Rightarrow (b)$ follows from 2.1 and the Lie-Kolchin theorem once it is verified that $T_n(k)$ is a semilattice of Archimedean semigroups.

4. Regular and reductive monoids

Throughout this section S denotes an irreducible algebraic monoid. S is *reductive* if $G(S)$ is a reductive, algebraic group. Recall that a reductive group is one that contains no closed, normal subgroup of positive dimension isomorphic to a subgroup of $U_n(k)$ (the group of upper triangular matrices with ones on the diagonal). The most familiar example $G\ell_n(k)$. The monoid S is *regular* if for any $x \in S$ there exists $a \in S$ such that $xax = x$. It is easy to show that any irreducible regular monoid is actually *unit regular*, namely we may choose $a \in G(S)$.

The following result is perhaps the most fundamental general result in the theory of algebraic monoids.

Theorem 4.1. ([14], Theorem 7.3) *Let S be an irreducible monoid with zero. The following are equivalent.*
(a) *S is regular.*
(b) *$G(S)$ is reductive.*
(c) *S contains no nonzero nil ideals.* □

The original proof [22] (due jointly to PUTCHA and the author) requires Weyl's Theorem on complete reducibility, the algebraic version of Hartog's Lemma, and Zariski's Main Theorem. Since then PUTCHA has found another proof that uses a more domestic blend of semigroup theory and algebraic group theory [14].

Using 4.1 one can readily obtain a characterization of regular monoids. First recall the *radical*, $R(G)$, of G. This is the unique, maximal, connected, solvable, normal subgroup of G [11].

Theorem 4.2. *Let S be an irreducible algebraic monoid. The following are equivalent.*

(a) S *is regular.*
(b) $\overline{R(G)}$ *is completely regular.*
(c) G_e *is reductive for any minimal idempotent* e *of* S. □

Here $G_e = \{g \in G(S) \mid eg = ge = e\}^0$. We also let $S_e = \overline{G_e} \subseteq S$. It is known [14] that $E(S_e) = \{f \in E(S) \mid fe = ef = e\}$. In Section 8 we describe the classification of reductive monoids and also the classification of completely regular monoids with solvable unit group. By Theorem 4.2, these are the two extremes of the general classification for regular, irreducible monoids. We end this section with a list of some of the important properties of regular monoids with zero. Proofs can be found in [14].

Theorem 4.3. *Let* S *be a regular monoid with zero, and let* $\Gamma \subseteq E(S)$ *be a chain,* $e \in \Gamma$.
(a) $C_G^r(\Gamma)$ *and* $C_G^\ell(\Gamma)$ *are opposite parabolic subgroups.*
(b) $C_G(\Gamma)$ *is reductive.*
(c) eSe *and* S_e *are regular monoids with zero.*
(d) *If* Γ *is a maximal chain, then* $C_G^r(\Gamma)$ *and* $C_G^\ell(\Gamma)$ *are Borel subgroups.* □

Property (a), in particular, is an important axiom in PUTCHA's theory of monoids with unit groups with BN-pairs. See Section 9.1.

5. The system of idempotents and Tits buildings

The purpose of this section is to describe the system of idempotents of a reductive, algebraic monoid. According to Nambooripad's theory of biordered sets [13], this goes a long way toward describing a regular semigroup.

The two basic ingredients in our description are 3.4 and 4.3(a).

Let S be a reductive monoid and let $e \in E(S)$. By 4.3(a), $C_G^r(e)$ and $C_G^\ell(e)$ are opposite parabolic subgroups of G. Also, $e \in J$, the $\mathcal{J}$-class of e. By 3.2(e), $E(J)$ forms but one conjugacy class under G, and so $C_G^r(e)$ and $C_G^r(f)$ are conjugate if $e\mathcal{J}f$. We say that parabolic subgroups P and Q are of the same *type*, and we write type(P) = type(Q), if P and Q are conjugate. It is a standard fact that the set of types, type(G), is indexed by the set of subsets 2^S of the set of nodes S of the associated Dynkin diagram of G. In any case we can associate a type to each $\mathcal{J}$-class, namely type(J) = type$(C_G^r$ (any $e \in E(J)$)). Define $\widehat{E}(S) = \{(P, Q, J) \in \Delta \times \Delta \times \mathcal{U}(S) \mid P$ and Q are opposite and type(P) = type$(J)\}$, where $\Delta = \{P \subseteq G \mid P$ is parabolic$\}$.

Theorem 5.1. ([19]) *Define* $\gamma\colon E(S) \to \widehat{E}(S)$ *by the rule* $\gamma(e) = (C^r_G(e), C^\ell_G(e), J_e)$. *Then*
(a) γ *is a bijection.*
(b) $E(S)$ *is determined (up to isomorphism) as a biordered set, by the type map* $\lambda\colon \mathcal{U}(S) \to \text{type}(G)$, $\lambda(J) = \text{type}(J)$, *and the poset structure of* $\mathcal{U}(G)$. □

Remarks. (1) It is straightforward to define the quasiorderings $\leq_r$ and $\leq_\ell$ on $\widehat{E}(S)$ that correspond to those of $E(S)$ under γ. It is however a nontrivial result (due to PUTCHA) to show that $\widehat{E}(S)$ is determined as a biordered set by $\leq_r$ and $\leq_\ell$.

(2) Proposition 5.1(b) motivates Putcha's theory of (abstract) monoids with unit group with BN-pair. See Section 9.1 for some discussion of this theory.

Theorem 5.1 makes it clear that the lattice of $\mathcal{J}$-classes is of central importance to the structure theory of reductive monoids. In [19] PUTCHA and the author produced an algorithm that computes $\mathcal{U}(S)$ and γ in case S has a zero, and only one minimal, non-zero $\mathcal{J}$-class. This is called the *$\mathcal{J}$-irreducible* case. This terminology is motivated by the following basic result.

Proposition 5.2. ([23], Corollary 8.33) *Let S be reductive with zero. The following are equivalent.*
(a) *S is $\mathcal{J}$-irreducible.*
(b) *There exists $\rho\colon S \to M_n(k)$ such that ρ is idempotent separating and irreducible as a representation.* □

One of the consequences of the algorithm in [19] is that, for $\mathcal{J}$-irreducible monoids, λ embeds $\mathcal{U}(S)$ into type(G). Furthermore, $\mathcal{U}(S)$ is determined by type(J_0), the minimal non-zero $\mathcal{J}$-class.

We illustrate the result by picturing $\mathcal{U}(S)$ for three classes of $\mathcal{J}$-irreducible monoids. By 5.2, we may describe each class in terms of the type of irreducible representation it can be associated with via 5.2(b). See Figures 1, 2 and 3.

6. Bruhat decompositions

One of the basic theorems about reductive groups (or more generally groups with BN-pair) yields a double coset decomposition

$$G = \bigcup_{w \in W} BwB$$

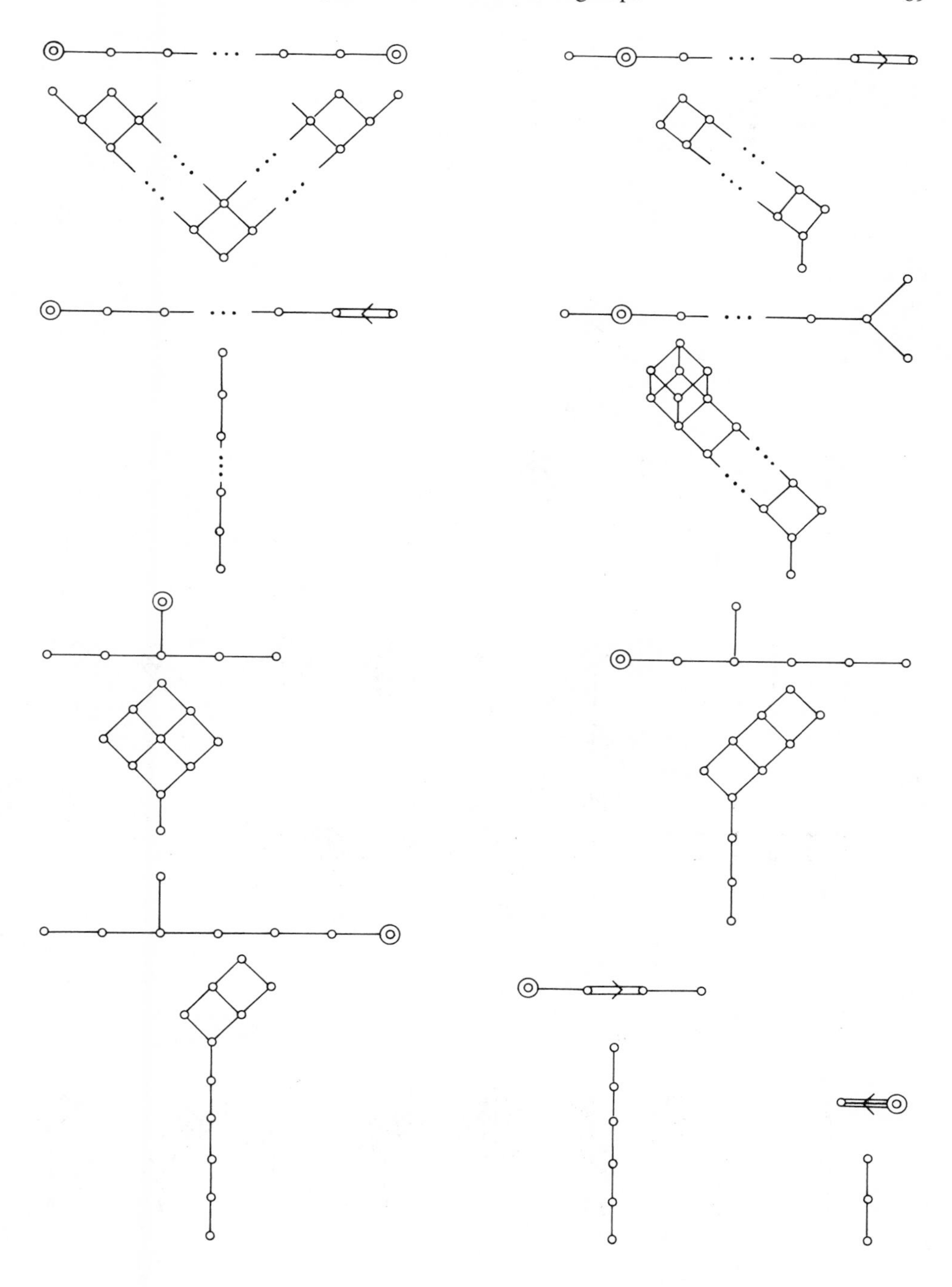

Fig. 1. The adjoint representations of simple algebraic groups. Here the lattice also represents the lattice of centers of unipotent radicals of standard parabolics.

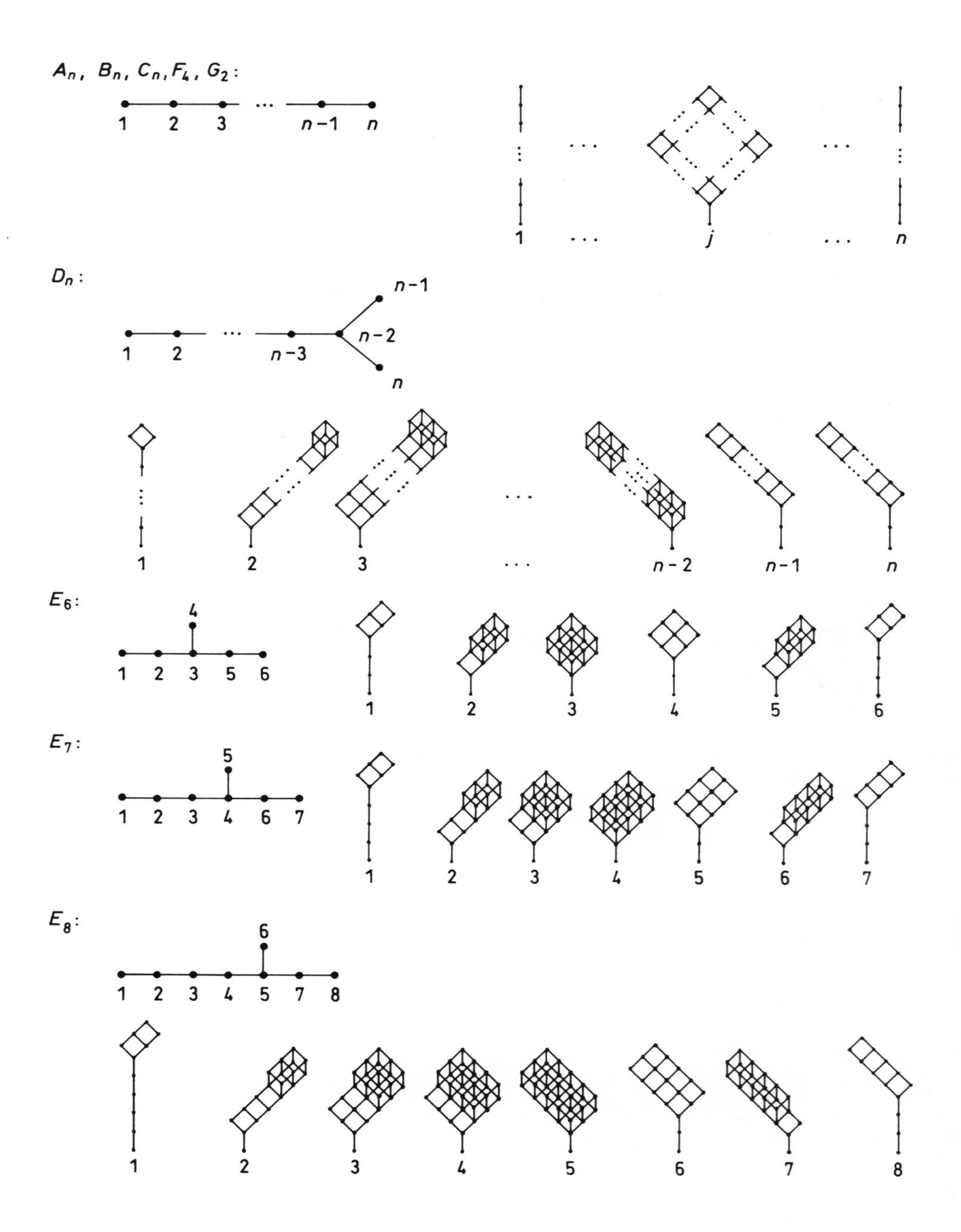

Fig. 2. Representations with fundamental high weight. For each simple group G there are rank(G) such possibilities. Recall that these correspond to the $\mathcal{J}_2$-irreducible monoids.

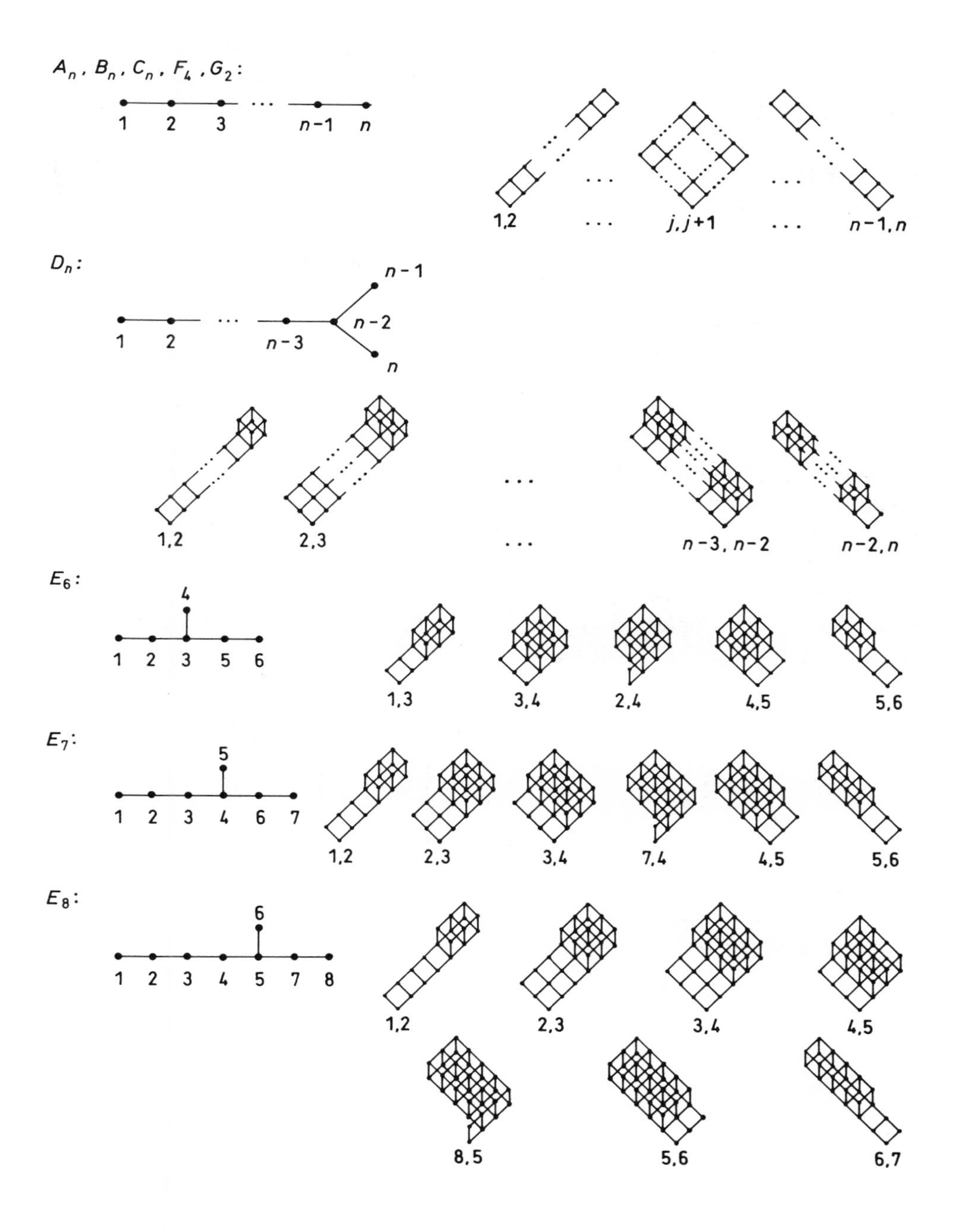

Fig. 3. Representations with high weight of the form $\lambda = a\lambda_1 + b\lambda_2$, with λ_1 and λ_2 fundamental and $\sigma_{\alpha_1}\sigma_{\alpha_2} \neq \sigma_{\alpha_2}\sigma_{\alpha_1}$. These correspond to the $\mathcal{J}$-irreducible monoids for which the $\mathcal{H}$-class of every idempotent is nearly simple.

where B is a Borel subgroup containing the maximal torus T and $W = N_G(T)/T$ is the Weyl group. A similar result holds for reductive monoids S. Let $T \subseteq G$ be a maximal torus and let $R = \{x \in S \mid xT = Tx\} = \overline{N_G(T)} \subseteq S$. Then R is an inverse monoid and $\mu = \{(x,y) \in R\times R \mid Tx = Ty\}$ is a fundamental congruence on R. We let $\mathcal{R} = R/\mu$. As above, let $W = N_G(T)/T \subseteq \mathcal{R}$, and let B be a Borel subgroup containing T.

Theorem 6.1. ([24]) *For the monoid $\mathcal{R}$, the following propositions hold:*
(a) *$\mathcal{R}$ is generated by W and $E(\mathcal{R}) \cong E(\overline{T})$.*
(b) *$S = \cup_{r\in\mathcal{R}} BrB$, disjoint union.*
(c) *For each $e \in E(\mathcal{R})$ there is a unique $r \in We$ such that $Br \subseteq rB$.*
(d) *If $\rho \in W$ is a simple reflection (relative to B) then $\rho Bx \subseteq BxB \cup B\rho xB$ for any $x \in \mathcal{R}$.* □

The reader is referred to [24] for the proof of these results. Putcha has obtained a more semigroup theoretic version of 6.1(b) in Chapter eleven of [14]. Part (c) of 6.1 is the starting point for a generalized Gauss–Jordan algorithm. These and other results are worked out in [24].

$\mathcal{R}$ can be partially ordered by the relation $x \le y$ if $BxB \subseteq \overline{ByB}$, generalizing the classical Bruhat order on W. If $S = M_2(k)$, then $\mathcal{R} = \mathcal{R}_2$ can be identified with the set $\{0, e, f, n, \overline{n}, 1, \rho\}$, where

$$0 = \begin{pmatrix} 0 & 0 \\ 0 & 0 \end{pmatrix}, \quad e = \begin{pmatrix} 1 & 0 \\ 0 & 0 \end{pmatrix}, \quad f = \begin{pmatrix} 0 & 0 \\ 0 & 1 \end{pmatrix},$$
$$n = \begin{pmatrix} 0 & 1 \\ 0 & 0 \end{pmatrix}, \quad \overline{n} = \begin{pmatrix} 0 & 0 \\ 1 & 0 \end{pmatrix}, \quad 1 = \begin{pmatrix} 1 & 0 \\ 0 & 1 \end{pmatrix},$$
$$\rho = \begin{pmatrix} 0 & 1 \\ 1 & 0 \end{pmatrix},$$

and the poset structure can be pictured as in Fig. 4.

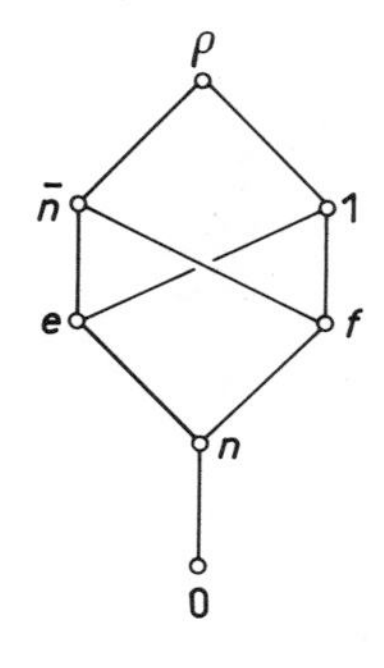

Fig. 4

For $S = M_n(k)$, $\mathcal{R} = \mathcal{R}_n$ can be identified with the inverse semigroup of injective partial transformations on n letters. If we let $r_n = |\mathcal{R}_n|$, then we obtain a recursive formula for r_n as follows:

$$r_0 = 1, \ r_1 = 2$$
$$r_n = 2nr_{n-1} - (n-1)^2 r_{n-2} \quad \text{for} \quad n \geq 2.$$

If we let $r(x) = \sum (r_n/n!)x^n$ then $r(x)$ converges for $|x| < 1$ to the function $\frac{e^{x/(1-x)}}{1-x}$.

If we linearly order the set of n elements (corresponding to a choice of Borel subgroup containing T) we may consider the submonoid $\mathcal{B}_n \subseteq \mathcal{R}_n$ consisting of order decreasing, injective, partial transformations. Alternatively, $\mathcal{B}_n = \{r \in \mathcal{R}_n \mid BrB \subseteq \overline{B}\}$. If we let $b_n = |\mathcal{B}_n|$, and consider the function

$$b(x) = \sum_{n \geq 1} \frac{b_n}{(n_1)!} x^{n+1}$$

then we obtain the curious result that

$$b(x) = e^{e^x - 1} - 1.$$

From this we obtain the formula $b_n = \frac{1}{e} \sum_{m=1}^{\infty} \frac{m^{n+1}}{m!}$.

It appears that there are many other infinite families of algebraic monoids that would yield enumerative problems similar to the above. See [2] for a detailed discussion of the above results.

7. Irreducible representations and conjugacy classes

In this section, S is an irreducible, reductive monoid. We assume also that S is normal as an algebraic variety. Let $T \subseteq G = G(S)$ be a maximal torus and let $Z = \overline{T} \subseteq S$. The following extension principle is fundamental for many of the deeper results about algebraic monoids. Let $\alpha: G(S) \to M$ and $\beta: Z \to M$ be morphisms of algebraic monoids.

Theorem 7.1. ([23], Corollary 4.6) *Let M be an algebraic monoid and assume $\alpha|T = \beta|T$. Then there exists a unique morphism $\varphi: S \to M$ of algebraic monoids such that $\varphi|G(S) = \alpha$ and $\varphi|Z = \beta$.* □

The basic idea behind 7.1 is to extend α to a morphism $\alpha' : U \to M$ for some open set $U \subseteq W_S$ satisfying $\mathrm{codim}_S(S \backslash U) \geq 2$. Once this is done, "Hartogs Lemma" ([9], Lemma 5.1) implies that α can be extended over all of S. Since $G \subseteq S$ is dense, the extension is automatically a morphism of algebraic monoids.

Using 7.1 it is relatively straightforward to deduce the following theorem. Here $X(T)$ is the character group of T, and
$X(Z) = \{\chi \in X(T) \mid \chi : T \to k \text{ extends over } Z\}$.

Theorem 7.2. ([25]) *There is a one-to-one correspondence betwen irreducible representations of S and the set $X(Z)/W$, where W is the Weyl group of T.* □

Just as in the case of groups, one may index the elements of $X(Z)/W$ by choosing the unique representative in the dominant chamber. For example, if $S = M_2(k)$ then the picture is as in Fig. 5.

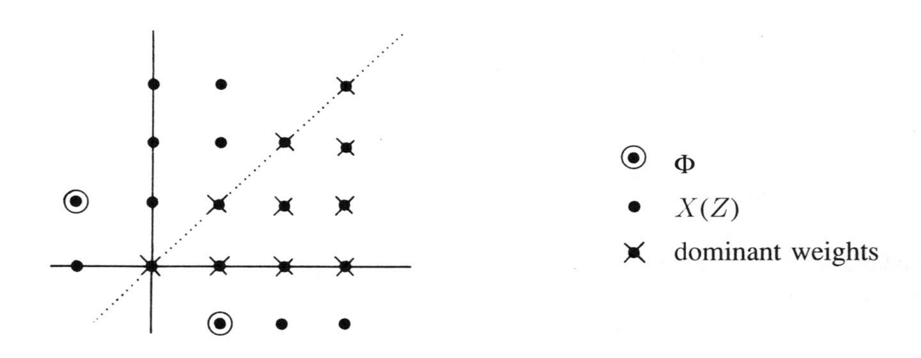

Fig. 5.

Here of course $W = \mathbb{Z}/2$, and the nontrivial element of W exchanges the axes.

Just as in the case of groups, one may consider the trace, $\mathrm{tr}(\rho) : S \to k$, of an irreducible representation $\rho : S \to M_n(k)$. Furthermore, $\mathrm{tr}(\rho)$ is an invariant function on S for the action $\mathrm{int} : G \times S \to S$, $(g, x) \mapsto gxg^{-1}$. Thus, $\mathrm{tr}(\rho)$ is a function on S/int, the geometric invariant theory quotient of S by G. Define an element $x \in S$ to be *semisimple* if $x \in \overline{T}$ for some maximal torus T of G.

Theorem 7.3. ([25]) *Let $x \in S$. The following are equivalent.*
(a) *x is semisimple.*
(b) *$\rho(x)$ is diagonalizable in every rational representation of S.*

(c) $x \in H_e$, *the* $\mathcal{H}$*-class of some idempotent, and* x *is a semisimple element of this algebraic group.*
(d) $C\ell_G(x)$, *the conjugacy class of* x, *in* S *is closed.* □

Thus, we have a characterization of semisimple elements in terms of semigroup theory, representation theory, and geometric invariant theory.

The fundamental theorem of Geometric Invariant Theory [12] identifies closed orbits with the points of the quotient variety $S/\text{int} = \text{Spec}(k[S]^G)$ via the canonical morphism $\pi: S \to S/\text{int}$. Because of Haboush's Theorem (formerly the Mumford Conjecture) [10], the fundamental theorem applies uniformly in all characteristics. From 7.3 it is fairly easy to deduce the following result.

Theorem 7.4. ([25]) *In the following commutative diagram*

$$\begin{array}{ccc} Z & \hookrightarrow & S \\ \downarrow & & \downarrow{\scriptstyle \pi} \\ Z/W & \underset{\theta}{\longrightarrow} & S/\text{int} \end{array}$$

θ *is an isomorphism of algebraic varieties.* □

Putcha has looked into the problem of conjugacy classes in complete generality for reductive monoids [17]. The basic result is roughly as follows: Let T be a maximal torus, and let $\Gamma = W \times E(T)$. For each $\alpha \in \Gamma$, Putcha defines a subvariety $M_\alpha \subseteq S$ and a surjective morphism (of varieties) $\zeta_\alpha: M_\alpha \to G_\alpha$, where G_α is a reductive group with a certain automorphism θ_α. Putcha's Theorem now reads as follows:

Theorem 7.5. ([17]) *Any element of* S *is conjugate to an element of some* M_α. *Two elements* $a, b \in M_\alpha$ *are conjugate in* S *if and only if there exists* $g \in G_\alpha$ *such that* $g\zeta_\alpha(a)\theta_\alpha(g(^{-1}) = \zeta_\alpha(b)$. □

See [17] for details. In my opinion this is a remarkable piece of work, and the above sketch is only a pale attempt to convey the basic idea. However, the one point I wish to make here is that conjugacy in reductive monoids reduces to twisted conjugacy in reductive groups via the above mechanism. There is definitely something new here that is not suggested by the classical Jordan canonical form.

8. Classification theorems

Because of Theorem 7.1 it is possible to classify reductive normal monoids using numerical data and root systems. For simplicity, we assume that S is reductive, normal, dim $ZG = 1$ and S has a zero element. Associated with S is its *polyhedral root system* (X, ϕ, C). This is obtained as follows:

Fix a maximal torus $T \subseteq G$. Let $X = X(T)$, the characters of T, and let $C = X(\overline{T}) = \{\chi \in X | \chi: T \to k \text{ extends over } \overline{T}\}$. $\phi \subseteq X(T)$ is the set of roots (i.e., the weights of T for the adjoint representation). For example, if $S = M_n(k)$, then

$$(X, \phi, C) = \Big(\bigoplus_{i=1}^{n} \mathbb{Z}e_i, \{e_i - e_j\}_{i \neq j}, \quad \langle e_i \rangle_{i=1}^{n}\Big)$$

where $\langle \ldots \rangle$ denotes "the monoid generated by". In section three of [23] this notion is axiomatized precisely. Basically, one starts with an integral root system (X, ϕ) in the classical sense, except that ϕ spans a subgroup of X of rank $X - 1$. One then chooses a rational polyhedral cone $C \subseteq X$ such that C is invariant under the Weyl group, C spans X, and $C \cap -C = \{0\}$.

Theorem 8.1. ([23]) *There is a one-to-one correspondence between isomorphism classes of semisimple algebraic monoids and isomorphism classes of polyhedral root systems.* □

The proof of 8.1 starts with a theorem of Chevalley which implies that any isomorphism of root systems is induced by an isomorphism of the corresponding groups. Putting this together with 7.1 yields one part of 8.1. The other part of 8.1 requires us to construct a semisimple monoid from a polyhedral root system (X, ϕ, C). To do this, one first takes the reductive group G with root system (X, ϕ) and then chooses enough characters $\lambda_1, \lambda_2, \ldots, \lambda_s$ such that $C = \langle \lambda_1, \ldots, \lambda_s \rangle$. For each λ_i, there exists an irreducible representation $\rho_i: G \to G\ell(V_i)$ of G such that $\lambda_i \in \text{Weights}(\rho_i) \subseteq \text{Convex Hull}(W \cdot \lambda_i)$, where W is the Weyl group. Then let $\rho = \bigoplus_{i=1}^{s} \rho_i$. It is then a routine matter to check that the normalization of $\overline{\rho(G)} \subseteq \prod_{i=1}^{s} \text{End}_k(V_i)$ is a semisimple monoid with polyhedral root system (X, ϕ, C)

A number of other results about reductive monoids are worked out in [23]:

(a) Theorem 8.1 first gives a criterion for the existence of a birational morphism between semisimple monoids S and T. Such morphisms are then classified in terms of polyhedral root systems.

(b) Theorem 8.2 proves that any semisimple monoid has a canonical, period two antiautomorhism which fixes pointwise the elements of a given maximal torus. In particular, any normal reductive monoid is isomorphic to its opposite.

(c) Theorem 8.4 proves that any smooth semisimple monoid is isomorphic to $M_n(k)$, for some n. This may be regarded as a generalization of Wedderburn's Theorem.

At the other extreme (see 4.2(b)) are the completely regular monoids with solvable unit group. Here there is also a discrete invariant that classifies each monoid (once the unit group has been fixed), but unlike the case of reductive groups there is no discrete classification of solvable groups. (In fact, there is probably no convincing general classification of connected, solvable groups.) Nonetheless, let $G = TU = UT$ be a connected, solvable, algebraic group with unipotent part U and maximal torus T. Let $\mathcal{L}$ be the Lie algebra of U. Then T acts on $\mathcal{L}$ via the adjoint representation $\mathrm{Ad}\colon T \to \mathrm{Aut}(\mathcal{L})$. Let $\psi \subseteq X(T)$ be the weights of T for the adjoint representation. So $\mathcal{L} = \bigoplus_{\alpha\in\psi} \mathcal{L}_\alpha$, where $\mathcal{L}_\alpha = \{x \in \mathcal{L} \mid \mathrm{Ad}(t)(x) = \alpha(t)x$ for all $t \in T\} \neq (0)$ for each $\alpha \in \psi$.

The purpose of this subsection is to classify all completely regular, irreducible, normal (NCR) monoids S with unit group G and the property that $\overline{T}$ has a zero element (for convenience). Let $C = X(\overline{T})$.

Theorem 8.2. ([26]) *Let S be an* NCR *monoid with unit group $C = TU$. Then $\psi \subseteq C \cup -C$.* □

The proof of 8.2 is a little tricky and will not be discussed here. But there is a revealing converse.

Let $T \subseteq Z = \overline{T}$ be a normal D-monoid with zero element and unit group T. Assume that

$$\psi \subseteq X(Z) \cup -X(Z)\ . \qquad (*)$$

So we can write $\psi = \psi_- \cup \psi_0 \cup \psi_+$ where $\psi_0 = \{0\}$ if $0 \in \psi$ and $\psi_0 = \emptyset$ if $0 \notin \psi$. By the results of [1] we may write $U = U_+U_0U_-$ where

$$\mathcal{L}(U_+) = \mathcal{L}_+ = \bigoplus_{\alpha\in\psi_+} \mathcal{L}_\alpha,\ \mathcal{L}(U_-) = \mathcal{L}_- = \bigoplus_{\alpha\in\psi_-} \mathcal{L}_\alpha,\ \text{and } \mathcal{L}(U_0) = \mathcal{L}_0\ .$$

By (*), the action $T \times U_+ \to U_+$, $(t,u) \mapsto tut^{-1} = u^t$, extends to an action $\alpha_1\colon Z \to \mathrm{End}(U_+)$; and the action $T \times U_- \to U_-$, $(t,u) \mapsto tut^{-1} = u^{\bar{t}}$, extends to an action $\beta_1\colon Z \to \mathrm{End}(U_-)$. Notice that the action of T on U_0 is trivial. Now $U_-U_+ \subseteq U = U_+U_0U_- \cong U_+ \times U_0 \times U_-$. This defines variety morphisms $\zeta_+\colon U_- \times U_+ \to U_+$, $\zeta_-\colon U_- \times U_+ \to U_-$, and $\zeta_0\colon U_- \times U_+ \to U_0$.

Theorem 8.3. ([26]) *Given $T \subseteq Z$ as above with $\psi \subseteq X(Z) \cup -X(Z)$, let $S_0 = Z \times U_0$ (with the usual product structure as an algebraic monoid). Define $S = U_+ \times S_0 \times U_-$ with multiplication law*

$$(u, x, v)(a, y, b) = (u\zeta_+(v, a)^x, x\zeta_0(v, a)y, \zeta_-(v, a)^{\overline{y}}b) .$$

Then S is an NCR *monoid with unit group $G = TU$, and $\overline{T} \cong Z$.* □

9. Related developments

The purpose of this section is to introduce the reader to several other recent developments which are closely related to algebraic monoids.

9.1. *Monoids on Groups with* BN-*Pair*. Inspired by Theorem 5.1, Putcha went on to develop an abstract theory of monoids that have unit groups with BN-pair [18].

Let (G, B, N) be a group with BN-pair (or Tits system) [29], and let S be a monoid with unit group G. Now S is a *monoid on G* if

(a) S is generated by $E(S)$ and G.
(b) If $e, f \in E(S)$ and $e \geq f$, then $C_G^r(e, f)$ and $C_G^\ell(e, f)$ are opposite parabolic subgroups.
(c) If $e, f \in E(S)$ and $e\mathcal{R}f$ or $e\mathcal{L}f$, then $e = gfg^{-1}$ for some $g \in G$.

The main examples here are reductive algebraic monoids (by 4.1, 4.2(a) and 3.2). Nonetheless, it is possible to extend Theorem 5.1 to the more general case of monoids on groups with BN-pair. In this case, $\mathcal{U}(S) = E(S)/\sim$ where $e \sim f$ if there exists $g \in G$ such that $geg^{-1} = f$, and $[e] \geq [f]$ in $\mathcal{U}(S)$ if $e \geq f_1$ for some $f_1 \sim f$. $\lambda: \mathcal{U}(S) \to$ type(G) is defined as in 5.1. But there is also a converse. Putcha obtains this as follows:

Let $\mathcal{U}$ be a poset with maximal element $1 \in \mathcal{U}$, and assume $\lambda: \mathcal{U} \to$ type$(G) \cong 2^S$ a set map with $\lambda(1) = G$. Define

$$E(\mathcal{U}, \lambda) = \{(J, P, Q) \mid J \in \mathcal{U},\ P \text{ and } Q \text{ are opposite and } \lambda(J) = \text{type}(P)\}.$$

We say that λ is *transitive* if $J_1 \geq J_2 \geq J_3$ implies that any connected component of $\lambda(J_2)$ is contained in either $\lambda(J_1)$ or $\lambda(J_3)$. Since S is identified with the set of nodes of a Dynkin diagram, we may discuss the connectivity of subsets of S.

Theorem 9.1. ([18]) *$E(\mathcal{U}, \lambda)$ is the biordered set of idempotents of some monoid S on G if and only if λ is transitive.* □

We note that by a theorem of Easdown [8], any biordered set (as in [13]) occurs as the set of idempotents of some semigroup S.

9.2. *Monoids on Groups with* BN-*Pair*. Motivated by some problems in intersection theory, several authors ([6], [3]) have considered the structure and classification of embeddings of spherical homogeneous spaces. (See [5] for a survey). Let G be a reductive algebraic group. A closed subgroup $H \subseteq G$ is called *spherical*, and G/H is called a *spherical homogeneous space*, if a Borel subgroup B of G has a dense orbit on G/H for the action $G \times G/H \to G/H$, $(g, hH) \mapsto ghH$. For example, consider $G = K \times K$, and let $H = \{(g,h) \in G \mid g = h\}$. Then G/H can be identified with K with the action $G \times K \to K$, $((g,h),k) \mapsto gkh^{-1}$. The Bruhat decomposition implies that K is a spherical homogeneous space for this canonical two sided action.

An *embedding* of a homogeneous space G/H is a variety X together with a G-action $G \times X \to X$, such that X has a dense (open) orbit G-isomorphic to G/H. A *spherical embedding* is an embedding of a spherical homogeneous space.

The theory of spherical embeddings has been developed systematically by Brion, Luna and Vust (in characteristic zero). On the other hand, there is a fundamental connection with algebraic monoids.

Theorem 9.2. ([27]) *Let $X = \overline{G/H}$ be a quasiprojective spherical embedding, and assume H is also reductive (equivalently, G/H is affine). Then there exists an algebraic monoid S with unit group $G \times k^*$, and a right $H \times k^*$-invariant, open subset $U \subseteq S$ such that $X \cong U/H \times k^*$ (geometric invariant theory quotient* [12]). *In particular, X is the semigeometric quotient by $H \times k^*$ of some open subset of an algebraic monoid.* □

9.3. *Cohomology of Compactifications of Semisimple Groups.* Associated with any semisimple monoid S is a certain projective variety X, defined as follows:

Consider the action $k^* \times S \backslash \{0\} \to S \backslash \{0\}$, $(\alpha, x) \mapsto \alpha x$, where k^* is identified with the connected component of the center of $G_1 = G(S)$. Let $X = (S\backslash\{0\})/k^*$. It is easily verified that

(a) X is a projective variety,

(b) the semisimple group $G = G_1/k^*$ embeds naturally in X, say $j: G \hookrightarrow X$, as a dense open subset, and

(c) the two-sided action of G on itself extends over j.

Conversely, by the results of [28], any normal projective variety satisfying (a), (b) and (c) above is obtained as above for some semisimple algebraic monoid S.

Assume now that $k = \mathbb{C}$, the complex numbers, and that X is nonsingular (this condition can be easily determined from the polyhedral root system of the associated monoid S). Let $K \subseteq G$ be a maximal, compact subgroup (in the classical topology), and let $Z \subseteq X$ be the closure of a maximal torus T (via j).

Assume $T_K = T \cap K \subseteq K$ is a maximal compact torus of K. Since $KTK = G$ and KZK is compact, we obtain

$$X = KZK \ .$$

Thus, we get a surjective (real analytic) map

$$\pi\colon (K \times K) \times_{T_K \times T_K} Z \to Z \ ,$$

since Z is $T_K \times T_K$ stable. The Weyl group W acts naturally on $\mathcal{U} = (K \times K) \times_{T_K \times T_K} Z$, and π is constant on the W-orbits. Thus, W acts on $H^*(\mathcal{U}; \mathbb{Q})$ and we obtain a morphism $\pi^*\colon H(X; \mathbb{Q}) \to H^*(\mathcal{U}, \mathbb{Q})^W$. See [7] for details.

Theorem 9.3. ([7]) *The map $\pi^*\colon H(X; \mathbb{Q}) \to H^*(\mathcal{U}, \mathbb{Q})^W$ is an isomorphism of rings.* □

De Concini and Procesi also compute $H^*(\mathcal{U}, \mathbb{Q})$ in [7], by extending some work of Danilov [4]. Notice that there is a fibration $\mathcal{U} \to K/T_K \times K/T_K$ with fibre Z, so that $\mathcal{U}$ may be regarded as a "relative torus embedding".

References

[1] Bialynicki-Birula, A., Some theorems on actions of algebraic groups, Annals of Math. 98 (1973), 480–497

[2] Borwein, D., S. Rankin and L. Renner, Ennumeration of injective partial transformations, Discrete Math. 73 (1989), 291–296

[3] Brion, M., D. Luna and Th. Vust, Espaces homogènes sphèriques, Invent Math. 84 (1986), 617–632

[4] Danilov, V. I., The geometry of toric varieties, Russian Math. Surveys 33 (1978), 97–154

[5] DeConcini, C., Equivariant embeddings of homogeneous spaces, Proc. of the ICM, Berkeley, 1986, 369-377

[6] DeConcini, C. and C. Procesi, Complete symmetric varieties II, Adv. Studies in Pure Math. 6 (1985), 481–513

[7] —, Cohomology of compactifications of algebraic groups, Duke Math. Journal 53 (1986), 585–596

[8] Easdown, D., Biordered sets come from semigroups, J. of Alg. 96 (1973), 581–591

[9] Grosshans, F., Observable groups and Hilbert's fourteenth problem, Amer. J. Math. 95 (1973), 229–253

[10] Haboush, W., Reductive groups are geometrically reductive, Annals of Math. 102 (1975), 67–83

[11] Humphreys, J. E., "Linear Algebraic Groups", Springer-Verlag, New York, 1980

[12] Mumford, D. and J. Fogarty, "Geometric Invariant Theory", Springer-Verlag, New York, 1982

[13] Nambooripad, K. S. S., Structure of regular semigroups I, Memoirs of the Amer. Math. Soc. 224 (1979)

[14] Putcha, M. S., "Linear algebraic monoids", London Math. Soc. Lecture Notes 133, Cambridge University Press, 1988

[15] —, Green's relation on a connected algebraic monoid, Linear and Multilinear Algebra 12 (1982), 205–214

[16] —, Regular linear algebraic monoids, Trans. Amer. Math. Soc. 290 (1985), 615–626

[17] —, Conjugacy classes in algebraic monoids, Trans. Amer. Math. Soc. 303 (1987), 529-540

[18] Putcha, M. S., Monoids on gropus with BN-pairs, J. of Alg., to appear

[19] Putcha, M. S. and L. Renner, The system of idempotents and and the lattice of $\mathcal{J}$-classes of reductive algebraic monoids, J. of Alg. 116 (1988), 385–399

[20] Renner, L., "Algebraic monoids", Ph. D. Thesis, University of British Columbia, 1982

[21] —, Quasiaffine algebraic monoids, Semigroup Forum 30 (1984), 167–176

[22] —, Reductive monoids are Von Neumann regular, J. of Alg. 93 (1985), 237–245

[23] —, Classification of semisimple algebraic monoids, Trans. Amer. Math. Soc. 292 (1985), 193–223

[24] —, Analogue of the Bruhat decomposition for algebraic monoids, J. of Alg. 101 (1986), 303–338

[25] —, Conjugacy classes of semisimple elements and irreducible representations of algebraic monoids, Comm. in Alg. 16 (1988), 1933–1943

[26] —, Completely regular algebraic monoids, J. of Pure and Appl. Alg., to appear

[27] —, Reductive embeddings, Proceedings of the McGill Conference on Group Actions, to appear

[28] —, Classification of semisimple algebraic varieties, J. of Alg. 122 (1989), 275–287

[29] Tits, J., Buildings of spherical type and finite BN-pairs, Springer-Verlag, New York, 1974

[30] Waterhouse, W. C., The unit groups of affine algebraic monoids, Proc. Amer. Math. Soc. 85 (1982), 506–508

Part II

The compact case

Compact semilattices, partial orders and topology

Michael W. Mislove

Introduction

The structure theory of compact semigroups is an area which enjoyed a tremendous amount of research interest in the 1950's–1970's, but it is an area in which research interest has declined since that time. One sub-area where research interest has continued (and, in fact, where research interest has grown) is the area of compact semilattices and their generalizations. While modern research activity centers on more general objects, the basic seeds of the present theory were sown in the development of compact semilattices. Our goal in this survey paper is to show how the developments in the structure theory of compact semilattices and its more modern outgrowths fit within a broader context.

That broader context is the relationship between partially ordered spaces and topological spaces. We begin by giving a brief history of the development of the structure theory of partially ordered spaces from the work of Nachbin on normally ordered spaces through the work of Lawson and others in the theory of compact semilattices. In Section 2 we examine the question of intrinsic topologies on algebraic objects. The first goal is to present Lawson's result that the topology of a compact semilattice is intrinsic, i.e., is determined by the algebraic structure. Then we document the fundamental role the Scott topology plays in the structure theory of partially ordered topological spaces.

In the third section, we describe a number of duality theorems which have the common thread that they describe dualities between categories of partially ordered sets, on the one hand, and categories of lattices, on the other. We start with the duality between the category of sets and functions and the category of complete atomic Boolean algebras and complete homomorphisms, and culminate with a duality theorem for a category of algebraic semilattices and a category of conditional $\vee$-semilattices; this duality theorem generalizes the better-known HMS-duality between the category of compact zero-dimensional semilattices and compact semilattice maps and the category of $\vee$-semilattice monoids and semilattice maps. In the process, we show how adding structure to the categories of sets and functions corresponds to relaxing structure on the categories of lattices.

In the fourth section we again trace the development of several duality theorems, but this time we bring topology into the picture more directly. Beginning once again with the LINDENBAUM–TARSKI duality between sets and Boolean algebras, this time we follow MARSHALL STONE's lead and seek dualities which involve topological spaces (instead of partially ordered spaces) on the one hand, and lattices on the other. Most of this theory is amply described in [4] and in [9], and so our presentation here is rather brief. But the point we are trying to make is amply justified: a number of these duality theorems can be recast as isomorphism theorems between categories of compact semilattice monoids and categories of sober spaces.

In the fifth and final section, we briefly describe some of the applications of the theory we have described. These applications center on one central theme, which we denote "The search for $2^{\mathbb{N}}$." The applications described are to harmonic analysis, infinite combinatorics, and, finally, theoretical computer science. In each case, we give results which focus on the existence of a copy of $2^{\mathbb{N}}$ in a given object, and we describe how each particular application is affected by the existence of a copy of $2^{\mathbb{N}}$.

1. Topology and order updated

From a historical perspective, the study of the relation between topology and order has its roots in the work of NACHBIN [20]. In this seminal work, NACHBIN explores the way the order on a partially ordered space influences the topology of the space, and his results provide the basis for the fundamental work of LAWSON [10] and others who followed. NACHBIN's work was in turn motivated by URYSOHN's results [29] stated below.

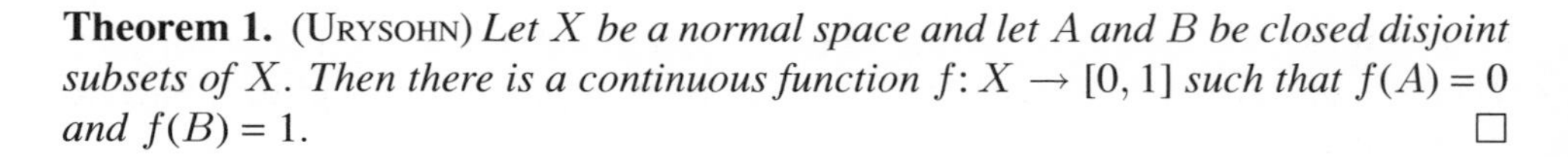

Theorem 1. (URYSOHN) *Let X be a normal space and let A and B be closed disjoint subsets of X. Then there is a continuous function $f\colon X \to [0,1]$ such that $f(A) = 0$ and $f(B) = 1$.* □

Theorem 2. (URYSOHN) *Let X be a normal space and $F \subseteq X$ a closed subset. Then each continuous function $f\colon F \to [0,1]$ has a continuous extension $f^*\colon X \to [0,1]$.* □

The proofs of these results proceed along now classical lines. For example, in the case of Theorem 1, given disjoint closed subsets A and B of the normal space X, one defines an order-dense chain of open sets $U_{p/2^n}$ satisfying $A \subseteq U_{p/2^n} \subseteq \overline{U_{p/2^n}} \subseteq U_{(p+1)/2^n}$, and $B \subseteq X \setminus U_{p/2^n}$ for each $n > 0$ and each $p \leq 2^n$. The desired continuous function $f\colon X \to [0,1]$ is then defined by $f(x) = \bigwedge\{p/2^n \leq 1 \mid x \in U_{p/2^n}\}$.

The prototypical normal spaces are the compact Hausdorff spaces, and Theorem 1 has the corollary that any second countable compact Hausdorff space is metrizable. NACHBIN saw that it would be possible to extend these results to the case of partially ordered spaces just in case the order and the topology were compatible in the following sense.

Definition 3. Let X be a topological space which is endowed with a partial order. Then X is *normally ordered* if given disjoint closed subsets A and B of X with A decreasing and B increasing, there are disjoint neighborhoods U of A and V of B with U decreasing and V increasing. □

The following results extend those of URYSOHN's to this setting; the proof of the first result amounts to choosing the appropriate open sets $U_{p/2^n}$ to be open decreasing sets.

Theorem 4. (NACHBIN) *The topological ordered space X is normally ordered if and only if, given two disjoint closed subsets A and B of X with A decreasing and B increasing, there is a continuous increasing function $f\colon X \to [0,1]$ such that $f(A) = 0$ and $f(B) = 1$.* □

Theorem 5. (NACHBIN) *Let X be a normally ordered space and let $F \subseteq X$ be a closed subset. Suppose that for every pair of closed subsets $A, B \subseteq F$ with $\downarrow A \cap \uparrow B \cap F = \emptyset$, we have $\downarrow A \cap \uparrow B = \emptyset$. Then each continuous increasing function $f\colon F \to [0,1]$ extends to a continuous increasing function $f^*\colon X \to [0,1]$.* □

In addition to being the canonical codomain for a bounded continuous real-valued function, the unit interval is also a topological semilattice under the min operation, and so it is natural to ask which compact semilattices admit enough continuous semilattice homomorphisms into the unit interval to separate the points. Using the idea of NACHBIN's approach, LAWSON solved this problem in his dissertation [10].

Definition 6. A topological semilattice S *has small semilattices* if each point of S has a neighborhood basis of subsemilattices. □

Theorem 7. (LAWSON) *A compact semilattice S has enough continuous semilattice homomorphisms into the unit interval I if and only if S has small semilattices.* □

The proof of this result involves choosing the sequence of open sets $U_{p/2^n}$ so that $U_{p/2^n} = S\setminus\uparrow x_{p/2^n}$ for some $x_{p/2^n} \in S$ for each $n > 0$ and each $p \leq 2^n$. Actually, this result holds for much broader classes of topological semilattices; e.g., for the class of locally compact finite dimensional topological semilattices (cf. [11]), but our focus is on compact semilattices. LAWSON [12] also produced the first example of a compact semilattice without small semilattices, and thus showed that the category of semilattices with small semilattices does not include all compact semilattices.

Nachbin's second theorem (Theorem 5) entered into the domain of compact semilattices in [7], where it was shown that epics in the category of compact semilattices with small semilattices are surmorphisms.

2. Intrinsic topologies for compact semilattices

The next issue in the development of the theory was the question whether the topology of a compact semilattice is *intrinsic*, i.e., whether it is somehow determined by the algebraic structure. Lawson again solved this problem by first showing that the topology of any compact topological lattice is intrinsic, and then applying this result to the case of compact semilattices.

First we note that the desired result does not hold for normally ordered spaces. Indeed the results in Section I imply that a compact Hausdorff space endowed with a closed partial order has the family of open upper sets and the family of open lower sets as a subbase for its topology. In particular, the topology is *locally convex*. However, this topology is not unique, as the following simple example shows.

Example 1. Let X_0 be an infinite set, and endow X_0 with the discrete topology and the discrete order. Let $X = X_0 \cup \{0, 1\}$, where 0 and 1 are not elements of X_0. Order X by letting 0 be the least element of X and 1 the greatest element of X. We can make X into a compact ordered space either by letting $X_0 \cup \{1\}$ be the one-point compactification of X_0 and adding 0 as a discrete point, or by making $X_0 \cup \{0\}$ the one-point compactification of X_0 and adding 1 as a discrete point. □

However, a semilattice S can have at most one compact Hausdorff topology relative to which it is a topological semilattice. For instance, if we want the space X of Example 1 to have a continuous inf-operation, then we must choose the topology relative to which 0 is the limit-point of X_0.

Definition 2. Let P be a partially ordered set. The subset $\emptyset \neq F \subseteq P$ is *filtered* if for every $x, y \in F$, there is some $z \in F$ such that $z \leq x, y$. Dually, the subset $\emptyset \neq D \subseteq P$ is *directed* if, for every $x, y \in D$ there is some $z \in D$ with $x, y \leq z$.□

It is an easy exercise to show that if S is a compact topological semilattice, then every filtered subset of S converges to its infimum, and every directed subset of S converges to its supremum. In particular, a compact semilattice monoid is a complete lattice.

Definition 3. Let L be a lattice.

i) A subset $C \subseteq L$ is *order-convex* (or, simply *convex*) if, given $x, y \in C$ and $z \in L$ with $x \leq z \leq y$, we also have $z \in C$.

ii) The net $\{x_\alpha\}$ *order-converges* to $x \in S$ if there are an ascending net $\{u_\alpha\}$ and a descending net $\{t_\alpha\}$ in S with $u_\alpha \leq x_\alpha \leq t_\alpha$ for all α, and $\bigvee u_\alpha = x = \bigwedge t_\alpha$.

iii) A subset $U \subseteq L$ is open in the *order topology* if $x \in U$ and $\{x_\alpha\}$ order converges to x imply x_α is eventually in U.

iv) The *convex-order topology* on L has for a basis those convex subsets U of L which are also open in the order topology. □

It is easy to show that the convex-order topology on a lattice is always locally convex and is contained in the order topology (obviously), and that it agrees with the order topology just in case the latter is locally convex.

Theorem 4. (Lawson [13]) *If the lattice L has a compact Hausdorff topology relative to which it is a topological lattice, then that topology is the convex-order topology.* □

Definition 5. For a lower set I in the semilattice S, define the subset $I^+ = \{\bigvee D \mid D \subseteq I \text{ is directed}\}$. □

A fundamental result about the topology of a compact semilattice is the following.

Proposition 6. Lawson [13]) *If I is a lower set in the compact semilattice S, then $\bar{I} = I^{++}$.* □

Lawson used Theorem 4 and Proposition 6 to prove that the topology of a compact semilattice is *intrinsic* (i.e., that it is defined by the algebraic structure) in the following way. First note that the lattice $\mathcal{I}(S)$ of all closed lower sets from S is a compact topological lattice when endowed with the Vietoris topology. By Theorem 4, this must be the convex-order topology on $\mathcal{I}(S)$, and so it is intrinsic. Next, the closed lower sets in S are defined intrinsically, since Proposition 6 shows they are exactly those lower sets which are closed under directed suprema. Finally, the map $x \mapsto \downarrow x \colon S \to \mathcal{I}(S)$ is an isomorphism of S onto its image in $\mathcal{I}(S)$, and so the topology of S is determined by the algebraic structure.

It is important to note that this result applies to *all* compact semilattices, not just to those having small semilattices. In fact, one of the most vexing problems in the theory is to give a direct algebraic description of the topology of a compact semilattice. Proposition 6 gives the required description of the closed lower sets, and so it remains to give analogous description of the closed upper sets; Nachbin's results then show that the topology is the one generated by the closed upper sets and the closed lower sets.

Actually, Proposition 6 contains the germ of one of the most fundamental ideas in topologizing partially ordered sets. It says that a lower set in a compact semilattice is closed if and only if it is closed under the formation of sups of directed sets. Dually, an upper set is open if and only if it is *inaccessible by directed suprema.*

Definition 7. Let P be a partially ordered set. A subset $U \subseteq P$ is *Scott open* if
1) $U = \uparrow U$, and
2) if $D \subseteq P$ is directed and $\bigvee D \in U$, then $D \cap U \neq \emptyset$. □

Since $P \backslash \downarrow x$ is Scott open for every $x \in P$, the Scott topology is always T_0. In this generality, nothing more can be said. But, with the addition of further hypotheses, the Scott topology becomes a rich structure whose study yields many rewarding results. We return to this general theme later on; for now, we limit our discussion to the case of semilattices, since our goal is to determine the topology of a compact semilattice from the algebraic structure.

Given a compact semilattice S, our comment following Definition 2 above shows that every directed set in S has a supremum and every filtered subset in S has an infimum, and so we make the following definition.

Definition 8. Let S be a semilattice. S is *complete* if every directed subset of S has a supremum and if every filtered subset of S has an infimum. □

If X is any non-empty subset of the complete semilattice S, then $\bigwedge F$ exists for all finite non-empty subsets $F \subseteq X$, and the family $\{\bigwedge F \mid \emptyset \neq F \subseteq X\}$ is a filtered subset of S. Since a filtered set of a complete semilattice has an infimum, and since this infimum is obviously $\bigwedge X$, the subset X has an infimum. Thus a complete semilattice S has the property that every non-empty subset has an infimum.

Now, given two points x, y in a complete semilattice S, if x and y have an upper bound in S, then the family B of upper bounds of x, y is a non-empty subsemilattice of S. Hence $\bigwedge B$ exists, and this is clearly the least upper bound of x and y. Thus, a complete semilattice S is also a *conditional sup-semilattice*, i.e., every pair of elements having an upper bound has a supremum.

So, a partial algebraic description of a compact semilattice is that it is a complete semilattice. Note that such a semilattice always has a least element (since S is non-empty), and it has an identity if and only if every subset has an infimum, if and only if S is a (complete) lattice.

Now, if S is a compact semilattice with small semilattices, then each point x of S has a neighborhood basis of subsemilattices. Since S is compact and Hausdorff, we can assume these subsemilattices are compact, and so each has an infimum. So, we know that $x = \bigvee\{y \in S \mid x \in (\uparrow y)^\circ\}$ for each $x \in S$, where $(\uparrow y)^\circ$ denotes the interior of the set $\uparrow y$. Moreover, the family $\{y \in S \mid x \in (\uparrow y)^\circ\}$ is directed, since x has a basis of compact semilattice neighborhoods. So, if we can find an algebraic description of the property that $x \in (\uparrow y)^\circ$, we have an algebraic description of the open upper sets in S.

Recalling Proposition 6, one condition y must satisfy for $x \in (\uparrow y)^\circ$ to hold is that for every directed subset D of S, if $x \leq \bigvee D$, then $D \cap \uparrow y \neq \emptyset$. That is, x must be in the Scott interior of $\uparrow y$. It turns out that this condition is also sufficient.

Definition 9. Let S be a complete semilattice, and let $x, y \in S$. Then we say *y is way-below x* if x is in the Scott interior of the set $\uparrow y$. If y is way-below x, then we write $y \ll x$. S is a *continuous semilattice* if for every $x \in S$ we have
a) $\Downarrow x \equiv \{y \in S \mid y \ll x\}$ is directed, and
b) $x = \bigvee \Downarrow x$. □

Theorem 10. *If S is a compact semilattice with small semilattices, then the following hold.*
1) S is a continuous semilattice.
2) The upper set $U \subseteq S$ is open in S if and only if U is Scott open.
3) A basis for the Scott topology on S is the family $\Uparrow y = \{x \in S \mid y \ll x\}$ for $y \in S$. □

But, we still need the other half of the topology for such an S, i.e., we need a description of the open lower sets of S. It turns out that this is quite simple to give. As motivation, consider the fact that in any partially ordered topological space X, the set $\uparrow x$ is closed for each point $x \in X$.

Definition 11. For a partially ordered set P, the *lower topology* on P has the family $\{P \setminus \uparrow x \mid x \in P\}$ as a subbase for the closed sets. The Lawson topology (or λ-topology, as it is also called) on P is the common refinement of the Scott topology and the lower topology. □

Theorem 12. *If S is a continuous semilattice, then the λ-topology is compact and Hausdorff, and S is a topological semilattice in the λ-topology. A basis for the λ-topology is the family $\{\Uparrow x \setminus \uparrow F \mid x \in S,\ F \subseteq S$ finite$\}$.* □

Actually, it is easy to show that the λ-topology is Hausdorff on any continuous semilattice S, and that such a semilattice is a topological semilattice in the λ-topology. The hard part is to show the topology is also compact. The proof of this goes back to the work of MARSHALL STONE [25, 26], which we describe in Section 4. The result we have been aiming for is the following.

Theorem 13. *Let S be a semilattice. S is a compact semilattice with small semilattices if and only if S is a continuous semilattice endowed with the λ-topology.* □

We close this section by singling out a very important special case of these results. If we are given a compact semilattice S which is totally disconnected, then an early result of WALLACE [30] shows that S is embeddable in a product of finite semilattices, and any one of these is embeddable in a power of the two point semilattice $\{0,1\}$. Thus, a compact totally disconnected semilattice S is embeddable in a power of $\{0,1\}$. Now, the Scott topology on a power of $\{0,1\}$ is easy to describe. In $\{0,1\}$, the set $\uparrow 1 = \{1\}$ is open, and, in the product topology on $\{0,1\}^X$ (for X any set), the same is true of the set $\uparrow (y_x)_{x\in X}$ if $y_x = 0$ for almost all $x \in X$, since the Scott topology is productive on powers of $\{0,1\}$. Moreover, every element $(z_x) \in \{0,1\}^X$ is the supremum of those elements $(y_x)_{x\in X} \leq (z_x)_{x\in X}$ satisfying $y_x = 0$ for almost all $x \in X$. Elements with this property deserve a special name.

Definition 14. If P is a partially ordered set, then the element $k \in P$ is *compact* if $\uparrow k$ is Scott open in P. The set of compact elements of P is denoted $K(P)$. The complete semilattice S is *algebraic* if the set $\downarrow s \cap K(S)$ is directed and $s = \bigvee(\downarrow s \cap K(S))$ for all $s \in S$. □

So, $\{0,1\}^X$ is an algebraic (semi)lattice for every set X. Furthermore, if S is any compact zero-dimensional semilattice and we regard S as a embedded in $\{0,1\}^X$, then given any compact element $k \in \{0,1\}^X$, the set $\uparrow k \cap S$ is a clopen subsemilattice of S. As such, it must have an infimum, k_S, and this element satisfies $\uparrow k_S = \uparrow k \cap S$ is clopen in S. Thus, $k_S \in K(S)$. Moreover, given two compact elements, $k, k' \in K(S)$, the set $\uparrow k \cap \uparrow k'$ is a clopen filter in S if it is non-empty, and so it has an infimum, $k'' \in K(S)$ in this case. Clearly $k'' = k \vee k'$ in S, so $K(S)$ is a conditional sup-semilattice. Finally, if $s \in S$, then this also shows that the set $\downarrow s \cap K(S)$ is directed.

Theorem 15. *The compact semilattice S is zero-dimensional if and only if S is an algebraic semilattice. In this case, the λ-topology has the family $\uparrow k \setminus \uparrow F$ for a basis, where $k \in K(S)$ and $F \subseteq K(S)$ is a finite set.* □

The results we have surveyed in the first two sections are all contained in the work [4]. Historical notes are provided there for all of these results, and the many generalizations which are also provided therein.

3. Duality theory: a different approach

So far we have given a few highlights of the development of the structure theory of compact semilattices, with an emphasis on semilattices with small semilattices. There are generalizations of this theory to the realm of partially ordered sets which we now want to describe. Rather than simply present the more general results without further motivation, we proceed along completely different lines.

Almost all undergraduates in mathematics have encountered the notion of a Boolean algebra, that is, a distributive complemented lattice. A fundamental result is that a Boolean algebra is complete and atomic if and only if it is the power set $\mathbb{P}(A)$ of some set A (this result is due to LINDENBAUM and TARSKI [27]). In fact, given a complete atomic Boolean algebra B, the set $A_B = \{a \in B \mid \downarrow a \backslash \{0\} = \{a\}\}$ is the set of *atoms* of B, and the isomorphism is the map $x \mapsto \downarrow x \cap A_B : B \to \mathbb{P}(A_B)$ sending each element of B to the set of atoms which the element dominates. Moreover, if $\phi: B \to B'$ is a complete homomorphism of Boolean algebras, then ϕ induces a function $\bar{\phi}: A_B \to A_{B'}$ defined by $\bar{\phi}(a) = \wedge \phi^{-1}(a)$. Dually, given a function $f: X \to Y$ between sets, the map $f^{-1}: \mathbb{P}(Y) \to \mathbb{P}(X)$ preserves all unions and all intersections, and so it is a complete homomorphism of Boolean algebras. Thus we have a duality between the category of complete atomic Boolean algebras and complete homomorphisms, on the one hand, and the category of sets and functions on the other.

Theorem 1. (LINDENBAUM & TARSKI) *The category of complete atomic Boolean algebras and complete homomorphisms is dual to the category of sets and functions under the functor which sends a complete atomic Boolean algebra B to its set A_B of atoms of B, and the functor which sends a set to its power set.* □

If we seek to generalize this result, a natural first step is to consider the category of partially ordered sets and monotone maps. After all, we can regard a set X as a partially ordered set endowed with the *discrete* order, wherein $x \leq y$ if and only if $x = y$. Then every subset of X is a lower set in this order, and every function between sets is monotone. If complete atomic Boolean algebras are dual to sets, then what are dual to partially ordered sets?

Definition 2. Let P be a partially ordered set.

1) The subset $X \subseteq P$ is a *lower set* in P if $X = \downarrow X$. The family of all lower sets of P is denoted $\mathcal{I}(P)$. Dually, the subset $U \subseteq P$ is an *upper set* of P if $U = \uparrow U$, and the family of upper sets of P is denoted $\mathcal{U}(P)$.
2) The subset $I \subseteq P$ is an *order ideal* of P if I is a directed lower set. The family of order ideals of P is denoted Id(P). Dually, the subset $F \subseteq P$ is a *filter* in P if F is a filtered upper set. The family of filters of P is denoted $Filt(P)$. □

So, the family $\mathcal{I}(P)$ is the analogue of $\mathbb{P}(X)$ for partially ordered sets. Clearly the intersection and union of any family of lower sets is another such, so $\mathcal{I}(P)$ is a complete sublattice of $\mathbb{P}(P)$. In fact, $\mathcal{I}(P)$ is a complete ring of sets, and as such, it is *completely distributive*. This means that the most general distributive law holds for $\mathcal{I}(P)$.

Definition 3. The complete lattice L is *completely distributive* if

$$\bigvee_{i\in I}\bigwedge_{j\in J} x_{i,j} = \bigwedge_{f\in J^I}\bigvee_{i\in I} x_{i,f(i)}$$

for every family $\{x_{i,j} \mid i \in I, j \in J\} \subseteq L$. □

Clearly $\mathbb{P}(X)$ is completely distributive for any set X, and since $\mathcal{I}(P)$ is closed in $\mathbb{P}(P)$ under all intersections and all unions, $\mathcal{I}(P)$ is completely distributive as well. The elements $\downarrow x$ for $x \in P$ play a special role in $\mathcal{I}(P)$: every element of $\mathcal{I}(P)$ is the union of such elements, and whenever $\downarrow x \subseteq \bigcup_i X_i$ in $\mathcal{I}(P)$, then $\downarrow x \subseteq X_i$ for some $i \in I$. That is, the elements $\downarrow x$ are *complete* $\cup$*-primes* in $\mathcal{I}(P)$.

Definition 4. Let L be a complete lattice L, and let $p \in L$

1) The element p is a *complete* $\vee$*-prime* of L if, for every subset $X \subseteq L$ with $p \leq \bigvee X$, there is some $x \in X$ with $p \leq x$. The family of complete $\vee$-primes of L is denoted $\mathrm{CSpec}_\vee(L)$. We say that $\mathrm{CSpec}_\vee(L)$ *order generates* L if $x = \bigvee(\downarrow x \cap \mathrm{CSpec}_\vee(L))$ for all $x \in L$. The concept of a *complete* $\wedge$*-prime* of L is defined dually.
2) The element p is a $\vee$*-prime* of L if for every finite subset $X \subseteq L$ with $p \leq \bigvee X$, there is some $x \in X$ with $p \leq x$. The family of $\vee$-primes of L is denoted $\mathrm{Spec}_\vee(L)$. We say that $\mathrm{Spec}_\vee(L)$ *order generates* L if $x = \bigvee(\uparrow x \cap \mathrm{Spec}_\vee(L))$ for all $x \in L$. The concept of a $\wedge$*-prime* of L is defined dually. □

Theorem 5. *A complete lattice L is of the form $\mathcal{I}(P)$ for some partially ordered set P if and only if* $\mathrm{CSpec}_\vee(L)$ *order generates L.*

Proof. As we commented above, $\mathrm{CSpec}_\cup(\mathcal{I}(P)) = \{\downarrow x \mid x \in P\}$ order generates $\mathcal{I}(P)$, and so we only need show that the converse holds. Suppose that L is a complete lattice in which $\mathrm{CSpec}_\vee(L)$ order generates L. Then $P = \mathrm{CSpec}_\vee(L)$ is a partially ordered set in the inherited order from L, and we define the map $\phi\colon L \to \mathcal{I}(P)$ by $\phi(x) = \downarrow x \cap P$. Since P order generates L, it follows that ϕ is one-to-one. Also, since P consists of complete $\vee$-primes of L, given any subset $X \subseteq P$, $\downarrow \bigvee X \cap P = X$, so $\bigvee X$ is the unique element x of L satisfying $\phi(x) = X$, which means ϕ is also surjective. Further use of the property that P consists of complete $\vee$-primes of L also shows that ϕ preserves all suprema and all infima. Thus ϕ is an isomorphism. □

Consider the category $\mathcal{CAL}$ whose objects are those complete lattices L in which $\mathrm{CSpec}_\vee(L)$ order generates and whose morphisms are complete lattice maps (i.e., those maps preserving all infima and all suprema), and the category $\mathcal{PO}$ of partially ordered sets and monotone maps. It is routine to verify that given a morphism

$f: L \to L'$ in $\mathcal{CAL}$, the map $\bar{f}: L' \to L$ defined by $\bar{f}(x) = \bigwedge f^{-1}(\uparrow x)$ preserves complete $\vee$-primes. Thus we have a monotone map $\bar{f}: \mathrm{CSpec}_\vee(L') \to \mathrm{CSpec}_\vee(L)$. We define the functor $\mathbf{CS}: \mathcal{CAL} \to \mathcal{PO}$ which associates to the complete lattice L the partially ordered set $\mathrm{CSpec}_\vee(L)$, and to each morphism $f: L \to L'$ in $\mathcal{CAL}$ the map $\bar{f}: \mathrm{CSpec}_\vee(L') \to \mathrm{CSpec}_\vee(L)$ in $\mathcal{PO}$. This functor is dual to the functor $\mathcal{I}: \mathcal{PO} \to \mathcal{CS}$ which associates to a partially ordered set P the lattice $\mathcal{I}(P)$ and to a monotone map $f: P \to P'$ the complete homomorphism $f^{-1}: \mathcal{I}(P') \to \mathcal{I}(P)$.

Theorem 6. *The category $\mathcal{CAL}$ of complete lattices which are order generated by complete $\vee$-primes and complete lattice maps is dual to the category $\mathcal{PO}$ of partially ordered sets and monotone maps.* □

Now every object L in $\mathcal{CAL}$ is completely distributive, and the complete $\vee$-primes order generate L. But complete $\vee$-primes are clearly compact elements of L, so $\mathrm{CSpec}_\vee(L) \subseteq K(L)$. Hence L is in fact a completely distributive *algebraic* lattice (recall Definition 2-14). Conversely, each completely distributive algebraic lattice is order generated by complete $\vee$-primes (cf., e.g., Proposition 4.2 of [18]), and so each is of the form $\mathcal{I}(P)$ where P is the set of complete $\vee$-primes of the lattice. Thus, an alternative statement of Theorem 6 is the following.

Theorem 6′. *The category $\mathcal{CDA}$ of completely distributive algebraic lattices and complete homomorphisms is dual to the category $\mathcal{PO}$.* □

Now that we are firmly entrenched within the realm of algebraic lattices, we can try to generalize this duality to one involving only algebraic lattices. We already have a subset which $\vee$-generates the algebraic lattice L, namely the family $K(L)$ of compact elements of L. Moreover, it is a simple exercise to show that the supremum of any two compact elements of L is another such, and clearly 0 is compact. So, $K(L)$ is a $\vee$-subsemilattice of L which contains 0 for any algebraic lattice L. Moreover, each point $x \in L$ gives rise to an order ideal of $K(L)$ via the map $x \mapsto \downarrow x \cap K(L)$ (since $K(L)$ is closed under finite suprema). Conversely, given any order ideal I of $K(L)$, the element $x = \bigvee I$ satisfies $\downarrow x \cap K(L) = I$ since I consists of compact elements. Thus, L is isomorphic to the lattice $\mathrm{Id}(K(L))$ of order ideals of the $\vee$-semilattice $K(L)$ under the map $x \mapsto \downarrow x \cap K(L)$.

Conversely, given a $\vee$-semilattice S with least element 0, the family $\mathrm{Id}(S)$ of all non-empty order ideals of S is a complete semilattice under the operations of intersection and directed union. Moreover, this semilattice has a largest element, the order ideal S itself, and so $\mathrm{Id}(S)$ is a complete lattice. Furthermore, any order ideal I of S is the union of the principal ideals $\downarrow x$ for $x \in I$, and clearly $\downarrow x \in K(\mathrm{Id}(S))$ for each $x \in S$. Thus $\mathrm{Id}(S)$ is an algebraic lattice in the usual order. Moreover, the map $x \mapsto \downarrow x: S \to \mathrm{Id}(S)$ maps S onto the family $K(\mathrm{Id}(S))$.

This gives the basics of a duality on the object level, but what maps should be used to complete the duality? The obvious maps on the semilattice side are those which preserve finite suprema (including 0, the supremum of the empty set). The obvious map to associate to a map $f\colon S \to S'$ is $f^{-1}\colon \mathrm{Id}(S') \to \mathrm{Id}(S)$. Since ideals are closed under the formation of arbitrary intersections (infima) and directed unions (directed suprema), the natural candidates for maps between algebraic lattices are those which preserve all infima and all directed suprema. In fact, if we consider a map $\phi\colon L \to L'$ between algebraic lattices which preserves all infima and all directed suprema, the map $\bar{\phi}\colon L' \to L$ defined by $\bar{\phi}(x) = \bigwedge \phi^{-1}(\uparrow x)$ takes compact elements to compact elements.

Proposition 7. *Let $\phi\colon L \to L'$ be a map between algebraic lattices which preserves all infima and all directed suprema. Then the map $\bar{\phi}\colon L' \to L$ defined by $\bar{\phi}(x) = \bigwedge \phi^{-1}(\uparrow x)$ takes $K(L')$ to $K(L)$ and preserves finite suprema.*

Proof. If $k \in K(L')$, then let $D \subseteq L$ be a directed set with $\bar{\phi}(k) \leq \bigvee D$. Since ϕ is monotone, the set $\phi(D)$ is directed in L', and $\phi(\bigvee D) = \bigvee \phi(D)$ since ϕ preserves directed suprema. Since ϕ also preserves all infima, we know that $\phi(\bar{\phi}(k)) \in \uparrow x$, so that $k \leq \phi(\bar{\phi}(k)) \leq \phi(\bigvee D)$. Thus, $k \leq \bigvee \phi(D)$, and so there is some $d \in D$ with $k \leq \phi(d)$. But then $\bar{\phi}(k) \leq \bar{\phi}(d) \leq d$. This shows that $\bar{\phi}(K(L')) \subseteq K(L)$. Since ϕ preserves all infima, it follows that $\bar{\phi}$ preserves all suprema, and so $\bar{\phi}$ preserves finite sups of compact elements. □

So, we have a duality between the category $\mathcal{AL}$ of algebraic lattices and maps preserving all infima and all directed suprema, and the category $\mathcal{SEM}$ of $\vee$-semilattice monoids and maps preserving finite suprema.

Theorem 8. *The category $\mathcal{AL}$ of algebraic lattices and morphisms preserving all infima and all directed suprema is dual to the category $\mathcal{SEM}$ of $\vee$-semilattice monoids and maps preserving finite suprema. The duality associates to an algebraic lattice L the $\vee$-semilattice $K(L)$, and to a $\vee$-semilattice S the algebraic lattice* $\mathrm{Id}(S)$. □

Using Theorem 2-15, this duality can be recast into a duality involving compact semilattices. Indeed, if $\mathcal{CZ}_0$ denotes the category of compact zero-dimensional semilattice monoids, then Theorem 2-15 implies that each such semilattice is an algebraic lattice. Furthermore, the Scott topology of an algebraic semilattice has family $\{\uparrow k \mid k \in K(S)\}$ as a basis. Now, the following result becomes crucial.

Proposition 9. *Let S and S' be a algebraic semilattices.*

1) *The map $f\colon P \to P'$ is Scott continuous if and only if f preserves directed suprema.*
2) *The homomorphism $f\colon P \to P'$ is λ-continuous if and only if f preserves all infima and all directed suprema.*

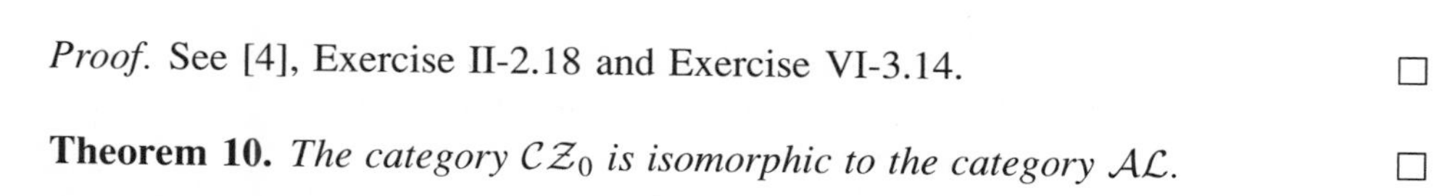
Proof. See [4], Exercise II-2.18 and Exercise VI-3.14. □

Theorem 10. *The category $\mathcal{CZ}_0$ is isomorphic to the category $\mathcal{AL}$.* □

Corollary 11. *(HMS-duality) The category $\mathcal{CZ}_0$ of compact zero-dimensional semilattice monoids and continuous semilattice maps is dual to the category $\mathcal{SEM}$ under the functors*

$\mathbb{K}: \mathcal{CZ}_0 \to \mathcal{AL}$ which sends a compact semilattice S to the $\vee$-semilattice $K(S)$ and sends a morphism $f: S \to S'$ of compact semilattice monoids to the $\vee$-semilattice map $\mathbb{K}(f): K(S') \to K(S)$ defined by $\mathbb{K}(f)(k) = \bigwedge f^{-1}(\uparrow k)$, and

$\mathbb{C}: \mathcal{SEM} \to \mathcal{CZ}_0$ which sends a $\vee$-semilattice S to the algebraic lattice Id(S) *endowed with the λ-topology, and the $\vee$-semilattice map $f: T \to T'$ to the $\mathcal{CZ}_0$-map* $\mathrm{Id}(f): \mathrm{Id}(T') \to \mathrm{Id}(T)$ *by* $\mathrm{Id}(f)(I) = f^{-1}(I)$. □

This duality was first discovered by Austin [1], but it was dubbed HMS-duality because of the extensive treatise [7] devoted to its study.

We now generalize this duality to the category $\mathcal{AS}$ of algebraic semilattices and maps preserving all infima and all directed suprema. According to Theorem 2-14 and Proposition 9, this is the same as the category of compact 0-dimensional semilattices and continuous semilattice maps. Of course, the object we want to associate with each algebraic semilattice S is the family $K(S)$ of compact elements of S, but $K(S)$ is no longer a $\vee$-semilattice. Indeed, without a universal upper bound, there may be compact elements of S with no common upper bound. But, if two compact elements $k, k' \in S$ have an upper bound, then they do have a supremum, and that supremum is a compact element.

Definition 12. A *conditional $\vee$-semilattice* is a partially ordered set T with least element in which every pair of elements having an upper bound has a least upper bound. □

Now it is easy to see that the set $K(S)$ is a conditional $\vee$-semilattice for S an algebraic semilattice. Conversely, for any conditional $\vee$-semilattice T, the family Id(T) of order-ideals of T is a complete semilattice; i.e., Id(T) is closed under all intersections (infima) and all directed unions (directed suprema). Moreover, Id(T) is algebraic, since the family $\{\downarrow x \mid x \in T\} \subseteq K(\mathrm{Id}(T))$ satisfies $I = \bigvee\{\downarrow x \mid x \in I\}$, and this set is directed (since I is an ideal) for every ideal $I \in \mathrm{Id}(T)$.

However, the morphisms on the algebraic semilattice side are somewhat problematical. One approach is to simply adjoin a discrete identity to each algebraic semilattice S to form the algebraic lattice S^1, extending the maps between algebraic semilattices to those between the corresponding algebraic lattices by preserving the identity, and then apply HMS-duality, and single out the objects and morphisms which arise in this manner. Then one must restrict the morphisms to be only those maps $f: S^1 \to S'^1$ between algebraic lattices which satisfy $f^{-1}(1) = \{1\}$, since these are the only maps which arise from maps between the algebraic semilattices S and S'. But there is another way to accomplish the same goal.

Consider what is true of a λ-continuous semilattice homomorphism $f: S \to S'$ for algebraic semilattices S and S'. The lower set $\downarrow f(S)$ is closed in S' since S and S' are compact Hausdorff spaces in the λ-topology. Moreover, the map f is retrievable from this lower set and the family $\downarrow f(S) \cap K(S')$, since $f(x) = \bigvee(\downarrow f(x) \cap K(S'))$ for each $x \in S$. So, what we want is the family $\downarrow f(S) \cap K(S')$, which is simply a lower set of S' intersected with $K(S')$. From the viewpoint of $K(S')$, this is a *lower set* in $K(S')$.

Definition 13 (cf. Definition 2). Let T be a conditional $\vee$-semilattice. The subset $R \subseteq T$ is a *lower set* of T if $R = \downarrow R = \{t \in T \mid (\exists r \in R)\ r \vee t = r\}$. □

Now, given a morphism of algebraic semilattices $f: S \to S'$, we define the lower set $T_f = \downarrow f(S) \cap K(S')$; this is again a conditional $\vee$-semilattice whose $\vee$-operation agrees with that of $K(S')$ where it is defined (since T_f is a lower set). So, we associate to f the morphism of conditional $\vee$-semilattices $\bar{f}: T_f \to K(S)$ by $\bar{f}(k) = \bigwedge f^{-1}(\uparrow k)$. Since $k \in \downarrow f(S)$, it follows that $f^{-1}(\uparrow k) \neq \emptyset$, and so we can show that $\bar{f}(k) \in K(S)$ and that $\bar{f}$ is a morphism of conditional $\vee$-semilattices.

Conversely, given a lower set R of $K(S')$ and a morphism of conditional $\vee$-semilattices $\phi: R \to K(S)$, the map $\bar{\phi}: S \to S'$ by $\bar{\phi}(s) = \bigvee \phi^{-1}(\downarrow s)$, $(\forall s \in S)$ is a morphism of algebraic semilattices; i.e., $\bar{\phi}$ preserves all infima and all directed suprema. We summarize these results in the following theorem.

Theorem 14. *The category $\mathcal{AS}$ of algebraic semilattices and maps preserving all infima and all directed suprema is dual to the category $\mathcal{CSEM}$ of conditional $\vee$-semilattices and maps defined on lower sets preserving all finite suprema which exist.* □

This concludes the development of the algebraic portion of the theory. We do have another purely algebraic duality theorem to obtain, but it is more convenient to do this using spectral theory than using the duality theory approach we have so far described.

4. Spectral theory: bringing topology to bear

Let's return for a moment to Theorem 3-1. It says there is a duality between the category of complete atomic Boolean algebras and complete homomorphisms, and the category of sets and functions. An obvious question is whether this duality extends to one involving arbitrary Boolean algebras. This was exactly the question Marshall Stone considered while studying families of commuting projections defined on a Hilbert space. Stone needed two ideas to obtain the desired generalization. First, he realized that instead of looking within the Boolean algebra B for a space to represent it, he should look at the space Spec(B) of prime ideals of B. Second, the set Spec(B) should be given a topology, namely the *hull-kernel topology*, in order to generalize the Lindenbaum-Tarski duality. In this topology, a closed set is formed by the family of all prime ideals containing a fixed ideal I of B (this is the *hull* of I), and the ideal I is the intersection of this family of prime ideals (each ideal is the *kernel* of its hull). Moreover, the clopen subsets of Spec(B) exactly correspond to the principal ideals, and so we can recover B from the space Spec(B). The final piece of the puzzle was the realization that the spaces which arise in this fashion from Boolean algebras are compact, Hausdorff and totally disconnected, or *Stone spaces* as they are now called. Given any Stone space X, the family $\mathcal{CO}(X)$ of all compact, open subsets of X is a Boolean algebra whose spectrum is homeomorphic to X.

This correspondence can be extended to a duality of categories. Any Boolean algebra homomorphism $\phi: B \to B'$ gives rise to a continuous function $\bar{\phi}$: Spec(B') $\to$ Spec(B) by $\bar{\phi}(P) = \phi^{-1}(P)$ between the spectra, and any continuous function $f: X \to Y$ between Stone spaces gives rise to a Boolean algebra homomorphism $f^{-1}: \mathcal{CO}(Y) \to \mathcal{CO}(X)$.

Theorem 1. (Stone [25, 26]) *The category of Stone spaces and continuous maps is dual to the category of Boolean algebras and Boolean homomorphisms.* □

Stone went on to single out the complete Boolean algebras under this duality as those whose spectrum is *extremely disconnected*, i.e., those spaces in which the closure of every open set is open.

In considering possible generalizations of Theorem 1, it is tempting to continue the progression we have seen so far. Namely, complete atomic Boolean algebras are dual to sets, complete Boolean algebras are dual to extremely disconnected spaces, and Boolean algebras are dual to Stone spaces. So, as we weaken the structure on the algebra side, we increase the structure on the spatial side.

Stone's insight was to substitute "topological space" for "set" in order to generalize the Lindenbaum-Tarski result to incomplete Boolean algebras. Turning this around, we can seek to find dualities for categories of topological spaces and continuous functions between them. To develop such a duality, the obvious ordered structure to consider as a dual object to a topological space is the lattice of open subsets of the space. For a topological space X, the lattice $\mathcal{O}(X)$ of open subsets of X is closed in $\mathbb{P}(X)$ under all unions and finite intersections. So, $\mathcal{O}(X)$ is a complete lattice. In fact, $\mathcal{O}(X)$ is a *complete Heyting algebra*.

Definition 2. The complete lattice L is a *complete Heyting algebra* if the following infinite distributivity law holds for every $x \in L$ and every $Y \subseteq L$:

$$x \wedge \bigvee Y = \bigvee_{y \in Y} x \wedge y \qquad \square$$

While it is not apparent from this definition, each complete Boolean algebra is in fact a Heyting algebra (cf., Exercise I-3.19 and Definition I-3.20 of [4]). But it is clear that the lattice $\mathcal{O}(X)$ is a complete Heyting algebra for each topological space X. Moreover, given a continuous map $f: X \to Y$ between topological spaces, the map $f^{-1}: \mathcal{O}(Y) \to \mathcal{O}(X)$ preserves all unions and all finite intersections, so we take as morphisms between complete Heyting algebras those maps which preserve all suprema and all finite infima.

But, we must make some additional assumptions on the space X if we are to recover X from $\mathcal{O}(X)$. The obvious map from X into $\mathcal{O}(X)$ is $x \mapsto X \setminus \overline{\{x\}}$, and this is one-to-one if and only if X is a T_0-space. Furthermore, the sets $X \setminus \overline{\{x\}}$ are prime in $\mathcal{O}(X)$ under intersection, as is easily shown. So, a natural candidate to identify X with is the family $\mathrm{Spec}_\cap(\mathcal{O}(X))$ of $\cap$-primes of $\mathcal{O}(X)$. To insure that every element of this set is of the form $X \setminus \overline{\{x\}}$ for some $x \in X$, we need an additional assumption.

Definition 3. The topological space X is called *sober* if each open set $U \in \mathrm{Spec}_\cap(\mathcal{O}(X))$ has the form $U = X \setminus \overline{\{x\}}$ for a unique $x \in X$. $\square$

Not only does this definition imply that every $\cap$-prime of $\mathcal{O}(X)$ arises from some point of X, it also guarantees that X is T_0. So, sober spaces are the natural candidates for developing duality theories involving topological spaces and continuous maps and lattices and lattice maps, since these spaces are in one-to-one correspondence with the spectrum of their lattice of open sets.

Now, given an element $U \in \mathcal{O}(X)$, we want to associate to U an open set in $\mathrm{Spec}_\cap(\mathcal{O}(X))$, and this is done in the following way. Consider the set $\uparrow U \cap \mathrm{Spec}_\cap(\mathcal{O}(X))$. We have

$$\uparrow U \cap \mathrm{Spec}_\cap(\mathcal{O}(X)) = \{X \setminus \overline{\{x\}} \mid U \subseteq X \setminus \overline{\{x\}}\} = \{x \in X \mid x \notin U\},$$

so that we can identify U with $\mathrm{Spec}_\cap(\mathcal{O}(X)) \setminus \uparrow U$ (cf. Definition 3-4).

Definition 4. For a complete lattice L, we define the *hull-kernel topology* on $\mathrm{Spec}_\wedge(L)$ to have a basis of open sets of the form $\mathrm{Spec}_\wedge(L) \setminus \uparrow x$, where $x \in L$. $\square$

This is always a sober topology on $\mathrm{Spec}_\wedge(L)$, and our remarks above show that if X is T_0, then the map $x \mapsto X \setminus \overline{\{x\}}: X \to \mathrm{Spec}_\cap(\mathcal{O}(X))$ embeds X homeomorphically into $\mathrm{Spec}_\cap(\mathcal{O}(X))$; this map is a homeomorphism just in case X is a sober space. Now, given a map $f: X \to Y$ between sober spaces, the morphism we associate to f is the map $f^{-1}: \mathcal{O}(Y) \to \mathcal{O}(X)$. This map preserves all suprema (unions) and all finite infima (intersections). But given a morphism $\phi: L \to L'$ of complete Heyting algebras, what is not so clear is how to find a continuous map $\bar{\phi}: \mathrm{Spec}_\wedge(L') \to \mathrm{Spec}_\wedge(L)$. To understand this, we need the following notion.

Definition 5. Let L and L' be complete lattices, and let $\phi: L \to L'$ and $\psi: L' \to L$ be order preserving maps. The pair (ϕ, ψ) is a *Galois adjunction* if $\phi \circ \psi \leq 1_L$ and $\psi \circ \phi \leq 1_{L'}$. In this case, ϕ preserves all infima and is called the *upper adjoint* of ψ, and ψ preserves all suprema and is called the *lower adjoint* of ϕ. □

The theory of Galois adjunctions between complete lattices in summarized in Section 0-3 of [4]. For complete lattices L and L', the following results are easily derived:

Each $\phi: L \to L'$ preserving all infima has a uniquely defined lower adjoint $\psi: L' \to L$ given by $\psi(x) = \bigwedge \phi^{-1}(\uparrow x)$, $\forall x \in L'$.
Each $\psi: L' \to L$ preserving all suprema has a uniquely defined upper adjoint $\phi: L \to L'$ given by $\phi(x) = \bigvee \psi^{-1}(\downarrow x)$, $\forall x \in L$.

For example, given a continuous map $f: X \to Y$ between sober spaces, the map $f^{-1}: \mathcal{O}(Y) \to \mathcal{O}(X)$ is a lower adjoint (since it preserves all suprema). Its upper adjoint is given by the formula $F: \mathcal{O}(X) \to \mathcal{O}(Y)$ given by

$$F(U) = \bigvee (f^{-1})^{-1}(\downarrow U) = \bigvee \{W \in \mathcal{O}(Y) \mid f(U) \supseteq W\},$$

and this map preserves all infima.

Proposition 6. (Hofmann & Lawson [6]) *Let L and L' be complete lattices in which the $\wedge$-primes order generate. Then the upper adjoint $\phi: L \to L'$ preserves $\wedge$-primes if and only if its lower adjoint $\psi: L' \to L$ preserves finite infima.* □

It is clear that the lower adjoint $f^{-1}: \mathcal{O}(Y) \to \mathcal{O}(X)$ preserves finite infima (intersections), and so its upper adjoint $F: \mathcal{O}(X) \to \mathcal{O}(Y)$ preserves $\wedge$-primes. Now, for each $x \in X$, we calculate

$$F(X \setminus \overline{\{x\}}) = \bigvee \{W \in \mathcal{O}(Y) \mid f(X \setminus \overline{\{x\}}) \supseteq W\} = Y \setminus \overline{\{f(x)\}}.$$

This is the same as the map $f: X \to Y$ with which we began, if we identify the spaces X and Y with $\mathrm{Spec}(\mathcal{O}(X))$ and $\mathrm{Spec}(\mathcal{O}(Y))$, respectively .

Theorem 7. *The category $\mathcal{SOB}$ of sober spaces and continuous maps is dual to the category $\mathcal{HEYT}$ of complete Heyting algebras in which the primes order* *generate, and morphisms preserving all suprema and all finite infima. This duality is implemented by the functor*

$$\mathcal{O}\colon \mathcal{SOB} \to \mathcal{HEYT}$$

which associates to a sober space X its lattice of open sets $\mathcal{O}(X)$, and to a continuous map $f\colon X \to Y$ the Heyting algebra map $f^{-1}\colon \mathcal{O}(Y) \to \mathcal{O}(X)$, on the one hand, and the functor

$$\mathrm{Spec}\colon \mathcal{HEYT} \to \mathcal{SOB}$$

which associates to a complete Heyting algebra L is spectrum Spec(L) *in the hull-kernel topology, and to a Heyting algebra map $\phi\colon L \to M$ the restriction and corestriction of ϕ's upper adjoint $\psi|_{\mathrm{Spec}\, M}\colon \mathrm{Spec}(M) \to \mathrm{Spec}(L)$.* □

This duality theorem is the basis upon which a whole range of duality theorems have arisen. These theorems show that, for various categories of topological spaces and continuous maps, there correspond dual categories of Heyting algebras and suitable morphisms between them. A listing of the dual categories is provided at the end of Section V-5 of [4]; a development more along the lines of Stone's approach is available in [9]. Rather than attempting to recite these theorems one-by-one, we confine ourselves to providing two examples. The first brings compact semilattice monoids into the picture.

Example 8. Let X be a locally compact Hausdorff space. Then $\mathcal{O}(X)$ is a complete Heyting algebra. If $U \in \mathcal{O}(X)$, then for each $x \in U$, there is an open set V_x satisfying $x \in V_x \subseteq \overline{V_x} \subseteq U$ and $\overline{V_x}$ is compact. It is easy to characterize the fact that V_x is contained in a compact subset of U within $\mathcal{O}(X)$: if $\mathcal{U}$ is any open cover of U, then there is a finite subcover of $\overline{V_x}$. But this is just the definition of the way-below relation in this particular situation (cf., Definition 2-9). Moreover, each open set $U \in \mathcal{O}(X)$ satisfies the property that U is the union of open subsets V which are way-below U in $\mathcal{O}(X)$. □

So, this example shows that continuous lattices arise as the lattice of open sets of a locally compact Hausdorff space. This generalizes to the category of T_0-spaces, where we must reinterpret the notion of local compactness.

Definition 9. A sober space X is *locally compact* if each point has a basis of compact neighborhoods. □

Theorem 10. (Hofmann & Lawson [6]) *For a sober space X, the lattice $\mathcal{O}(X)$ is a continuous lattice if and only if X is locally compact.* □

We can thus invoke the duality between sober spaces and complete Heyting algebras with enough primes to conclude the following corollary.

Corollary 11. (HOFMANN & LAWSON [6]) *The category of locally compact sober spaces and continuous maps is dual to the category of distributive continuous lattices and maps preserving all suprema and all finite infima.* □

Now, if we want to regard a continuous lattice as a compact semilattice (by endowing it with the Lawson topology), then the maps of interest are those preserving all infima and directed suprema. A Heyting algebra map $\phi: L \to M$ between continuous lattices has an upper adjoint $\psi: M \to L$ which preserves all infima. The following result characterizes when ψ also preserves directed suprema.

Proposition 12. *Let L and M be continuous lattices, and let $\phi: L \to M$ be the lower adjoint of the map $\psi: M \to L$. Then ψ is Scott-continuous if and only if ϕ preserves the way-below relation; i.e., if $x \ll y \in L$, then $\phi(x) \ll \phi(y) \in M$.* □

For a map $f: X \to Y$ between locally compact sober spaces, the Heyting algebra map $f^{-1}: \mathcal{O}(Y) \to \mathcal{O}(X)$ is the lower adjoint of the map which is the candidate for a morphism of compact semilattices. For the upper adjoint of f^{-1} to be Scott-continuous, the Proposition says f^{-1} must preserve the way-below relation. This can actually be characterized in terms of the map f from which f^{-1} is derived. First, call a subset of a topological space *saturated* if it is the intersection of the open sets which contain it. Then, a map $f: X \to Y$ is *proper* if $f^{-1}(A)$ is compact for every saturated compact subset A of Y.

Proposition 13. *Let $f: X \to Y$ be a continuous function between locally compact sober spaces. The Heyting algebra map $f^{-1}: \mathcal{O}(Y) \to \mathcal{O}(X)$ preserves the way-below relation if and only if f is a continuous proper map.* □

So, we can summarize these results as the following theorem:

Theorem 14. (HOFMANN & LAWSON [6]) *There is an isomorphism between the category of locally compact sober spaces and continuous proper maps, and the category of distributive continuous lattices and Scott-continuous maps preserving all infima and prime elements. This isomorphism associates to a sober space X the lattice $\mathcal{O}(X)$ and to the map $f: X \to Y$ the upper adjoint $\psi: \mathcal{O}(X) \to \mathcal{O}(Y)$ of the Heyting algebra map $f^{-1}: \mathcal{O}(Y) \to \mathcal{O}(X)$. Dually, the isomorphism associates to a distributive continuous lattice L the space* $\mathrm{Spec}(L)$ *in the hull-kernel topology, and to the map $\psi: L \to M$ preserving primes, the restriction and co-restriction* $\psi|_{\mathrm{Spec}(L)}: \mathrm{Spec}(L) \to \mathrm{Spec}(M)$. □

This is really a theorem about compact semilattices. It states that the category of compact Lawson semilattice monoids (i.e., those having a basis of subsemilattices at each point) which are distributive lattices and compact semilattice maps which preserve primes is isomorphic to the category of locally compact sober spaces and proper maps.

Actually, the statement of the Theorem tells us a bit more. Let $f: X \to Y$ be a continuous proper map between locally compact sober spaces, and let $\psi: \mathcal{O}(X) \to \mathcal{O}(Y)$ be the upper adjoint of f^{-1}. Then the natural isomorphisms $\eta_X: X \to \mathrm{Spec}(\mathcal{O}(X))$ and $\eta_Y: Y \to \mathrm{Spec}(\mathcal{O}(Y))$ satisfy the property that $\eta_Y \circ f = \eta_Y^{-1} \circ \psi|_{\mathrm{Spec}(\mathcal{O}(X))}$. So, the map ψ extends the map induced by f between the spectra of $\mathcal{O}(X)$ and $\mathcal{O}(Y)$ to a morphism of compact semilattices.

This is one of two theorems involving spectral theory which we wanted to point out. The other involves the algebraic theory we described in the last section. Since Theorem 14 presents an isomorphism involving continuous lattices, one might wonder which spaces give rise to distributive algebraic lattices under this isomorphism. We show that such spaces must have a basis of compact-open subsets. If L is a distributive algebraic lattice, then each compact element $k \in K(L)$ satisfies the property that $\mathrm{Spec}(L) \setminus \uparrow k$ is compact and open in $\mathrm{Spec}(L)$. Indeed, the set $\mathrm{Spec}(L) \setminus \uparrow k$ is open since each element of L gives rise to an open set in $\mathrm{Spec}(L)$ in this way. But the set is also compact, since any open cover of $\mathrm{Spec}(L) \setminus \uparrow k$ of the form $\{\mathrm{Spec}(L) \setminus \uparrow k_i \mid i \in I\}$ has the property that $k \leq \bigvee_{i \in I} k_i$, so there is a finite subset $F \subseteq I$ with $k \leq \bigvee_{i \in F} k_i$. But then

$$\bigcap_{i \in F} \mathrm{Spec}(L) \setminus \uparrow k_i = \mathrm{Spec}(L) \setminus (\bigcup_{i \in F} \uparrow k_i) \supseteq \mathrm{Spec}(L) \setminus \uparrow k,$$

since $\mathrm{Spec}(L)$ order-generates L and $k \leq \bigvee_{i \in I} k_i$.

Conversely, if X is a sober space with a basis of compact-open subsets, then X is necessarily locally compact, so $\mathcal{O}(X)$ is a continuous lattice. But each compact-open subset $C \subseteq X$ is a compact element of $\mathcal{O}(X)$, and so $\mathcal{O}(X)$ is an algebraic lattice.

Theorem 15. *There is a isomorphism between that category of sober spaces having a basis of compact-open subsets and continuous proper maps and the category of distributive algebraic lattices and maps preserving all infima, all directed suprema and prime elements.* □

Of course, there is the subcategory of *arithmetic lattices*, which are those algebraic lattices L which have the property that $K(L)$ is a sublattice (as opposed to being only a sup-subsemilattice) of L. On the spatial side, the property that $K(\mathcal{O}(X))$ is a sublattice is precisely that the compact-open subsets of X are closed under both union and intersection. These are the *coherent spaces*, and so we conclude the following corollary.

Corollary 16. *There is a isomorphism between the category of coherent spaces and continuous proper maps and the category of distributive arithmetic lattices and maps preserving all infima, all directed suprema and prime elements.* □

In terms of compact semilattices, this states that the category of coherent spaces and continuous proper maps is isomorphic to the category of compact 0-dimensional semilattice monoids which are distributive arithmetic lattices, and compact semilattice maps preserving prime elements. There are many further isomorphism theorems which can be derived in this setting; we refer to the comprehensive treatments in Chapter V of [4] or in [9] for further details. But our basic point clearly has been made: compact semilattices (in a suitable guise) play a central role in the general theory of the relationship between categories of topological spaces and categories of complete lattices.

5. Applications

In this the final section, we describe applications of the structure theory of compact semilattices to three distinct areas. Our presentation is necessarily somewhat sketchy, with few details of proofs provided. Moreover, we do not pretend to present a complete overview of this area. There are, for example, several other applications described in detail in [4], and further references there to still other applications. We focus on one particular question which has arisen in several distinct areas: when does an object contain a copy of the complete lattice $2^{\mathbb{N}}$? In each case, we also document the implications of the existence of such a copy for the theory we are describing. We begin with what was historically the first area where this question was considered.

Harmonic Analysis: Each locally compact abelian group G admits a translation invariant regular measure, called *Haar measure*, which is unique up to multiplicative constant. The algebra $L^1(G)$ of all measurable functions f satisfying $\int_G |f(x)|dx < \infty$ completely determines G, in that the character group of G, $\hat{G}$ arises as the space of complex homomorphisms of $L^1(G)$. This is one path to obtaining the Pontryagin Duality Theorem, which states that the category of locally compact abelian groups and continuous homomorphisms is self-dual under the functor which associates to an lca group G the group $\hat{G} = \mathrm{Hom}(G, \mathbb{R}/\mathbb{Z})$, and to the continuous homomorphism $f: G \to H$ the map $\hat{f}: \hat{H} \to \hat{G}$ by $\hat{f}(\phi) = \phi \circ f$.

There is also a larger algebra associated with G, the algebra $M(G)$ of all finite regular Borel measures on G. This is a commutative Banach algebra with identity, δ_0, and which admits the algebra $L^1(G)$ as a subalgebra (in fact, as an L-ideal) under the map $f \mapsto f dx$. As a commutative Banach algebra, $M(G)$ has a maximal ideal space, $\Delta M(G)$, which is a compact Hausdorff space in the weak *-topology. Moreover, $\Delta M(G)$ is a semitopological semigroup in this topology, and $\hat{G}$ forms the group of units of this semigroup. One obvious question is, "When is $\hat{G} = \Delta M(G)$?" If G is discrete, then $L^1(G) = M(G)$, and so $\hat{G} = \Delta M(G)$. To obtain the converse requires more work.

The measure algebra $M(G)$ admits a natural involution: given $\mu \in M(G)$, define $\mu^*(f) = \int_G \overline{f(-x)} d\mu(x)$. Moreover, each character ϕ of G respects this involution, in that $\int_G \phi(x) d\mu^*(x) = \overline{\int_G \phi(x) d\mu(x)}$. If a Banach algebra M with involution satisfies the property that every complex homomorphism respects its involution, then it is called *symmetric*; otherwise it is called *asymmetric*. The celebrated HEWITT-KAKUTANI Theorem states that any non-discrete locally compact abelian group G contains a copy of the Cantor set C which is *independent*, and such a subset supports a self-adjoint measure μ of unit norm so that the measure $\delta_0 - \mu^2$ has spectral norm 2. Hence there is a complex homomorphism h of $M(G)$ satisfying $h(\mu^2) = -1$, which means h cannot respect the self-adjointness of μ, and so $M(G)$ is asymmetric. In particular, h does not arise from a character of G.

In the constellation of compact semigroups, groups and semilattices can be thought of as residing at antipodal points. But it is still interesting to take this same question and apply it in the setting of compact semilattices. Namely, a compact semilattice monoid S has a measure algebra $M(S)$ consisting of all finite regular Borel measures on S, and the maximal ideal space $\Delta M(S)$ is a compact semitopological semigroup in the weak *-topology. Once again, each universally measurable semicharacter $\phi\colon S \to D$ of S into the unit disk in the plane naturally gives rise to a complex homomorphism of $M(S)$ via the formula $\mu \mapsto \int_S \phi(x) d\mu(x)$. The question then becomes whether all complex homomorphisms of $M(S)$ arise in this fashion.

The first work on this question was due to HEWITT and ZUCKERMAN [5], who showed that $\Delta M([0,1])$ consists precisely of complex homomorphisms induced by semicharacters of $[0,1]$, the unit interval. This work was then advanced independently by BAARTZ [2] and NEWMAN [21], who showed that $\Delta M(S)$ consists of semicharacters if S is a locally compact semilattice which is embeddable in a finite product of locally compact chains. Moreover, they also both showed that this result fails for $2^{\mathbb{N}}$. In fact, NEWMAN's proof of this result consists of showing that the HEWITT-KAKUTANI proof carries over to this setting, since the Cantor set can be embedded in the compact semilattice $2^{\mathbb{N}}$. This utilizes the fact that there is a natural involution on $M(S)$: for $\mu \in M(S)$, define $\mu^*(f) = \int_S \overline{f(s)} d\mu(s)$.

The definitive results along this line were obtained in [14] and [17]. First, we say a locally compact semilattice has *compactly finite breadth* if each compact subset $C \subseteq S$ has a finite subset $F \subseteq C$ with $\bigwedge C = \bigwedge F$. The following result then becomes crucial.

Theorem 1. (LAWSON, LIUKKONEN & MISLOVE) *For a locally compact semilattice S, the following are equivalent:*
i) *S has compactly finite breadth.*
ii) *S admits no copy of $2^{\mathbb{N}}$ as a subsemilattice.*
iii) *Each complex homomorphism of $M(S)$ is given as integration against a semi-character of S_d, the discrete semilattice S.* □

As a corollary of this result, the idempotent measures on S can be completely described, and it can also be shown that the only invertible measures on S are those which are exponential measures. Since any locally compact semilattice which contains a copy of $2^{\mathbb{N}}$ must have an asymmetric measure algebra, this theorem describes the only semilattices which satisfy the property that all complex homomorphisms of $M(S)$ arise from semicharacters of S.

Infinite Combinatorics: The second area of application of compact semilattices is that of infinite combinatorics. For a poset P, the structure of P is often understood in terms of the structure of the lattice $\mathcal{I}(P)$ of all lower sets of P, and the poset Id(P) of all order-ideals of P (i.e., directed lower sets of P). Typical of the type of result which is forthcoming in this area is the following.

Let K denote the poset which consists of all pairs (i, j) of positive integers i, j with $i \leq j$, and which is ordered by

$$(i, j) \leq (k, l) \quad iff \quad i = k \ \& \ j \leq l \ \ or \ \ j < k.$$

Let K^d denote K with the opposite order, and let ω denote the natural numbers in the usual order.

Theorem 2. (DUFFUS, POUZET & RIVAL [3]) *Let P be a partially ordered set. Then the following are equivalent:*
i) *$\mathcal{I}(P)$ contains an infinite antichain.*
ii) *P contains an infinite antichain, a copy of K, of K^d, or of $\omega \oplus \omega^d$.* □

It was the goal of work in [16] to dismantle this result, so that the occurrance of the posets K, K^d, and $\omega \oplus \omega^d$ could be better understood. The fundamental result turned out to be clarifying when P has an infinite antichain in terms of $\mathcal{I}(P)$. That result relies on the observation that the natural map $p \mapsto \downarrow p : P \to \mathcal{I}(P)$ maps P onto the complete $\cup$-primes of $\mathcal{I}(P)$. Thus, if P contains an infinite antichain, then $\mathcal{I}(P)$ contains a copy of $2^{\mathbb{N}}$ closed under all infima and all suprema. The converse also holds, as the following result shows:

Theorem 3. (LAWSON, MISLOVE & PRIESTLEY [16]) *For a partially ordered set P, the following are equivalent:*
i) *P contains an infinite antichain.*

ii) *$\mathcal{I}(P)$ contains a copy of $2^{\mathbb{N}}$ under a map preserving all infima and all suprema.*
iii) *There is some element I of $\mathcal{I}(P)$ which is not the union of finitely many order-ideals of P.* □

The third equivalent statement can be regarded as the restatement of the compactly finite breadth of $\mathcal{I}(P)$. Then the definitive occurrances of K, of K^d or of $\omega \oplus \omega^d$ in P arise in the case where every element of $\mathcal{I}(P)$ is the union of finitely many order-ideals of P. In this case, we can still classify when each of these posets arise in P, as the following theorem shows:

Theorem 4. (LAWSON, MISLOVE & PRIESTLEY [16]) *If the poset P has no infinite antichain, then the following results hold:*
i) *P has a copy of K (respectively, of K^d) if and only if $(\mathcal{I}(P), \cup)$ (respectively, $(\mathcal{I}(P), \cap)$) contains a copy of the free semilattice on countably many generators.*
ii) *If P has no copy of K or of K^d, then P has a copy of $\omega \oplus \omega^d$ if and only if $\mathcal{I}(P)$ has an infinite antichain, but this antichain does not generate a free $\cup$- or a free $\cap$-subsemilattice of $\mathcal{I}(P)$.* □

If one focuses on Theorem 3, then the existence of an infinite antichain within P forces $\mathcal{I}(P)$ to contain a copy of $2^{\mathbb{N}}$, since P resides among the elements of $\mathcal{I}(P)$ which have the strongest generation property with respect to $\cup$. More generally, one might ask about the existence of infinite antichains in compact semilattices, and whether they must generate copies of $2^{\mathbb{N}}$. The results in [15] address exactly this question, and the answer lies in how the infinite antichain resides within the semilattice S, with respect to the order and the topology of S. Namely, an infinite antichain $A = \{x_n\}$ within a complete semilattice S *converges upwards* to $x \in S$ if x is an upper bound for A, and if there exists an increasing sequence $\{z_n\}$ with $z_n \leq x_n$ for each n, and $x = \bigvee z_n$. An antichain A in the complete semilattice S is *completely free* if $\bigwedge C = \bigwedge D$ implies $C = D$ for all subsets C, D of A.

Theorem 5. (LAWSON, MISLOVE & PRIESTLEY [15]) *For a continuous semilattice S, the following are equivalent:*
i) *S contains a copy of the compact semilattice $2^{\mathbb{N}}$.*
ii) *S contains a infinite antichain A which is completely free and converges upwards to some element of S.*
iii) *S does not have compactly finite breadth.* □

If the continuous semilattice S has no copy of $2^{\mathbb{N}}$, then any infinite antichain within S must either not be completely free, or else it must not converge upwards. The alternative to the latter is that it converge *downwards* or *sideways* (cf. [15] for the precise definitions). It is an open question whether there is a finite catalogue of continuous semilattices $\{S_i \mid n = 1, \ldots, n\}$ such that any continuous semilattice which contains a completely free infinite antichain which does not converge upwards must contain a copy of one of the S_i's.

Theoretical Computer Science: The final area of application we consider is theoretical computer science. The semantics of high-level programming languages studies methods of assigning meanings to programs so that they can be subjected to analytical reasoning. A compiler or interpreter for such a language has a built-in syntax checker called the *parser*, and this checks each program to be sure that it complies with the grammatical rules for writing programs which are given by the syntax of the language. But even if a program makes it through the compiler, it is not clear what that program will do. In an effort to understand the complicated code which can be generated in writing a long program, theoretical computer scientists have developed methods to assign meanings to programs which are *modular* and *mathematical*. That the methods are modular means that the meaning of a complicated construct is made up from the meanings of its simpler constituent parts.

The untyped λ-calculus is often regarded as a prototypical programming language (without assignment), and Dana Scott used continuous lattices to give the first models of this calculus [24]. What he needed was a cartesian closed category and an object which was isomorphic to its function space in the category. The category consisted of continuous lattices and Scott continuous maps, and the object was our old friend $2^{\mathbb{N}}$, this time recognized as the inverse limit of the semilattices 2^n in the obvious way. Gordon Plotkin pointed out that more general objects than lattices are needed as mathematical models for programming languages, and the notion of a *domain* was formulated. In the terminology of this paper, a domain is an algebraic poset P satisfying $K(P)$ is countable. Plotkin proposed the truth value domain $T = \{tt, ff, \perp\}$ as a basis for modeling languages, where tt and ff are incomparable, but $\perp$ lies below both of them. In T the element tt stands for "true", ff for "false", and $\perp$ for the least element of the poset. Plotkin showed that $D = T^{\mathbb{N}}$ satisfies the equation $D \simeq [D \to D]$, where $[D \to D]$ denotes the domain of Scott continuous selfmaps defined on D (cf. [22]).

Now, it is clear that $2 \subseteq T \subseteq 2^2$, and so each of $2^{\mathbb{N}}$ and $T^{\mathbb{N}}$ contains a copy of the other. The question then arises whether any domain D satisfying $D \simeq [D \to D]$ must contain a copy of $2^{\mathbb{N}}$.

Theorem 6. (Plotkin (cf. [19])) *If a domain D satisfies $D \simeq [D \to D]$, then D contains a copy of $2^{\mathbb{N}}$ under a map preserving all suprema.* □

This result has been generalized somewhat, in that it is now known that any domain D which contains a copy of $[D \to D]$ under a Scott continuous injection must contain a copy of $2^{\mathbb{N}}$ under a map preserving all suprema. Moreover, it can also be shown that an algebraic semilattice S which contains a copy of $[S \to S]$ must contain a copy of $2^{\mathbb{N}}$ as a subsemilattice. This means that any domain which is large enough to accommodate a model of the untyped λ-calculus must contain a copy of $2^{\mathbb{N}}$, and so this is a minimal model.

References

[1] Austin, C. W., Duality theorems for some commutative semigroups, Trans. Amer. Math. Soc. 109 (1963), 245–256

[2] Baartz, A., The measure algebra of a lcoally compact semigroup, Pac. J. Math. 21 (1967), 199–214

[3] Duffus, D., M. Pouzet and I. Rival, Complete ordered sets with no infinite antichains, Discrete Mathematics 35 (1981), 31–52

[4] Gierz, G., K. H. Hofmann, K. Keimel, J. D. Lawson, M. Mislove and D. S. Scott, "A Compendium of Continuous Lattices," Springer-Verlag, Berline etc., 1980

[5] Hewitt, E. and H. Zuckerman, Structure theory for a class of convolution measure algebras, Pac. J. Math. 7 (1957), 913–941

[6] Hofmann, K. H. and J. D. Lawson, The spectral theory of distributive continuous lattices, Trans. Amer. Math. Soc. 246 (1978), 285–310

[7] Hofmann, K. H., and M. Mislove, Epics of compact Lawson semilattices are surjective, Arch. d. Math. (Basel) 26 (1975), 337–345

[8] Hofmann, K. H., M. W. Mislove and A. R. Stralka, "The Pontryagin Duality of Compact 0-Dimensional Semilattices and its Applications," Lecture Notes in Mathematics 396 (1974), Springer-Verlag

[9] Johnstone, P. T., "Stone Spaces," Cambridge University Press, Cambridge, 1982

[10] Lawson, J. D., Vietoris mappings and embeddings of topological lattices, Dissertation, University of Tennessee, Knoxville, 1967

[11] —, Topological semilattice with small semilattices, J. London Math. Soc. 2 (1969), 719–724

[12] —, Lattices with no interval homomorphisms, Pac. J. Math. 32 (1970), 459–465

[13] —, Intrinsic topologies in topological lattices and semilattices, Pac. J. Math. 44 (1973), 593–602

[14] Lawson, J. D., J. R. Liukkonen and M. W. Mislove, Measure algebras of semilattices of finite breadth, Pac. J. Math. 69 (1977), 125–139

[15] Lawson, J. D., M. W. Mislove and H. A. Priestley, Ordered sets with no infinite antichains, Discrete Mathematics 63 (1987), 225–230

[16] —, Infinite antichains in semilattices, Order 2 (1985), 275–290

[17] Liukkonen, J. R. and M. W. Mislove, Measure algebras of locally compact semilattices, in: Lecture Notes in Mathematics (Springer-Verlag) 998 (1983), 202–214

[18] Mislove, M. W., When are order scattered and toplogically scattered the same?, in: "Orders: Descriptions and Roles", Annals of Discrete Mathematics 23, (1985), 39–60

[19] —, Detecting local finite breadth in continuous lattices and semilattices, in: Lecture Notes in Computer Science (Springer- Verlag) 239 (1986), 205–214

[20] Nachbin, L., "Topology and Order," D. Van Nostrand Company, Inc., Princeton, New Jersey, 1954

[21] Newman, S. E., Measure algebras on idempotent semigroup, Pac. J. Math. 31 (1969), 161–169

[22] Plotkin, G. D., T^{∞} as a universal domain, J. Computing and Systems Science 17 (1978), 209–236

[23] Rudin, W. "Fourier Analysis on Groups," John Wiley & Sons, New York, 1962

[24] Scott, D. S., Data types as lattices, SIAM J. of Computing 5 (1976), 522–587

[25] Stone, M., The theory of representations for Boolean algebras, Trans. Amer. Math. Soc. 40 (1936), 37–111

[26] —, Applications of the theory of Boolean rings to general topology, Trans. Amer. Math. Soc. 41 (1937), 375–481

[27] Tarski, A., Zur Grundlegung der Boole'schen Algebra, Fund. Math. 24 (1935), 177–198

[28] Taylor, J., "Measure Algebras," CBMS Regional Conference Series in Mathematics 16 (1972)

[29] Urysohn, P., Über die Mächtigkeit der zusammenhängenden Mengen, Math. Annalen 94 (1925), 262–295

[30] Wallace, A. D., Acyclicity of compact connected semigroups, Fund. Math. 50 (1961), 99–105

Compact semitopological semigroups

W. A. F. Ruppert

1. Semigroups with separately continuous multiplication

In topological algebra, the natural and straightforward way to define objects is to endow an algebraic object with a topology such that all operations involved are continuous, and it is in this obvious way that the important concepts of a topological group and of a topological semigroup are derived. However when it comes to applications often a broader concept is needed. For instance, in many situations we want to use a compact topology on a certain semigroup but cannot garantee that under this topology the multiplication of the semigroup will be jointly continuous. In particular, in contexts where operator semigroups act on a locally convex vector space one typically either has to use a non-compact topology or to abandon joint continuity.

We are therefore led to define *semitopological semigroups*, i.e., semigroups S where only the translations $x \mapsto sx$ and $x \mapsto xs$ are continuous (but multiplication as a map $S \times S \to S$ can be discontinuous). Let us look first at some examples.

Example 1.1. The most simple (but nonetheless very instructive) example of a semitopological semigroup is the one-point compactification $\mathbb{R}_\omega = \mathbb{R} \cup \{\omega\}$ of the reals, where ω acts as a zero element. Note that the multiplication of $\mathbb{R}_\omega$ is jointly continuous at all points of $\mathbb{R}_\omega \times \mathbb{R}_\omega$ *except* (ω, ω). Also, the induced action of the topological subsemigroup $\mathbb{R}_+ \cup \{\omega\}$ on the whole of $\mathbb{R}_\omega$ is separately, but not jointly continuous.

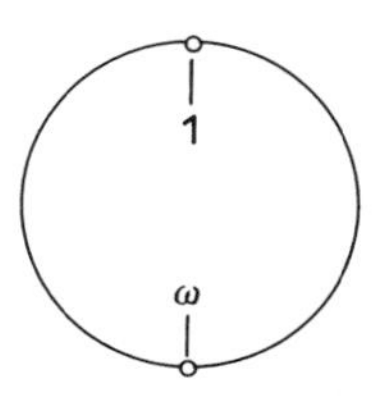

Fig. 1

(This example is universal in that it can be shown ([44]) that any separately, but not jointly continuous action of $\mathbb{R}_+ \cup \{\omega\}$ on a compact space X contains a subaction which maps homomorphically onto $(\mathbb{R}_\omega, \mathbb{R}_+ \cup \{\omega\})$.)

Example 1.2. The above construction can be carried through also for general locally compact topological groups G. Still more generally, if S_0 is a locally compact cancellative topological semigroup then the one-point compactification $S = S_0 \cup \{\omega\}$ becomes a compact semitopological semigroup if ω is defined as a zero element.

Example 1.3. The semigroups of type $P(g)$. Using examples 1.1 and 1.2 it is easy to find more non-group semitopological monoids on a manifold, e. g. the sphere $\mathbb{R}^2 \cup \{\omega\}$ and the direct products $\mathbb{R}^k_\omega$. Less trivial examples are the semigroups of type $P(g)$, which are exactly the semitopological monoids living on a compact surface of genus g. We do not give the details of their construction (which can be found in [43], chapter IV), but only list the essential properties. Let S be a semigroup of type $P(g)$. Then

(i) S is homeomorphic with a compact orientable surface of genus g;
(ii) S is a finite semilattice of groups; (that is, S is the union of finitely many maximal subgroups and the idempotents of S are central in S;)
(iii) the group of units $H(\mathbf{1})$ of S is dense in S and isomorphic with $\mathbb{R} \times \mathbb{R}$;
(iv) S has exactly $2g + 2$ idempotents, namely the identity $\mathbf{1}$, a zero element $\mathbf{0}$, and $2g$ idempotents $e_1, e_2, \ldots, e_{2g}$ with $e_i e_j = \mathbf{0}$ if $i \neq j$;
(v) every maximal subgroup $H(e)$ is open in the corresponding principal ideal eS.

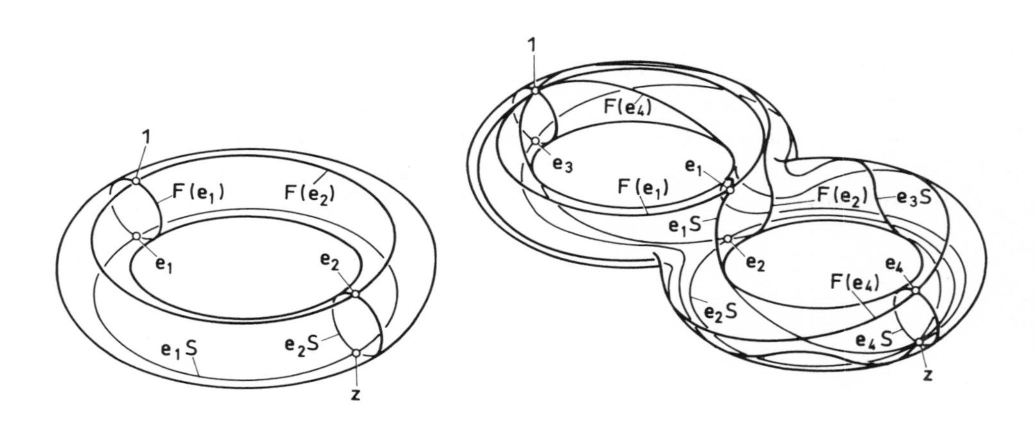

Fig. 2.

Figure 2 illustrates the cases $g = 1$ and $g = 2$. The semigroups of type $P(g)$ are typical examples of semitopological monoids on compact manifolds (cf. 4.3.1).

Example 1.4. The direct product $S = \prod_{i \in J} \mathbb{R}_\omega$, where J is any index set. Note that for infinite index sets J the group of units of S is not locally compact and every neighborhood of the identity contains card J idempotents.

Before presenting the next example we recall some definitions.

Definition 1.5. Let S be a semitopological semigroup and suppose that $f: S \to \mathbb{C}$ is a bounded continuous function on S. We write O^f for the set of all functions $S \to \mathbb{C}, x \mapsto f(sx); s \in S$. Then f is called

(i) *almost periodic* if O^f is precompact with respect to the norm topology on $\mathcal{C}(S)$;

(ii) *weakly almost periodic* if O^f is precompact with respect to the weak topology on $\mathcal{C}(S)$.

The almost periodic and the weakly almost periodic functions form norm-closed subalgebras of $\mathcal{C}(S)$.

Example 1.6. It is well known that the Bohr compactification G^b of a topological group G can be obtained as the Gelfand space of the C*-algebra of all almost periodic functions on G. If we replace the almost periodic by the weakly almost periodic functions then we also get a compactification of G; however the induced multiplication is in general only separately continuous. The "*weak almost periodic compactification*" (G^w, i_G) so obtained is universal among semitopological compactifications: If $f: G \to S$ is any continuous homomorphism into a compact semitopological semigroup then there is a morphism $f^w: G^w \to S$ such that $f = f^w \circ i_G$. This example of a compact semitopological semigroup actually was the starting point for all later research in this area. Obviously, the group G can be replaced by any semitopological semigroup S.

Example 1.7. Let V be an infinite dimensional Hilbert space and write S for the semigroup of all contractions on V, that is, $S = \{U \in \mathcal{B}(V) \mid \|U\| \leq 1\}$ (where $\mathcal{B}$ denotes the set of all bounded operators on V). Endow S with the topology of weak convergence. Then S is a compact semitopological semigroup with respect to composition. The subsemigroup of unitary operators on V is exactly the group of units $H(1)$ of S, and $H(1)$ is dense in S. The group $H(1)$ is a topological group which by an old result of DIEUDONNÉ cannot be embedded into a complete topological group. (The above properties of S are taken from HELMER's paper [28, 1983], which also contains further references.) The structure of this semigroup is of particular interest for the theory of unitary representations of topological groups and semigroups (cf. Section 5.2 on the Mautner phenomenon).

Example 1.8. Let G be a locally compact abelian topological group. Then the maximal ideal space Δ_G of the measure algebra $M(G)$ of G is a compact semitopological monoid with respect to the Gelfand topology. The group of units of Δ_G can be identified in a natural way with the dual group $\hat{G}$. Also, Δ_G can be represented as the set of all continuous homomorphisms $B \to D$ mapping a compact topological semigroup B, the so-called *structure semigroup of* G, into the unit disk $D = \{z \in \mathbb{C} \mid |z| \leq 1\}$ under multiplication. The semigroup Δ_G was studied extensively by J. L. TAYLOR in [51], [52], we also recommend the very readable short account in [20].

Example 1.9. Let μ be a measure on some measure space X. Then the unit ball $B = \{f \in L^\infty(X, \mu) \mid \|f\| \leq 1\}$ is weakly compact and forms a semigroup under the usual (pointwise) multiplication, which is weakly separately continuous. Morphisms into such compact semitopological semigroups are called semicharacters by DUNKL and RAMIREZ [24] and are the basis on which they build a Representation Theory for commutative semitopological semigroups. Note also that TAYLOR's work on the structure semigroup can be subsumed under this approach.

The above examples show the characteristic feature of the theory of semitopological semigroups: its close connection, one might say, its symbiotic existence, with "applied" subjects, such as harmonic analysis and representation theory. Much of the dynamics in its development is due to the permanent exchange of questions and answers between structural research and research in applied fields. In the beginning of the theory, forty years ago, compact semitopological semigroups were considered mainly as a device permitting the formulation and interpretation of results about weakly almost periodic functions. At that time some of the major structural results about semitopological compactifications could be obtained only by translating them into properties of weakly almost periodic functions. At present the intrinsic structure theory of compact semitopological semigroups has become a full-fledged theory of its own, which in turn contributes new insights and powerful proof techniques to various other fields of research in analysis.

2. A panoramic view of the subject

After the motivating examples of the first section let us now take an overall view of the scenery. The accompanying diagram groups the topics of the theory into three main branches: basic theory, structure theory, and applications. This division results from practical considerations: we do not claim to offer a neat partition of the field into tidily separated compartments. On the contrary, there is considerable overlap between different topics and many connecting wires run back and forth between them. Also, "application" is not to be understood as a "unidirectionally flow", conveying information from semitopological semigroup theory to some other field, but as a channel transferring results and problems in both directions. Because of their central importance, three of the topics in the diagram are given special prominence: Joint Continuity in the group of Basic tools, Compactifications in the Structure Theory group, and Representations in the group collecting applications. These topics also constitute the principal landmarks on which "Research and Development" in the theory is oriented.

Topics in the Theory of
Compact Semitopological Semigroups

Basic Tools	Structure Theory	Applications
Elementary Properties of Green's Relations	General Structure Theory	**Representations**
Grothendieck's Criterion	**Compactifications**	Periodicity Properties of Functions
Joint Continuity	Semigroups on Special Spaces	Group Topologies
Examples and General Constructions	Special Algebraic Properties	Combinatorics

While admiring the scenery let us also have a look at the guide-books describing it. The first comprehensive discussion of compact semitopological semigroups was given in 1961 by K. DeLeeuw and I. Glicksberg [18], [19], their presentation was centered on weakly almost periodic functions and the associated compactification. This pioneering work was followed in 1967 by the Lecture Notes [6] by

J. F. BERGLUND and K. H. HOFMANN, which combined a categorical treatment of the compactification machinery with basic structural assertions and elements from the theory of semigroups acting affinely on convex sets in Banach spaces.

A few years later, in 1971, C. F. DUNKL and D. E. RAMIREZ published their inspiring book [20] on Harmonic Analysis, which, among other things, contains a short introduction to TAYLOR's work (cf. Example 1.8). Later on, DUNKL and RAMIREZ developed the theory of L^∞-representations of commutative semitopological semigroups (cf. Ex. 1.9), their results are collected in [24], 1975. On a less profound level R. BURCKEL's book ([13], 1971) summarized the then known properties of weakly almost periodic functions.

The categorical study of the lattice of compactifications was pursued further by BERGLUND and his co-authors H. D. JUNGHENN and P. MILNES, in [7] (1978).

In 1974, J. D. LAWSON's paper on Joint continuity opened the door to the investigation of compact semitopological semigroups along strictly "structural" lines; in [43] (1984) the present author tried to give a fairly complete account of this approach. This book also contains a collection of 42 research problems, other sets of problems were published earlier by LAWSON [33] (1976), [34] (1983) and BERGLUND [3] (1980).

Right during the preparation of the present survey a further book by the three authors BERGLUND, JUNGHENN and MILNES came from the press ([8] 1989). This new monograph offers a general treatment of several important topics concerning analysis on semigroups (with one-sided as well as with two-sided continuity of the multiplication). Notably the authors dwell upon function spaces and the periodictiy behavior ("dynamical properties") of functions defined on semigroups, invariant means and weakly almost periodic semigroups of operators.

3. Basic tools

3.1. Green's relations

It is obvious that when starting the investigation of certain topologized algebraic objects one will first look at the algebraic theory, hoping to find useful cues for a suitable conceptual framework on which a vocabulary for structural descriptions can be based. On the other hand, some help is expected to be gained from the existing algebraic machinery. In the case of topologized semigroups (with varying degrees of continuity of the multiplication) the most valuable device inherited from the algebraic theory is the partition of the semigroup in question into "eggboxes", according to Green's relations. In particular this device is a very effective implement in unravelling the structure of the minimal ideal in a compact semigroup (the "First Fundamental Theorem"; cf. 4.1.1 below).

Other results following from elementary properties of Green's relations deal with the pre-orders $\leq_L$ and $\leq_R$, induced by Green's relations on the set of idempotents, and the existence of maximal and minimal idempotents. (For any two idempotents e, f the relation $e \leq_L f$ means $ef = e$, and similarly $e \leq_R f$ means $fe = e$.) Also, if the idempotents in a compact semitopological semigroup commute then they form a complete meet-continuous semilattice ([43], I.2.13).

3.2. Grothendieck's criterion

The basic device for the transfer of problems and ideas between the structural theory of compact semitopological semigroups and applications in harmonic analysis is GROTHENDIECK's double limit criterion [26]. This criterion is also a standard tool for the construction of examples of weakly almost periodic functions.

Its outstanding importance is due mostly to the fact that it allows to replace general nets by sequences; basically it makes "countable arguments" and "countable constructions" available to the general theory of compact semitopological groups.

Grothendieck's Criterion 3.2.1. *Let $f: S \to \mathbb{C}$ be a continuous function on a semitopological semigroup. Then the following assertions are equivalent:*
(i) *f is weakly almost periodic on S.*
(ii) *Let $\langle x_m \mid m \in \mathbb{N} \rangle$ and $\langle y_n \mid n \in \mathbb{N} \rangle$ be two sequences in S. Then the iterated limits $\lim_m (\lim_n f(x_m y_n))$ and $\lim_n (\lim_m f(x_m y_n))$ are equal whenever both of them exist.*

Problem 1. Find other (non-trivial) necessary and sufficient conditions for a bounded function on $\mathbb{N}$ or $\mathbb{Z}$ (or on similar semigroups) to be weakly almost periodic.
(J. F. Berglund)

Combinatorial conditions of the above type are available for finitely-valued functions on discrete semigroups, by a suitable adaption of the results in Theorem 5.4.1 below (cf. also the assertions in Theorem 4.2.13 which imply that on certain connected locally compact groups every weakly almost periodic function must be the sum of a constant with a function vanishing at infinity). BERGLUND and HINDMAN [5] list several conditions which are either necessary or sufficient, but none of them is both.

3.3. Joint continuity

In dealing with knotty problems concerning the structure of semitopological semigroups we fortunately have one major aid at our disposal: the Joint Continuity Theorem. Actually, this theorem has developed into a theory of its own, laying in our hands a bundle of "magic wands" which can be used comfortably in many different situations. Although it provides important structural information as well we include it here because of its character as a central tool in the theory.

The most widely employed Joint Continuity assertion says that, under fairly general assumptions, in a separately continuous action at least the units must act in a jointly continuous way.

The Joint Continuity Theorem of Ellis-Lawson 3.3.1. *Let S be a locally compact semitopological semigroup with identity* **1** *and*

$$\pi: X \times S \to X, (x, s) \mapsto x \cdot s,$$

a separately continuous action of S on a compact space X with $x \cdot \mathbf{1} = x$ for all $x \in X$. Then the action π is jointly continuous at every point (x_0, h_0) with $h_0 \in H(\mathbf{1}), x_0 \in X$.

This result has a long history. Its early precursors were assertions securing "categorically many" points of joint continuity for separately continuous maps on metrizable spaces— some of these statements go back to René Baire himself. In case that S is a group it already suffices to show that for every $x \in X$ the action is jointly continuous at *one* of the points (x, g), $g \in S$; the rest of the theorem can then be inferred by applying suitable translations which "transport" the given continuity point (x, g) to any other point (x, h) (cf. [43], chapter II). Using projective limits and the above transport principle Robert Ellis ([25], 1957) extended the original Baire-type arguments to the case where X is an arbitrary compact space and S is a separable locally compact (semitopological) group; he then applied a result in Weil's book [57] (which allows the deduction of compactness from countable compactness in spaces of continuous selfmaps), to get the above theorem for actions of general locally compact groups. No further significant progress was made until 1973, when Jimmie D. Lawson [31] extended the Transport Principle in an ingenious way to the case where the units do not form an open subsemigroup of S. Later on Isaku Namioka ([39]) found a simplified and generalized formulation of the process involving the transition from the separable case to the general one; this formulation uses a non-Hausdorff version of Baire's Theorem. In [16] J. P. R. Christensen generalized Namioka's results to phase spaces X which belong to the larger class of "α-favorable spaces" (a special type of Baire spaces, cf. also [8]).

Many other variants and proofs of Joint Continuity assertions (also for non-compact phase spaces and non-locally compact semigroups) were added by Lawson himself [32], [34], [35] and several other authors. A very promising line of research branching off from the basic papers by Ellis and Lawson is the "approche nouvelle" to weak almost periodicity (as well as to general compact semitopological semigroups) which was initiated by Jean Pierre Troallic and his co-workers. This approach combines the above mentioned ideas with convexity arguments, by passing to appropriate affine actions on convex sets. We especially refer to TROALLIC's paper [53] and his survey article in this volume [54].

By courtesy of the Theorem of Ellis-Lawson we have the following two fundamental structural assertions [31].

Theorem 3.3.2. (i) *Every maximal subgroup $H(e)$ of a compact semitopological semigroup is a topological group which is complete with respect to the two-sided uniformity.*
(ii) *Every compact semitopological semilattice is a topological semigroup.*

Note that assertion 3.3.2(i) only states that $H(e)$ is complete *with respect to the two-sided uniformity* (i.e. the join of the left and the right uniformity). In general, $H(e)$ need not be complete with respect to a one-sided uniformity, as is shown by Example 1.7 above.

The reader is also warned that, in contrast to group actions, a separately continuous action of a compact topological semilattice on a compact space need not be jointly continuous (cf. [43], p. 239; it must be jointly continuous, however, if S is a compact totally ordered space under the min-multiplication).

We illustrate the use of the Joint Continuity Theorem 3.3.1 by proving the following Corollary (its abelian version was given in [44]; for the general version compare TROALLIC's contribution to this volume, [54], 4.3(1)).

Corollary 3.3.3. *Let S be a compact semitopological semigroup and suppose that s is an element of S with $s \in sS \cap Ss$. Furthermore, suppose that $\pi: X \times S \to X$, $(a,b) \mapsto a.b$, is a separately continuous action of S on a compact space X. Then the restriction of π to $X \times (sS \cap Ss)$ is jointly continuous at every point (x, s) with $x \in X$.*

Proof. Let $\langle x_n \rangle$, $\langle s_n \rangle$ be nets in X, $sS \cap Ss$, respectively, with $\lim x_n = x$, $\lim s_n = s$. We may assume that $s_n \neq s$ for all n, so that we can find elements t_n, t'_n with $s_n = st_n = t'_n s$. Passing over to suitable subnets we enforce that the limits $z = \lim x_n.s_n$, $z' = \lim x_n.t'_n$, $t = \lim t_n$, exist. Then $s = st = t's$ and $z = z'.s \in X.s$. Let e be a minimal idempotent in the compact semigroup $\{u \in S \mid su = s\}$. Then ete is a unit in eSe, hence by 3.3.1 we have

$$\begin{aligned}\lim x_n.s_n &= z = z'.s = z'.se = z.e = \lim x_n.s_n e = \lim x_n{\cdot}st_n e \\ &= \lim(x_n.se).(et_n e) = (\lim x_n.se).(\lim et_n e) = (x.se).ete = x.s\end{aligned}$$

and the assertion follows. □

Transport techniques often also work in a non-compact setting, yielding statements like the following.

Corollary 3.3.4. *If H is a closed submonoid of a locally compact group G and has dense interior in G then every separately continuous action of H on a compact space X is jointly continuous.*
(Cf. [34] for a much more general, but also more technical version of this statement.)

The second of the two central Joint Continuity results essentially asserts that for separately continuous actions on compact spaces the units act equicontinuously on every subset on which they act minimally. This result forms the basis for a strictly "structural" approach to the famous Theorem of Ryll-Nardzewski on fixed points of distal actions; such an approach was given first in connection with the above-cited "approche nouvelle" by J. P. TROALLIC [53], [54]; results concerning semigroups with dense subsets of left (or right) units can be found in [27]. The proof in [43], chapter II, parallels Lawson's original proof of 3.3.1, replacing the Transport Principle for points of Joint Continuity by a Transport Principle for points of equicontinuity.

Theorem 3.3.5. *Let S be a compact semitopological semigroup with dense group of units and suppose that $\pi: X \times S \to X$ is a separately continuous action of S on a compact space X.*
(i) *If $x_0 \in X$ with $\pi(x_0, S) = X$ then π is jointly continuous at every point (x_0, s), $s \in S$.*
(ii) *If the action of S is minimal (i.e., $\pi(x, S) = X$ for every $x \in X$) then the maps $x \mapsto \pi(x, s)$; $s \in S$, form an equicontinuous family and hence π is actually a jointly continuous action.*

(For further Joint Continuity results and more references cf. [43,44,45].)

3.4. Examples and general constructions

Last but not least we want to emphasize the role of examples and general constructions as basic research tools. They are useful not only by helping our intuition and exhibiting limitations to possible generalizations, but often are also basic instruments in proofs.

For instance, in order to show that on a locally compact group every function vanishing at infinity is weakly almost periodic one only has to look at Example 1.2. To show that every weakly almost periodic function on a closed central subgroup Z of a locally compact topological group can be extended to a weakly almost periodic function on G the easiest way is to construct a semitopological compactification

(S, φ) of G such that the induced compactification $(\overline{\varphi(Z)}, \varphi|Z)$ is equivalent to the weak almost periodic compactification of Z.

In the description of universal semitopological compactifications, examples and the associated constructions are often the most adequate way of expressing mathematical insights—the bits and chunks of such structural information as we have.

4. Structure theory

4.1. General structure theory

Virtually the only structural feature compact semitopological and compact topological semigroups have in common is the existence and algebraic description of the minimal ideal; following the traditions of Topological Semigroup Theory this group of statements is usually summarized under the title "First Fundamental Theorem". Actually the essential parts of this Theorem also hold more generally for compact semigroups with only one-sided continuous multiplication. Its proof combines elementary properties of Green's relations with the fact that compact semigroups with one-sided continuous multiplication must contain idempotents as well as minimal left ideals (or minimal right ideals, depending on which side the multiplication is continuous).

In concise form the theorem reads as follows.

The First Fundamental Theorem 4.1.1. *Let S be a compact semigroup such that the translations $x \mapsto sx, (s \in S)$, are continuous. Then S contains a minimal ideal $M(S)$, which is a paragroup. In particular, $M(S)$ contains idempotents, and for each idempotent $e \in M(S)$ we have*

(i) *eS is a minimal right ideal, Se is a minimal left ideal;*

(ii) *$eS \cap Se = eSe$ is a group;*

(iii) *$e_1e_2 = e_1$ for any pair of idempotents e_1, e_2 in Se; similarly, $f_1f_2 = f_2$ for any pair of idempotents f_1, f_2 in eS.*

If S is a semitopological semigroup then every minimal left and every minimal right ideal (but not necessarily the minimal ideal) is closed. In the semitopological case the minimal ideal $M(S)$ is a topological semigroup if and only if $M(S)$ as well as the set of idempotents in $M(S)$ are closed (cf. [6], further details and bibliographical references are given in [43]).

The second result recorded in this subsection is a rather straightforward consequence of Theorem 3.3.5 above. Its assertion (iii) implies that, in particular, the minimal ideal of the weak almost periodic compactification of a group is a compact topological group. This special case is called the "Ryll-Nardzewski Theorem". The Ryll-Nardzewski Theorem was one of the first highlights of the theory, and several different proofs of it are available now. Apart from RYLL-NARDZEWSKI's original proof in [50] we want to mention the "geometric proof" by ASPLUND and NAMIOKA [1], Veech's "fixed-point free approach" [55], TROALLIC's "approche nouvelle" [53] and the author's proof in [43]. The proofs in [50] and [1] were formulated as fixed point theorems, while [55] uses properties of distal actions; these methods do not yield 4.1.2 in full generality.

Theorem 4.1.2. *Let S be a compact semitopological semigroup with dense group of units $H(\mathbf{1})$. Then the following assertions hold:*

(i) *For every $x \in S$ the map*

$$xS \times S \to S, (a, b) \mapsto ab,$$

is jointly continuous at all points (x, s), with $s \in S$.

(ii) *For every pair of idempotents e, f in S we have $ef = e$ if and only if $fe = e$. Thus the idempotents are partially ordered with respect to the relation $e \leq f \iff ef = e$. Every maximal chain K of idempotents in S is compact and the order topology on K coincides with the topology inherited from S.*

(iii) *The minimal ideal $M(S')$ of any compact subsemigroup S' of S is a compact topological group.*

Translated into the language of compactifications, assertion (iii) above amounts to the classical result that the minimal ideal of the weak almost periodic compactification of a topological group G is canonically isomorphic with the Bohr compactification of G (cf. [6], chap. III, or [43], p. 105).

4.2. Semitopological compactifications

Because of its universal applicability to structural questions we recorded Theorem 4.1.2 as belonging to the category "general structure theory". However from the conceptual point of view it could have been subsumed as well under the heading "Compactifications", our next theme.

Compactifications play a dominant role in the theory, and virtually all recent research efforts are in some way related to this central topic. One reason for this dominance is the historic origin of the theory, which can be traced back to two sources: the study of weakly almost periodic functions and the study of group actions on compact spaces. Both of these subjects are naturally connected with compactifications. The other reason lies in the fact that compactifications also arise as universal structure elements in the description of compact semitopological semigroups. For instance, the closure of the set $\{x^k \mid k = 1, 2, \ldots\}$ of all positive powers of an element x in a compact semitopological semigroup S amounts to a semitopological compactification of $\mathbb{N}$; similarly, the pointwise closure of the

Schuetzenberger group $G_{H(x)}$ of an $\mathcal{H}$-class $H(x)$ in S defines a semitopological compactification of $G_{H(x)}$.

Notation. Following the notation in Example 1.6 we henceforth write (S^w, i_S) for the *weak almost periodic compactification* of a semitopological semigroup S.

Weak almost periodic compactifications of discrete semigroups

Whereas in the topological theory of semigroups we have very detailed structural descriptions of the universal (=Bohr) compactifications of such household semigroups as the positive integers and the positive reals, cones in finite-dimensional vector spaces, or more generally, infinitesimally generated semigroups like "Heisenberg's Beak", the "Affine Triangle" and the "Whirlpot", life becomes much harder in the semitopological case. To begin with, troubles already arise with the weak almost periodic compactification of the positive integers, the semitopological analogue of the universal compact monothetic semigroup. Moreover, the known structural features of $\mathbb{N}^w$ and $\mathbb{Z}^w$ suggest that it is hopeless to search for "explicit constructions" like those available in the topological theory. Studying the Bohr compactification of semigroups like $\mathbb{N}$ is comparable to the disassembling and reassembling of a mechanical device, while studying the corresponding weak almost periodic compactifications is something like digging in a mine. Let us illustrate our point by collecting the major morsels of information we have.

Theorem 4.2.1. *We write $(\mathbb{N}^w, i_{\mathbb{N}})$ for the weak almost periodic compactification of $\mathbb{N}$ and $(\mathbb{Z}^w, i_{\mathbb{Z}})$ for the weak almost periodic compactification of $\mathbb{Z}$. The following assertions hold true.*

(i) a) *Denote with j^w the homomorphism $\mathbb{N}^w \to \mathbb{Z}^w$ canonically associated with the inclusion map $j: \mathbb{N} \to \mathbb{Z}$. Then the restriction of j^w to $E(\mathbb{N}^w)\mathbb{N}^w$ induces an algebraic isomorphism $E(\mathbb{N}^w)\mathbb{N}^w \to \big(E(\mathbb{Z}^w) \setminus \{\mathbf{1}\}\big)\mathbb{Z}^w$. In particular, j^w gives rise to an algebraic isomorphism between the semilattice $E(\mathbb{N}^w)$ of idempotents in $\mathbb{N}^w$ and the semilattice of non-identity idempotents in $\mathbb{Z}^w$.*
b) *However j^w is not injective on the whole of $\mathbb{N}^w$.*

(ii) a) *The minimal ideals of $\mathbb{Z}^w$ and $\mathbb{N}^w$ are canonically isomorphic with the Bohr compactification $\mathbb{Z}^b$.*
b) *The minimal ideals of $\mathbb{Z}^w$ and $\mathbb{N}^w$ are not prime ideals. In fact, there are non-minimal idempotents $e, f \in E(\mathbb{Z}^w) \setminus \{\mathbf{1}\} \simeq E(\mathbb{N}^w)$ whose product is the minimal idempotent in $E(\mathbb{Z}^w)$.*

(iii) a) *The semilattice $E(\mathbb{Z}^w)$ is complete and meet-continuous.*
b) *The semilattice $E(\mathbb{Z}^w)$ contains a subsemilattice which is isomorphic with the semilattice $\{0,1\}^{2^{\aleph_0}}$.*

(iv) *$\mathbb{Z}^w$ contains an algebraic copy of the free abelian semigroup in $2^{\aleph_0}$ generators which is discrete in itself.*

(v) *$\mathbb{Z}^w$ contains a homeomorphic image of the Stone-Čech compactification $\beta\mathbb{N}$.*

Part a) of assertion (i) will be a consequence of the remarks in 4.2.3 below; part b) is a result of BERGLUND and HINDMAN [5]. Statement a) of (ii) is a consequence of 4.1.2 (iii), while (ii) b) is the answer to a question by J. BAKER (a proof is given in [49]). Part a) of (iii) follows from the general observations recorded in the last remarks of 3.1, part b) of (iii) goes back to T. WEST [58] (1968) and to the subsequent paper by BROWN and MORAN [11] (1971; cf. also [12]). In fact, West produced an example of a semitopological compactification of $\mathbb{N}$ with at least two idempotents; Brown and Moran showed that this same example actually contains uncountably many idempotents and gave a variant of West's construction which yields examples with exactly two idempotents. For assertion (iv), as well as for a "structural" proof of (iii) b), we refer to the material in Section 5.2 below. Assertion (v) was known already to WALTER RUDIN ([41], 1959) and D. E. RAMIREZ (1968, [40]); cf. Theorem 5.4.2 below.

Proposition 4.2.2 (cf. [49], section 4). *Let $\varphi\colon \mathbb{N} \to S$ be a homomorphism into a compact semitopological semigroup S. If $\{\varphi(1)\}$ is not open in $\overline{\varphi(\mathbb{N})}$ then there exists a homomorphism $\psi\colon \mathbb{Z} \to S$ which extends φ, i.e., $\psi(n) = \varphi(n)$ for all $n \in \mathbb{N}$.*

Remarks. The following discussion provides a proof of 4.2.1(i), part a).

(i) If e is an idempotent in $\mathbb{N}^w$ then $ei_{\mathbb{N}}(1)$ cannot be open in $ei_{\mathbb{N}}(\mathbb{N})$. To see this, take a net $\langle n_\alpha \rangle$ in $\mathbb{N}$ with $\lim i_{\mathbb{N}}(n_\alpha) = e$. The map $\mathbb{N} \to e\mathbb{N}^w, x \mapsto ei_{\mathbb{N}}(x+1)$ is injective since otherwise $e\mathbb{N}^w = \overline{ei_{\mathbb{N}}(\mathbb{N})}$ would be finite and hence the minimal ideal of $\mathbb{N}^w$ would be a finite group, a contradiction to 4.2.1(ii)a). It follows that the net $\langle ei_{\mathbb{N}}(n_\alpha + 1) \rangle$ cannot be eventually constant. Hence its limit $ei_{\mathbb{N}}(1)$ cannot form an open singleton in $ei_{\mathbb{N}}(\mathbb{N})$.

(ii) Combining observation (i) with 4.2.2 and the unversality of $\mathbb{Z}^w$, we find to every idempotent $e \in \mathbb{Z}^w$ a morphism $k_e\colon \mathbb{Z}^w \to e\mathbb{N}^w$ such that $k_e(i_{\mathbb{Z}}(n)) = ei_{\mathbb{N}}(n)$, for every $n \in \mathbb{N}$. Let $\langle n_\alpha \rangle$ be a net in $\mathbb{N}$ with $e = \lim i_{\mathbb{N}}(n_\alpha)$. Then $k_e(j^w(ei_{\mathbb{N}}(n))) = \lim k_e(i_{\mathbb{Z}}(n_\alpha + n)) = \lim ei_{\mathbb{N}}(n_\alpha + n) = ei_{\mathbb{N}}(n)$ for all $n \in \mathbb{N}$, and therefore $k_e(j^w(es)) = es$ for all $s \in \mathbb{N}^w$.

(iii) We now show that the restriction of j^w to $E(\mathbb{N}^w)\mathbb{N}^w$ is injective. Suppose that $s, s' \in E(\mathbb{N}^w)\mathbb{N}^w$ with $j^w(s) = j^w(s')$, and let e, e' be the minimal idempotents of the semigroups $\{t \in \mathbb{N}^w | ts = s\}$, $\{t' \in \mathbb{N}^w | t's' = s'\}$, respectively. Then by (ii) $es' = k_e(j^w(es')) = k_e(j^w(e)j^w(s')) = k_e(j^w(e)j^w(s)) = k_e(j^w(es)) = es = s$. In the same fashion we see that $e's = s'$. Thus $s = es' = ee's$ and $s' = ee's'$. By the minimality of e, e' this implies $e = ee' = e'$, hence $s = es' = e's' = s'$.

(iv) Finally we show that $j^w(E(\mathbb{N}^w)\mathbb{N}^w) = (E(\mathbb{Z}^w) \setminus \{\mathbf{1}\})\mathbb{Z}^w$. To this end it suffices to prove that $fi_{\mathbb{Z}}(z)$ is contained in the closure of $i_{\mathbb{Z}}(\mathbb{N})$ for every idempotent $f \in \mathbb{Z}^w \setminus \{\mathbf{1}\}$ and every $z \in \mathbb{Z}$. Suppose that f is the limit of a net $\langle i_{\mathbb{Z}}(z_\beta) \rangle$ with $z_\beta \in \mathbb{Z}$ and $\lim |z_\beta| = \infty$. Then $fi_{\mathbb{Z}}(z) = \lim i_{\mathbb{Z}}(z_\beta + z)$. By passing to a suitable subnet we enforce that the limit $s = \lim i_{\mathbb{Z}}(-z_\beta)$ exists and that all of the integers z_β, $z_\beta + z$ have the same sign. Obviously, we may suppose that z_β, $z_\beta + z \in -\mathbb{N}$ for every β. By the Joint Continuity Theorem of Ellis-Lawson (3.3.1) we have $fs = \lim f i_{\mathbb{Z}}(z_\beta) i_{\mathbb{Z}}(-z_\beta) = f$. Let U be an open subset of $\mathbb{Z}^w$ with $f i_{\mathbb{Z}}(z) \in U$. Then $s i_{\mathbb{Z}}(z_{\beta_0} + z) \in U$ for some index β_0 and therefore $i_{\mathbb{Z}}(-z_\beta + z_{\beta_0} + z) \in U$ for all large indexes β. Since $-z_\beta \to \infty$ this implies that U must meet $i_{\mathbb{Z}}(\mathbb{N})$, which establishes our claim.

More Remarks. (v) It is not known whether the above algebraic embedding $j^w \mid E(\mathbb{N}^w)N^w$ is a *homeomorphic* embedding (i.e., has a continuous inverse). Our discussion only shows that the restriction of j^w to any closed subsemigroup of $E(\mathbb{N}^w)N^w$ is also a homeomorphic embedding.

(vi) The idempotent semigroup $E(\mathbb{N}^w) \simeq E(\mathbb{Z}^w) \setminus \{\mathbf{1}\}$ can be described also in terms of group topologies on $\mathbb{Z}$, cf. sect. 5.2 below.

(vii) There are a few cases where we can explicitely compute the weak almost periodic compactification S^w of a discrete semigroup S.

For instance, it is easy to construct the weak almost periodic compactifications of left or right zero semigroups and of semigroups with constant multiplication ($xy = \mathbf{0}$ for all $x, y \in S$).

If S is a totally ordered set under the max-multiplication (and with the discrete topology), then S^w must be a chain of idempotents and hence is a topological semigroup (cf. 3.3.2(ii) above). Moreover S^w can be represented as $(U \cup L)/\sim$, where U denotes the set of all upper subsets and L the set of all lower subsets of S, $U \cup L$ is ordered by the naturally induced order relation ($a \leq b$ in each of the following cases a)–d): a) $a \in L$, $b \in U$, $a \cap b = \emptyset$; b) $a \in L$, $b \in L$, $a \subseteq b$; c) $a \in U$, $b \in L$, $a \cap b \neq \emptyset$; d) $a \in U$, $b \in U$, $b \subseteq a$), and $\sim$ identifies for every $s \in S$ the elements $\uparrow s = \{t \in S \mid s \leq t\}$ and $\downarrow s = \{t \in S \mid t \leq s\}$ in $U \cup L$; the topology of S^w is the usual order topology.

Using the above observation it is also not difficult to describe the structure of S^w *in terms of the weak almost periodic compactification of* $\mathbb{N}$ if S is, say, the semidirect product of a certain chain of idempotents with $\mathbb{N}$ (notably if S is the bicyclic semigroup $(\mathbb{N}, \wedge) \rtimes (\mathbb{N}, +)$).

Problem 2. We noted above that the inclusion $j: \mathbb{N} \to \mathbb{Z}$ induces a morphism $j^w: \mathbb{N}^w \to \mathbb{Z}^w$ which is not injective. On the other hand, Proposition 4.2.2 shows that the restriction of j^w to $E(\mathbb{N}^w)\mathbb{N}^w$ is an injection. Is it true that j^w is injective on $\left(\mathbb{N}^w \setminus i_w(\mathbb{N})\right)^2$? More generally, describe the kernel congruence of j^w. (Or equivalently, characterize the weakly almost periodic functions on $\mathbb{N}$ which can be extended to weakly almost periodic functions on $\mathbb{Z}$. Cf. also section 5.4 below and [5]).

Problem 3. Is one of the idempotent sets $E(\mathbb{Z}^w)$, $E(\mathbb{N}^w)$ compact?

As we shall see in section 5.3 below, there is a Galois correspondence between the set of idempotents in $\mathbb{Z}^w$ and the set of group topologies on $\mathbb{Z}$. Thus Problem 3 is closely connected with the following.

Problem 4. Is it true that for every group topology τ on $\mathbb{Z}$ the Hausdorff completion $\mathbb{Z}_\tau$ of $(\mathbb{Z}, \tau)$ can be embedded into a compact semitopological semigroup?

Semitopological compactifications of connected groups

Let us now turn to semitopological compactifications of *connected* locally compact groups G. Here the salient fact is that the semitopological compactifications of such a group G must be "nearly abelian", and that all "structural complexity" can be described in terms of compactifications of abelian groups. The following theorem comprises the most remarkable features, more details and proofs for the general case can be found in [43], chapter III.

Theorem 4.2.3. *Let (S, φ) be a semitoplogical compactification of a locally compact connected topological group G. Then the following statements hold:*
(i) $s\varphi(G) = \varphi(G)s$ *(hence $sS = Ss$) for all $s \in S$.*
(ii) *Every idempotent in S is central.*
(iii) *The idempotents of S form a complete meet-continuous lattice.*
(iv) *For every element $s \in S$ the left and right isotropy groups coincide,* $\{g \in G \mid s\varphi(g) = s\} = \{g \in G \mid \varphi(g)s = s\}$.

Remarks. (i) Note that 4.2.3(i) implies in particular that in S all of Green's relations coincide, since every left or right ideal in S must be an ideal.

(ii) Actually, a little bit more is true: If G is a connected subgroup of *any* locally compact topological group H and $\alpha: H \to S$ is a continuous homomorphism into a compact semitopological semigroup S then $\alpha(G)s = s\alpha(G)$ for all $s \in S$.

(iii) Assertion 4.2.3(i) also shows that the isotropy groups are normal in G. Indeed, if $s\varphi(n) = s$ and $g \in G$ then $s\varphi(g) = \varphi(g_*)s$ for some $g_* \in G$ and hence $s\varphi(gng^{-1}) = \varphi(g_*)s\varphi(n)\varphi(g^{-1}) = \varphi(g_*)s\varphi(g^{-1}) = s\varphi(gg^{-1}) = s$.

(iv) Theorem 4.2.3 does not hold for non-connected locally compact groups. If G is discrete and contains an infinite subgroup H and an element g such that $g^{-1}Hg \cap H$ is finite then the minimal idempotent e of $\overline{\varphi(H)}$ does not commute with $\varphi(g)$ (cf. [43], p. 141).

There are also results which ensure that certain isotropy groups are non-trivial. In computing weak almost periodic compactifications such results are always most welcome. Since every connected locally compact group is the projective limit of Lie groups, we may restrict our attention to the case where G is a connected Lie group. The following Proposition is not only an assertion about isotropy groups, it is also the main argument in the proof of Theorem 4.2.3, as given in [43], p.133ff.

Proposition 4.2.4. *Let (S, φ) be a semitopological compactification of a connected Lie group G with Lie algebra $\mathfrak{g}$ and let $\langle g_n \rangle$ be a net in G such that $\langle \varphi(g_n) \rangle$ converges to some $s \in S$. We write $\mathfrak{f}(s)$ for the ideal in $\mathfrak{g}$ which corresponds to the isotropy group $F(s)$ of s. In order to simplify notation we endow $\mathfrak{g}$ with a Euclidean norm $\|.\|$.*

(i) *Suppose that for some $X \in \mathfrak{g}$ we have $\lim g_n.X = 0$, where "." denotes the adjoint action of G on $\mathfrak{g}$. Then $X \in \mathfrak{f}(s)$.*
(ii) *If $\lim \|g_n.X\| = \infty$ for some $X \in \mathfrak{g}$ then $\mathfrak{f}(s) \neq \{0\}$.*

For the next Corollary recall that a vector $X \in \mathfrak{g}$ is called ad-compact if $\operatorname{ad} X$ is a semisimple operator with purely imaginary spectrum. This Corollary will be needed in the section concerned with the Mautner Phenomenon.

Corollary 4.2.5. *In 4.2.4, pick an element $X \in \mathfrak{g}$ and take for s the minimal idempotent e in the closure of $\varphi(\exp \mathbb{R}\cdot X)$. Then the quotient morphism $G \to G/F(e)$ induces a quotient morphism $q: \mathfrak{g} \to \mathfrak{g}/\mathfrak{f}(e)$ between Lie algebras; this morphism maps X onto an ad-compact element.*

Proof. Let f be the minimal idempotent in the closure of $\varphi(F(e))$. Then $fe = e$ and the closure N of $f\varphi\big(F(e)\big)$ is a compact normal subgroup of the maximal subgroup $H(f)$. We form the quotient $S_1 = fS/N$ (in the same way as with groups) and define $\varphi_1: G/F(e) \to S_1, g \mapsto f\varphi(g)N$. The pair (S_1, φ_1) is a semitopological compactification of $G/F(e)$; by construction, the isotropy group of the idempotent eN (with respect to (S_1, φ_1)) is trivial. Thus it suffices to show the assertion under the additional assumption that $F(e) = \{1\}$. (Note that the assertion is obvious if $X \in \mathfrak{f}(e)$.)

Now let $\langle t_n | n \in D \rangle$ be an unbounded net of real numbers such that the net $\langle \varphi(\exp t_n \cdot X) \rangle$ converges to e. From Proposition 4.2.4 we infer that for every $Y \in \mathfrak{g}$ there is a subnet $\langle \exp t_m \cdot X \rangle$ of $\langle \exp t_n \cdot X \rangle$ such that $\lim(\exp t_m \cdot X).Y$ exists. Since $\mathfrak{g}$ is finite-dimensional, we therefore may suppose that the limit of the operators $\exp t_n \cdot \operatorname{ad} X: Y \mapsto (\exp t_n \cdot X).Y$ exists. But this means that the set of operators $\exp t \cdot \operatorname{ad} X$, $t \in \mathbb{R}$ is contained in a torus subgroup of $\mathrm{Gl}(\mathfrak{g})$ (otherwise, by the well-known result of ANDRÉ WEIL about locally compact solenoidal groups, $\exp \mathbb{R} \cdot \operatorname{ad} X$ would be an isomorphic copy of the reals). The assertion follows. □

Although at present we still lack a systematic and complete description in the general case, for many important types of connected Lie groups G the available information already provides all essential ingredients for the explicit construction of G^w *in terms of weak almost periodic compactifications of abelian groups.* To keep within the limits of this article we only present some carefully selected examples.

Example 4.2.6. Let G be the "ax+b"-group, $G = (\mathbb{R}, +) \rtimes (\mathbb{R}_+, \cdot)$. Then $N \stackrel{\text{def}}{=} (\mathbb{R}, +) \times \{1\}$ is a closed normal subgroup of G and $\overline{i_G(N)} = i_G(N) \cup \{e\}$, where e

is an idempotent with $G^w \backslash i_G(G) = eG^w$. Moreover, eG^w is isomorphic with the weak almost periodic compactification of $\mathbb{R}$. (More details, especially about the way the semigroup eG^w is embedded in G^w, can be found in [43], p. 121.)

Example 4.2.7. Let G be the Heisenberg group, written as the set of all triples $(a, b, c) \in \mathbb{R}^3$, endowed with the sandwich multiplication $(u, v, w)(x, y, z) = (u + x, v + wx + y, w + z)$. Denote the center $\{(0, c, 0) \mid c \in \mathbb{R}\}$ by Z. Then G^w is the union $i_G(G) \cup \overline{i_G(Z)} i_G(G) \cup S_1$, where S_1 is isomorphic with the weak almost periodic compactification of $\mathbb{R}^2$ (cf. [43], p. 147).

Example 4.2.8. Suppose that G is the semidirect product $G = W \rtimes K$, where W is a vector group and K is a compact connected Lie group which acts irreducibly on W. Then G^w is the union $i_G(G) \cup M(G^w)$, where $M(G^w)$ is the minimal ideal of G^w. Moreover, we have $\overline{i_G(W)} = i_G(W) \cup \{m\}$ and $M(G^w) = m i_G(K) \simeq K$, where m is the minimal idempotent of G^w.

(The first examples of this kind were studied by Chou [14]; the more general case, where K is not necessarily compact, is treated in [43], p. 145ff. It is also shown in [43] that for connected non-compact locally compact topological groups G the formula $G^w = i_G(G) \cup M(G^w)$ holds if and only if G contains a compact normal subgroup N such that G/N is either a simple Lie group with finite center or isomorphic with a group $V \rtimes K$ as described above.)

Example 4.2.9. Let G be a non-compact connected simple Lie group with finite center. Then G^w is isomorphic with the one-point compactification $G \cup \{\mathbf{0}\}$, with a zero element added, as described in Example 1.2.

Example 4.2.10. Suppose that in Example 4.2.9 the center Z of G is infinite. Then $G^w = \overline{i_G(Z)} i_G(G) \cup \{\mathbf{0}\}$, where $\overline{i_G(Z)}$ is isomorphic with the weak almost periodic compactification of Z. (An inspection of a table of simple Lie groups shows that Z must be isomorphic to a direct product $\mathbb{Z} \times \mathbb{Z}/m\mathbb{Z}$ of the integers with some finite cyclic group.)

Example 4.2.11. If G is a connected semisimple Lie group with finite center then G^w can be described as follows: Being a connected semisimple Lie group, G can be written as the quotient $G = G_1 \times G_2 \times \cdots \times G_n / N$, where each G_i is a simple and simply connected Lie group and N is a (necessarily discrete) central subgroup of the direct product $G_1 \times G_2 \times \cdots \times G_n$. For every $i \in \{1, 2, \ldots, n\}$ let G_i^* be the weak almost periodic compactification of G_i (so $G_i^* = G_i$ if G_i is compact and $G_i^* = G_i \cup \{\mathbf{0}\}$ otherwise). Then G^w is isomorphic with the quotient $G_1^* \times G_2^* \times \cdots \times G_n^* / N$ (where as before "$/N$" means the congruence $x \sim y \iff xN = yN$).

The above computation of G^w was given first by Veech [56]. It is also not difficult to give similar descriptions in the case where the center of G is not assumed to be finite; cf. [43], p. 142.

There are several interesting characterizations showing that if the weak almost periodic compactification of G has a "nice" or "simple" structure then it must be highly "non-abelian." Let us associate a kind of mental image with this phenomenon, by way of physical analogy. According to this analogy, abelian groups G are "soft" and bendable, thus are fitted easily into intricate compact shapes. Compared with abelian G's, non-abelian ones are somehow "stiff;" they cannot be bent freely in all directions, hence are not embeddable into sophisticated compact structures. In fact, if G is not abelian, then there are non-trivial inner automorphisms which must extend to G^w as well, and since G is assumed to be connected this restricts the possible ways a compact semitopological semigroup can be built around G.

Theorem 4.2.12. *Let G be a connected locally compact group. Then the following assertions are equivalent:*

(i) *G^w is a semilattice of groups.*
(ii) *G^w has finitely many idempotents.*
(iii) *G^w is a regular semigroup.*
(iv) *If N is a closed normal subgroup of G then the center of G/N is compact.*
(v) *G is productible—that is, if S is a semitopological semigroup then the weak almost periodic compactification of $G \times S$ is naturally isomorphic with the direct product $G^w \times S^w$.*
(vi) *If G_1 is a locally compact topological group and $\alpha: G \to G_1$ a continuous homomorphism then $\alpha(G)$ is closed in G_1.*

Some of these equivalences probably also hold for non-connected locally compact groups G; the discrete case is mentioned in subsection 5.4 below. A proof for the above Theorem (and some additional material) can be found in [43], III, section 6, and [47]. (cf. also Chou [14], as well as [30] and the bibliography given in [43].) The following result of [43] is an extension of Veech's result on simple Lie groups with finite center.

Theorem 4.2.13. *Let G be a non-compact connected locally compact topological group. Then G^w is a group with zero* (as in Example 1.2) *if and only if the center of G is compact and there is a connected simple Lie group H and a morphism $\alpha: H \to G$ with $\overline{\alpha(H)} = G$.*

Conjecture 4.2.14. Theorem 4.2.13 remains true if we drop the assumption that G is connected.

It was shown in [46] that if G is discrete and non-finite then G^w cannot be a regular semigroup (cf. section 5.4 below). It is to be expected that the arguments given there can be extended to show the following conjecture.

Conjecture 4.2.15. Let G be a locally compact topological group. If G^w is a regular semigroup then G must be connected.

4.3. Semitopological semigroups on special spaces

In the early days of topological semigroup theory the following two types of problems were central to most investigations:
(1) Given a class $\mathcal{C}$ of topological spaces (say intervals, or manifolds, or n-cells), classify—or at least describe—all topological semigroups (or more specifically all monoids, or all semigroups S with $S^2 = S$) which can be defined on a member of this class $\mathcal{C}$.
(2) Given a class $\mathcal{S}$ of topological semigroups (say monoids, or semilattices, or divisible semigroups), classify—or at least describe—all topological spaces on which a semigroup of class $\mathcal{S}$ can be defined (or cannot be defined).

In a nutshell this is the program A. D. WALLACE presented in his 1953 address to the American Mathematical Society, which is still one of the major sources of research problems. Needless to say that this program still makes sense if the term "topological semigroup" is replaced throughout by "semitopological semigroup." It is a "natural" problem in the semitopological context too, but in the beginning it was not evident whether there was any chance at all to attack it successfully, even under restrictive additional assumptions. In order to get substantial results we either have to find very "elementary" arguments paralleling existing proofs in the topological theory or to devise new methods.

The first alternative has proved to be viable in the case of semigroups on particularly nice spaces. For instance, careful analysis has shown that the structure theory of interval semigroups actually can be developed equally well and with virtually the same effort for semitopological semigroups, and yields the same results. Thus every semitopological interval semigroup S with $S^2 = S$ is a topological semigroup (cf. BERGLUND's paper [4] and the more general treatment in [43], section II.5). Similarly, it was shown with elementary arguments that no semitopological monoid can be defined on the topological sine $\{(x,y) \in \mathbb{R}^2 |$ either $0 < x \leq 1$, and $y = \sin \frac{1}{x}$ or $x = 0$ and $-1 \leq y \leq 1\}$ ([42]).

The second alternative was adopted in investigating the structure of semitopological monoids on compact connected manifolds. This subject has no parallel in the topological case: in any topological semigroup S on a manifold the group of units must be open and hence must be all of S if S is compact connected. In the semitopological case, the Joint Continuity Theorem of Ellis-Lawson still implies that the group of units is open, but this does not mean that S must be a group, since maximal subgroups of compact semitopological semigroups need not be closed. In the next theorem we list the most important structure assertions about such semigroups; a detailed exposition of the subject is given in chapter IV of [43]. A low-dimensional illustration of the situation is provided by Example 1.3.

Theorem 4.3.1. *Let S be a compact semitopological semigroup with identity $\mathbf{1}$ and assume that S is embedded in a compact connected manifold E such that S is a neighborhood of $\mathbf{1}$ in E. If S is not a group then the following statements hold:*

(i) *$S = E$ and S is an orientable manifold with even Euler-Poincaré characteristic.*

(ii) *The group of units $H(\mathbf{1})$ is open and dense in S.*

(iii) *$H(\mathbf{1})$ contains a closed normal subgroup G, which is homeomorphic, but not necessarily isomorphic, with some Euclidean space $\mathbb{R}^n$, and a compact subgroup K such that S is isomorphic with the semidirect product $\overline{G} \rtimes K$, with multiplication $(a,b)(c,d) = (abcb^{-1}, bd)$.*

(iv) *S has only finitely many idempotents and the number of idempotents is even. More specifically, the number of idempotents can be computed in terms of the Betti numbers $\beta_q(\overline{G})$:*

$$\operatorname{card}\{e^2 = e \in S \mid \dim eG = n - q\} = \beta_q(\overline{G}) = \\ = \beta_{n-q}(\overline{G}) = \operatorname{card}\{e^2 = e \in S \mid \dim eG = q\}.$$

(v) *Every idempotent $e \in S$ is contained in the closure of its isotropy group $F(e) = \{g \in H(\mathbf{1}) \mid eg = e\}$ and the maximal subgroup $H(e)$ is open in eS. Furthermore, the isotropy groups $F(e)$ are homeomorphic with Euclidean spaces $\mathbb{R}^k$.*

(vi) *S is a semilattice of groups, that is, it is a union of groups and every idempotent is central.*

The key idea used in the proof of the above statements was to apply the machinery of Alexander-Spanier cohomology to conclude that the isotropy groups of elements outside the group of units have positive dimension and that the idempotents are just the zero elements in the closure of their isotropy groups.

In a similar vein one can study the structure of compact connected semitopological monoids where the identity is a local cutpoint or is "non-peripheral" in a rather strong sense. Unfortunately, the popular peripherality concept of LAWSON and MADISON no longer works in the semitopological context (cf. [43], p. 168ff.).

We conclude this discussion with some unsolved problems.

Problem 5. Find a concept of peripherality which ensures that if the identity is non-peripheral in the semigroup S then the closure of $H(\mathbf{1})$ is "thick" in S (say, is dense in an open subset of S, or is of second category).

Problem 6. As for topological semigroups, it is not difficult to show that every compact connected semitoplogical monoid contains a minimal connected submonoid connecting the identity with the minimal ideal. Traditionally a connected semigroup is called "irreducible" if no proper connected submonoid meets the minimal ideal. Is it true that every irreducible compact semitopological monoid must be abelian? *(J. F. Berglund)*

Problem 7. Describe those compact semitopological monoids on a manifold which cannot be decomposed into a semidirect product of non-trivial subsemigroups. *(K. H. Hofmann)*

Problem 8. Characterize those finite semilattices which occur as semilattice of idempotents in a compact semitopological monoid on a manifold.

Problem 9. Can the construction yielding the semigroups of type $P(g)$ be transferred to higher dimensions (in particular to dimension 4)?

4.4. Semitopological semigroups with special algebraic properties

When investigating the structure of certain objects in topological algebra one usually tries to single out special algebraic properties which ensure "nice" structure theorems. In the case of compact semitopological semigroups such nice structure results are available in each of the following cases:

(i) S is a group, or a semilattice, or a left [right] group, or has trivial (left or right zero, or constant) multiplication (since then S is a topological semigroup);

(ii) S is a group with zero (in this case S is either a topological semigroup or is of the type described in Example 1.3)

(iii) S is a semilattice of groups (i.e., it is the union of its maximal subgroups and all idempotents are central). In this case the idempotents form a closed subsemigroup with jointly continuous multiplication, inversion is continuous, and all ideals of S are closed (this case has been studied by Berglund [2] and the author; cf. [43], p. 200).

(iv) S does not contain proper (closed) ideals (This case is already settled by the First Fundamental Theorem), or S is zero-minimal (i.e., does not contain non-zero ideals; Berglund [2]).

It seems that the above examples are already exhaustive, at least if one does not combine special algebraic properties with special topological properties.

5. Applications

5.1. Representations: the Mautner Phenomenon

There are three main lines of research connecting semitopological semigroups and weakly almost periodic functions with Representation Theory: Convolution Semigroups, such as the structure semigroup of a locally compact abelian group (cf. Example 1.8), L^∞-representations of commutative semigroups (cf. Example 1.9), and Mautner's Phenomenon. For lack of space, and for methodical reasons we do not attempt a discussion of the first two topics. The interested reader is referred to the accounts in the existing literature, notably, to [51], [20] for the first one and to [24] for the second.

In the theory of unitary representations the label "Mautner Phenomenon" is attached to a group of observations of the following general form.

5.1.1. Mautner's Phenomenon. *Let V be a Hilbert space and let $\rho: G \to \mathcal{U}(V)$ be a unitary representation of a Lie group G with Lie algebra $\mathfrak{g}$. Suppose that $\langle \exp t{\cdot}X \mid t \in \mathbb{R} \rangle$ $(X \in \mathfrak{g})$ is a one parameter subgroup of G and that v is a vector in V which is invariant under $\rho(\exp t{\cdot}X)$, that is, $\rho(\exp t{\cdot}X)v = v$ for all $t \in \mathbb{R}$. Then v is fixed by the elements of a usually much larger subgroup H of G.*
(Of course it depends on the structure of G *how* much larger H is.)

The investigation of the Mautner Phenomenon originated in the ergodic theory of geodesic flows, with F. I. Mautner's paper [36] of 1957. Since then the theory has developed considerably and many interesting results have been obtained. As a particular example we recall the following theorem of Calvin C. Moore [38] (1969).

Theorem 5.1.2. *Let $\rho: G \to \mathcal{U}(V)$ be a unitary representation of a Lie group G on some Hilbert space V and suppose that a vector $v \in V$ is fixed under the action of some one-parameter subgroup $\exp \mathbb{R}{\cdot}X$, $X \in \mathfrak{g}$. Then v remains fixed under the smallest normal subgroup N of G such that the corresponding quotient Lie algebra morphism $\mathfrak{g} \to \mathfrak{g}/\mathfrak{n}$ maps X onto an* ad*-compact element.*

(It is easy to show that such a smallest normal subgroup always exists.)

We observed already in Example 1.7 that the contractions of a Hilbert space form a weakly compact set. Thus, writing S for the weak closure of $\rho(G)$ in $\mathcal{U}(V)$, the corestriction $\rho^*: G \to S$ of any unitary representation $\rho : G \to \mathcal{U}(V)$ defines a semitopological compactification of G, and it is therefore natural to ask whether the general structure theory of compact semitopological semigroups also yields assertions of the above type. This is indeed true; from the structural point of view statements like Moore's Theorem are just nice applications of the general properties of semitopological compactifications of connected Lie groups. More specifically, Mautner's Phenomenon can be considered as a group of corollaries derived from propositions about isotropy groups (which in turn were derived from Joint Continuity Statements). In the present notes we shall not offer an exhaustive treatment of the Theme "Applications of weak almost periodic compactifications to the representation theory of Lie groups"—instead we try to make clear the basic ideas and concepts by applying them to a few typical situations, such as the above Theorem 5.1.2.

Whereas the connection between the Mautner Phenomenon and the theory of weakly almost periodic functions was made evident rather early in the history of the subject, by several authors, it seems that we owe the first strictly "structural" approach to W. A. VEECH. The approach proposed in the following lines is published here for the first time, it is mostly based on and inspired by his papers [55] and [56].

Conventions. Throughout the rest of this section, G will always denote a connected Lie group with Lie algebra $\mathfrak{g}$ and $\rho: G \to \mathcal{C}(V)$ a representation of G by contraction operators (i.e. operators of norm ≤ 1) on a (complex) reflexive Banach space V. Then the weak closure of $\rho(G)$ in $\mathcal{C}(V)$ will be a compact semitopological semigroup which we denote by S. Since no ambiguities are to be feared we also write ρ for the associated compactification map $G \to S$. Note that the induced action $V \times S \to V, (v, s) \mapsto v.s$, is separately continuous and that every eigenvalue of an operator $\rho(g)$, $g \in G$ must have modulus one.

Lemma 5.1.3. *Let $X \in \mathfrak{g}$ and suppose that v is an eigenvector of $\rho(\exp X)$, so that $v.\rho(\exp X) = \lambda \cdot v$ for some scalar λ with $|\lambda| = 1$. Then the minimal idempotent e in the closure of $\{\rho(\exp z \cdot X) | z \in \mathbb{Z}\}$ leaves v invariant, $v.e = v$.*

(Note that this assertion also implies that v is left invariant under all elements of the isotropy group $F(e)$.)

Proof. Choose a net $\langle k_n | n \in D \rangle$ with $\lim \rho(\exp k_n \cdot X) = e$ and such that the limit $\lambda_e = \lim \lambda^{k_n}$ exists. Since the action of S on V is separately continuous, we have $v.e = \lim v.\rho(\exp k_n \cdot X) = \lambda_e.v$. Now $\lambda_e \cdot v = v.e = v.e^2 = \lambda_e^2.v$, hence $v = \lambda_e \cdot v = v.e$. □

Combining the above Lemma 5.1.3 with Corollary 4.2.5 we get the following version of MOORE's Theorem.

Theorem 5.1.4. *Let $\rho: G \to \mathcal{C}(V)$ be a representation of a Lie group G by contraction operators and suppose that $v \in V$ is an eigenvector of the operator $\rho(\exp X)$, where $X \in \mathfrak{g}$. Then v remains fixed under the smallest normal subgroup N of G*

such that the corresponding quotient Lie algebra morphism $\mathfrak{g} \to \mathfrak{g}/\mathfrak{n}$ *maps* X *onto an* ad-*compact element.*

Example 5.1.5. Let G, X, ρ be as in Theorem 5.1.4 and assume that $G = W \rtimes K$ is a group of the type described in Example 4.2.8 and and that $\exp X$ is not contained in a compact subgroup of G. Then v remains fixed under the action of W (and hence its orbit under G is compact).

The following Theorem also follows from our general results on weak almost periodic compactifications (cf. Example 4.2.11 above); it can be considered as a generalized version of MOORE's Egodicity Theorem in [37]. The formulation in its present form is due to VEECH [55].

Theorem 5.1.6. *Let* G *be a semisimple connected Lie group with no compact factors, and let* H *be a totally unbounded subgroup of* G*, that is, the adjoint representation of* H *on each of the simple ideals of the Lie algebra* $\mathfrak{g}$ *is non-compact. Furthermore, suppose that one of the following conditions* (i), (ii) *is satisfied:*

(i) *G has finite center and ρ is a bounded, strongly continuous representation of G on a reflexive Banach space B.*

(ii) *B is a Hilbert space and ρ is a unitary representation of G.*

Then we have:

(iii) *If V is a finite dimensional subspace of B which is invariant under $\rho(H)$, then $v.\rho(g) = v$ for all $g \in G$ and $v \in V$.*

The implication (i)$\Longrightarrow$(iii) is an immediate consequence of 4.26, in order to establish (ii)$\Longrightarrow$(iii) one first reduces the problem to the case where G is simple and ρ is irreducible and then applies Schur's Lemma to find a factorization ρ' of ρ over a locally compact group G' with compact center and containing a dense continuous image of G. In other words, we find a continuous homomorphism $\alpha: G \to G'$ with $\overline{\alpha(G)} = G'$ and a unitary representation $\rho': G' \to \mathcal{U}(B)$ of G' such that $\rho = \rho' \circ \alpha$. Now it suffices to recall Theorem 4.2.13.

5.2. Periodicity properties of functions

Periodicity and idempotents. One of the most important achievements in classical harmonic analysis is BOCHNER's discovery that a function $f: \mathbb{R} \to \mathbb{C}$ is almost periodic, according to BOHR's original definition, if and only if its orbit $O^f = \{x \mapsto f(r + x) \mid r \in \mathbb{R}\}$ under the action of $\mathbb{R}$ is precompact with respect to the sup-norm. It is exactly this characterization which extends naturally to continuous functions on arbitrary groups and semigroups, and thus leads to the present-day definitions of almost periodicity and weak almost periodicity as recalled above. However, especially in the general form of weak almost periodicity, this concept now deviates considerably from the intuitive notion of periodicity, which is concerned with pattern-recurrence rather than with compactness of orbit closures. A striking illustration is provided by the continuous functions on a locally compact group which vanish at infinity; such functions are obviously weakly almost periodic but are as "non-periodic in the common sense" as possible.

From the "pattern-recurrence" point of view, a "nearly periodic" function f, say on a group G, should possess a non-trivial period-within-ε, for every $\varepsilon > 0$. In other words, f should satisfy a condition like the following:

($*$) There exists a **1**-neighborhood U in G such that to every $\varepsilon > 0$ we can find an element $p_\varepsilon \in G \setminus U$ with

$$|f(p_\varepsilon g) - f(g)| < \varepsilon \text{ for all } g \in G.$$

To simplify the discussion we restrict our attention to abelian groups G. Following Harald Bohr ([9]) we call a continuous function $f: G \to \mathbb{C}$ *periodic-like* if it satisfies ($*$). Bohr himself noticed that the periodic-like functions on $\mathbb{R}$ do not form a vector space, and it can be shown that a periodic-like function need not be weakly almost periodic ([49]).

If f is weakly almost periodic then its periodicity behavior with respect to property ($*$) can be measured by idempotents in G^w.

Theorem 5.2.1. ([48]) *Let $f: G \to \mathbb{C}$ be a weakly almost periodic function on an abelian topological group G. Then f is periodic-like if and only if there exists an idempotent $e \neq \mathbf{1}$ such that the canonical extension $f^w: G^w \to \mathbb{C}$ satisfies $f^w(x) = f^w(ex)$ for all $x \in G^w$.*

Functions derived from signed a-adic expansions. We now present a class of functions $\mathbb{Z} \to [0, 1]$ which is very convenient as a testing ground and as a source of examples for various periodicity concepts. (Details will be given in the forthcoming paper [49].)

To construct these functions we use the following variant of **a**-adic expansions: Let $\mathbf{a} = \langle a_1, a_2, \ldots \rangle$ be a sequence of natural numbers > 2. We define $A_0 = 1$ and, for every $n \in \mathbb{N}$, A_n to be the product of all a_k's with $k \leq n$. Let x be any integer. Then there are integers u_i and x_i, $i = 0, 1, 2, \ldots$, with

$$x_0 = x, \tag{1}$$

$$x_i = u_i + x_{i+1} a_{i+1}, \quad \text{where} \tag{2}$$

$$\text{either} \quad |u_i| < \frac{a_{i+1}}{2} \quad \text{or} \quad u_i = \frac{a_{i+1}}{2}. \tag{3}$$

The numbers u_i, x_i are computed recursively; they are uniquely determined (note that all numbers a_i are supposed to be > 2). Combining the equations (2) we therefore get a unique representation of x as a sum

$$x = \sum_{i=0}^{\infty} u_i A_i, \tag{4}$$

where for every i either $|u_i| < \frac{a_{i+1}}{2}$ or $u_i = \frac{a_{i+1}}{2}$; and $u_i \neq 0$ for at most finitely many indexes i.

The representation (4) will be called the *signed* **a**-*adic expansion of* x.

Now let $\mathbf{c} = \langle c_i \,|\, i = 0, 1, 2, \ldots \rangle$ be a "tempering" sequence of non-negative reals with

$$c_i \leq a_{i+1}^{-1}, \tag{5}$$

so that $\frac{1}{2} \leq 1 - c_i|u_i| \leq 1$. Then we define a function $f\colon \mathbb{Z} \to [0,1]$ by

$$f(x) \stackrel{\text{def}}{=} \prod_{i=0}^{\infty} (1 - c_i|u_i|), \tag{6}$$

where $x = \sum_i u_i A_i$ is the signed **a**-adic expansion of x. The function f will be called an $(\mathbf{a}, \mathbf{c})$-*function* for short.

Admittedly, the so defined $(\mathbf{a}, \mathbf{c})$-functions look somewhat grim, so let us say something favorable about their periodicity behavior.

Theorem 5.2.2. *For the above defined functions* $f\colon \mathbb{Z} \to [0,1]$ *the following assertions hold:*

(i) *f is periodic-like if*

$$\lim_i c_i = 0 \quad \textit{and} \quad \sum_{a_{i+1} \equiv 0(2)} c_i < \infty.$$

Conversely, if f is periodic-like then $\liminf_i c_i = 0$.

(ii) *f is weakly almost periodic if the sequences* $\mathbf{a}, \mathbf{c}$ *satisfy one of the following conditions* (a), (b):

2(a) $\sum_{i=0}^{\infty} c_i$ *converges.*

2(b) *The sequence* $\mathbf{c}$ *is monotone non-increasing and only finitely many of the a_i's are even.*

Conversely, if f is weakly almost periodic then we have:

2(c) *If $\langle b_i \,|\, i \in \mathbb{N} \cup \{0\} \rangle$ is a sequence with $b_i \in \{0,1\}$ and such that the sum $\sum_{i=0}^{\infty} b_i c_i a_{i+1}$ converges then $\sum_{i=0}^{\infty} b_i c_{i+1}$ converges too.*

(iii) *The following assertions are equivalent:*

2(d) *f is almost periodic,*

2(e) $\sum_{i=0}^{\infty} c_i a_{i+1} < \infty$,

(iv) *f is periodic if and only if all but finitely many of the c_i's vanish. Moreover, if f is periodic then the primitive period of f must be a divisor of one of the numbers A_i, and hence of all but finitely many of them.*

The above theorem says, roughly speaking, that the sequence $\mathbf{c}$ controls the periodicity behavior of f: the "smaller" the c_i's the more will f show some degree of periodicity. Once this Theorem is established, the $(\mathbf{a}, \mathbf{c})$-functions are easy to handle. For instance, it is now a matter of straightforward elementary analysis to produce weakly almost periodic functions on $\mathbb{Z}$ (hence also on $\mathbb{N}$ and, by extension, on $\mathbb{R}$) which are, or are not, periodic-like, or of periodic-like functions which are not weakly almost periodic, and similar examples.

The main virtue of the weakly almost periodic $(\mathbf{a}, \mathbf{c})$-functions is that each such function generates a semitopological compactification whose structure is perfectly well understood, due to the fact that $(\mathbf{a}, \mathbf{c})$-functions are "near-homomorphisms": $f(x+y) = f(x)f(y)$ if x and y have disjoint supports with respect to their signed $\mathbf{a}$-adic expansions. (A similar idea of "near-homomorphism" lies behind the construction of large free subgroups in $\beta\mathbb{N}$ by HINDMAN and PYM and the associated concept of "oid"; cf. the chapter by J. PYM in this volume.) Before listing more details, let us agree to follow HEWITT and ROSS [29] in denoting the group of $\mathbf{a}$-adic integers with $\Delta_\mathbf{a}$. We view the elements of $\Delta_\mathbf{a}$ as infinite sequences $\langle u_i \rangle$ of integers which satisfy (3).

Theorem 5.2.3. *Suppose that the $(\mathbf{a}, \mathbf{c})$-function f is weakly almost periodic but not periodic. Then f generates a semitopological compactification (S_f, φ_f) of $\mathbb{Z}$ with the following properties:*

(i) *The group of units of S_f is isomorphic with the subgroup*

$$G \stackrel{\text{def}}{=} \{\langle u_i \rangle \in \Delta_\mathbf{a} : \sum_{i=0}^{\infty} c_i |u_i| < \infty\},$$

of $\Delta_\mathbf{a}$, endowed with a group topology τ which is not coarser than the topology inherited from $\Delta_\mathbf{a}$. (Actually the topology τ can be described explicitely by means of a "near-norm".)

(ii) *Let J be the set of all points $x \in S_f$ which can be written as a limit $x = \lim \varphi(x_n)$, where $\langle x_n \rangle$ is a sequence in $\mathbb{Z}$ which converges to 0 with respect to the $\mathbf{a}$-adic topology, that is, x_n lies eventually in $A_m\mathbb{Z}$ for every m. Then the natural extension of f to a function $\widehat{f}: S_f \to [0, 1]$ satisfies $\widehat{f}(xy) = \widehat{f}(x)\widehat{f}(y)$ for all $x \in J$, $\quad y \in S_f$. Moreover, $\widehat{f}$ maps J isomorphically onto a closed subsemigroup of $([0, 1], \cdot)$.*

(iii) *If f is not almost periodic then S_f has a zero element $\mathbf{0}$ and $\mathbf{0} \in J$. (Thus S_f has at most two idempotents.)*

(iv) *The action $S_f \times J \to S_f$, $(s, j) \mapsto sj$, of J on S_f is (jointly) continuous.*

(v) *If f is not almost periodic then the map*

$$\psi: J \times H(\mathbf{1})/\{\mathbf{0}\} \times H(\mathbf{1}) \to S_f,\ \psi(x,y) = \begin{cases} xy & \text{if } x \neq \mathbf{0}, \\ \mathbf{0} & \text{otherwise;} \end{cases}$$

is a continuous homomorphism and an algebraic isomorphism.
(However the inverse ψ^{-1} is *not* continuous.)

Addendum. *The restriction of $\widehat{f}$ to J is an isomorphism onto $I = ([0,1],\cdot)$ if there is a sequence $\langle i(k) : k \in \mathbb{N}\rangle$ of positive integers such that $\lim_k c_{i(k)} = 0$ and $\sum_{k\in\mathbb{N}} c_{i(k)} a_{i(k)+1} = \infty$.*

Combining compactifications S_f for different f's we get semitopological compactifications with many idempotents. This solves the problem of finding a "direct" proof of assertion 4.2.1(iii) (cf. Berglund's Problem in [3], recorded as Problem (28) in [43]). Actually it also yields some additional information.

Theorem 5.2.4. *To every cardinal m with $m \leq 2^{\aleph_0}$ there exists a semitopological compactification (S, φ) of $\mathbb{Z}$ with the following properties:*

(i) *S contains a closed subsemigroup J which is isomorphic with the cube I^m, where I denotes the semigroup $([0,1],\cdot)$.*

(ii) *The map*

$$J \times H(\mathbf{1}) \to S,\ (j,h) \mapsto jh,$$

is a continuous and surjective homomorphism.

Question. Is there any connection between the examples constructed here and the examples given by Brown and Moran in [11]?

5.3. Group topologies

The theory of group topologies is connected with compact semitopological semigroups in view of the following observations:
(1) It is natural to ask which topological groups can be embedded isomorphically into compact semitopological semigroups. More specifically, given a group G, one would like to have information about the topologies τ on G such that (G, τ) is isomorphically embedded into its weak almost periodic compactification and about the lattice of all such topologies (cf. Problem 4 and the discussion in connection with Theorem 4.2.1).
(2) We shall see below that there is a Galois Connection between group topologies on a given group G and idempotents in the weak almost periodic compactification of G_d, where G_d denotes the group G, endowed with the discrete topology. Thus

the semilattice of idempotents in ${G_d}^w$ can be described in terms of group topologies on G.
(3) In [21], [22] and [23], DUNKL and RAMIREZ studied idempotents in the semitopological semigroup Δ_G of Example 1.8 which are contained in a locally compact maximal subgroup. Such a maximal subgroup is naturally isomorphic with the Pontryagin dual of the abstract group G, endowed with a locally compact group topology which is stronger than the original topology on G.

We still do not have satisfactory answers to the questions in (1), but the author is confident that the recent success in the investigation of minimal group topologies will also lead to a better understanding of this problem. On the other hand, the group topologies arising from certain semitopological compactifications of the integers, such as the "generalized **a**-adic topologies" induced by weakly almost periodic (**a**,**c**)-functions, may also have some independent interest for the general theory of topological groups (cf. the survey article by COMFORT [17]). Two problems concerning integer topologies are appended at the end of this section.

Let us now summarize briefly the definition and essential properties of the Galois connection announced in (2).

Notation 5.3.1. In the following G denotes a fixed abelian group, which we shall endow with various group topologies. These topologies are not assumed to be Hausdorff. The term "completion", however, always means "Hausdorff completion."

If τ is a group topology on G then G_τ denotes the associated Hausdorff completion, and (G_τ^w, i_τ) the weak almost periodic compactification of G_τ. If τ is the discrete topology then we omit subscripts and simply write (G^w, i). As before we write $E(G^w)$ for the set of all idempotents in G^w.

A group topology τ on G is called *weakly almost periodic*, or a *wap topology* for short, if i_τ is a homeomorphic embedding. The set of all wap-topologies on G is denoted by $\mathbf{T}(G)$.

Also, we write

(i) $\mathcal{T}$ for the map $E(G^w) \to \mathbf{T}(G)$ assigning to every idempotent e in G^w the topology induced on G by the homomorphism

$$G \to eG^w,\ g \mapsto ei_G(g),$$

and, in the other direction,

(ii) $\mathcal{E}$ for the map $\mathbf{T}(G) \to E(G^w)$ assigning to every wap-topology τ the minimal idempotent in the kernel $j_\tau^{-1}(\mathbf{1})$ of the canonical morphism $j_\tau \colon G^w \to G_\tau^w$ making the diagram

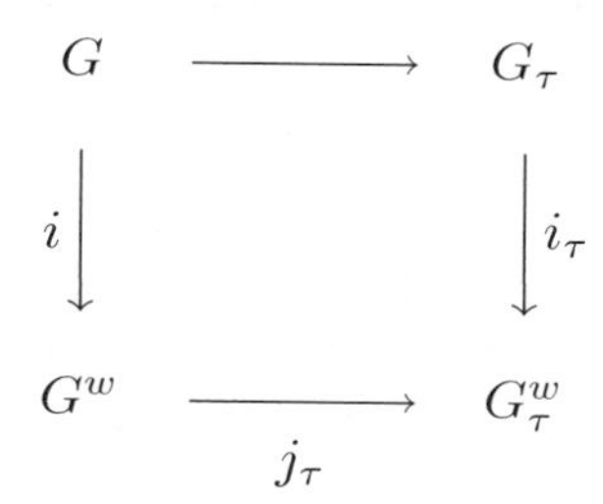

commutative.

Note that for every idempotent $e \in G^w$ we have $\mathcal{E}\mathcal{T}(e) = e$, and for every topology $\tau \in \mathbf{T}(G)$ the inclusion $\tau \subseteq \mathcal{T}\mathcal{E}(\tau)$ holds.

Theorem 5.3.2. ([48]) *Let τ_1, τ_2 be two wap-topologies on G and set $e_1 = \mathcal{E}(\tau_1)$, $e_2 = \mathcal{E}(\tau_2)$. Furthermore, write $\varphi_1: G_d \to G_{\tau_1}$, $\varphi_2: G_d \to G_{\tau_2}$ for the natural homomorphisms into the Hausdorff completions G_{τ_1}, G_{τ_2}.*

Then the following assertions (i), (ii), (iii) *are equivalent:*

(i) $e_1 \leq e_2$

(ii) *Every τ_2-Cauchy filter in G can be refined to a τ_1-Cauchy filter.*

(iii) *G_{τ_1} contains a compact normal subgroup K, such that the quotient morphism $\kappa: G_{\tau_1} \to G_{\tau_1}/K$ can be combined with the maps φ_1, φ_2 and a surjection $f: G_{\tau_2} \to G_{\tau_1}/K$ to form a commutative diagram*

$$\begin{array}{ccc} G_d & \xrightarrow{\varphi_1} & G_{\tau_1} \\ {\scriptstyle \varphi_2}\downarrow & & \downarrow{\scriptstyle \kappa} \\ G_{\tau_2} & \xrightarrow[f]{} & G_{\tau_1}/K \end{array}$$

of continuous homomorphisms.

We henceforth write $\tau_1 \leq_c \tau_2$ if one of the equivalent statements (i), (ii) holds, and we write $\tau_1 =_c \tau_2$ if both $\tau_1 \leq_c \tau_2$ and $\tau_2 \leq_c \tau_1$. (So the subscript "c" stands both for "Cauchy" and "compact.") Note that the pre-order $\leq_c$ is compatible with the usual order by inclusion.

Corollary 5.3.3. ([48]) *In the notation of 5.3.2, the two topologies τ_1 and τ_2 are c-equivalent, and hence $\mathcal{E}(\tau_1) = \mathcal{E}(\tau_2)$, if and only if there are compact normal subgroups K_1 and K_2 of G_{τ_1} and G_{τ_2}, respectively, such that there is an isomorphism $f: G_{\tau_2}/K_2 \to G_{\tau_1}/K_1$ rendering the diagram*

$$\begin{array}{ccccc} G_{\tau_2} & \xleftarrow{\varphi_2} & G_d & \xrightarrow{\varphi_1} & G_{\tau_1} \\ \kappa_2 \downarrow & & & & \downarrow \kappa_1 \\ G_{\tau_2}/K_2 & & \xrightarrow{f} & & G_{\tau_1}/K_1 \end{array}$$

commutative.

Remarks. (i) The set $\mathbf{T}(G)$ is a lattice with respect to inclusion. The maximum of $\mathbf{T}(G)$ is the discrete topology, and the minimum is the lump topology of G (i.e. the topology $\{\emptyset, G\}$). Also, $\bigvee \mathbf{T}_1 \in \mathbf{T}(G)$, for every subset $\mathbf{T}_1$ of $\mathbf{T}(G)$. (In particular, to every idempotent $e \in G^w$ there is always a finest wap-topology τ with $\mathcal{E}(\tau) = e$, of course $\tau = \mathcal{T}(e)$.)
(ii) The map $\mathcal{T}$ is injective and order preserving.
(iii) The map $\mathcal{E}$ is order preserving (*it is in general not injective*) and we have $\mathcal{E}\mathcal{T}(e) = e$ for every idempotent e.

The above observations already show that the pair $(\mathcal{E}, \mathcal{T})$ is a Galois connection. To have it at hand, let us recall the definition.

Definition 5.3.4. Let X and Y be two ordered sets. A pair (g, d) of order morphisms $g: X \to Y$ and $d: Y \to X$ is called a *Galois connection* or an *adjunction* between X and Y if for $(x, y) \in X \times Y$ the assertions
(i) $g(x) \geq y$ and
(ii) $x \geq d(y)$
are equivalent. The map d is called the *lower adjoint*, the map g is called the *upper adjoint* of the given Galois connection. The composition dg is called the *kernel operator*, and gd is called the closure operator associated with (d, g).

Proposition 5.3.5. (i) *The pair $(\mathcal{T}, \mathcal{E})$ is a Galois connection between the set of idempotents $(E(G^w, \leq)$ and the ordered set $(\mathbf{T}(G), \leq)$ of wap topologies on G.*
(ii) *The kernel operator $\mathcal{E}\mathcal{T}$ of this Galois connection is the identity on $E(G^w)$.*
(iii) *The closure operator $\mathcal{T}\mathcal{E}$ is the map assigning to each wap topology $\tau \in \mathbf{T}(G)$ the finest wap topology which is c-equivalent with τ.*

We close this section with two problems concerning integer topologies.

Problem 10. Find an example of a weakly almost periodic integer topology τ such that $\mathcal{E}(\tau)$ is a maximal idempotent in $E(\mathbb{Z}^w) \setminus \{\mathbf{1}\}$. (Such topologies must exist since $E(\mathbb{Z}^w) \setminus \{\mathbf{1}\}$ contains maximal elements.)

Problem 11. Is it true that to every weakly almost periodic non-precompact integer topology τ there is another one, say τ', such that $\tau' \subset \tau$ but $\tau' \neq \tau$?

5.4. Combinatorics

Thanks to the work of Neil Hindman and his school we have learned that compact semigroups can be used profitably for the proof (and the understanding) of combinatorial facts, and that, conversely, combinatorial methods are a powerful means for structural research on compact semigroups. Most of these relations to combinatorics concern semigroups with one-sided continuous multiplication, but there are also some interesting connections with compact semitopological semigroups.

Let us start with a very natural problem. Suppose that A is a subset of a discrete semigroup S. When is the indicator function c_A (defined to be 1 on A and 0 outside) weakly almost periodic? Following a widespread practice we shall call such a set A *weakly almost periodic*.

Applying Grothendieck's Criterion we find a combinatorial answer to this question.

Theorem 5.4.1. ([46]) *A subset A of a discrete semigroup S is weakly almost periodic if and only if every infinite subset $B \subseteq S$ contains a finite subset F such that the sets*

$$D_R(B,F) = \Big(\bigcap_{b \in F} Ab^{-1}\Big) \setminus \Big(\bigcap_{b \in B \setminus F} Ab^{-1}\Big)$$

and

$$D_L(B,F) = \Big(\bigcap_{b \in F} b^{-1}A\Big) \setminus \Big(\bigcap_{b \in B \setminus F} b^{-1}A\Big)$$

are finite.

Here, as elsewhere, an expression like Ux^{-1} $[x^{-1}U]$ means $\{u \in U \mid ux \in U\}$ $[\{u \in U \mid xu \in U\}]$. Of course, the above condition can be rewritten using the fact that $S \backslash A$ is weakly almost periodic if and only if A is weakly almost periodic.

Next we ask for weakly almost periodic subsets satisfying the stronger condition that every bounded function supported in A is weakly almost periodic. The answer also provides some information about the structure of S^w. (Recall that a continuous map $f: X \to Y$ is called proper if the inverse image $f^{-1}(y)$ of each point $y \in f(X)$ is compact.)

Theorem 5.4.2. *Let A be a subset of a discrete semigroup S and consider the statements* (i)–(v) *below:*

(i) *Every bounded function which vanishes off A is weakly almost periodic.*

(ii) *Every infinite subset B of S contains a finite subset F such that the sets $\bigcap_{b \in F} Ab^{-1}$ and $\bigcap_{b \in F} b^{-1}A$ are finite.*

(iii) *$\overline{i_S(A)}$ is open in S^w and does not intersect $\left(S^w \backslash i_S(S)\right)^2$.*

(iv) *A is weakly almost periodic and the induced compactification $(\overline{i_S(A)}, i_S \mid A)$ is equivalent to the Stone-Čech compactification of A.*

Then the following implications hold: (ii)$\Longrightarrow$(iii)$\Longrightarrow$(iv) $\Longleftrightarrow$(i). *If the left and right translations in S are proper then the assertions* (i)–(iv) *are pairwise equivalent.*

Remarks 5.4.3. (i) It seems that the first results in the direction of 5.4.2 are due to Walter Rudin ([41], 1959) and Donald E. Ramirez ([40], 1968). Ramirez used the ideas in Rudin's paper to introduce the concept of *T-sets*, which is closely related to statement 5.4.2(ii):

(t) A subset E of a group G is called a T-set if $(E \cap Eg) \cup (E \cap gE)$ is finite for every $g \in G$.

Ramirez observed that every T-set satisfies the above condition 5.4.2.(iv), and concluded that $\mathbb{Z}^w$ contains a copy of $\beta\mathbb{N}$. The present form of Theorem 5.4.2 is taken from [46] (1985), in this paper the author continued the terminology of Ramirez by calling a set A *translation-finite* if it satisfies condition 5.4.2.(ii). Clearly, every finite union of T-sets is translation-finite, but, as shown by Chou Ching in in his study of "thin" subsets of groups [15], every infinite discrete group G contains translation-finite subsets which are not of this type. Chou also noted that statement 5.4.2.(ii) can be given the following equivalent form:

(c) A does not contain infinite triangles $\{x_i y_j \mid 1 \le i \le j, j \in \mathbb{N}\}$ or $\{x_i y_j \mid 1 \le j \le i, i \in \mathbb{N}\}$, where $\langle x_i \mid i \in \mathbb{N}\rangle$ and $\langle y_j \mid j \in \mathbb{N}\rangle$ are sequences of distinct elements in S.

(In [15] Translation-finite sets are called R_W-sets.)

(ii) As the author realized only after [46] was published, the implications (ii$'$)$\Longrightarrow$(iii)$\Longrightarrow$(iv) in 5.4.2 were deduced earlier by Berglund and Hindman [5] (1985) from their construction of S^w as a set of equivalence classes of ultrafilters on S.

(iii) If a countable discrete semigroup has proper translations and if the coincidence sets $\{x \in S \mid sx = tx \text{ or } xs = xt\}$ are finite for every pair of distinct elements $s, t \in S$ then each infinite subset of S contains an infinite translation-finite subset of S. As a corollary we conclude from 5.4.2.(iv) that G^w cannot be a union of groups if G is an infinite discrete group (cf. Theorem 4.2.12 above).

Now let us specialize to the case where $S = \mathbb{N}$ or $S = \mathbb{Z}$. In this case it is very easy to find examples of T-sets and of weakly almost periodic sets which are not translation-finite in S. For instance, the elements a_i of any sequence in $\mathbb{N}$ satisfying a Fibonacci-growth condition $a_{i+2} \geq a_i + a_{i+1} > 0$ form a T-set in $\mathbb{Z}$ as well as in $\mathbb{N}$. Indeed, this claim is a consequence of the fact that for such sequences the equation $a_i = a_j + d$, with $d \in \mathbb{N}$ implies $d = a_i - a_j \geq a_{j+1} - a_j = a_{j-1} \geq j - 1$. Applying this argument repeatedly it can be shown that also the set of pairwise sums

$$PS(A) \stackrel{\text{def}}{=} \{x + y \mid x \neq y,\ x, y \in A\}$$

is weakly almost periodic in $\mathbb{N}$. (A direct proof can be found in [5].) Similarly, but with slightly more technical apparatus, it can be shown that the set of finite differences $PD(A) = \{a_i - a_j \mid j < i\}$ is translation-finite in $\mathbb{N}$ (but not even weakly almost periodic in $\mathbb{Z}$). In [5], BERGLUND and HINDMAN have shown that the indicator function of the set

$$U = \{2^{2^n(2r+1)} - 2^m \mid m, n \in \mathbb{N},\ m \leq n\}$$

cannot be extended to a weakly almost periodic function on $\mathbb{Z}$. Since c_U is weakly almost periodic on $\mathbb{N}$ this means that not every weakly almost periodic function on $\mathbb{N}$ extends to a weakly almost periodic function on $\mathbb{Z}$, or in other words, that the injection $\mathbb{N} \to \mathbb{Z}$ does *not* induce an injective morphism $\mathbb{N}^w \to \mathbb{Z}^w$.

References

[1] Asplund, E., and I. Namioka, A geometric proof of Ryll-Nardzewski's fixed point theorem, Bull. Amer. Math. Soc. 73 (1967), 443–445

[2] Berglund, J. F., Compact completely 0-simple semitopological semigroups, Proc. Amer. Math. Soc. 50 (1975), 483–488

[3] Berglund, J.F., Problems about semitopological semigroups, Semigroup Forum 14 (1980), 373–380

[4] Berglund, J. F., Compact connected ordered semitopological semigroups, J. London Math. Soc. 4 (1972), 533-540

[5] Berglund, J. F., and N. Hindman, Filters and the weak almost periodic compactification of a discrete semigroup, Trans. Amer. Math. Soc. 284 (1984), 1–38

[6] Berglund, J. F., and K. H. Hofmann, "Compact semitopological semigroups and weakly almost periodic functions," Springer Lecture Notes in Math. 42 (1967)

[7] Berglund, J. F., H. D. Junghenn, and P. Milnes, "Compact right topological semigroups and generalizations of weak almost periodicity," Springer Lecture Notes in Math. 663 (1978)

[8] Berglund, J. F., H. D. Junghenn, and P. Milnes, "Analysis on Semigroups: Function Spaces, Compactifications, Representations," Wiley-Interscience Publications, New York etc. 1989

[9] Bohr, H., On the definition of almost periodicity, J. Anal. Math. 1 (1951), 11–27

[11] Brown, G., and W. Moran, Idempotents of compact monothetic semitopological semigroups, Proc. London Math. Soc. (3) 22 (1971), 203–216

[12] —, The idempotent semigroups of compact monothetic semigroups, Proc. Roy. Irish Acad. A 72 (1972), 17–33

[13] Burckel, R. B., " Weakly almost periodic functions on semigroups," Notes on mathematics and its applications, Gordon and Breach, New York 1971

[14] Chou Ching, Minimally weakly almost periodic groups, J. Funct. Anal. 36 (1980), 1–17

[15] —,Weakly almost periodic functions and thin sets in discrete groups, (forthcoming)

[16] Christensen, J. P. R, Joint continuity of separately continuous functions, Proc. Amer. Math. Soc. 82 (1981), 455–461

[17] Comfort, W. W., Topological Groups, in: Kunen, K., and J. E. Vaughan (ed.), Handbook of set-theoretic topology. North-Holland, Amsterdam, etc. 1984

[18] DeLeeuw, K., and I. Glicksberg, Applications of almost periodic compactifications, Acta Math. 105 (1961), 63–97

[19] —, The decomposition of certain group representations, J. d'Analyse Math. 15 (1965), 135–192

[20] Dunkl, C. F., and D. E. Ramirez, "Topics in harmonic analysis," Meredith, New York, 1971

[21] —, Locally compact subgroups of the spectrum of the measure algebra, Semigroup Forum 3 (1971), 95–107

[22] —, Locally compact subgroups of the spectrum of the measure algebra II, Semigroup Forum 3 (1971), 267–269

[23] —, Locally compact subgroups of the spectrum of the measure algebra III, Semigroup Forum 5 (1972), 65–76

[24] —, Representations of commutative semitopological semigroups, Springer Lecture Notes in Math. 435 (1975)

[25] Ellis, R., Locally compact transformation groups, Duke Math. J. 24 (1957), 119–126

[26] Grothendieck, A., Critères de compacité dans les espaces fonctionnels généraux, Amer. J. Math. 74 (1952), 168–186

[27] Hansel, G., and J. P. Troallic, Points de continuité à gauche d'une action de semigroupe, Semigroup Forum 26 (1983), 20–27

[28] Helmer, D., On weakly almost periodic compactifications, in: Hofmann, K. H., (ed.), Almost periodic compactifications, continuity, and compact semigroups, THD Preprint Series No. 781, Darmstadt 1983, 24–36

[29] Hewitt, E., and K. A. Ross, "Abstract harmonic analysis," Springer Grundlehren 115, Berlin-Heidelberg-New York 1963

[30] Junghenn, H. D., and B. T. Lerner, Semigroup compactifications of semidirect products, Trans. Amer. Math. Soc. 265 (1981), 393–404

[31] Lawson, J. D., Joint continuity in semitopological semigroups, Ill. J. Math. 18(1974), 275–285

[32] —, Additional notes on continuity in semitopological semigroups, Semigroup Forum 12(1976), 265–280

[33] —, Some research problems concerning joint continuity of actions of compact topological monoids, Semigroup Forum 12, (1976), 281

[34] —, Points of continuity for semigroup actions, in: Hofmann, K. H., (ed.), Almost periodic compactifications, continuity, and compact semigroups, THD Preprint Series No. 781, Darmstadt 1983, 46–88

[35] —, Points of continuity for semigroup actions, Trans. Amer. Math. Soc. 284 (1984), 183–202

[36] Mautner, F. I., Geodesic flows on symmetric Riemannian spaces, Ann. of Math. 65 (1957), 416–431

[37] Moore, C. C., Ergodicity of flows on homogeneous spaces, Amer. J. Math. 88 (1968), 154–178

[38] —, The Mautner Phenomenon for general unitary representations, Pac. J. Math. 86 (1980), 155–169

[39] Namioka, I., Separate continuity and joint continuity, Pac. J. Math. 51(1974), 515–531

[40] Ramirez, D. E., Uniform approximation by Fourier-Stieltjes transforms, Proc. Cambridge Phil. Soc. 64 (1968), 615–624

[41] Rudin, W., Weak almost periodic functions and Fourier-Stieltjes transforms, Duke Math. J. 26 (1959), 215–220

[42] Ruppert, W., No semitopological monoid can be defined on the closure of the set $\{(x, \sin \frac{1}{x}) \mid x \in \mathbb{R}\backslash\{0\}\}$, Semigroup Forum 22 (1981), 283–284

[43] —, "Compact Semitopological Semigroups: An Intrinsic Theory," Springer Lecture Notes in Math. 1079, Berlin etc., 1984

[44] —, On semitopological actions of generalized I-semigroups, Semigroup Forum 31 (1985), 153–180

[45] —, A note on the location of joint discontinuity, Semigroup Forum 32 (1985), 173–182

[46] —, On weakly almost periodic sets, Semigroup Forum 32 (1985) 267–281

[47] —, A note on the WAP-compactification of groups, Semigroup Forum 33 (1986), 103–109

[48] —, On group topologies and idempotents in weak almost periodic compactifications, Semigroup Forum (to appear)

[49] —, On signed a-adic expansions and weakly almost periodic functions, (forthcoming)

[50] Ryll-Nardzewski, C., On fixed points of semigroups of endomorphisms of linear spaces, Proc. of the Fifth Berkeley Sympos. on Math. Statistics and Probability II(1965)

[51] Taylor, J. F., " Measure algebras," Amer. Math. Soc., Reg. Conf. Ser. 16 (1973)

[52] —, The structure of convolution measure algebras, Trans. Amer. Math. Soc. 119 (1965), 150–166

[53] Troallic, J. P., Une approche nouvelle de la presque-périodicité faible, Semigroup Forum 22(1981), 247–255

[54] —, Semigroupes affines semitopologiques compacts, this volume

[55] Veech, W. A., A fixed point theorem free approach to weak almost periodicity, Trans. Amer. Math. soc. 177(1973), 353–362

[56] —, Weakly almost periodic functions on semisimple Lie groups, Monatshefte f. Math. 88(1979), 55–68

[57] Weil, A., "Sur les espaces à structure uniforme et sur la topologie générale," Hermann, Paris 1937

[58] West, T. T., Weakly compact monothetic semigroups of operators in Banach spaces, Proc. Roy. Irish Acad. A 67 (1968), 27–37

Semigroupes affines semitopologiques compacts

Jean-Pierre Troallic

1. Introduction

Ce travail a pour but de présenter quelques aspects de la théorie des semigroupes affines semitopologiques compacts. Dans les sections 4 et 5, deux types de problèmes sont abordés qui ont trait, les uns à l'existence de points de continuité à gauche pour des actions séparément affines et séparément continues de semigroupes affines semitopologiques compacts (section 4), les autres à l'extrémalité des points de continuité à gauche obtenus (section 5). Ces deux sections ont ceci en commun qu'elles s'appuient l'une et l'autre sur une variante d'un résultat de I. NAMIOKA ([24] théorème 2.3 complété par [25] théorème 4.3) établie en 2.2. Le théorème 2.2 est l'outil principal de ce travail; il joue un rôle comparable à celui que joue le théorème de point fixe de C. RYLL-NARDZEWSKI dans l'approche originale de la presque-périodicité faible. (Le théorème de point fixe de C. RYLL-NARDZEWSKI n'est pas utilisé ici; il convient cependant de noter que la démonstration de ce théorème de point fixe donnée par E. ASPLUND et I. NAMIOKA s'appuie essentiellement sur le théorème 2.3 de [24] signalé ci-dessus).

Les sections 6, 7 et 8 sont consacrées à des applications des résultats établis dans les sections 4 et 5 (théorèmes du type "L. Alaoglu, G. Birkhoff, C. Ryll-Nardzewski", ou encore du type "K. DeLeeuw, I. Glicksberg, C. Ryll-Nardzewski"...).

2. Points de continuité à gauche extrémaux

L'énoncé principal de cette section est le théorème 2.2. Il s'agit d'une variante d'un résultat de I. NAMIOKA ([24] Théorème 2.3 complété par [25] théorème 4.3). Sa démonstration est une adaptation aisée des méthodes utilisées dans [24] et [25]. Nous nous appuierons sur ce théorème:

– dans la section 4, pour établir l'existence de points de continuité à gauche pour des actions séparément affines et séparément continues de semigroupes affines semitopologiques compacts.

– dans la section 5, pour étudier l'extrémalité des points de continuité à gauche obtenus dans la section 4.

Notations, définitions et rappels 2.1. 1) Soient X, Y et Z des espaces topologiques et soit $f: X \times Y \to Z$ une application.

α) Pour tout $x \in X$ [resp. $y \in Y$], on note f_x [resp. f_y] l'application de Y dans Z [resp. de X dans Z] définie par $f_x(y) = f(x, y)$ pour tout $y \in Y$ [resp. $f_y(x) = f(x, y)$ pour tout $x \in X$]. On dit que f est *séparément continue* si pour tout $x \in X$ et pour tout $y \in Y$, f_x et f_y sont continues.

β) On dit que $x \in X$ est un *point de continuité à gauche de* f si pour tout $y \in Y$, (x, y) est un point de continuité de f.

Supposons que f_x soit continue pour tout $x \in X$, que Y soit compact et que $Z = \mathbb{R}$. Alors, pour qu'un point $x \in X$ soit un point de continuité à gauche de f, il faut et il suffit que la condition suivante soit vérifiée: pour tout $\varepsilon > 0$, il existe un voisinage V de x dans X tel que pour tout $x' \in V$, l'on ait

$$\sup_{y \in Y} |f(x, y) - f(x', y)| \leq \varepsilon.$$

De plus, si X est compact, l'ensemble des points de continuité à gauche de f est un G_δ de X. (On trouvera dans [25] (voir théorème 1.2) les démonstrations de ces propriétés.)

2) Soit E un espace localement convexe séparé (réel). Pour toute partie A de E, on note $\text{co}(A)$ [resp. $\overline{\text{co}}(A)$] l'enveloppe convexe [resp. l'enveloppe convexe fermée] de A dans E. Soit K un convexe compact de E et soit $\text{ext}(K)$ l'ensemble des points extrémaux de K. Alors:

α) On a $\overline{\text{co}}(\text{ext}(K)) = K$ (théorème de Krein-Milman [18]).

β) Soit X une partie compacte de K. Si $\overline{\text{co}}(X) = K$, $\text{ext}(K)$ est inclus dans X (théorème de Milman [18]).

Théorème 2.2. *Soient K un convexe compact d'un espace localement convexe séparé et Y un espace compact. Soit $f: K \times Y \to R$ une application affine par rapport à la première variable et séparément continue. Soit A le G_δ de K constitué par les points de continuité à gauche de f. Alors, $A \cap \text{ext}(K)$ est dense dans* $\text{ext}(K)$.

Démonstration. Soit $C(Y, \mathbb{R})$ l'espace vectoriel réel des applications continues de Y dans $\mathbb{R}$ et soit $C_s(Y, \mathbb{R})$ [resp. $C_u(Y, \mathbb{R})$] l'espace localement convexe séparé [resp. l'espace de Banach] obtenu en munissant $C(Y, \mathbb{R})$ de la topologie de la convergence simple [resp. de la norme $\| \,.\, \|_u$ de la convergence uniforme]. Soit $\Phi: K \to C(Y, \mathbb{R})$ l'application affine définie par $\Phi(x)(y) = f(x, y)$ pour tout $x \in K$ et tout $y \in Y$; $\Phi: K \to C_s(Y, \mathbb{R})$ est une application continue et d'après 2.1, A coïncide avec l'ensemble des points de continuité de l'application $\Phi: K \to C_u(Y, \mathbb{R})$.

Désignons par X l'adhérence de ext(K) dans K et par ϕ la restriction de Φ à X ; l'application $\phi: X \to C_s(Y, \mathbb{R})$ est continue, par conséquent d'après le théorème 2.2 de [25], le G_δ R de X constitué par les points de continuité de l'application $\phi: X \to C_u(Y, \mathbb{R})$ est dense dans X ; le sous-espace ext(K) de K étant un espace de Baire (théorème de G. Choquet [7]), $R \cap \text{ext}(K)$ est un G_δ dense de ext(K). Montrons que $R \cap \text{ext}(K)$ est inclus dans A (ce qui achèvera la démonstration).

Soit $x \in R \cap \text{ext}(K)$ et soit $\varepsilon > 0$; notons V_ε l'image réciproque par Φ de la boule ouverte de $C_u(Y, \mathbb{R})$ de centre $\Phi(x)$ et de rayon ε; nous devons prouver que $x \in \text{Int}(V_\varepsilon)$. Soit ω un voisinage ouvert de x dans X dont l'image par ϕ soit incluse dans la boule fermée de $C_u(Y, \mathbb{R})$ de centre $\phi(x)$ et de rayon $\varepsilon/2$. Posons $K_1 = \overline{\text{co}}(X \setminus \omega)$, $K_2 = \overline{\text{co}}(\omega)$ et

$$C_r = \{\alpha x_1 + (1 - \alpha)x_2 : x_i \in K_i,\ r \leq \alpha \leq 1\}$$

pour tout $r \in [0, 1]$. Les C_r sont des convexes compacts, on a $C_0 = K$ et pour tout $r \in]0, 1]$, $x \notin C_r$ (cf. le 2) de 2.1). Le convexe compact $\Phi(K) - \Phi(K)$ de $C_s(Y, \mathbb{R})$ est absorbé par la boule unité fermée de $C_u(Y, \mathbb{R})$ (cf. [18] page 90) ; il existe donc $t \in]0, 1]$ tel que $\Phi(K) - \Phi(K)$ soit contenu dans la boule fermée de $C_u(Y, \mathbb{R})$ de centre 0 et de rayon $\varepsilon/2t$. On vérifie aisément que $K \setminus C_t$ est inclus dans l'image réciproque V_ε par Φ de la boule ouverte de $C_u(Y, \mathbb{R})$ de centre $\phi(x)$ et de rayon ε (rappelons que toute boule fermée de $C_u(Y, \mathbb{R})$ est un fermé de $C_s(Y, \mathbb{R})$). On en déduit que $\overline{\text{co}}(K \setminus V_\varepsilon) \subset C_t$. Il en résulte que $x \notin \overline{\text{co}}(K \setminus V_\varepsilon)$ et par conséquent que $x \in \text{Int}(V_\varepsilon)$. □

Soient K un convexe compact d'un espace localement convexe séparé, $A(K, R)$ l'espace vectoriel des applications affines et continues de K dans $\mathbb{R}$ et S un sous-ensemble uniformément borné de $A(K, R)$. Dans [19], S. S. Khurana démontre que S est relativement compact dans $A(K, R)$ muni de la topologie de la convergence simple sur ext(K) si et seulement si S est relativement compact dans $A(K, R)$ muni de la topologie de la convergence simple sur K (ce qui généralise un résultat de J. Bourgain et M. Talagrand [5]). La proposition suivante est essentiellement une reformulation de cette propriété.

Proposition 2.3. *Soient K un convexe compact d'un espace localement convexe séparé et Y un espace compact. Soit X un compact de K tel que $\overline{\text{co}}(X) = K$. Soit $f : K \times Y \to \mathbb{R}$ une application affine par rapport à la première variable. Alors, les conditions suivantes sont équivalentes:*

1) *f est séparément continue.*

2) *f est bornée ; f est continue par rapport à la première variable et pour tout $x \in X$, l'application partielle $f_x : Y \to \mathbb{R}$ est continue.*

3) *f est bornée ; f est continue par rapport à la première variable et pour tout $x \in \text{ext}(K)$, l'application partielle $f_x : Y \to \mathbb{R}$ est continue.*

Démonstration. Supposons la condition 1) vérifiée ; pour tout $x \in X$, l'application $f_x : Y \to \mathbb{R}$ est alors bien sûr continue ; montrons que f est bornée. Soit $C(Y, \mathbb{R})$ l'espace vectoriel des applications continues de Y dans $\mathbb{R}$ et soit $C_s(Y, \mathbb{R})$ [resp. $C_u(Y, \mathbb{R})$] l'espace localement convexe séparé [resp. l'espace de Banach] obtenu en munissant $C(Y, \mathbb{R})$ de la topologie de la convergence simple [resp. de la norme $\| \,.\, \|_u$ de la convergence uniforme]. Soit $\Phi_1 : K \to C(Y, \mathbb{R})$ l'application affine définie par $\Phi_1(x)(y) = f(x, y)$ pour tout $x \in K$ et tout $y \in Y$. L'application $\Phi_1 : K \to C_s(Y, \mathbb{R})$ est affine et continue, par conséquent $\Phi_1(K)$ est une partie convexe compacte de $C_s(Y, \mathbb{R})$; il résulte donc du théorème d'absorption (cf. [18] page 90) que $\Phi_1(K)$ est une partie bornée de $C_u(Y, \mathbb{R})$, autrement dit que f est bornée.

Nous avons montré que 1) implique 2) ; le fait que 2) implique 3) résulte de l'inclusion $\operatorname{ext}(K) \subset X$; supposons la condition 3) vérifiée et démontrons 1). Soit $A(K, \mathbb{R})$ l'espace vectoriel des applications affines et continues de K dans $\mathbb{R}$ et soit $A_s(K, \mathbb{R})$ [resp. $A_e(K, \mathbb{R})$] l'espace [vectoriel] topologique [localement convexe séparé] obtenu en munissant $A(K, \mathbb{R})$ de la topologie de la convergence simple sur K [resp. sur $\operatorname{ext}(K)$]. Les espaces topologiques $A_s(K, \mathbb{R})$ et $A_e(K, \mathbb{R})$ sont séparés et la topologie de $A_e(K, \mathbb{R})$ est moins fine que celle de $A_s(K, \mathbb{R})$. Soit $\Phi_2 : Y \to A(K, \mathbb{R})$ l'application définie par $\Phi_2(y)(x) = f(x, y)$ pour tout $x \in K$ et tout $y \in Y$. L'application $\Phi_2 : Y \to A_e(K, \mathbb{R})$ est continue, donc $\Phi_2(Y)$ est un sous-ensemble compact de $A_e(K, \mathbb{R})$; ce sous-ensemble de $A(K, \mathbb{R})$ est uniformément borné car f est par hypothèse bornée ; il résulte par conséquent du théorème de S.S. Khurana rappelé ci-dessus que $\Phi_2(Y)$ est une partie compacte de $A_s(K, \mathbb{R})$; en particulier, les topologies induites sur $\Phi_2(Y)$ par $A_s(K, \mathbb{R})$ et $A_e(K, \mathbb{R})$ coïncident et par conséquent $\Phi_2 : Y \to A_s(K, \mathbb{R})$ est continue. On en déduit que f est séparément continue. □

3. Semigroupes affines semitopologiques: propriétés générales

Définitions 3.1. 1) Un *semigroupe semitopologique* [resp. *topologique*] est un espace topologique non vide muni d'une multiplication associative séparément continue [resp. continue].

Soit S un semigroupe semitopologique [resp. topologique]. Un *sous-semigroupe* de S est une partie non vide de S stable pour la multiplication. Nous regarderons toujours un sous-semigroupe de S comme étant lui-même un semigroupe semitopologique [resp. topologique] en le munissant de la multiplication et de la topologie induites par celles de S.

2) Soit E un espace localement convexe séparé (réel). Un *semigroupe affine semitopologique* [resp. *topologique*] dans E est un sous-espace topologique convexe non vide de E muni d'une multiplication associative séparément affine et séparément continue [resp. continue].

Soit S un semigroupe affine semitopologique [resp. topologique] dans E. Un *sous-semigroupe affine* de S est une partie convexe non vide de S stable pour la multiplication. Nous regarderons toujours un sous-semigroupe affine de S comme étant lui-même un semigroupe affine semitopologique [resp. topologique] dans E en le munissant de la multiplication et de la topologie induites par celles de S.

La proposition suivante est immédiate.

Proposition 3.2. 1) *Soit S un semigroupe semitopologique et soit A un sous-semigroupe de S. Alors, l'adhérence $\overline{A}$ de A dans S est un sous-semigroupe de S. Si A est commutatif, $\overline{A}$ est commutatif.*

2) *Soit E un espace localement convexe séparé et soit S un semigroupe affine semitopologique dans E fermé dans E. Soit A un sous-semigroupe de S. Alors, l'enveloppe convexe* $\text{co}(A)$ *[resp. l'enveloppe convexe fermée $\overline{\text{co}}(A)$] de A dans E est un sous-semigroupe affine de S. Si A est commutatif,* $\text{co}(A)$ *et* $\overline{\text{co}}(A)$ *sont commutatifs.*

Le résultat suivant est une conséquence de 2.3. Notons que l'équivalence entre 1) et 2) est établie dans [3], par une autre méthode, dans le cas particulier où X est un sous-semigroupe fermé de S.

Proposition 3.3. *Soit E un espace localement convexe séparé. Soit S un sous-espace topologique convexe compact non vide de E muni d'une multiplication associative séparément affine. Pour tout $s \in S$, soit λ_s [resp. ρ_s] l'application de S dans S définie par $\lambda_s(z) = sz$ [resp. $\rho_s(z) = zs$] pour tout $z \in S$. Soit X un compact de S tel que $\overline{\text{co}}(X) = S$. Alors, les conditions suivantes sont équivalentes:*

1) *S est un semigroupe affine semitopologique compact dans E.*

2) *ρ_s est continue pour tout $s \in S$ et λ_x est continue pour tout $x \in X$.*

3) *ρ_s est continue pour tout $s \in S$ et λ_x est continue pour tout $x \in$* $\text{ext}(S)$.

Démonstration. Il est évident que 1) implique 2) ; le fait que 2) implique 3) résulte de l'inclusion $\text{ext}(S) \subset X$; supposons la condition 3) vérifiée et établissons 1). Soit $g : S \to \mathbb{R}$ une application affine et continue et soit $f : S \times S \to \mathbb{R}$ l'application définie par $f(a, b) = g(ab)$; f est bornée (car g est bornée) ; f est affine et continue par rapport à la première variable et pour tout $x \in \text{ext}(S)$, $f \circ \lambda_x : S \to \mathbb{R}$ est continue. Il résulte donc de la proposition 2.3 que f est séparément continue. On en déduit que la multiplication de S est séparément continue (en effet, les applications affines et continues de S dans $\mathbb{R}$ engendrent la topologie de S). □

Exemples 3.4. Rappelons deux exemples bien connus de semigroupes affines semitopologiques compacts.

α) **Exemple 1** (I. Glicksberg [14], [15]).

Soient S un semigroupe semitopologique compact et $C_u(S, R)$ l'espace de Banach, pour la norme uniforme, des applications continues de S dans $\mathbb{R}$. Désignons par $C_u(S, \mathbb{R})^*$ l'espace localement convexe séparé obtenu en munissant le dual topologique de $C_u(S, \mathbb{R})$ de la topologie faible. Soit $\widetilde{S}$ le sous-espace topologique convexe compact de $C_u(S, \mathbb{R})^*$ constitué par les fonctionnelles linéaires non négatives et normalisées sur $C_u(S, \mathbb{R})$. (Si $\mu \in \widetilde{S}$ et si $f \in C_u(S, \mathbb{R})$, $\mu(f)$ est encore noté $\int_S f(s)d\mu(s)$, notation justifiée par le théorème de représentation de F. Riesz [11]). Soient μ et $\nu \in \widetilde{S}$; on voit facilement (en utilisant 2.3) que pour tout $f \in C_u(S, \mathbb{R})$, l'application de S dans $\mathbb{R}$ qui à y associe $\int_S f(xy)d\mu(x)$ est continue. L'application de $C_u(S, \mathbb{R})$ dans $\mathbb{R}$ qui à f associe $\int_S \int_S f(xy)d\mu(x)d\nu(y)$ $(= \int_S \int_S f(xy)d\nu(y)d\mu(x)$ [15]) est un élément de $\widetilde{S}$ noté $\mu * \nu$ et appelé le produit de convolution de μ et ν. Munissons $\widetilde{S}$ du produit de convolution $*$; $\widetilde{S}$ est alors un semigroupe affine semitopologique compact dans $C_u(S, \mathbb{R})^*$. Si S est commutatif, $\widetilde{S}$ est également commutatif. Notons que l'application de S dans $\widetilde{S}$ qui à s associe l'élément δ_s de $\widetilde{S}$ défini par $\delta_s(f) = f(s)$ pour tout $f \in C_u(S, \mathbb{R})$, permet d'identifier (algébriquement et topologiquement) S au sous-semigroupe $\{\delta_s : s \in S\}$ de $\widetilde{S}$. Notons aussi que l'on a $\widetilde{S} = \overline{\text{co}}\{\delta_s : s \in S\}$ [11]. On pourra consulter [2] ou [3] (ou encore [14], [15]) pour une démonstration détaillée de ces propriétés.

β) **Exemple 2** (K. DeLeeuw et I. Glicksberg [9]).
Soit B un espace de Banach (réel) et soit S un semigroupe d'opérateurs (linéaires continus) sur B (autrement dit, un ensemble non vide d'opérateurs sur B stable pour la composition des opérateurs). Supposons S faiblement presque-périodique, c'est-à-dire que $\{s(x): s \in S\}$ est une partie faiblement relativement compacte de B pour tout $x \in B$. On rencontre bien sûr ce type de situation lors de l'étude des fonctions faiblement presque-périodiques sur un semigroupe [9], [12]. Désignons par $L_w(B)$ l'espace localement convexe séparé obtenu en munissant l'espace vectoriel $L(B)$ des opérateurs sur B de la topologie faible des opérateurs ; alors, l'enveloppe convexe fermée $\overline{\text{co}}(S)$ de S dans $L_w(B)$, munie de la composition des opérateurs, est un semigroupe affine semitopologique compact dans $L_w(B)$ (auquel nous nous intéresserons tout particulièrement dans la section 8). L'étude de (B, S) se trouve facilitée par l'introduction de $\overline{\text{co}}(S)$.

Définition 3.5. Soient S un semigroupe semitopologique et Y un espace topologique. Une *action* de S dans Y est une application de $S \times Y$ dans Y qui à (s, y) associe un point noté sy telle que pour tous $s, t \in S$ et tout $y \in Y$, l'on ait $(st)y = s(ty)$. Les actions de S dans Y considérées seront séparément continues ; en fait, S sera ici un semigroupe affine semitopologique compact dans un espace localement convexe séparé E, Y sera un sous-espace topologique convexe compact d'un espace localement convexe séparé F et les actions de S dans Y considérées seront séparément affines et séparément continues.

La proposition suivante se démontre comme 3.3.

Proposition 3.6. *Soient S un semigroupe affine semitopologique compact dans un espace localement convexe séparé E, K un sous-espace topologique convexe compact d'un espace localement convexe séparé F et ϕ une action séparément affine de S dans K. Soit X un compact de S tel que $\overline{\text{co}}(X) = S$. Alors, les conditions suivantes sont équivalentes:*

1) *ϕ est séparément continue.*

2) *ϕ_y est continue pour tout $y \in K$ et ϕ_x est continue pour tout $x \in X$.*

3) *ϕ_y est continue pour tout $y \in K$ et ϕ_x est continue pour tout $x \in \text{ext}(S)$.*

Remarque 3.7. Nous avons en (3.4,α) identifié tout semigroupe semitopologique compact à un sous-semigroupe d'un semigroupe affine semitopologique compact; on peut, de la même façon, prolonger toute action séparément continue ϕ d'un semigroupe semitopologique compact S dans un espace compact Y en une action séparément affine et séparément continue $\widetilde{\phi}$ d'un semigroupe affine semitopologique compact $\widetilde{S}$ dans un convexe compact $\widetilde{Y}$. Nous omettons les détails (cf. (3.4,α)).

4. Points de continuité à gauche d'une action séparément affine

Dans toute cette section, on désigne par S un semigroupe affine semitopologique compact dans un espace localement convexe séparé E, par K un sous-espace topologique convexe compact d'un espace localement convexe séparé F et par $\phi: S \times K \to K$ une action de S dans K séparément affine et séparément continue. On pose

$$T = \{s: s \in S,\ Ss = S\} \text{ et } U = \{s: s \in S,\ sS = S\}.$$

La proposition suivante est immédiate.

Proposition 4.1. 1) *Si T [resp. U] n'est pas vide, alors T [resp. U] est un sous-semigroupe de S.*

2) *Si S admet un élément neutre, T [resp. U] n'est autre que l'ensemble des éléments de S inversibles à gauche [resp. à droite] et $T \cap U$ est l'ensemble des éléments inversibles de S.*

3) *S admet un élément neutre si et seulement si $T \cap U$ est non vide.*

L'énoncé de base de cette section est le théorème 4.2. Ce théorème comporte deux parties:

– la première de ces parties est démontrée dans [36] en supposant K égal à S et en prenant pour action ϕ la multiplication dans S (restriction qui ne simplifie d'ailleurs en rien la démonstration). Cette première partie de 4.2 est essentiellement une application de 2.2. Plusieurs résultats connus obtenus dans [13], [16], [29], [33], [34], [36] sont retrouvés ici comme corollaires de (4.2,1).

– la seconde partie du théorème 4.2 est une extension simple d'un théorème bien connu de J. D. Lawson rappelé en 4.9. Pour obtenir cette extension, nous nous appuierons à nouveau sur le théorème 2.2 ; signalons toutefois que le théorème de I. Namioka [25] concernant l'existence de points de continuité jointe pour une fonction réelle séparément continue définie sur le produit de deux compacts, suffisait à sa démonstration.

Théorème 4.2. *Soit $s \in S$. Alors:*

1) Si $s \in \overline{\mathrm{co}}(sU)$, s est un point de continuité à gauche de la restriction de ϕ à $\overline{\mathrm{co}}(sU) \times K$. En particulier, si $\overline{\mathrm{co}}(U) = S$, s est un point de continuité à gauche de la restriction de ϕ à $(sS) \times K$.

2) Si $s \in sS \cap Ss$, s est un point de continuité à gauche de la restriction de ϕ à $(sS \cap Ss) \times K$. En particulier, si S est commutatif, s est un point de continuité à gauche de la restriction de ϕ à $(sS) \times K$.

Démonstration. 1) Soit f une application affine et continue de K dans $\mathbb{R}$ et soit $\varepsilon > 0$. Montrons qu'il existe un voisinage V de s dans $\overline{\mathrm{co}}(sU)$ tel que pour tout $s' \in V$, l'on ait

$$\sup_{y \in K} |f(sy) - f(s'y)| \leq \varepsilon.$$

Il en résultera que s est un point de continuité à gauche de la restriction de $f \circ \phi$ à $\overline{\mathrm{co}}(sU) \times K$ (cf. (2.1,1,β)), donc que s est un point de continuité à gauche de la restriction de ϕ à $\overline{\mathrm{co}}(sU) \times K$ (en effet, les applications affines et continues de K dans $\mathbb{R}$ engendrent la topologie de K). Soit $x \in \overline{\mathrm{co}}(sU)$ un point de continuité à gauche extrémal de la restriction de $f \circ \phi$ à $\overline{\mathrm{co}}(sU) \times K$ (cf. 2.2) et soit V_1 un voisinage de x dans $\overline{\mathrm{co}}(sU)$ tel que pour tout x' $\in V_1$, l'on ait

$$\sup_{y \in K} |f(xy) - f(x'y)| \leq \varepsilon/2$$

(cf. (2.1,1,β)). Comme x est un point extrémal de $\overline{\mathrm{co}}(sU)$, x appartient à $\overline{sU}$ (cf. (2.1,2,β)). Soit $u \in U$ tel que $su \in V_1$. Les translations à droite dans S sont affines et continues ; par conséquent, $(\overline{\mathrm{co}}(sU))u$ est inclus dans $\overline{\mathrm{co}}(sU)$ et il existe un voisinage V de s dans $\overline{\mathrm{co}}(sU)$ tel que $Vu \subset V_1$. Pour tout $s' \in V$, on a alors

$$\begin{aligned}\sup_{y\in K}|f(sy)-f(s'y)| &= \sup_{y\in K}|f(suy)-f(s'uy)| \\ &\leq \sup_{y\in K}|f(suy)-f(xy)| + \sup_{y\in K}|f(xy)-f(s'uy)| \\ &\leq \varepsilon\end{aligned}$$

2) Soit $y \in K$; montrons que la restriction de ϕ à $(sS\cap Ss)\times K$ est continue en (s,y). Soit (s_α) une suite généralisée de points de $sS \cap Ss$ telle que $\lim s_\alpha = s$ et soit (y_α) une suite généralisée de points de K telle que $\lim y_\alpha = y$; montrons que $\lim s_\alpha y_\alpha = sy$. Soit z une valeur d'adhérence de $(s_\alpha y_\alpha)$ dans K et soit $f : K \to \mathbb{R}$ une application affine et continue ; il suffit de montrer que $f(z) = f(sy)$ (en effet, d'une part K est compact et d'autre part les applications affines et continues de K dans $\mathbb{R}$ séparent les points de K). Soit A l'ensemble des points de continuité à gauche de la restriction de $f \circ \phi$ à $(Ss) \times K$ (Ss est un convexe compact de E car les translations dans S sont affines et continues); d'après le théorème 2.2, $A \cap \mathrm{ext}(Ss)$ est dense dans $\mathrm{ext}(Ss)$. Soit $a \in A$ et soit $r \in S$ tel que $a = rs$; la suite généralisée (rs_α) converge vers a dans Ss et la suite généralisée (y_α) converge vers y dans K, par conséquent, la suite généralisée $(f(rs_\alpha y_\alpha))$ converge vers $f(ay)$; $f(rz)$ étant une valeur d'adhérence de $(f(rs_\alpha y_\alpha))$, on a $f(ay) = f(rz)$. Le point z appartient à sK ; en effet, sK est un fermé de K, z est une valeur d'adhérence de $(s_\alpha y_\alpha)$ et pour tout α, $s_\alpha y_\alpha$ appartient à sK car s_α appartient à sS. Soit $y_1 \in K$ tel que $z = sy_1$; on a

$$f(ay_1) = f(rsy_1) = f(rz) = f(ay)$$

pour tout a ∈A. L'application $f \circ \phi$ étant affine et continue par rapport à la première variable et $\overline{\mathrm{co}}(A)$ étant égal à Ss, on a $f(by_1) = f(by)$ pour tout $b \in Ss$; en particulier, puisque s appartient à Ss (par hypothèse, s appartient à $sS \cap Ss$), on a $f(sy_1) = f(sy)$ c'est à dire $f(z) = f(sy)$. □

On a bien sûr le corollaire suivant de 4.2:

Corollaire 4.3. *Soit $s \in S$. Alors:*

1) Si $s \in \overline{\mathrm{co}}(sU)$, s est un point de continuité à gauche de la restriction à $\overline{\mathrm{co}}(sU)\times S$ de la multiplication de S. En particulier, si $\overline{\mathrm{co}}(U) = S$, s est un point de continuité à gauche de la restriction à $(sS) \times S$ de la multiplication de S.

2) Si $s \in sS \cap Ss$, s est un point de continuité à gauche de la restriction à $(sS \cap Ss) \times S$ de la multiplication de S. En particulier, si S est commutatif, s est un point de continuité à gauche de la restriction à $(sS) \times S$ de la multiplication de S. (Ce cas particulier est signalé dans [31] *par* W. Ruppert).

Démonstration. S agit dans lui-même par translations à gauche de manière séparément affine et séparément continue ; 4.3 découle donc de 4.2. □

Remarque 4.4. 1) On a bien sûr un énoncé symétrique de 4.3 en considérant $S\times\overline{\mathrm{co}}(Ts)$ et $S\times(sS\cap Ss)$ au lieu de $\overline{\mathrm{co}}(sU)\times S$ et $(sS\cap Ss)\times S$. Pour l'obtenir, il suffit d'introduire sur S la loi interne qui à (s,t) associe ts et de s'appuyer sur 4.3. Cette remarque s'applique à bon nombre des énoncés qui vont suivre.

2) Signalons que dans [29] apparaît le résultat suivant: soit A un semigroupe semitopologique compact contenant un groupe dense et soit a ∈A. Alors, a est un point de continuité à gauche de la restriction à $(aA)\times A$ de la multiplication de A. Ce résultat peut être déduit de (4.3,1) (cf. (3.4,α)), (mais ce n'est pas la démonstration la plus directe).

Définitions 4.5. Soit A un semigroupe semitopologique. On dit que A est *topologiquement simple à droite* [resp. *à gauche*] si pour tout $a\in A$, aA [resp. Aa] est dense dans A. On dit que A est *simple à droite* [resp. *à gauche*] si pour tout $a\in A$, on a $aA = A$ [resp. $Aa = A$].

Le résultat suivant provient de [16].

Proposition 4.6. *Tout sous-semigroupe de S topologiquement simple à droite (ou à gauche) est un semigroupe topologique. En particulier, tout sous-semigroupe de S simple à droite (ou à gauche) est un semigroupe topologique.*

Démonstration. Soit A un sous-semigroupe de S topologiquement simple à droite. L'enveloppe convexe fermée $\overline{\mathrm{co}}(A)$ de A dans E est un semigroupe affine semitopologique compact (cf. (3.1,2) et (3.2,2)). Pour tout $a\in A$, $a(\overline{\mathrm{co}}(A))$ est égal à $\overline{\mathrm{co}}(A)$ donc, d'après (4.3,1), a est un point de continuité à gauche de la multiplication de $\overline{\mathrm{co}}(A)$. On en déduit que A est un semigroupe topologique. □

Corollaire 4.7. *Tout sous-groupe G de S est un groupe topologique compact.*

Démonstration. La multiplication dans G est continue d'après 4.6. On vérifie aisément que l'inversion dans G est continue. En effet, désignons par e l'élément neutre de G, par g un élément de G et par (g_α) une suite généralisée d'éléments de G telle que $\lim g_\alpha = g$. Soit $(g_{\alpha_\beta}^{-1})$ une sous-suite généralisée de (g_α^{-1}) convergeant vers un élément h de S. On a

$$gh = (\lim g_{\alpha_\beta})(\lim g_{\alpha_\beta}^{-1}) = \lim g_{\alpha_\beta} g_{\alpha_\beta}^{-1} = e;$$

de même on a $hg = e$; il en résulte que l'inverse de g dans G est la seule valeur d'adhérence de (g_α^{-1}) dans G et par conséquent que (g_α^{-1}) converge dans G vers g^{-1}. □

Remarque 4.8. Le résultat précédent est dû à J. D. Lawson [20]. On peut en déduire le résultat plus classique de R. Ellis [13].

On suppose désormais, et jusqu'à la fin de cette section, que S admet un élément neutre e. Les deux théorèmes suivants (auxquels nous ferons appel à diverses reprises) sont des résultats de J. D. Lawson [20], [21].

Théorème 4.9. *Tout point s de $T \cap U$ est un point de continuité à gauche de ϕ.*

Démonstration. C'est une conséquence de (4.2,2) ; en effet, un point s de S appartient à $T \cap U$ si et seulement si $sS \cap Ss = S$. □

Théorème 4.10. *Tout point de $\{(T \cap U) \times S\} \cup \{S \times (T \cap U)\}$ est un point de continuité de la multiplication dans S.*

Démonstration. Il résulte de (4.3,2) que la multiplication de S est continue en tout point de $(T \cap U) \times S$. Par symétrie, elle est aussi continue en tout point de $S \times (T \cap U)$. □

Le théorème 4.12, qui généralise 4.9, provient de [16]. Nous ferons appel à ce théorème dans la section 8. Sa démonstration utilise le lemme suivant qui provient également de [16].

Lemma 4.11. *Soit $s \in S$. Alors, les conditions suivantes sont équivalentes:*
1) *e appartient à $\overline{sU}$.*
2) *s appartient à $\overline{T} \cap U$.*

Démonstration. Supposons que $e \in \overline{sU}$; montrons que $s \in \bar{h}T \cap U$. Soit (u_α) une suite généralisée d'éléments de U telle que $\lim su_\alpha = e$. Pour tout α, soit $t_\alpha \in S$ tel que $u_\alpha t_\alpha = e$. On peut supposer que (u_α) et (t_α) convergent dans S (en effet, S est compact). Il résulte de 4.10 que

$$s = \lim s(u_\alpha t_\alpha) = \lim(su_\alpha)t_\alpha = \lim(su_\alpha) \lim t_\alpha = \lim t_\alpha;$$

par conséquent, $s \in \overline{T}$. On a d'autre part $e = s(\lim u_\alpha)$, donc $s \in \overline{T} \cap U$.

Supposons inversement que $s \in \overline{T} \cap U$; montrons que $e \in \overline{sU}$. Soit (t_α) une suite généralisée d'éléments de T telle que $\lim t_\alpha = s$. Pour tout α, soit $u_\alpha \in S$ tel que $u_\alpha t_\alpha = e$. On peut supposer que (u_α) converge dans S (en effet, S est compact). soit $t \in S$ tel que $e = st$. Il résulte de 4.10 que

$$e = st = s \lim(u_\alpha t_\alpha)t = s \lim u_\alpha(t_\alpha t) = s \lim u_\alpha \lim t_\alpha t = s \lim u_\alpha = \lim su_\alpha.$$

Par conséquent, $e \in \overline{sU}$. □

Théorème 4.12. *Tout point de $\overline{T} \cap U$ est un point de continuité à gauche de ϕ.*

Démonstration. Soit $s \in \overline{T} \cap U$, soit f une application affine et continue de K dans $\mathbb{R}$ et soit $\varepsilon > 0$. Montrons qu'il existe un voisinage V de s dans S tel que pour tout $s' \in V$, l'on ait

$$\sup_{y \in K} |f(sy) - f(s'y)| \leq \varepsilon.$$

Il en résultera que s est un point de continuité à gauche de $f \circ \phi$ (cf. (2.1,1,β)), donc que s est un point de continuité à gauche de ϕ (en effet, les applications affines et continues de K dans $\mathbb{R}$ engendrent la topologie de K). D'après 4.9, e est un point de continuité à gauche de $f \circ \phi$; soit V_1 un voisinage de e dans S tel que pour tout $x \in V_1$, l'on ait

$$\sup_{y \in K} |f(ey) - f(xy)| \leq \varepsilon/2.$$

D'après 4.11, $e \in \overline{sU}$; soit $u \in U$ tel que $su \in V_1$. Les translations à droite dans S étant continues, il existe un voisinage V de s dans S tel que $Vu \subset V_1$. Soit $t \in S$ tel que $ut = e$; alors, pour tout $s' \in V$, on a

$$\begin{aligned}\sup_{y \in K} |f(sy) - f(s'y)| &= \sup_{y \in K} |f((su)(ty)) - f((s'u)(ty))| \\ &\leq \sup_{y \in K} |f((su)(ty)) - f(e(ty))| + \sup_{y \in K} |f(e(ty)) - f((s'u)(ty))| \\ &\leq \varepsilon.\end{aligned}$$

Corollaire 4.13. *Tout point de $\{(\overline{T} \cap U) \times S\} \cup \{S \times (T \cap \overline{U})\}$ est un point de continuité de la multiplication dans S.*

Démonstration. S agit dans lui même par translations à gauche de manière séparément affine et séparément continue donc, d'après 4.12, la multiplication de S est continue en tout point de $(\overline{T} \cap U) \times S$. Par symétrie, elle est aussi continue en tout point de $S \times (T \cap \overline{U})$. □

L'énoncé 4.15 est essentiellement le théorème 4.1 de [34] (cet énoncé sera complété en 5.6). Il s'agit d'un corollaire du lemme 4.11 (ou plus précisément de la reformulation suivante de 4.11 pour ce qui concerne l'équivalence entre (4.15,1) et (4.15,2)).

Lemma 4.14. *Soient u, $t \in S$ tels que $ut = e$. Alors, u appartient à $\overline{T}$ si et seulement si $t \in \overline{U}$.*

Démonstration. Si $t \in \overline{U}$, alors $ut(= e)$ appartient à $\overline{uU}$ et d'après 4.11, $u \in \overline{T}$. Par symétrie, si $u \in \overline{T}$ alors $t \in \overline{U}$. □

Théorème 4.15. *Les conditions suivantes sont équivalentes:*

1) *U est inclus dans $\overline{T}$.*
2) *T est inclus dans $\overline{U}$.*
3) *$\overline{T}$ est égal à $\overline{U}$.*
4) *Le sous-semigroupe T de S est topologiquement simple à gauche.*
5) *Le sous-semigroupe U de S est topologiquement simple à droite.*

Démonstration. Les conditions 1) et 2) sont équivalentes d'après 4.14 et ces conditions sont bien sûr équivalentes à 3). Il est immédiat que 4) est vérifiée si et seulement si $e \in \overline{Tt}$ pour tout $t \in T$; on en déduit que 1) implique 4) ; on en déduit également, par application de 4.11, que 4) implique 2) ; les quatre premières conditions sont donc équivalentes. L'équivalence entre 3) et 4) implique, par symétrie, l'équivalence entre 3) et 5). □

5. Points extrémaux dans un semigroupe affine

Dans toute cette section, on désigne par S un semigroupe affine semitopologique compact dans un espace localement convexe séparé E. On pose

$$T = \{s: s \in S, Ss = S\} \text{ et } U = \{s: s \in S, sS = S\}$$

comme dans la section précédente.

L'énoncé principal de cette section est le théorème 5.1. Il comporte deux parties (comme le théorème 4.2 de la section 4):

– dans la première partie nous considérons, comme en (4.2,1), le convexe compact $\overline{\text{co}}(sU)$ de E (s étant un point fixé de S) et montrons que s en est un point extrémal (si $s \in \overline{\text{co}}(sU)$). Cette propriété est prouvée dans [36] en supposant que $\overline{\text{co}}(U) = S$ (restriction qui ne simplifie en rien la démonstration). Signalons que I. Namioka [26] a obtenu un résultat de ce type, mais pour une autre classe d'actions affines.

– dans la seconde partie du théorème 5.1 nous considérons, comme en (4.2,2), le convexe compact $sS \cap Ss$ de E (s étant un point fixé de S) et montrons que s en est un point extrémal (si $s \in sS \cap Ss$).

Dans le cas particulier où $sS = Ss = S$ (autrement dit, dans le cas particulier où S admet un élément neutre et où s est inversible (cf.4.1)), ce résultat est dû à H. Cohen et H. S. Ccollins [8] (cf. également K. DeLeeuw et I. Glicksberg [9] ainsi que J. G. Wendel [38]).

Dans le cas particulier où S est commutatif, (5.1,2) peut être déduit d'un théorème prouvé en 1966 par S. Dubuc [10]. Notre démonstration de (5.1,2) est d'ailleurs voisine de celle du théorème de S. Dubuc.

Théorème 5.1. *Soit $s \in S$. Alors:*

1) *Si $s \in \overline{\text{co}}(sU)$, s est un point extrémal de $\overline{\text{co}}(sU)$. En particulier, si $\overline{\text{co}}(U) = S$, s est un point extrémal de sS.*

2) *Si $s \in sS \cap Ss$, s est un point extrémal de $sS \cap Ss$. En particulier, si S est commutatif, s est un point extrémal de sS.*

Démonstration. 1) Supposons que $s \in \overline{\text{co}}(sU)$. Soient $a, b \in \overline{\text{co}}(sU)$ et $\gamma \in]0,1[$ tels que $s = \gamma a + (1-\gamma)b$; nous devons montrer que $a = b$. Soit $f : S \to \mathbb{R}$ une application affine et continue ; montrons que $f(a) = f(b)$; il en résultera que $a = b$, car les applications affines et continues de S dans $\mathbb{R}$ séparent les points de S. Soit $z \in \overline{\text{co}}(sU)$ un point de continuité à gauche extrémal de l'application de $\overline{\text{co}}(sU) \times S$ dans $\mathbb{R}$ qui à (x, y) associe $f(xy)$ (cf. 2.2) et soit $v \in \overline{U}$ tel que $z = sv$ (cf.(2.1,2,β)). On a $z = \gamma av + (1-\gamma)bv$ (car les translations à droite dans S sont affines) donc on a $z = av = bv$ (en effet, z est un point extrémal de $\overline{\text{co}}(sU)$ et $av, bv \in \overline{\text{co}}(sU)$). Soit (u_α) une suite généralisée d'éléments de U convergeant vers v dans S et pour tout α, soit $t_\alpha \in T$ tel que $u_\alpha t_\alpha = e$. On peut supposer que (t_α) converge dans S (car S est compact). On a alors:

$$\begin{aligned} f(a) = f(\lim a u_\alpha t_\alpha) &= f(\lim a u_\alpha \lim t_\alpha) = f(\lim b u_\alpha \lim t_\alpha) \\ &= f(\lim b u_\alpha t_\alpha) = f(b). \end{aligned}$$

2) Supposons que $s \in sS \cap Ss$. Soient $a, b \in sS \cap Ss$ et $\alpha \in]0,1[$ tels que $s = \alpha a + (1-\alpha)b$; nous devons montrer que $a = b$. Soient $a_1, b_1 \in S$ tels que $a = a_1 s$ et $b = b_1 s$; on a $s = \alpha a_1 s + (1-\alpha)b_1 s$. Soit $z \in \text{ext}(sS)$ et soit $z_1 \in S$ tel que $z = sz_1$; les translations à droite dans S sont affines, donc on a

$$z = sz_1 = (\alpha a_1 s + (1-\alpha)b_1 s)z_1 = \alpha a_1 s z_1 + (1-\alpha)b_1 s z_1 = \alpha a_1 z + (1-\alpha)b_1 z.$$

On a $a_1 z = a_1 s z_1 = a z_1$, donc $a_1 z \in sS$ (car $a \in sS$). De même,$b_1 z \in sS$. On a par conséquent $z = a_1 z = b_1 z$.

Pour tout $z \in \text{ext}(sS)$, on a $a_1 z = b_1 z$; les translations à droite dans S étant affines et continues, on a $a_1 x = b_1 x$ pour tout $x \in \overline{\text{co}}(sU)$ par application du théorème de Krein-Milman ; en particulier, on a $a_1 s = b_1 s$ c'est-à-dire $a = b$. □

On suppose désormais, et jusqu'à la fin de cette section, que S admet un élément neutre e.

Corollaire 5.2. ([8], [9], [38]) *Tout élément inversible de S est un point extrémal de S.*

Démonstration. En effet, un point s de S est inversible si et seulement si $sS \cap Ss = S$. □

Le théorème suivant généralise 5.2 (de la même façon que 4.12 généralisait 4.9). Nous ferons appel à ce théorème dans la section 8.

Théorème 5.3. *Tout élément de $(\overline{T} \cap U) \cup (T \cap \overline{U})$ est un point extrémal de S.*

Démonstration. Soit $s \in \overline{T} \cap U$; montrons que s est un point extrémal de S. (Par symétrie, tout point de $T \cap \overline{U}$ est alors également un point extrémal de S). Soient $a, b \in S$ et $\gamma \in]0, 1[$ tels que $s = \gamma a + (1 - \gamma)b$; nous devons montrer que $a = b$. Soit $t \in T$ tel que $st = e$; on a $e = \gamma at + (1 - \gamma)bt$ (car les translations à droite dans S sont affines) ; on a donc $e = at = bt$ d'après 5.2. Le point s appartient à $\overline{T}$, par conséquent d'après 4.14, t appartient à $\overline{U}$. Soit (u_α) une suite généralisée d'éléments de U convergeant vers t dans S et pour tout α, soit $t_\alpha \in T$ tel que $u_\alpha t_\alpha = e$. On peut supposer que (t_α) converge dans S (car S est compact). On a alors:

$$a = \lim au_\alpha t_\alpha = \lim au_\alpha \lim t_\alpha = \lim bu_\alpha \lim t_\alpha = \lim bu_\alpha t_\alpha = b$$

(ceci, compte-tenu de 4.10). □

Le lemme suivant complète le lemme 4.14.

Lemma 5.4. *Soient $u, t \in S$ tels que $ut = e$. Alors, les conditions suivantes sont équivalentes:*

1) *u appartient à $\overline{T}$.*
2) *t appartient à $\overline{U}$.*
3) *u appartient à $\overline{\mathrm{co}}(T)$.*
4) *t appartient à $\overline{\mathrm{co}}(U)$.*

Démonstration. L'équivalence entre 1) et 2) est prouvée en 4.14. Si u appartient à $\overline{T}$, alors u appartient à $\overline{\mathrm{co}}(T)$. Inversement, supposons que u appartienne à $\overline{\mathrm{co}}(T)$; le point e appartient alors à $\overline{\mathrm{co}}(Tt)$ (car les translations à droite dans S sont affines et continues). Il résulte donc de 5.2 et du théorème de Milman que e appartient à $\overline{Tt}$, donc que t appartient à $\overline{U}$ (cf.4.11). On en déduit que u appartient à $\overline{T}$ (cf.4.14). Les conditions 1) et 3) sont donc équivalentes. Par symétrie, 2) est alors équivalente à 4). □

Remarque 5.5. L'équivalence entre les conditions 1) et 3) [resp. 2] et 4)) de 5.4 signifie que l'on a $\overline{T} \cap U = \overline{\mathrm{co}}(T) \cap U$ [resp. $T \cap \overline{U} = T \cap \overline{\mathrm{co}}(U)$].

Le théorème suivant complète le théorème 4.15.

Théorème 5.6. *Les conditions suivantes sont équivalentes:*

1) *U est inclus dans $\overline{T}$.*
2) *T est inclus dans $\overline{U}$.*
3) *$\overline{T}$ est égal à $\overline{U}$.*
4) *Le sous-semigroupe T de S est topologiquement simple à gauche.*
5) *Le sous-semigroupe U de S est topologiquement simple à droite.*
6) *U est inclus dans $\overline{\mathrm{co}}(T)$.*
7) *T est inclus dans $\overline{\mathrm{co}}(U)$.*
8) *$\overline{\mathrm{co}}(T)$ est égal à $\overline{\mathrm{co}}(U)$.*
9) *On a $\overline{\mathrm{co}}(Tt) = \overline{\mathrm{co}}(T)$ pour tout $t \in T$.*
10) *On a $\overline{\mathrm{co}}(uU) = \overline{\mathrm{co}}(U)$ pour tout $u \in U$.*

Démonstration. Les cinq premières conditions sont équivalentes d'après 4.15. L'inclusion $U \subset \overline{T}$ équivaut à $\overline{T} \cap U = U$, donc à $\overline{\mathrm{co}}(T) \cap U = U$ (cf. 5.5), donc à $U \subset \overline{\mathrm{co}}(T)$; les conditions 1) et 6) sont donc équivalentes ; par symétrie, 2) équivaut à 7). Les conditions équivalentes 6) et 7) sont bien sûr équivalentes à 8). Il est immédiat que 4) [resp. 9)] est vérifiée si et seulement si $e \in \overline{Tt}$ [resp. $e \in \overline{\mathrm{co}}(Tt)$] pour tout $t \in T$; par conséquent, compte-tenu de 5.2 et du théorème de Milman, 4) et 9) sont équivalentes ; par symétrie, 5) et 10) sont également équivalentes. □

Le corollaire suivant de 5.6 est un résultat de [37].

Théorème 5.7. *Les conditions suivantes sont équivalentes:*

1) *$\overline{\mathrm{co}}(T)$ est égal à S.*
2) *$\overline{\mathrm{co}}(U)$ est égal à S.*

Démonstration. La condition 1) implique la condition 2). En effet, si $\overline{\mathrm{co}}(T) = S$ on a $U \subset \overline{\mathrm{co}}(T)$ et d'après 5.6, on a $\overline{\mathrm{co}}(T) = \overline{\mathrm{co}}(U)$; on a donc $\overline{\mathrm{co}}(U) = S$. Par symétrie, 2) implique 1). □

6. Semigroupes affines ergodiques en moyenne

Dans toute cette section,on désigne par S un semigroupe affine semitopologique compact dans un espace localement convexe séparé E. On pose

$$T = \{s: s \in S,\ Ss = S\} \text{ et } U = \{s: s \in S,\ sS = S\}.$$

Definitions 6.1. 1) Soit $a \in S$. On dit que a est un *zéro à gauche* [resp. *à droite*] de S si l'on a $as = a$ [resp. $sa = a$] pour tout $s \in S$. On dit que a est un *zéro* de S si a est à la fois un zéro à gauche et un zéro à droite de S.

2) On dit que S est *ergodique en moyenne* s'il existe un zéro dans S (un tel zéro est alors, bien sûr, unique).

Nous nous appuierons en 6.3 sur le théorème suivant.

Théorème 6.2. *Supposons que tout point s de S soit un point extrémal des convexes compacts Ss et sS de E, dès lors que $s \in Ss$ et $s \in sS$. Alors, S est ergodique en moyenne.*

Démonstration. Ordonnons par inclusion l'ensemble des convexes fermés non vides K' de S tels que $SK' \subset K'$. Le lemme de Zorn assure l'existence, dans cet ensemble ordonné, d'éléments minimaux. Soit K un tel élément minimal; pour tout $x \in K$, on a $Sx = K$, par conséquent x appartient à ext(K); il en résulte que K est réduit à un point, autrement dit que S admet un zéro à droite u. Par symétrie, S admet aussi un zéro à gauche v. Posons $a = uv$; alors, a est un zéro de S. □

Divers résultats obtenus successivement par J. VON NEUMANN [27], S. K. AKUTANI [17], A. MARKOFF et S. KAKUTANI [17], L. ALAOGLU et G. BIRKHOFF [1], W. F. EBERLEIN [12], J. E. L. PECK [28], H. COHEN et H. S. COLLINS [8], K. DELEEUW et I. GLICKSBERG [9], C. RYLL-NARDZEWSKI [32], J. F. BERGLUND et K. H. HOFMANN[2], ont conduit en 1967 ([2] et [9]) à l'énoncé suivant (que nous retrouvons ici par application de 5.1 et 6.2).

Théorème 6.3. *S est ergodique en moyenne dans chacun des deux cas suivants:*

1) *S admet un élément neutre et $\overline{\text{co}}(T) = S$.*

2) *S est commutatif.*

Démonstration. 1) on a $\overline{\text{co}}(T) = S$, donc on a $\overline{\text{co}}(U) = S$ (cf. 5.7) ; par conséquent, d'après (5.1,1), pour tout $s \in S$, s est un point extrémal de Ss et de sS. Il résulte donc de 6.2 que S est ergodique en moyenne.

2) Pour tout $s \in S$, s est d'après (5.1,2) un point extrémal de Ss (= sS). Il résulte donc à nouveau de 6.2 que S est ergodique en moyenne. □

Remarque 6.4. 1) Le corollaire suivant de (6.3.,1) est prouvé dans [8] (en supposant de plus la multiplication de S jointement continue): *s'il existe un sous-groupe fermé G de S tel que $\overline{\mathrm{co}}(G) = S$, S est ergodique en moyenne (notons que l'élément neutre de G est alors élément neutre de S).*

On peut donner, de cet énoncé, la démonstration relativement simple suivante: soit $s \in S$; montrons que $s \in \mathrm{ext}(Ss)$. (Il en résultera, par symétrie, que $s \in \mathrm{ext}(sS)$ et par application de 6.2 que S est ergodique en moyenne). Soient $a, b \in Ss$ et $\alpha \in]0, 1[$ tels que $s = \alpha a + (1 - \alpha)b$; nous devons montrer que $a = b$. Soit $z \in \mathrm{ext}(Ss)$; on a $\overline{\mathrm{co}}(Gs) = S$, donc $z \in Gs$ (Gs est un fermé de Ss) ; soit $g \in G$ tel que $z = gs$; on a $z = \alpha ga + (1 - \alpha)gb$, donc $z = ga = gb$ et par conséquent $a = b$ (car g est inversible dans S).

2) si $\overline{\mathrm{co}}(T) = \overline{\mathrm{co}}(U) = S$, alors S est encore ergodique en moyenne. Cette propriété (un peu plus générale que (6.3,1) d'après 5.7) découle, comme (6.3,1), de (5.1,1) et 6.2.

7. Extension du théorème 6.3

On reprend les notations des précédentes sections; autrement dit, on désigne par S un semigroupe affine semitopologique compact dans un espace localement convexe séparé E et on pose

$$T = \{s: s \in S,\ Ss = S\} \text{ et } U = \{s: s \in S,\ sS = S\}.$$

Definitions 7.1. 1) Un *idempotent* de S est un élément u de S tel que $u^2 = u$.

2) Soit A un sous-semigroupe fermé de S. Le *noyau* de A est l'intersection $N(A)$ des parties non vides I de A telles que $AI \subset I$ et $IA \subset I$.

La proposition suivante est immédiate.

Proposition 7.2. *Soit A un sous-semigroupe affine fermé de S et soit $a \in A$. Alors,on a $N(A) = \{a\}$ si et seulement si A admet a pour* 0.

On a également la propriété simple suivante [9].

Proposition 7.3. *Soit A un sous-semigroupe affine fermé de S. Alors, $N(A)$ est un groupe si et seulement si $N(A)$ est réduit à un point.*

Démonstration. Si $N(A)$ est réduit à un point, alors $N(A)$ est bien sûr un groupe. Inversement, supposons que $N(A)$ soit un groupe et désignons par x un élément de $N(A)$; on a $N(A) = Ax$ (immédiat), par conséquent $N(A)$ est un semigroupe affine semitopologique compact dans E ; $N(A)$ étant un groupe, il résulte de 5.2

que tous les points de $N(A)$ sont des points extrémaux de $N(A)$ et donc que $N(A)$ est réduit à un point. □

Nous nous abstiendrons, dans la démonstration de la proposition suivante, de prouver en détail le fait que la condition 4) implique la condition 1) (implication remarquée par W. Ruppert [29]). Signalons toutefois que la démonstration de ce fait s'appuie sur le théorème 2.3 de [9] concernant la structure du noyau d'un semigroupe semitopologique compact.

Proposition 7.4. [23]. *Les conditions suivantes sont équivalentes:*

1) *Le noyau de tout sous-semigroupe fermé de S est un groupe.*
2) *Le noyau de tout sous-semigroupe fermé de S est un groupe topologique compact.*
3) *Tout sous-semigroupe affine fermé de S est ergodique en moyenne.*
4) *Pour tous idempotents u et v de S , on a $uv = u$ si et seulement si on a $vu = u$.*

Démonstration. Si A est un sous-semigroupe fermé de S et si $N(A)$ est un groupe, alors $N(A)$ est un fermé de S (en effet, si $x \in N(A)$, on a $N(A) = Ax$); $N(A)$ est donc, d'après 4.7, un groupe topologique compact ; par conséquent, la condition 1) implique la condition 2). La condition 2) implique la condition 3) d'après 7.3 et 7.2. Soient u et v deux idempotents de S tels que $uv = u$; $\overline{\mathrm{co}}\{u, v, vu\}$ est un sous-semigroupe affine fermé de S car $\{u, v, vu\}$ est un sous-semigroupe de S ; on a $ux = u$ pour tout $x \in \{u, v, vu\}$, par conséquent on a $ux = u$ pour tout $x \in \overline{\mathrm{co}}\{u, v, vu\}$ (les translations à gauche dans S sont affines et continues) ; si la condition 3) est vérifiée et si a désigne le zéro de $\overline{\mathrm{co}}\{u, v, vu\}$, on a les égalités $ua = u = a$, donc $vu = va = a = u$. Par symétrie, si $vu = u$ on a $uv = u$. La condition 3) implique donc la condition 4). □

Nous nous appuierons en 7.6 sur le théorème suivant (qui, compte-tenu de 7.3, généralise le théorème 6.2).

Théorème 7.5. [23]. *Supposons que tout point s de S soit un point extrémal des convexes compacts Ss et sS de E, dès lors que $s \in Ss$ et $s \in sS$. Alors, le noyau de tout sous-semigroupe fermé de S est un groupe topologique compact.*

Démonstration. Montrons que la condition 4) de 7.4 est vérifiée. Soient u et v des idempotents de S ; supposons que l'on ait $uv = u$; montrons que l'on a $vu = u$. (Par symétrie, si $vu = u$, alors $uv = u$). On a

$$v(\frac{1}{2}vu + \frac{1}{2}u) = \frac{1}{2}v^2u + \frac{1}{2}vu = \frac{1}{2}vu + \frac{1}{2}vu = vu$$

et

$$u(\frac{1}{2}vu + \frac{1}{2}u) = \frac{1}{2}uvu + \frac{1}{2}u^2 = \frac{1}{2}u^2 + \frac{1}{2}u^2 = u^2 = u$$

par conséquent, vu et u appartiennent à $S(\frac{1}{2}vu+\frac{1}{2}u)$. Le point $\frac{1}{2}vu+\frac{1}{2}u$ du convexe $S(\frac{1}{2}vu + \frac{1}{2}u)$ étant par hypothèse extrémal, on a $vu = u$. □

L'énoncé suivant (qui, compte-tenu de 7.3, généralise le théorème 6.3) est l'énoncé principal de cette section. La première des deux parties qu'il comporte est un résultat de [23] ; la seconde provient de [9].

Théorème 7.6. *Le noyau de tout sous-semigroupe fermé de S est un groupe topologique compact dans chacun des deux cas suivants:*
1) *S admet un élément neutre e et $\overline{\text{co}}(T) = S$.*
2) *S est commutatif.*

Démonstration. 1) On a $\overline{\text{co}}(T) = S$, donc on a $\overline{\text{co}}(U) = S$ (cf. 5.7) ; par conséquent, d'après (5.1,1), pour tout $s \in S$, s est un point extrémal de Ss et de sS. Il résulte donc de 7.5 que le noyau de tout sous-semigroupe fermé de S est un groupe topologique compact.

2) C'est une conséquence de ce que dans 7.4, la condition 4) implique la condition 2). □

Remarque 7.7. 1) Si $\overline{\text{co}}(T) = \overline{\text{co}}(U) = S$, alors le noyau de tout sous-semigroupe fermé de S est encore un groupe topologique compact. Cette propriété (un peu plus générale que (7.6,1), d'après 5.7) découle,comme (7.6,1),de (5.1,1) et 7.5.

2) Soit A un semigroupe semitopologique compact et soit $N(A)$ le noyau de A. Supposons que A admette un élément neutre et que l'ensemble des éléments de A inversibles à gauche soit dense dans A. Alors:

1)α) $N(A)$ est un groupe topologique compact [9], [32].

β) Plus généralement, le noyau N(B) de tout sous-semigroupe fermé B de A est un groupe topologique compact [22].

2) Il existe sur A une probabilité régulière invariante par translations [9], [32].

Il est aisé de retrouver ces résultats par application de (7.6,1) (ou encore de (6.3,1) pour ce qui concerne la propriété 2)) ; il suffit pour cela d'introduire le semigroupe affine semitopologique compact $\tilde{A}$ dans $C_u(A, \mathbb{R})^*$ défini en (3.4,α).

Rappelons que la démonstration originale de la propriété 2) s'appuie sur le théorème de point fixe de C. Ryll-Nardzewski [32] ; celle de (1,α) s'appuie, d'une part sur la propriété 2) et d'autre part sur le résultat suivant de K. DeLeeuw et I. Glicksberg [9]:

le noyau d'un semigroupe semitopologique compact est un groupe topologique compact si et seulement s'il existe sur ce semigroupe semitopologique compact une probabilité régulière invariante par translations.

Rappelons également que l'on peut obtenir la propriété 1) par voie purement topologique (cf. [29], [33] pour (1,α) et [22] pour (1,β)).

8. Semigroupes faiblement presque-périodiques d'opérateurs

Cette section a pour objet de présenter quelques applications des résultats établis dans les sections précédentes à l'étude des semigroupes faiblement presque-périodiques d'opérateurs sur un espace de Banach (réel).

Préliminaires 8.1. 1) Soit $(B, \|\,.\,\|)$ un espace de Banach et soit $L(B)$ l'espace vectoriel des opérateurs (linéaires continus) sur B. On note $L_w(B)$ [resp. $L_s(B)$] l'espace localement convexe séparé obtenu en munissant $L(B)$ de la topologie faible [resp. forte] des opérateurs. Rappelons que l'enveloppe convexe fermée d'une partie de $L(B)$ est la même dans $L_w(B)$ et dans $L_s(B)$. Rappelons également que l'enveloppe convexe fermée d'une partie relativement compacte de $L_w(B)$ [resp. $L_s(B)$] est une partie compacte de $L_w(B)$ [resp. $L_s(B)$]. Rappelons enfin qu'une partie A de $L_w(B)$ [resp. $L_s(B)$] est relativement compacte si et seulement si pour tout $x \in B$, $\{a(x): a \in A\}$ est une partie faiblement relativement compacte [resp. relativement compacte] de B.

2) Soit S un semigroupe d'opérateurs sur B (c'est à dire un ensemble non vide d'opérateurs sur B stable pour la composition des opérateurs). On dit que S est *faiblement presque-périodique* [resp. *presque-périodique*] si pour tout $x \in B$, $\{s(x): s \in S\}$ est une partie faiblement relativement compacte [resp. relativement compacte] de B.

3) $(L_w(B), \circ)$ et $(L_s(B), \circ)$ sont des semigroupes affines semitopologiques (dans eux-mêmes). Soit S un semigroupe d'opérateurs sur B; alors, l'enveloppe convexe fermé $\overline{\text{co}}(S)$ de S dans $L_w(B)$ [resp. $L_s(B)$] est un sous-semigroupe affine fermé de $L_w(B)$ [resp. $L_s(B)$] (cf. 3.2). Si S est faiblement presque-périodique [resp. presque-périodique], alors $\overline{\text{co}}(S)$, muni de la composition des opérateurs, est un semigroupe affine semitopologique compact dans $L_w(B)$ [resp. $L_s(B)$].

Notation 8.2. Dans tous les énoncés qui suivent, S désigne un semigroupe faiblement presque-périodique d'opérateurs sur un espace de Banach réel $(B, \|\,.\,\|)$. On suppose que S admet un élément neutre e.

Le théorème suivant est du même type que le théorème 5.2 de [34]. Sa démonstration s'appuie sur les théorèmes 4.2 et 4.12 de la même façon que celle du théorème 5.2 de [34] s'appuyait sur le théorème de J. D. Lawson rappelé en 4.9.

Théorème 8.3. *Supposons que l'on ait* $\overline{\mathrm{co}}(S) = S$. *Posons*

$$T = \{s : s \in S, S \circ s = S\} \text{ et } U = \{s : s \in S, s \circ S = S\}.$$

Soit $s \in S$. *Alors:*

1) *Les voisinages faibles et forts de* s *dans* $\overline{\mathrm{co}}(T \circ s)$ *coïncident.*
2) *Les voisinages faibles et forts de* s *dans* $s \circ S \cap S \circ s$ *coïncident.*
3) *Si* $s \in T \cap \overline{U}^{w.o.}$, *les voisinages faibles et forts de* s *dans* S *coïncident* ($\overline{U}^{w.o.}$ *étant l'adhérence de* U *dans* $L_w(B)$).

Démonstration. On considère S comme un semigroupe affine semitopologique compact dans $L_w(B)$ en adoptant pour loi interne sur S la loi notée '.' définie par $s.t = t \circ s$. Soit A un sous-semigroupe affine fermé de S. Munissons le dual topologique B^* de B de la topologie faible et désignons par K l'enveloppe convexe fermée dans B^* du sous-ensemble

$$Y = \{\gamma \circ s \colon \gamma \in B^*, \parallel \gamma \parallel \leq 1, s \in A\}$$

de B^*. Le semigroupe A est équicontinu (d'après le théorème de Banach-Steinhaus), par conséquent Y est une partie équicontinue de B^* ; K est donc un convexe compact de B^*. Soit ϕ: $A \times K \to K$ l'application définie par $\phi(s, \theta) = \theta \circ s$; alors, ϕ est une action séparément affine et séparément continue de A dans K. Soit $a \in A$ un point de continuité à gauche de ϕ ; montrons que les voisinages faibles et forts de a dans A coïncident: tout voisinage faible de a dans A est bien sûr un voisinage fort de a dans A. Soit (a_α) une suite généralisée d'éléments de A convergeant vers a pour la topologie faible des opérateurs; montrons qu'elle converge vers a pour la topologie forte des opérateurs; il en résultera que tout voisinage fort de a dans A est un voisinage faible de a dans A. Soit $x \in B$. Le point a est un point de continuité à gauche de ϕ donc, compte-tenu de (2.1,1), on a

$$\lim_\alpha \sup_{\theta \in K} |(\theta \circ a_\alpha)(x) - (\theta \circ a)(x)| = 0.$$

On en déduit que

$$\lim_\alpha \sup_{\substack{\gamma \in B^* \\ \parallel \gamma \parallel \leq 1}} |\gamma(a_\alpha(x) - a(x))| = 0,$$

autrement dit, que $\lim_\alpha \parallel a_\alpha(x) - a(x) \parallel = 0$ (d'après le théorème de Hahn-Banach).

Pour obtenir la propriété 1) [resp. 2)], [resp. 3)], il suffit de prendre $A = \overline{\mathrm{co}}(T \circ s)$ [resp. $A = s \circ S \cap S \circ s$], [resp. $A = S$], $a = s$ et de faire appel au théorème (4.2,1) [resp. (4.2,2)], [resp. 4.12]. □

Si S est un groupe et si S est compact pour la topologie faible des opérateurs, il est bien connu que les topologies faible et forte sur S coïncident [9]. Le théorème (8.3,1) permet d'obtenir l'énoncé plus général suivant.

Proposition 8.4. *Si S est topologiquement simple à gauche pour la topologie faible des opérateurs, alors les topologies faible et forte sur S coïncident.*

Démonstration. Considérons le semigroupe affine semitopologique compact $(\overline{\text{co}}(S), \circ)$ dans $L_w(B)$. Pour tout $s \in S$, on a

$$\overline{\text{co}}(S) \circ s = \overline{\text{co}}(S \circ s) = \overline{\text{co}}(S)$$

donc, d'après (8.3,1), les voisinages faibles et forts de s dans $\overline{\text{co}}(S)$ coïncident. On en déduit que les topologies faible et forte sur S coïncident. □

Le théorème 8.6 est du même type que le théorème 5.6 de [34]. Il s'agit d'une application des théorèmes 5.1, 5.3 et 8.3.

Définition 8.5. Soit A une partie non vide d'un espace localement convexe séparé E. Un point a de A est appelé un *point denté* de A dans E si pour tout voisinage V de 0 dans E, $a \notin \overline{\text{co}}(A \setminus (a + V))$.

Théorème 8.6. *Supposons que l'on ait $\overline{\text{co}}(S) = S$. Posons*

$$T = \{s : s \in S,\ S \circ s = S\} \text{ et } U = \{s : s \in S,\ s \circ S = S\}.$$

Soit $s \in S$. Alors:

1) s est un point denté de $\overline{\text{co}}(T \circ s)$ dans $L_s(B)$.

2) s est un point denté de $s \circ S \cap S \circ s$ dans $L_s(B)$.

3) Si $s \in T \cap \overline{U}^{w.o.}$, s est un point denté de S dans $L_s(B)$.

Démonstration. Soit A une partie convexe compacte non vide de $L_w(B)$ et soit a un point de A. Supposons que $a \in \text{ext}(A)$ et que les voisinages faibles et forts de a dans A coïncident ; montrons que a est alors un point denté de A dans $L_s(B)$. Soit V un voisinage de 0 dans $L_s(B)$; $(a + V) \cap A$ est un voisinage de a dans A pour la topologie faible des opérateurs, donc a n'appartient pas à l'adhérence de $A \setminus (a + V)$ dans $L_w(B)$; il en résulte que a n'appartient pas à $\overline{\text{co}}(A \setminus (a + V))$; en effet, dans le cas contraire, a serait un point extrémal de $\overline{\text{co}}(A \setminus (a + V))$, donc a appartiendrait à l'adhérence de $A \setminus (a + V)$ dans $L_w(B)$ (d'après le théorème de Milman). Le point a est par conséquent un point denté de A dans $L_s(B)$.

Pour obtenir la propriété 1) [resp. 2)], [resp.3)], il suffit de prendre $A = \overline{\text{co}}(T \circ s)$ [resp. $A = s \circ S \cap S \circ s$], [resp. $A = S$], $a = s$ et de faire appel aux théorèmes (5.1,1) et (8.3,1) [resp. (5.1,2) et (8.3,2)], [resp. 5.3 et (8.3,3)]. □

Une application importante de 6.3 est le théorème ergodique suivant [1], [9], [32].

Théorème 8.7. 1) *Si S est topologiquement simple à gauche (ou à droite) pour la topologie faible des opérateurs, ou si S est commutatif, il existe une et une seule application p de B dans B vérifiant les conditions suivantes:*

α) *p est un opérateur sur B.*

β) *Pour tout $x \in B$, $p(x) \in \overline{\text{co}}\{s(x)\colon s \in S\}$.*

γ) *Pour tout $s \in S$, $p \circ s = s \circ p = p$.*

2) *L'opérateur p appartient à $\overline{\text{co}}(S)$ et pour tout $x \in B$, $p(x)$ est le seul point fixe par rapport à S appartenant à $\overline{\text{co}}\{s(x)\colon s \in S\}$.*

Démonstration. Considérons le semigroupe affine semitopologique compact $(\overline{\text{co}}(S), \circ)$ dans $L_w(B)$. Supposons S topologiquement simple à gauche pour la topologie faible des opérateurs, ou commutatif ; il existe alors, d'après 6.3, un zéro dans $\overline{\text{co}}(S)$ (dans le premier cas, S est inclus dans l'ensemble T des éléments de $\overline{\text{co}}(S)$ inversibles à gauche dans $\overline{\text{co}}(S)$, donc on a $\overline{\text{co}}(T) = \overline{\text{co}}(S)$). Désignons par p le zéro de $\overline{\text{co}}(S)$. Il est immédiat que p vérifie les conditions (1,α), (1,β) et (1,γ).

Soit $x \in B$; d'après (1,β) et (1,γ), $p(x)$ est un point fixe par rapport à S dans $\overline{\text{co}}\{s(x)\colon s \in S\}$; montrons que c'est le seul (ce qui achèvera la démonstration). Soit y un point fixe quelconque par rapport à S dans $\overline{\text{co}}\{s(x)\colon s \in S\}$; il résulte de (1,$\beta$) que $p(y) = y$ et de (1,α) et (1,γ) que p est une application constante sur $\overline{\text{co}}\{s(x)\colon s \in S\}$; on a donc $y = p(x)$. □

Si S est topologiquement simple à gauche (ou à droite) pour la topologie faible des opérateurs, ou si S est commutatif, les résultats établis dans la section 7 permettent d'obtenir des théorèmes de décomposition du type "W.F. Eberlein, K. DeLeeuw et I. Glicksberg, C. Ryll-Nardzewski". On trouvera dans [9] et [33] (cf. également [2], [3], [6], [34]) des informations sur cette question qui n'est pas développée ici.

References

[1] Alaoglu, L., and G. Birkhoff, General ergodic theorems, Annals of Math. 41 (1940), 293–309

[2] Berglund, J. F., and K. H. Hofmann, "Compact semitopological semigroups and weakly almost periodic functions", Lectures Notes in Mathematics 42, Springer-Verlag, Berlin, Heidelberg, New York (1967)

[3] Berglund, J. F, H. D. Junghenn and P. Milnes, "Compact right topological semigroups and generalizations of almost periodicity", Lecture Notes in Mathematics, 663, Springer- Verlag, Berlin, Heidelberg, New York (1978)

[4] Bourbaki, N., "Topologie générale", Chap. 10. Paris, Hermann, 1967 (Act. Scient. et ind., 1084)

[5] Bourgain, J., et M. Talagrand, Compacité extrémale, Proc. Amer. Math. Soc. 80 (1980), 68–70

[6] Burckel, R. B., "Weakly almost periodic functions on semigroups", Notes on mathematics and its applications, Gordon and Breach, New York, 1971

[7] Choquet, G., "Lectures on analysis", Vol.2, New York, W.A. Benjamin, 1969

[8] Cohen, H., and H. S. Collins, Affine semigroups, Trans. Amer. Math. Soc. 93 (1959), 97–113

[9] DeLeeuw, K., and I. Glicksberg, Applications of almost periodic compactifications, Acta Math. 105 (1961), 63–97

[10] Dubuc, S., Dissertation, Cornell University, 1966

[11] Dunford, N., and J. T. Schwartz, "Linear operators, I: General theory", Interscience, New York, 1958

[12] Eberlein, W. F., Abstract ergodic theorems and weakly almost periodic functions, Trans. Amer. Math. Soc. 67 (1949), 217–240

[13] Ellis, R., Locally compact transformation groups, Duke Math. J. 24 (1957), 119–125

[14] Glicksberg, I., Convolution semigroups of measures, Pac. J. Math. 9 (1959), 51–67

[15] —, Weak compactness and separate continuity, Pac. J. Math. 11 (1961), 205–214

[16] Hansel, G., et J. P. Troallic, Points de continuité à gauche d'une action de semigroupe, Semigroup Forum 26 (1983), 20–27

[17] Kakutani, S., Two fixed-point theorems concerning bicompact convex sets, Proc. Imp. Acad. Tokyo 14 (1938), 242–245

[18] Kelley, J., and I. Namioka, "Topological Spaces", D. Van Nostrand Co., Inc., 1963

[19] Khurana, S. S., Pointwise compactness on extreme points, Proc. Amer. Math. Soc. 83 (1981), 347–348

[20] Lawson, J. D., Joint continuity in semitopological semigroups, Illinois J. Math. 18 (1974), 275–285

[21] —, Additional notes on continuity in semitopological semigroups, Semigroup Forum 12 (1976), 265–280

[22] —, Points of continuity for semigroup actions, Trans. Amer. Math. Soc. 284 (1984), 183–202

[23] Ménard, E., et J. P. Troallic, Points extrémaux dans un monoïde affine semitopologique, Semigroup Forum 31 (1985), 19–24

[24] Namioka, I., Neighborhoods of extreme points, Israel J. Math. 5 (1967), 145–152

[25] —, Separate continuity and joint continuity, Pac. J. Math. 51 (1974), 515–531

[26] —, Affine flows and distal points, Math. Z. 184 (1983), 259–269

[27] von Neumann, J., Almost periodic functions in groups, Trans. Amer. Math. Soc. 36 (1934), 445–492

[28] Peck, J. E. L., An ergodic theorem for non commutative semigroup of linear operators, Proc. Amer. Math. Soc. 2 (1951), 414–421

[29] Ruppert, W. A. F., On structural methods and results in the theory of compact semitopological semigroups, Recent developments in the algebraic, analytical and topological theory of semigroups, Lectures Notes in Mathematics 998 (1983), Springer-Verlag, Berlin, Heidelberg, New York, Tokyo, 215–238

[30] —, "Compact semitopological semigroups: an intrinsic theory", Lectures Notes in Mathematics 1079, Springer-Verlag, Berlin, Heidelberg, New-York, (1984)

[31] —, On semitopological actions of generalized I-semigroups, Semigroup Forum 31 (1985), 153–180

[32] Ryll-Nardzewski, C., On fixed points of semigroups of endomorphisms of linear spaces, Proc. Fifth Berkeley Sympos. Math. Stat. and Prob. (Berkeley, 1965/66), Vol. II, Part 1, 55-61

[33] Troallic, J. P., Une approche nouvelle de la presque-périodicité faible, Semigroup Forum 22 (1981), 247–255

[34] —, Semigroupes semitopologiques et presque-périodicité, Recent developments in the algebraic, analytical and topological theory of semigroups, Lecture Notes in Mathematics 998 (1983), Springer-Verlag, Berlin, Heidelberg, New York, Tokyo, 239–251

[35] —, Notes sur la structure des semigroupes semitopologiques compacts, Séminaire Initiation à l'Analyse, G. Choquet–M. Rogalski–J. Saint Raymond 54 (1981–1982), Exposé 14

[36] —, Points de continuité jointe dans un monoïde affine semitopologique, Séminaire Initiation à l'Analyse, G. Choquet–M. Rogalski–J. Saint Raymond 66 (1983–1984), Exposé 10

[37] —, Flots affines et points extrémaux, Séminaire Initiation à l'Analyse, G. Choquet–M. Rogalski–J. Saint Raymond 78 (1984–1985), Exposé 17

[38] Wendel, J. G., Haar measure and the semigroup of measures on a compact group, Proc. Amer. Math. Soc. 5 (1954), 923–929

Compact semigroups with one-sided continuity

John S. Pym

This survey must begin by drawing attention to competing accounts of the subject. First, there is WOLFGANG RUPPERT's book [55]. This advertises itself as being about separately continuous semigroups but in fact contains a lot of material about semigroups with multiplication continuous on one side only. Then there is NEIL HINDMAN's survey [24]. This, even more improbably for an article which contains a substantial amount on compact right topological semigroups, claims to be about ultrafilters and Ramsey theory. Finally there is the book by JOHN BERGLUND, HUGO JUNGHENN and PAUL MILNES – not the familiar volume of 1978 in the Lecture Notes series, but the entirely new book which will appear in 1989. This offers a superb elegant account of the whole theory of topological semigroups and their compactifications, not as deep as RUPPERT's masterful summary of his special area, but providing a real headache for anyone attempting to produce an alternative over-view of the field. I shall, of course, try to say a little about all aspects of the subject, but in order to avoid too much overlap I shall concentrate on those areas which are more familiar to me and about which I may, at this moment, have more information than other authors.

Any reviewer of this field faces another, much more irritating problem: terminology. I shall be dealing with semigroups in which the multiplication $(s, t) \mapsto st$ is continuous in one of the variables. Obviously it does not matter which is chosen, and different authors have made different choices. If we decide on continuity in s, we might describe this as 'left-continuity' because s is the left-hand variable. But we might equally observe that it is the operation of translation on the right by the element t which is continuous and use the term 'right-continuity'. Again, individual authors have followed their own inclinations. What is worse, if you find a writer's terminology unpalatable, you cannot simply ask your word processor to go through a paper interchanging 'left' and 'right', because everyone agrees what (for example) a left ideal is. It is obviously time that a decision was made on a uniform terminology. In spite of the consequences for some of my own preferences, I hereby declare that BERGLUND, JUNGHENN and MILNES [6] will become the standard reference work in this area; I shall follow their terminology and notation and I recommend all other writers to do the same.

There is one further aspect of notation about which an innocent reader should be warned. In this subject, commutative semigroups can be dense in non-commutative ones. For this reason the practice has grown up of sometimes using the expression $s + t$ to denote a non-commutative product of s and t, and we shall do this when convenient. An addition sign does not necessarily imply commutativity.

1. Introduction and examples

Let S be a semigroup with a topology. For $s, t \in S$ with product st we write

$$\rho_t(s) = st = \lambda_s(t).$$

Then S is called *right topological* if ρ_t is continuous for each t. All our semigroups will be right topological (except where the contrary is stated). An important role in the theory is played by the subsemigroup

$$\Lambda(S) = \{s : \lambda_s \text{ is continuous}\}.$$

Obviously the algebraic centre

$$Z(S) = \{s : st = ts \text{ for all } t \text{ in } S\}$$

satisfies $Z(S) \subseteq \Lambda(S)$. If $Z(S)$ is dense in S, then in fact $Z(S) = \Lambda(S)$, as is easy to check.

Compactness for a right topological semigroup is enough to ensure a rich algebraic structure. In 1958, Ellis [14] had already proved the existence of idempotents under these conditions. Ruppert [52] established the following algebraic structure theorem in 1973.

Theorem 1.1. *A compact right topological semigroup has a minimal ideal K and*

$$K \cong E(Se) \times eSe \times E(eS)$$

where e is any idempotent in K and $E(T)$ denotes the set of idempotents in T. $E(Se)$ is a left-zero semigroup, $E(eS)$ right-zero, and eSe is a group, and K is not necessarily an algebraic direct product but is a paragroup. □

Topologically K exhibits worse behaviour as we shall see.

This algebraic structure theorem holds under even weaker hypotheses. The space 2^X of all closed subsets of a compact space X has a natural topology called the Vietoris or finite topology (see Michael [36]). Forouzanfar [15] took a closed subset S of this space and assumed it was a semigroup under some multiplication which was monotone and 'upper-semicontinuous' in the left-hand variable, meaning that if $s_\alpha \to s$ then lim $\sup_\alpha s_\alpha t \leq st$ for each t in S. Such a semigroup contains idempotents and has a minimal ideal with the algebraic structure just described. Of course, 2^X is a particular compact lattice, and there should be an abstract formulation of FOROUZANFAR's work.

We have seen that we shall be able to say something substantial about compact right topological semigroups if they exist, but do they in any worthwhile sense? Are there any interesting examples? Their early appearances were in connection with transformation groups and semigroups.

Example 1.2. Take a compact space X. The set $S = X^X$ can be considered as consisting of all maps from X to itself in which case it is a semigroup under composition of functions, and also as a direct product of compact spaces, so that it is compact; the topology is the same as the pointwise topology – $s_\alpha \longrightarrow s$ in S iff $s_\alpha(x) \longrightarrow s(x)$ for each X. This semigroup is right topological since

$$s_\alpha \circ t(x) = s_\alpha(t(x)) \longrightarrow s(t(x)) = s \circ t(x).$$

In the same way it can be seen that if s is a continuous function then $s \in \Lambda(S)$. On the other hand, if $s \in \Lambda(S)$ and $y_\alpha \longrightarrow y$ in X, then we take the constant functions $t_\alpha(x) = y_\alpha, t(x) = y$ for all x, and we find

$$s(y_\alpha) = s \circ t_\alpha(x) \longrightarrow s \circ t(x) = s(y),$$

so that s is continuous. Thus in this case $\Lambda(S)$ is precisely the set of continuous functions. It is easy to see that $\Lambda(S)$ is dense in S here. In this example, $\Lambda(S)$ is clearly an important subsemigroup.

A way of producing a large number of examples is by constructing universal compactifications of a semigroup S. Some care is needed. If a semigroup compactification is defined to be a pair (ψ, X) consisting of a compact right topological semigroup X and a continuous homomorphism ψ with $\psi(S)$ dense in X, then maximal compactifications (meaning compactifications of which every other compactification is a quotient) need not exist; a counterexample was produced by J. W. BAKER (see [5], V.1.11). However, if the additional condition $\psi(S) \subseteq \Lambda(X)$ is added then satisfactory theories can be obtained, a fact recognised as early as 1973 by MILNES [37].

The best approach to semigroup compactifications appears to me to be the one using subdirect products presented in Chapter 3 of [6], following JUNGHENN and PANDIAN [29]. A P-compactification is a pair (ψ, X) which has the above properties together with a further property P. For a large class of properties P, maximal (or universal) compactifications exist. For example, if P is 'X has separately continuous multiplication' then the universal compactification is the weakly almost periodic compactification of S, denoted by $S^{\mathcal{WAP}}$. Of the many important properties P we mention three more:

(1) P is the joint continuity property, that the map $(s,x) \mapsto \psi(s)x$ of $S \times X$ into X is jointly continuous;
(2) P is the property that $xey = xy$ for all x, y and all idempotents e in X;
(3) P is the empty property (so that no additional conditions are imposed on (ψ, X) beyond being a compactification of X).

The universal compactification in case (1) is denoted by $S^{\mathcal{LC}}$; in case (2) by $S^{\mathcal{D}}$; and in case (3) by $S^{\mathcal{LMC}}$.

The reason for these notations is as follows. If $(\psi, S^{\mathcal{F}})$ is a compactification of S and $\mathcal{C}(S^{\mathcal{F}})$ denotes the space of bounded continuous (complex-valued) functions on $S^{\mathcal{F}}$, then

$$\mathcal{F}(S) = \{f \circ \psi : f \in \mathcal{C}(S^{\mathcal{F}})\}$$

is a Banach algebra of continuous functions on S whose maximal ideal space is $S^{\mathcal{F}}$. The universal compactifications are thus determined by function spaces. The set $\mathcal{LC}(S)$is called the space of *left norm continuous functions* (sometimes previously called left uniformly continuous functions, as in [5]); $\mathcal{D}(S)$ is the space of *distal* functions; and $\mathcal{LMC}(S)$ the space of *left multiplicatively continuous* functions. These function spaces can be characterized in other ways (see [6]). When S has an identity, $S^{\mathcal{D}}$ is a group, and is the universal compactification for the property P 'X is a group' ([6], Theorem 4.6.5).

When S is locally compact, $\mathcal{LC}(S) = \mathcal{LMC}(S)$ ([6], Theorem 4.5.7). If in addition S is discrete, then $\mathcal{LC}(S) = \mathcal{C}(S)$. The maximal ideal space of $\mathcal{C}(S)$ is then the Stone-Čech compactification βS, so in this case $\beta S = S^{\mathcal{LC}}$ is a semigroup. It is easy to use the continuity properties of multiplication in βS to give a formula for the product in βS: given x, y in βS, take nets $s_\alpha \longrightarrow x, t_\beta \longrightarrow y$ in S, and then

$$xy = \lim_\alpha \lim_\beta s_\alpha t_\beta.$$

In this iterated limit, the order is vital. The first explicit mention of βS as a semigroup appears to be in Civin and Yood [9] in 1961 (though they, unnecessarily, take S to be a group).

In the last few paragraphs we have been vague about the relationship between the semigroup S and its topology, though in [6], for example, it is explicitly required that multiplication in S be separately continuous. In fact, no connection between the topology and the multiplication is necessary; this was shown by Hindman and Milnes [26] and was further explained by Papazyan [43] who produced a universal separately continuous semigroup starting from any semigroup with a topology.

Civin and Yood [9] were in fact studying the second duals of Banach algebras, and these algebras produce another class of examples. Any Banach algebra A can be embedded as a weak* dense subspace of its second dual A^{**}. Given x, y in A^{**}. there are nets $(s_\alpha), (t_\beta)$ in A with $s_\alpha \to x, t_\beta \to y$ in the weak* topology.

Theorem 1.3. *The formula*

$$xy = w^* - \lim_{\alpha} \; w^* - \lim_{\beta} s_\alpha t_\beta$$

defines a Banach algebra multiplication in A^{**}. In the weak topology, A^{**} is right topological and $A \subseteq \Lambda(A^{**})$. If the norm in A satisfies $\|st\| \leq \|s\|.\|t\|$, the unit ball of A^{**} is a weak* compact right topological semigroup.* □

These multiplications were constructed by Arens [1] and have been widely studied since (see Duncan and Hosseniun [12]) though very little from a semigroup viewpoint.

Example 1.4. If S is discrete and A is taken to be the semigroup algebra

$$\ell^1(S) = \{f : f \text{ maps } S \text{ to } \mathbb{C} \text{ and } \sum_{s \in S} \|f(s)\| < \infty\}$$

with convolution as multiplication, then $\{F \in A^{**} : F \geq 0, \|F\| = 1\}$ with the weak* topology is the universal affine compactification of S (definition 3.4.1 of [6]) as is easily seen.

2. Ideal structure

In this section, we look at some properties of compact right topological semigroups which are concerned with the existence of ideals. To begin with, we shall consider universal semigroup compactifications (ψ, X) of a semigroup S (as in case (3) above), and to simplify matters we suppose that ψ is injective, so that we may assume $S \subseteq X$, and also that S has the topology induced by X. Since $S = \psi(S) \subseteq \Lambda(X)$, this means that S has a separately continuous multiplication. These conditions are satisfied in particular when S is a discrete semigroup and $X = \beta S$.

Take S to be locally compact but not compact. We first ask when $X \setminus S$ is a left ideal. In this case, $X \setminus S$ is compact and for each $s \in S, s(X \setminus S) \subseteq X \setminus S$. Since $s \in \Lambda(X)$, for each open neighbourhood U of $X \setminus S$, $\lambda_s^{-1}(U)$ is an open neighbourhood of $X \setminus S$. But open neighbourhoods of $X \setminus S$ are precisely the complements of compact subsets of S, so that a necessary (and, it is equally easy to see, a sufficient) condition is provided by the following Theorem.

Theorem 2.1. *$X \setminus S$ is a left ideal if and only if for each compact $K \subseteq S$ and each $s \in S$, $\lambda_s^{-1}(K)$ is compact.* □

Such semigroups S are called *left almost cancellative.*

PARSONS [45] was looking for conditions under which X could be written as a disjoint union $S \cup L_1 \cup L_2$ where L_1 and L_2 were both closed left ideals. Let U_1, U_2 be disjoint open neighbourhoods of L_1, L_2 respectively. Given $s \in S$, take an open neighbourhood V of L_1 with $sV \subseteq U_1$. Then $U_1 \setminus V$ is relatively compact in S, and $sU_1 \setminus U_1 = (s(U_1 \setminus V) \cup sV) \setminus U_1 \subseteq s(U_1 \setminus V)$ is relatively compact in S. Calling $U \subseteq S$ *left almost invariant* if $sU \setminus U$ is relatively compact in S we have found a necessary (and, equally easily, a sufficient) condition for our property.

Theorem 2.2. *X is the union of two disjoint closed left ideals if and only if S is the union of two disjoint left almost invariant sets with non-empty interiors.* □

This condition is trickier than might be supposed even for discrete semigroups. Infinite almost cancellative periodic semigroups satisfy it if and only if they are countable. Commutative groups satisfy it if and only if they are either torsion or the direct product of a finite group with Z (the theory for groups is connected with homology.) The simplest semigroup, $(N, +)$, does not satisfy it. PARSONS [46] gives a general necessary and sufficient condition, but it is too complicated to describe here and we refer to his papers for details.

If S is commutative, then PARSONS' work allows us to conclude that $\Lambda(X \setminus S) = S$ when $X = S \cup L_1 \cup L_2$, for under these conditions $\Lambda(X) = Z(X)$, and no element of L_1 can commute with any element of L_2. PARSONS was in fact concerned to find the centre of the second dual of a certain measure algebra on S. For a general locally compact group G, a most satisfactory result has been obtained by LAU and LOSERT [30].

Theorem 2.3. $\Lambda(L^1(G)^{**}) = L^1(G)$. □

A profitable area of research would be to investigate such questions for other algebras, perhaps in connection with the theory of affine compactifications touched on in Example 1.4 above.

HAMYE [18] offers another approach to closed left ideals. For simplicity, assume S has an identity. Call $U \subseteq S$ *large* with respect to the compactification X if the set

$$I(U) = \bigcap_{s \in S} \operatorname{cl}_X \lambda_s^{-1}(U)$$

is not empty. Such a set is a closed left ideal. HAMYE proves

Theorem 2.4. *Every closed left ideal is an intersection of sets $I(U)$ with U both large and a zero set for $\mathcal{C}(X)|_S$ (the space of functions on S which have continuous extensions to X).* □

Using these techniques, HAMYE is able to prove that $R^{\mathcal{LC}}$ (where R means the additive group of reals) has infinitely many disjoint closed left ideals. (The exact number here is 2^c – see BAKER and MILNES [4] for example.)

The most significant results in this area (as in so much of this whole field) have been obtained by NEIL HINDMAN. He is mostly concerned with compactifications of discrete semigroups, and in particular with βS for discrete S. HINDMAN prefers to work with ultrafilters which means that 'ordinary' semigroup theorists are often faced with a problem of translation. We shall present in this section some results from [23].

Denote by $U_\kappa(S)$ the set of points $p \in \beta S$ with the property

$$p \in \operatorname{cl}_{\beta S} A \text{ for } A \subseteq S \text{ implies card } A \geq \kappa.$$

Each point of βS is (representable as) an ultrafilter on S, and $p \in U_\kappa(S)$ if and only if each set belonging to the ultrafilter p has at least κ elements. Another alternative is to say that if V is a neighbourhood of p in βS, then $V \cap S$ has at least κ elements. When $\kappa = 1, U_1(S)$ is just βS. When $\kappa = \omega$, the first infinite ordinal, $U_\omega(S) = \beta S \setminus S$. HINDMAN gives conditions for $U_\kappa(S)$ to be a subsemigroup, a left ideal or a right ideal in βS. We now describe some of these.

Theorem 2.5. *For $U_\kappa(S)$ to be a left ideal, it is necessary and sufficient that $A \subseteq S$, card $A < \kappa$ and $s \in S$ imply card$(\lambda_s^{-1}(A)) < \kappa$.* □

This generalizes the result given in Theorem 2.1 in the discrete case, since 'compact' means 'finite' here. The result for right ideals is trickier.

Theorem 2.6. *$U_\kappa(S)$ is a right ideal if and only if for $A, B \subseteq S$ with card $A < \kappa$ and card $B = \kappa$ there is a finite set $\{s_1, \ldots, s_n\} \subseteq B$ with $\bigcap_{i=1}^{n} \lambda_{s_i}^{-1}(A) = \phi$.* □

If S is cancellative, then it can be seen from Theorems 2.5 and 2.6 that $U_\kappa(S)$ is always an ideal in S.

These methods enable HINDMAN to characterize the minimal ideal of $U_\kappa(S)$. We give the result only for $\beta S = U_1(S)$, where the condition is easier to state. First, considering $p \in \beta S$ as an ultrafilter, write $C_1(p) = \{A \subseteq S$: for every $s \in S, \lambda_s^{-1}(A) \in p\}$.

Theorem 2.7. *p is in the minimal ideal of βS if and only if for every $A \in p$ there is a finite set $\{s_1, \ldots, s_n\} \subseteq S$ with $\bigcup_{i=1}^{n} \lambda_{s_i}^{-1}(A) \in C_1(p)$.* □

In particular, HINDMAN is able to characterize the closure of the minimal ideal in βS, and is able to deduce that the closure of the minimal ideal is an ideal. This result should be related to an observation made by RUPPERT [56] (from which HINDMAN's follows only in the case in which S is commutative).

Theorem 2.8. *If T is a compact right topological semigroup and $\Lambda(T)$ is commutative and dense, then the closure of each right ideal of T is a (two-sided) ideal.* □

The formulation and proof of analogues of all of Hindman's results valid for general locally compact semigroups is likely to be a difficult task.

If $\beta S \setminus S$ is an ideal in βS, then so is $(\beta S \setminus S).(\beta S \setminus S)$, and this ideal must contain the minimal ideal M of βS. UMOH [59] showed that it is not in general true that $\mathrm{cl} M \subseteq (\beta S \setminus S).(\beta S \setminus S)$. He gives a sufficient condition for this to hold ('inflatability') which is too complicated to state here, but it is satisfied by $(N, +)$ and by the free semigroup on two generators.

Finally in this section we draw attention to the paper of HINDMAN and MILNES [25] on X^X, for any Hausdorff topological space X. They determine completely the left, right, closed left, closed right, minimal left and minimal right ideals of this semigroup. (The algebraic part of their results is in CLIFFORD and PRESTON [10]). The principal tools in their work are the spaces of all non-empty subsets of X and of all partitions of X, and certain non-Hausdorff topologies on these spaces. They find, in particular, that in this semigroup the closure of any left ideal is a left ideal and that this closure does not depend on the topology of X. It would be of interest to find other classes of semigroup for which such complete information could be obtained.

3. Detailed algebraic structure

We now turn to a consideration of more detailed algebraic properties of right topological semigroups. Analysts concerned with the second duals of Banach algebras have largely concentrated on the question of when, for a commutative algebra, the second dual is commutative (or to be more precise, the non-abelian version of this question). A related question for us is whether βS is commutative if S is. The answer is (almost always) no, in a spectacular way.

Naturally I believe the easiest approach here is by a method used in one of my own papers [50]. Let $x = (x_i)$ be a sequence of 0's and 1's. Define supp $x = \{i : x_i = 1\}$. Write $S = \{x : \text{supp } x \text{ is finite}\}$. Define $x + y$ in S if and only if supp $x \cap$ supp $y = \phi$ to be $(x_i + y_i)$. Then S with this minimal algebraic structure is called a *standard oid*. The direct sum of two-element groups, $\bigoplus_{i=1}^{\infty}\mathbb{Z}_2$, has a standard oid structure. So does $(\mathbb{N},+)$, as is seen by writing integers in binary form.

We introduce a semigroup structure on the compact subset H of βS defined by

$$H = \{x \in \beta S : x = \lim_{\alpha} x_\alpha \text{ with } \inf(\text{supp } x_\alpha) \to \infty\}$$

by writing $x + y = \lim_\alpha \lim_\beta (x_\alpha + y_\beta)$ if $x = \lim x_\alpha, y = \lim y_\beta$. Then H is a right topological semigroup. (And when $S = \mathbb{N}$, for example, this is the same semigroup operation as that induced by $\beta\mathbb{N}$.) Although we write the operation in H additively, it need not be commutative.

We shall now produce a continuous homomorphism h from H to any compact topological group G_2 which contains the free group on two generators a_1, a_2. Partition $I = \{1, 2, \ldots\}$ arbitrarily into two infinite sets I_1, I_2. Define $h : S \to G_2$ by

$$h(x) = h(x_1, x_2, x_3, \ldots) = \tilde{h}(x_1).\tilde{h}(x_2).\tilde{h}(x_3)\ldots$$

where $\tilde{h}(x_i) = 1$ (the identity of G_2) if $x_i = 0, \tilde{h}(x_i) = a_1$ if $x_i = 1$ and $i \in I_1$, and $\tilde{h}(x_i) = a_2$ if $x_i = 1$ and $i \in I_2$; thus the apparently infinite product on the right is essentially finite. Since G_2 is compact, h extends to a continuous map from βS to G_2, so that h is defined on H. It is easy to use the relation $h(x + y) = h(x)h(y)$ if $x, y \in S$ and supp $x <$ supp y to deduce that h is a homomorphism on H. The fact that G_2 contains a free group on two generators implies that the minimal ideal of H also contains a free group on two generators.

Now in fact much more is true. Similar, elementary, arguments show

Theorem 3.1. *The minimal ideal of H contains a free group on 2^c generators, 2^c minimal left ideals and 2^c minimal right ideals.* □

These results were obtained for the case of $(\mathbb{N},+)$ by HINDMAN and PYM [27], when the minimal ideal of H is contained in the minimal ideal of $\beta\mathbb{N}$. The basic idea of considering a map h which is a homomorphism on a subsemigroup H of $\beta\mathbb{N}$ instead of requiring it to be a homomorphism on the whole of $\beta\mathbb{N}$ is HINDMAN's.

This theory raises the question: Which semigroups contain oids? Before we answer this, let us reflect on how HINDMAN was drawn into semigroup theory. He was concerned with sets of finite sums. In a general commutative semigroup S, the set $FS(x_n)$ of finite sums of a sequence (x_n) with distinct terms consists of all elements which can be written in the form $x_{i_1}+\ldots+x_{i_r}$ for some finite collection of distinct suffices $i_1,\ldots,i_r$. We say that the finite sums are distinct if the relation $x_{i_1}+\ldots+x_{i_r} = x_{j_1}+\ldots+x_{j_s}$ implies $r = s$ and $\{i_1,\ldots,i_r\} = \{j_1,\ldots,j_r\}$ (i. e. the expression for a finite sum is unique). Now here is an important observation of

Papazyan [44]. Write $e_n = (0,\ldots,0,1,0,0,\ldots)$ (with 1 in the nth place); in a standard oid, $FS(e_n)$ makes sense and is the whole oid. Then if the finite sums $FS(x_n)$ are distinct, the natural map which extends $e_n \mapsto x_n$ is an isomorphism of the standard oid onto $FS(x_n)$. This is obvious (once pointed out) and shows that oids and finite sums amount to the same thing.

Now we can give some answers to our question about semigroups containing oids. Hindman [23] Lemma 5.4 (see also van Douwen [11] Theorem 4.5) gives this important theorem.

Theorem 3.2. *Let S be cancellative and let $p \in \beta S$ be idempotent. Then for each neighbourhood V of p, $V \cap S$ contains an oid.* □

It is then a consequence of Theorem 3.1 that V contains a free group on 2^c generators, and 2^c idempotents. Papazyan [44] has shown that almost every commutative semigroup contains an oid (though not all do, for example ($\mathbb{N}$, max)). She has also noticed that every oid has many sub-oids, for example

$$\{(x_1, x_2, \ldots) : x_{2r} = 0 \text{ for all } r\},$$

and using this she has obtained the following generalization of a result of Lisan [34].

Theorem 3.3. *H contains c copies of its minimal ideal M which do not intersect* cl M. □

Strauss [57] has pointed out some remarkably simple ways of using oids to produce results about Stone-Čech compactifications. We have space only for one illustration. Given a standard oid S, define $c\colon S \to \mathbb{N}$ by $c(x) = \text{card}(\text{supp } x)$. Then if supp $x \cap$ supp $y = \phi$, so that $x + y$ is defined, we have $c(x + y) = c(x) + c(y)$. It follows easily that c extends to a homomorphism from H into the one-point compactification $N \cup \{\infty\}$. If we put

$$H_n = \{x \in H : c(x) = n\}$$

then $H_1 = \{x \in H : x \in \text{cl}\{e_1, e_2, e_3, \ldots\}\}$ (where e_n is as above) is the closure of a discrete countable set, so has cardinal 2^c. We shall now see that, given $y, z \in H_1$, we can determine y and z from $y + z$. Indeed, we can express any $x \in S$ in the form $x = e_{i_1} + \ldots + e_{i_r}$. We define $\sigma_f(x) = e_{i_1}, \sigma_\ell(x) = e_{i_r}$ (f for 'first', ℓ for 'last'). Both σ_f and σ_ℓ extend to maps of βS into itself, and it is easy to see that $\sigma_f(y + z) = y,\ \sigma_\ell(y + z) = z$. These methods show

Theorem 3.4. *H_1 generates a free semigroup in H. This free semigroup misses the minimal ideal and in fact is not contained in any free group.* □

The algebraic structure of βS, or of its minimal ideal, is far from being understood, even in the case $S = \mathbb{N}$. Can we find any other groups of which βS contains copies? Can the group in the minimal ideal of βS be determined (up to algebraic isomorphism)?

The theory described above is all in the discrete case. Little is known about the general case. BAKER and MILNES [4] have shown that for a locally compact (non-compact) group G, the compactification $G^{\mathcal{LC}}$ contains 2^c minimal left and 2^c minimal right ideals, which includes HAMYE's result quoted as Theorem 2.4 above. LAU and PYM [31] know how to obtain $\mathbb{R}^{\mathcal{LC}}$ from $\beta\mathbb{N} \times [0, 1]$, and so can find all the pathology of $\beta\mathbb{N}$ in $\mathbb{R}^{\mathcal{LC}}$.

Finally in this section, we should mention a specific problem about $\beta\mathbb{N}$. The set $\mathbb{N}$ has two natural operations, addition and multiplication, and hence $\beta\mathbb{N}$ has too. Can there be found p, q, r, s in $\beta\mathbb{N} \setminus \mathbb{N}$ with $p + q = r.s$? (This question was asked by VAN DOUWEN [11]). HINDMAN [21] has proved that this is not possible with the single additional restriction $p \in \bigcap_n \mathrm{cl}_{\beta\mathbb{N}}(n\mathbb{N})$ (a result which has been obtained by methods I find simpler by STRAUSS [57]). Using the fact that $\mathbb{N}$ is an oid, this result implies that if the theory is extended to the oid situation, then $p + q = r.s$ is impossible in the semigroup H constructed at the beginning of this section.

4. Topological properties

So far we have considered properties which placed the emphasis on algebra rather than topology. We now turn to questions of a topological kind. First we have two results which show that some algebraically significant sets are topologically small. STRAUSS [57] proves

Theorem 4.1. *For any countable discrete semigroup S and any $x \in \beta S \setminus S$, the set $\{y \in \beta S \setminus S : y + x = x + y\}$ of elements commuting with x is of first category in $\beta S \setminus S$.* □

VAN DOUWEN [11] has shown

Theorem 4.2. $(\beta\mathbb{N} \setminus \mathbb{N}) + (\beta\mathbb{N} \setminus \mathbb{N})$ *is of first category in $\beta\mathbb{N} \setminus \mathbb{N}$.* □

By way of contrast, HINDMAN [22] has shown

Theorem 4.3. *If S is countable and cancellative, then the set $\{x \in \beta S \setminus S : y + x = z + x \text{ implies } y = z\}$ of right cancellative elements contains a dense open set.* □

These last two results are in fact essentially the same: STRAUSS [57] also shows that each $x \notin (\beta\mathbb{N} \setminus \mathbb{N}) + (\beta\mathbb{N} \setminus \mathbb{N})$ is right cancellative, a conclusion contained in the earlier thesis of BUTCHER [7] but not published. Her techniques enable her to generalize her result given as Theorem 3.4: with

$$X = (\beta\mathbb{N} \setminus \mathbb{N}) \setminus ((\beta\mathbb{N} \setminus \mathbb{N}) + (\beta\mathbb{N} \setminus \mathbb{N})),$$

the set X/Z generates a free semigroup in $(\beta\mathbb{N} \setminus \mathbb{N})/Z$ (where the quotient sign means identifying for each $x \in \beta\mathbb{N} \setminus \mathbb{N}$, all elements $x + n$ $(n \in Z)$).

We next turn to a collection of results of a different kind. These assert that at certain points of a compact right topological semigroup, multiplication must be continuous. They derive from the ELLIS-LAWSON Theorem ([13], [33]). The simplest proof of the basic theorem is due to CHRISTENSEN [8] and an account is given in [6]. The semigroup applications are excellently described by RUPPERT [55] and so we only give the principal theorem as an illustration.

Theorem 4.4. *Let S be a locally compact right topological semigroup with identity 1, which has a separately continuous action $(s, x) \mapsto sx$ on a compact space X, with 1 acting as the identity on X. Then $(s, x) \mapsto sx$ is continuous at $(1, y)$ for each $y \in X$.* □

If additional assumptions are made, continuity results can be improved. The most striking is that a compact *metrizable* right topological group S with $\Lambda(S)$ dense is in fact a topological group; this was proved independently by NAMIOKA [40] and RUPPERT [52]. With stronger assumptions come stronger conclusions: any right topological group defined on the real line R or the torus T is a topological group (RUPPERT [53]).

We now return to our main example of βS for a discrete semigroup S. We take S to be commutative, so that S is in the algebraic centre of βS. If $x \in \beta S$, then $\mathrm{cl}_{\beta S}(S + x) = \beta S + x$. We shall now prove an unpublished result due to BAKER [2].

Theorem 4.5. *Let $e \in \beta S$ be idempotent and give $S + e$ the topology induced by βS; then $\beta(S + e) = \beta S + e$.*

Proof. Define $\overline{f}$ to be the unique continuous extension to βS of the function $\overline{f}$ defined on S by $\overline{f}(s) = f(s + e)$. It is enough to show that $\overline{f}$ agrees with f on $S + e$. Choose (t_α) in S with $t_\alpha \to e$ in βS. Then, for $s \in S$,

$$\begin{aligned}\overline{f}(s + e) = \overline{f}(e + s) = \lim_\alpha \overline{f}(t_\alpha + s) = \lim_\alpha f(t_\alpha + s + e) = f(e + s + e)\\ = f(s + e + e) = f(s + e),\end{aligned}$$

as required. □

This result is more than a curiosity. If L is a minimal left ideal of βS, and $e \in L$, then $\beta S+e = L$, and $S+e$ is a subsemigroup of the group $e+L = e+\beta S+e$. Thus, for each idempotent $e \in L$, we have $\beta(S+e) = \beta(eL) = L$. Now the topologies induced on $S+e$ by βS are distinct in the sense that the map of $S+e$ to $S+e'$ given by $s+e \mapsto s+e'$ is never continuous if $e \neq e'$ (though whether any other map is a homeomorphism is not known – see PYM [47] and so, taking $S = N$, we find 2^c different topologies on N for which the Stone-Čech compactifications are the same.

The semigroups $S+e$ just described are also examples of non-discrete semigroups for which $\beta(S+e)$ are naturally semigroups. The work of BAKER and BUTCHER [3] provides some general conditions under which this property holds. Is a more comprehensive theory possible?

A minimal left ideal L in βS is algebraically a direct product $E(L) \times G$, and if e is any idempotent of $L, e+L$ is algebraically isomorphic with G. We have just seen that $e+L$ is topologically dense in L. How can this be? The first transparent example of this phenomenon (for a much simpler semigroup) was discussed in this context by BAKER and MILNES [4].

Example 4.6. Write $S_o = \{0, e_1, e_2\}$ with $0+e_i = e_i+0 = e_i, e_i+e_j = e_i (1 \leq i, j \leq 2)$, and consider the direct product semigroup $S = S_o \times \mathbb{T}$, where $\mathbb{T}$ is the circle group; S is algebraically the union of three circles. The singletons $\{(0,t)\}$ for $t \in \mathbb{T}$ are to be open sets. A base for the other open sets is to consist of all the sets of the form

$$(\{0\} \times (t_1, t_2)) \cup (\{e_1\} \times [t_1, t_2)) \cup (\{e_2\} \times (t_1, t_2])$$

(where the intervals (t_1, t_2) etc in $\mathbb{T}$ have an obvious sense). Any neighbourhood of a point in $\{e_2\} \times \mathbb{T}$ meets both $\{0\} \times \mathbb{T}$ and $\{e_1\} \times \mathbb{T}$. It is therefore easy to see that $\Lambda(S) = \{0\} \times \mathbb{T}$ is dense in S and that, for $i = 1, 2, \{e_i\} \times \mathbb{T}$ is dense in $L = \{e_1, e_2\} \times \mathbb{T}$. This S is compact but not metrizable.

More complex examples of the same type are possible.

Example 4.7. Let G be a compact group with identity 1 and take a set E such that $(E \times \{1\}) \cup (G \setminus \{1\})$ is a compactification of $G \setminus \{1\}$ (for example, E could be $\beta(G \setminus \{1\}) \setminus (G \setminus \{1\})$, or if $G = \mathbb{T}, (E \times \{1\}) \cup (\mathbb{T} \setminus \{1\})$ could be the two-point compactification of the circle). Make $E \times G$ into a left group. Give it the coarsest topology which is invariant under the action of G and for which the obvious projection from $E \times G$ onto $(E \times \{1\}) \cup (G \setminus \{1\})$ is continuous. Then $E \times G$ has the property that $\{e\} \times G$ is dense for each $e \in E$. Moreover, it is possible to add on a discrete copy of G just as in Example 4.6 (and if we start with the two-point compactification of the circle, we obtain exactly Example 4.6).

Examples like 4.7 originally came from an investigation of second dual algebras (IŞIK, PYM and ÜLGER [28]). PYM [49] showed that they were the maximal ideal spaces of the C^*-algebra of bounded functions f on G which have the property that osc (f, x) (the oscillation of f at x) exceeds any $\varepsilon > 0$ at only a finite number of points. The simple construction in Example 4.7 is due to HELMER and IŞIK [19]. They also point out a curious fact: the construction is only of interest for metrizable G, since if G is not metrizable the only compactification of $G \setminus \{1\}$ is G itself. (This contrasts strangely with a remark made after Theorem 4.4: the most interesting compact right topological groups must be non-metrizable.) A simple proof of the last result is given by HELMER and PYM [20].

Example 4.7 showed us a way of deriving the BAKER-MILNES example from the circle group. Here is another.

Example 4.8. Start with the neighbourhood (t_1, t_2) of 1 in $\mathbb{T}$, and split it into three parts $(t_1, 1) \cup \{1\} \cup (1, t_2)$. Divide the set $\{0, e_1, e_2\} \times (t_1, t_2)$ into neighbourhoods of $(0, 1), (e_1, 1), (e_2, 1)$ respectively as follows:

$$\{(0,1)\};\ \{(e_1,1)\} \cup (\{0,e_1,e_2\} \times (1,t_2));\ \{(e_2,1)\} \cup (\{0,e_1,e_2\} \times (t_1,1)).$$

It turns out that this process of splitting neighbourhoods can be repeated, though the technical details are unwelcome and we must refer to PYM [48] for those. An interesting feature is that this method can produce semigroups $E \times \mathbb{T}$ not obtainable by the procedure of Example 4.7. Whether this construction is possible for groups other than $\mathbb{T}$ is not known.

The measure theoretic properties of the three-circle semigroup discussed as Example 4.6 are of interest, especially as far as its minimal ideal $L = \{e_1, e_2\} \times \mathbb{T}$ is concerned. Every open set in L is measurable for the product topology on $\{e_1, e_2\} \times \mathbb{T}$. It therefore appears that Lebesgue measure on $\{e_i\} \times \mathbb{T}$ for $i = 1, 2$ will provide two different right invariant measures on L. However, $\{e_i\} \times \mathbb{T}$ is not measurable in the topology of Example 4.6, and in fact we can construct only one invariant measure by this method. The measure theory of the spaces described in Examples 4.7 and 4.8 has not been considered.

We return briefly to a Stone-Čech compactification, specifically to $\beta\mathbb{Z}$. We showed in Theorem 4.5 that for any idempotent $e \in \beta\mathbb{Z} \setminus \mathbb{Z}$, we had $\beta(\mathbb{Z}+e) = \beta\mathbb{Z}+e$. Which topologies on $\mathbb{Z} + e$ can occur here, or in other words, which topologies on $\mathbb{Z}$ make it homeomorphic with a subspace $\mathbb{Z} + e$ of $\beta\mathbb{Z}$? Obviously any such topology must be translation invariant and totally disconnected. PAPAZYAN [44] has shown that any extreme topology with these properties must occur, but that not every topology with the properties does.

5. Transformation semigroups

We close with three short sections on topics which are adequately covered elsewhere or are marginal to the present account. We begin by looking at semigroups of transformations for which we refer to [6] for more information. Let S be a semigroup with a topology which makes multiplication separately continuous, let X be a compact space, and let there be an action $(s, x) \mapsto sx$ of S on X which is also separately continuous; the pair (S, X) is called a *flow*. Then there is a natural mapping of S into X^X which in fact sends S into a set of continuous functions, or in other words into $\Lambda(X^X)$. We shall assume this map is injective for simplicity. The closure Σ of S in X^X is called the *enveloping semigroup* of S. It is a famous theorem of Ellis [14] that Σ is a group (called the *Ellis group*) whose identity is the identity mapping on X if and only if (S, X) is *distal*, that is, if x, y are in X and (s_α) in S with $\lim\limits_\alpha s_\alpha x = \lim\limits_\alpha s_\alpha y$, then $x = y$. Thus, by starting with distal flows, it is possible to find examples of compact right topological groups G with $\Lambda(G)$ dense.

One such group is calculated by Namioka [42]; the original transformation group was considered by Furstenberg [16] and further discussed by Milnes [38].

Example 5.1. Take the semigroup S to be $(\mathbb{N}, +)$ and take X to be the torus $\mathbb{T}^2$. It is convenient here to identify $\mathbb{T}$ with $\mathbb{R}/\mathbb{Z}$ and to write its group operation additively. Since 1 generates $\mathbb{N}$, we need only describe the action of 1 on $\mathbb{T}^2$; this is given by fixing an irrational number ξ in $\mathbb{T}$ and writing $1.(x, y) = (\xi + x, x + y)$. To describe the Ellis group Σ, we first remark that the set of endomorphisms $E(\mathbb{T})$ of the torus is in fact a ring (one operation is addition, the other composition of mappings) and is a compact subspace of $\mathbb{T}^\mathbb{T}$. The compact right topological group Σ turns out to be homeomorphic and isomorphic to $E(\mathbb{T}) \times \mathbb{T}$ with the multiplication given by

$$(f, x).(g, y) = (f + g, x + y + f \circ g(\xi)).$$

Namioka also shows that $\Lambda(E(\mathbb{T}) \times \mathbb{T}) = \mathbb{Z} \times \mathbb{T}$.

Further examples are discussed by Milnes [39].

When X is in addition a convex subset of a topological vector space and each $s \in S$ acts as an affine map, the pair (S, X) is called an *affine flow*. For this case, Namioka [41] provides the following interesting theorem (from which he deduces easily the Ryll-Nardzewski fixed point theorem).

Theorem 5.2. *Let I be a minimal left ideal of Σ and let $C \subseteq X$ be minimal subject to the conditions that it is convex, compact and closed under the action of S; then each s in I maps C to one point.* □

This result prompts the question of how affine flows might be obtained. One obvious way is to take a flow (S, X), denote by $P(X)$ the space of probability measures on X with its usual (weak*) topology, and then note that $(S, P(X))$ is an affine flow. However, the relationship between the two enveloping semigroups Σ, one obtained from (S, X) and the other from $(S, P(X))$ is not clear to me.

Yet another result of NAMIOKA [40] says that a compact right topological group G has a left invariant measure μ, but a little care is needed in interpreting this result: invariance is only under the action of elements x in $\Lambda(G)$, *not* of all elements in G. (Full details of the proof are given in [6]). This raises the question of whether the only elements of G under which μ is invariant are in $\Lambda(G)$. The problem is even worse than it seems at first sight; if $x \notin \Lambda(G)$, then λ_x may not even have a sensible extension to $P(G)$. Is $\{x \in G : \lambda_x \text{ is measurable}\}$ the same as $\Lambda(G)$?

The question just asked makes sense for general compact right topological semigroups: is there anything to be said about the measurable analogue of $\Lambda(S)$?

6. Universal semigroups

We saw in section 1 that universal constructions provided some of the most important examples of compact right topological semigroups, and the bulk of this survey has been concerned with a particular semigroup of this kind, βS. Here we draw attention to three different results belonging to this area.

The first is from JUNGHENN and PANDIAN [29]. It provides a structure theorem for compactifications of a certain type which is analogous to FURSTENBERG's structure theorem for distal flows [16] (see also NAMIOKA [40]) and whose form is very suggestive: could there be similar decompositions for other compactifications? Recall from section 1 that a compactification of S is a pair (ϕ, X) with X compact and right topological and $\phi(S) \subseteq \Lambda(X)$. Write $(\phi, X) \geq (\psi, Y)$ if there is a continuous homomorphism $\pi : X \to Y$ with $\psi = \pi \circ \phi$.

Theorem 6.1. *Let (ϕ, X) be a compactification in which X is a right topological group and suppose either (i) that S is σ-compact and $(s, x) \mapsto \phi(s)x$ is continuous or (ii) that S is a separable k-space. Then there is an ordinal κ and a family $\{(\phi_\xi, X_\xi) : 0 \leq \xi \leq \kappa\}$ of compactifications of S with the properties (i) (ϕ_0, X_0) is the trivial compactification and $(\phi_\kappa, X_\kappa) = (\phi, X)$, (ii) if $\xi \leq \eta$ then $(\phi_\eta, X_\eta) \geq (\phi_\xi, X_\xi)$, (iii) if $\xi < \kappa$ and $\pi : X_{\xi+1} \to X_\xi$ is the canonical quotient map, then multiplication in $X_{\xi+1}$ restricted to $X_{\xi+1} \times \pi^{-1}(1)$ (where 1 is the identity of X_ξ) is continuous; (iv) if η is a limit ordinal, then X_η is the projective limit of the family $\{X_\xi : \xi < \eta\}$.* □

In section 1 we mentioned four universal compactifications. In [6] many more are listed and discussed. Of course, these compactifications sometimes coincide:

we have already remarked that for locally compact S, $\mathcal{LC}(S)= \mathcal{LMC}(S)$ and for discrete S, $\mathcal{LC}(S)= \mathcal{C}(S)$. More cases of equality are described in [6], but there is plenty of scope for further investigation in this area. Many years ago, MACRI [35] found a condition for discrete semigroups under which $\mathcal{C}(S) = \mathcal{WAP}(S)$. More recently, PYM and VASUDEVA [51] considered the semigroup 2^S of all subsets of a discrete semigroup S (with the product $AB = \{xy : x \in A, y \in B\}$ for $A, B \subseteq S$ and the Vietoris topology). They showed that $\mathcal{C}(2^S) = \mathcal{WAP}(2^S)$ if and only if $\mathcal{C}(2^S) = \mathcal{LMC}(2^S)$ and that this is equivalent to S having either the property that for some partition $S = A_1 \cup \ldots \cup A_k, xA_i$ has one element for each i and each $x \in S$ or the property that for some partition $S = A_1 \cup \ldots \cup A_k, A_ix$ has one element for each i and each $x \in S$. The interest of this paper probably lies more in its method than its main result; the procedure involved in the proof of the most difficult implication is a complex passage to subsequences in an iterated limit condition.

One of the questions addressed by RUPPERT [54] is finding $\Lambda(G^{\mathcal{LMC}})$ for topological groups G.

Theorem 6.2. *If G is locally compact and abelian, or if G is discrete (in which case $G^{\mathcal{LMC}} = \beta G$ of course) and has no element of order 2, and in some other cases, $\Lambda(G^{\mathcal{LMC}}) = G$.* □

RUPPERT also shows by example that in general this conclusion does not hold. Another interesting example is provided by HAMYE [18]: the topological subgroup $\mathbb{Q}$ of $\mathbb{R}$ has the property that $\Lambda(Q^{\mathcal{LMC}}) = \mathbb{Q}$, but because $\mathbb{Q}^{\mathcal{LC}} = \mathbb{R}^{\mathcal{LC}}$, we have $\Lambda(\mathbb{Q}^{\mathcal{LC}}) = \mathbb{R}$. The problem of determining $\Lambda(S)$ for interesting compact semigroups S is a significant one.

7. Second duals

In Example 1.4, we mentioned the second duals of Banach algebras as related to affine compactifications of semigroups. The main question which has been addressed in the forty years' life of the theory of second duals is that of whether, or when, an algebra is *regular*. In our context, this means whether every continuous bounded affine function on the metrizable convex semigroup which is the unit ball of the Banach algebra is weakly almost periodic. Every such function is, in fact, always left multiplicatively continuous. For some algebras, such as C^*-algebras, these functions are indeed all weakly almost periodic. There may be a natural advantage in studying these semigroups as opposed to general ones because a Banach algebra has two natural topologies, the norm topology and the weak topology, but as far as I am aware no attempt has been made to exploit such a situation in semigroup theory. (Anyone wishing to investigate this area could consult the survey article by DUNCAN and HOSSENIUN [12], the account by GROSSER [17] of the most general theory, or the elegant introduction to the basic results by ÜLGER [58].)

Recently an investigation of the algebraic structure of a particular second dual produced a specific example of a compact right topological semigroup in much the same way as CIVIN and YOOD [9] found βS. To be specific, take the group algebra $L^1(G)$ of a compact group G. The dual $L^1(G)^*$ is $L^\infty(G)$ which is a natural C^*-algebra, and therefore isomorphic with $\mathcal{C}(X)$ for some compact space X. As usual, X can be identified with a weak* compact subset of $L^\infty(G)^*$. IŞIK, PYM and ÜLGER [28] prove

Theorem 6.1. *X is a subsemigroup of $L^1(G)^{**}$ and has the algebraic structure of a left group, $E \times H$. Here, card $E \geq 2^c$ if G is infinite and the group H is algebraically isomorphic with the original group G.* □

There is an obvious similarity here with the situation in the minimal left ideal of βS as described in section 3, but there are essential differences. For example, given $e \in E$, the closure of $\{e\} \times H$ must contain a large number of idempotents, but it is not necessarily the whole of $E \times H$. The topological structure of this semigroup is far from clear.

When G is only locally compact, the situation changes: the structure space X of $L^\infty(G)$ may not even be a semigroup (LAU and PYM [31]). However, there is a subspace $L_0^\infty(G)$ of functions in $L^\infty(G)$ which 'vanish at ∞' in a certain sense. The structure space of this C^*-algebra without identity has the algebraic structure $(E \times H) \cup \{0\}$, where $E \times H$ is a left group, H is algebraically isomorphic with G, and 0 is a zero; this work will be published by LAU and PYM [32]). In general, the algebraic structure of $L^1(G)^{**}$ remains obscure.

References

[1] Arens, R., Operations induced in function classes, Monatsh. Math. 55 (1951), 1–19

[2] Baker, J. W., University of Sheffield, personal communication (1979)

[3] Baker, J. W. and R. J. Butcher, , The Stone-Čech compactification of a topological semigroup, Proc. Camb. Phil. Soc. 80 (1976), 103–107

[4] Baker, J. W. and P. Milnes, The ideal structure of the Stone-Čech compactification of a group, Math. Proc. Camb. Phil. Soc. 82 (1977), 401–409

[5] Berglund, J. F., H. D. Junghenn and P. Milnes, "Compact Right Topological Semigroups and Generalizations of Almost Periodicity", Lecture Notes in Mathematics 663, Springer, Berlin, 1978

[6] Berglund, J. F., H. D. Junghenn and P. Milnes, "Analysis on Semigroups: Function Spaces, Compactifications, Representations", Wiley, New York, 1989

[7] Butcher, R. J., "The Stone-Čech compactification of a topological semigroup and its algebra of measures", Dissertation, University of Sheffield, 1975

[8] Christensen, J. P. R., Joint continuity of separately continuous functions, Proc. Amer. Math. Soc. 82 (1981), 455–461

[9] Civin, P. and B. Yood, The second conjugate space of a Banach algebra as an algebra, Pacific J. Math. 11 (1961), 847–870

[10] Clifford, A. H. and G. B. Preston, "The Algebraic Theory of Semigroups I", Math. Surveys 7, Amer. Math. Soc., Providence, 1961

[11] Douwen, E. K. van The Čech-Stone compactification of a discrete groupoid, Topology and its Applications, to appear

[12] Duncan, J. and S. A. R. Hosseniun, The second dual of a Banach algebra, Proc. Roy. Soc. Edinburgh A 84 (1979), 309–325

[13] Ellis R., Locally compact transformation groups, Duke Math. J. 24, (1957) 119–126

[14] Ellis R., Distal transformation groups, Pacific J. Math. 9 (1958), 401-405

[15] Forouzanfar, A. M., Compact right upper semi-continuous semigroups of closed sets, preprint, University of Sheffield (1988)

[16] Furstenberg, H., The structure of distal flows, Amer. J. Math. 85 (1963), 477-515

[17] Grosser, M., "Bidualräume und Vervollständigungen von Banachmoduln", Lecture Notes in Mathematics 717, Springer, Berlin, 1979

[18] Hamye, S. M., Some compactifications of a semitopological semigroup, Semigroup Forum 34 (1987), 341–357

[19] Helmer, D. and N. Işik, Construction of right topological compactifications for discrete versions of subsemigroups of compact groups, Semigroup Forum, 39 (1989), 65–73

[20] Helmer, D. and J. S. Pym, A short proof of $\beta(G\setminus\{1\}) = G$ for non-metrizable compact groups G, Semigroup Forum, to appear

[21] Hindman, N., Sums equal to products in $\beta\mathbb{N}$, Semigroup Forum 21 (1980), 221–255

[22] —, Minimal ideals and cancellation in $\beta\mathbb{N}$, Semigroup Forum 25 (1982), 291–310

[23] —, The ideal structure of the space of κ-uniform ultrafilters on a discrete semigroup, Rocky Mountain J. Math. 16 (1986), 685–701

[24] —,"Ultrafilters and Ramsey theory – an update", Proceedings of the conference 'Set theory and its applications, York 87', to appear

[25] Hindman, N. and P. Milnes, The ideal structure of X^X, Semigroup Forum 30 (1984), 41–51

[26] —, The LMC-compactification of a topologized semigroup, Czech. Math. J. 38 (1988), 103-119

[27] Hindman, N. and J. S. Pym, Free groups and semigroups in $\beta\mathbb{N}$, Semigroup Forum 30 (1984), 177–193

[28] Işik, N., J. S. Pym and A. Ülger, The second dual of the group algebra of a compact group, J. London Math. Soc. 35 (1987), 135–148

[29] Junghenn, H. D. and R. D. Pandian, Existence and structure theorems for semigroup compactifications, Semigroup Forum 28 (1984), 109–122

[30] Lau, A. T. and V. Losert, On the second conjugate algebra of $L_1(G)$ of a locally compact group, J. London Math. Soc. 37 (1988), 464–470

[31] Lau, A. T. and J. S. Pym, Universities of Alberta and Sheffield, personal communication

[32] —, Concerning the second dual of the group algebra of a locally compact group, preprint, Universities of Alberta and Sheffield, (1989)

[33] Lawson, J. D., Joint continuity in semitopological semigroups, Illinois J. Math. 18 (1974), 275–285

[34] Lisan, A., Free groups in $\beta\mathbb{N}$ which miss the minimal ideal, Semigroup Forum 37 (1988), 233–239

[35] Macri, N., The continuity of Arens' product on the Stone-Čech compactification of semigroups, Trans. Amer. Math. Soc. 191 (1974), 185–193

[36] Michael, E., Topologies on spaces of subsets, Trans. Amer. Math. Soc. 71 (1951), 152–182

[37] Milnes, P., Compactifications of topological semigroups, J. Australian Math. Soc. 15 (1973), 488–503

[38] —, Continuity properties of compact right topological groups, Math. Proc. Camb. Phil. Soc. 86 (1979), 427–435

[39] —, Ellis groups and group extensions, Houston J. Math. 12 (1986), 87–108

[40] Namioka, I., Right topological groups, distal flows and a fixed point theorem, Math. Systems Theory 6 (1972), 193–209

[41] —, Affine flows and distal points, Math. Z. 184, (1983), 259–269

[42] —, Ellis groups and compact right topological groups, Contemporary Mathematics 26 (1984), 295–300

[43] Papazyan, T., The LMC compactification of a topologized semigroup by a universal mapping construction, preprint, University of Sheffield (1988)

[44] —, University of Sheffield, personal communication

[45] Parsons, D., The centre of the second dual of a commutative semigroup algebra, Math. Proc. Camb. Phil. Soc. 95 (1984), 71–92

[46] —, Abelian semigroups whose Stone-Čech compactifications have left ideal decompositions, Math. Proc. Camb. Phil. Soc. 97 (1985), 473–479

[47] Pym, J. S., Footnote to a paper of Baker and Milnes, Math. Proc. Camb. Phil. Soc. 85 (1979), 315

[48] —, A topological construction in the theory of compact semigroups, J. Nigerian Math. Soc. 2 (1982), 39–53

[49] —, A construction of compact right topological semigroups from compact groups, Semigroup Forum 35 (1987), 207–225

[50] —, Semigroup structure in Stone-Čech compactifications, J. London Math. Soc. 36 (1988), 421–428

[51] Pym, P. S. and H. L. Vasudeva, Compactifications of some hyperspace semigroups, Semigroup Forum 30 (1984), 263–282

[52] Ruppert, W., Rechtstopologische Halbgruppen, J. reine angew. Math. 261 (1973), 123–133

[53] —, Rechtstopologische Intervallhalbgruppen und Kreishalbgruppen, Manuscripta Math. 14 (1974), 183–193

[54] —, On semigroup compactifications of topological groups, Proc. Roy. Irish Acad. A 79 (1979), 179–200

[55] —, "Compact Semitopological Semigroups: An Intrinsic Theory", Lecture Notes in Mathematics 1079, Springer, Berlin, 1984

[56] —, In a left topological semigroup with dense center the closure of any left ideal is an ideal, Semigroup Forum 36 (1987), 247

[57] Strauss, D., University of Hull, personal communication, (1988)

[58] Ülger, A. , Weakly compact bilinear forms and Arens regularity, Proc. Amer. Math. Soc. 101 (1987), 697–704

[59] Umoh, H., Ideals of the Stone-Čech compactification of semigroups, Semigroup Forum 32 (1985), 201–214

Part III

Functional analysis on semigroups

Measure algebras on semigroups

John W. Baker

1. Scope of survey

In deciding what to include in this survey article I have taken as a model the much earlier survey article by WILLIAMSON [64]. Broadly I have decided to cover the same range of topics as was covered there but, of course, excluding any of those topics which appear more relevant to any of the other surveys. My aim is also to cover those topics which I feel are of interest to mathematicians who claim an allegiance to 'topological semigroups'. Therefore, with certain exceptions, I have excluded material on topological groups or on algebraic (i.e., discrete) semigroups, even though both of these are technically topological semigroups. The main exception is that I shall discuss the Arens regularity (and related concepts) of certain spaces of measures on groups; this is certainly of interest to researchers in the field since the second duals of such algebras can be represented as algebras of measures on certain compactifications of these groups, and these compactifications are normally semigroups but not groups. The other exception is that I have discussed the convolution algebra $\ell^1(S)$ for an algebraic semigroup S. This topic tends to be more of interest to analysts rather than algebraists, so seems to be relevant here.

I have split the survey into a number of sections. After the present section there is an introductory section, and then the remainder of the contents is as follows. Section 3 is concerned with intrinsic properties of the algebra of bounded measures on a topological semigroups. The main interest is in the problem of doing analysis on such semigroups and their measure algebras. In particular we discuss the problems of determining the maximal ideals and representations of these algebras. In Section 4 we discuss recent work on the algebra $\ell^1(S)$; it turned out that this topic has been very little studied in recent years, but I have kept to my original plan for a chapter on this topic. The fifth section takes a look at problems on the second dual of certain measure algebras, including, of course, Arens regularity. More work on this topic occurs in Section 7. An account of the theory of multipliers of measure algebras is given in Section 6. The subject of Section 7 is that of weighted convolution algebras. Even though it is not strictly relevant to measure algebras, it seemed relevant to mention recent work on invariant measures, since the main role of such a measure would be to provide an analogue of the group algebra $L^1(G, m)$.

This work is discussed in Section 8. Section 9 has been reserved for any material not fitting naturally into Sections 3 to 8.

I have had to make the decision whether to attempt to offer a comprehensive account of the subject matter or to provide more detailed discussion of what appears to be the more significant parts of recent work. I have had to guess at the likely preference of possible readers and have elected to do the former of the two possibilities. This has meant that little will be said about methods of proof used in the research. Finally, it was necessary to decide on a meaning for the word 'recent'. Broadly I have taken this to mean that I should go back to about 1980, which roughly takes us back to the previous Oberwolfach conference. Of course, sometimes it has been necessary to go back earlier in order to set the scene.

2. Preliminaries. Convolution of measures

Throughout this article, unless otherwise stated, S will be a locally compact Hausdorff separately continuous semigroup, written multiplicatively. Occasionally, authors consider measure algebras on non-locally compact semigroups, but I do not feel this interests too many mathematicians as there are few examples of interest and usually there is little more than observing that, with a little extra care, the same arguments can be used in the more general case. For each x in S we have the *left translation* maps $y \mapsto xy$ and the *right translation* maps $y \mapsto yx$ of S to itself. So our hypothesis is that both of these maps are continuous, for all x in S. If we only know that each of the right translation maps is continuous we say that S is a *right topological semigroup*. If the right (resp. left) translation maps on S are injective we say that S is *right* (resp. *left*) *cancellative*; if these translation maps are proper we say that S is *topologically right* (resp. *left*) *cancellative*. (A mapping ϕ between topological spaces is *proper* if $\phi^{-1}(K)$ is compact for all compact subsets K of the codomain.)

Also, throughout this article, unless otherwise stated G will be a locally compact topological group. By $\mathbb{R}$ will always be understood the additive group $\mathbb{R}$ in its usual topology. By $\mathbb{R}_+$ will be understood the set of *positive* real numbers; unless otherwise stated $\mathbb{R}_+$ will be understood to have addition as its operation. We will denote by I the unit interval in its usual topology with the binary operation maximum. In many papers authors allow for the more general semigroup $[a, b]$, possibly with minimum instead of maximum. However as all such semigroups are isomorphic to I it seems preferable to make the slightly more simple formulation of the theory which one has if one takes S to be I.

We shall not normally require semigroups to have an identity. We use the standard notation listed below,

$$AB = \{ab : a \in A, b \in B\}, \quad A^{-1}B = \{x \in S : ax \in B \text{ for some } a \in A\}$$

for $A, B \subset S$, and then if $y \in S$ define $yB = \{y\}B, \quad y^{-1}B = \{y\}^{-1}B$ for $B \subset S$; other related conventions should then be clear. We denote by $C_b(S)$ the Banach space of bounded and continous, complex-valued functions on S with the supremum norm $\|\cdot\|_\infty$. For $f \in C_b(S)$ and $x \in S$ we define the translates of f by x to be the functions

$$f_x(y) = f(yx), \quad {}_xf(y) = f(xy) \quad (y \in S).$$

We shall reserve the term *measure* for either a positive or a complex regular Borel measure on a given locally compact space, as defined in [57]. We can then construct the Banach algebra $M(S)$ of bounded measures on S by giving $M(S)$ the total variation norm and the *convolution* product given by

$$(\mu * \nu)(f) = \int\int f(xy)d\mu(x)d\nu(y) \; (\mu, \nu \in M(S), \; f \in C_b(S)). \tag{1}$$

Equivalently we can define

$$(\mu * \nu)(A) = \int \mu(Ax^{-1})d\nu(x)$$

for $\mu, \nu \in M(S)$ and A a Borel subset of S. The algebra $M(S)$ is commutative if and only if S is commutative. As mathematicians were warned by GLICKSBERG [28], it is a mistake to take the definition and properties of convolution of measures too much for granted. However, as S is locally compact there is no problem if the measures being considered are bounded. It is only in Sections 7 and 8 where unbounded measures occur, and in Section 8 there are no convolution products. It turns out that the measures occuring in the weighted measure algebras of Section 7 are σ-finite, which means that there is no problem which cannot be overcome with a little care. However, as is pointed out by REJALI [54], various authors have defined the space $M(S, w)$ in such a way that, with the usual meaning attached to the term measure, the space is incomplete. Briefly, there is no way that we can construct the *vector space* of measures $M(S, w)$ where the elements of $M(S, w)$ can ascribe a value to every Borel subset of S. However, this is easily dealt with if we regard a measure as a linear functional on the space $C_{00}(S)$ of continuous functions on S with compact support. If S is jointly continuous we can define a convolution product by equation (1), taking f in $C_{00}(S)$. Any reader wanting to study the more general situation is referred to [54], which studies a very general situation which appears to include most previous work, and to [34].

For a locally compact group G we denote by m a left Haar measure on G. If x is an element of S then $\bar{x}$ will denote the measure consisting of the point mass at x.

If X is a locally compact space, $M(X)$ is the vector space of bounded measures on X with total variation norm and if μ is a positive measure on X (possibly unbounded) we shall regard the space $L^1(X, \mu)$ as a subspace of $M(X)$, as well as a space of (equivalence classes of) functions on X. It inherits the norm from $M(X)$ and if $X = S$ may inherit the convolution product if $L^1(S, \mu)$ is a subalgebra of $M(S)$. The norms $\|\cdot\|_p$ for $1 \leq p \leq \infty$ on a space of functions have their usual meanings.

Finally, I have stated some results in a formal way and others are given implicitly in the text. Those which have been stated formally are identified as 'theorems' without meaning to imply that they have equal significance or require a similar amount of proof.

During preparation of this article I made extensive use of Mathematical reviews. I have retained the MR numbers of each of the papers in the references in case it should be useful for readers.

3. The algebra of bounded measures on a topological semigroup

It is unreasonable to expect to find many (if any) non-trivial theorems applying to the algebra $M(S)$ for all semigroups S. Therefore most interest has been in obtaining theorems which apply for particular classes of semigroups. The three classes which have been the subject of interest are (a) 'totally ordered semigroups', such as I, with the order topology and the operation maximum on some totally ordered set, and finite products of such semigroups, (b) 'foundation semigroups' which possess an algebra of measures analogous to the group algebra $L^1(G, m)$, and (c) discrete semigroups. I shall stick to my original plan and discuss (c) in the next (short) section. In fact (c) is a special case of (b), but one hopes to obtain much more information in the case of (c). In this section I first look at (b) and then at (a).

A thorough account of the theory of foundation semigroups up to 1983 is given in Dzinotyiweyi's book [22]. The only significant work since 1984 seems to be that contained in two papers by Bami [7], [8]. Both for completeness and in order to discuss the work of Bami I will give a brief introduction to the topic. This work appears in papers by A. C. and J. W. Baker, [2], [3], Dzinotyiweyi [21] to [24] and Sleijpen [58] to [62].

Even for a commutative locally compact group G, the algebra $M(G)$ is rather large and unwieldy and has proved difficult to study. In harmonic analysis on such groups it is normal to make use, instead, of the subalgebra $L^1(G, m)$. As is explained in Section 8, it is exceptional for a locally compact semigroup to possess an invariant measure, which might play the role played by Haar measure on groups. An alternative possibility might be to look for a positive *quasi-invariant* measure μ (i.e., $\mu(A) = 0$ implies $\mu(x^{-1}A) = 0$), but then the algebra $L^1(S, \mu)$ is not, in general, very useful.

Let S be a *jointly continuous* semigroup. We define

$$M_a^\ell(S) = \{\mu \in M(S) \mid \text{ the map } x \mapsto |\mu|(x^{-1}K) \text{ of } S \text{ into } \mathbb{R} \text{ is continuous for every compact subset } K \text{ of } S\}.$$

We then define $M_a^r(S)$ similarly with Kx^{-1} replacing $x^{-1}K$ (ℓ = left, r = right). Finally define

$$M_a(S) = M_a^\ell(S) \cap M_a^r(S).$$

This definition appeared simultaneously in [2] and in the Ph. D. Thesis of HART [31]. An alternative definition, which is equivalent to the above, is that

$$M_a^\ell(S) = \{\mu \in M(S) \mid \text{the map } x \mapsto \bar{x} * |\mu| \text{ of } S \text{ into } M(S)\} \text{ is continuous for the weak topology of } M(S)\}.$$

One then has the following properties of $M_a(S)$, which are established in [2], [3] and can be found in, for example, [22].

(i) For a locally compact group G we have $M_a(G) = L^1(G, m)$.

(ii) $M_a(S)$ is a closed 2-sided ideal of $M(S)$, and is invariant under translation by elements of S (i.e., $\mu \in M_a(S)$, $x \in S$ implies $\bar{x} * \mu$ and $\mu * \bar{x} \in M_a(S)$.

(iii) $M_a(S)$ is an L-subspace of $M(S)$ (i.e., if $\mu \in M_a(S)$ and $\nu \ll |\mu|$ then $\nu \in M_a(S)$).

(iv) If T is a closed subsemigroup of S and $\mu \in M_a(S)$ then $\mu|_T$ (the restriction of μ to T) is in $M_a(T)$.

(v) If S is cancellative (but not in general) then $\mu \in M_a^\ell(S)$ iff the map $x \mapsto \mu(x^{-1}K)$ is continuous for all compact subsets K of S.

As is explained in [21], it has not yet proved possible to extend these results to the case in which S is only separately continuous; no suitable definition of $M_a(S)$ has been found which, in that case, will ensure (ii) and (iii).

We now define the *foundation* of S to be the closure of the union of the supports of all measures in $M_a(S)$. This set is either empty or is an ideal of S. We call S a *foundation semigroup* if its foundation is the whole of S. Clearly if S is discrete then $M_a(S) = \ell^1(S)$, so that S is a foundation semigroup if it is a discrete semigroup or is a topological group (by virtue of (i) above). If S is not a foundation semigroup but its foundation, T say, is not empty then $M_a(S)$ can only provide information about T. By virtue of (iv) above, we see that T certainly is a foundation semigroup; in fact $\{\mu|_T : \mu \in M_a(S)\} \subseteq M_a(T)$. So we concentrate attention on foundation semigroups.

Now every discrete semigroup and every open subsemigroup ('open' can be weakened) of a locally compact group is a foundation semigroup. This leads to a large class of such semigroups since any finite product of foundation semigroups and any image of a foundation semigroup under a continuous and proper homo-

morphism is of the same type. Unfortunately, in spite of some very hard work by Sleijpen, no intrinsic simple necessary and sufficient conditions for S to be a foundation semigroup have been found, and it would appear that in general there is no measure μ on S such that $M_a(S) = L^1(S,\mu)$. I shall describe Sleijpen's progress for the case where S has an identity. Strangely this does not help with the general case since no construction has been found for embedding an arbitrary foundation semigroup into one with an identity. We call S a *stip* (an acronym) if S has an identity 1 and if for each neighbourhood U of 1 and each x in S
(i) $U^{-1}(Ux) \cap (xU)U^{-1}$ is a neighbourhood of x, and
(ii) $U^{-1}v \cap w^{-1}U$ is a neighbourhood of 1 for all $v, w \in U$.

In his Ph.D. thesis [61], SLEIJPEN showed that if S is a foundation semigroup with identity then S is a stip, (for details see [22]). Partial converses to this result have been obtained, see [58], [59], [60] and [22]. The lack of work on this topic in recent years is, to some extent, due to a lack of characterisation of foundation semigroups. All that seems to be known, as is implied by the definition of a stip, is that a foundation semigroup needs to have a topology which is, in some sense, homogeneous.

I now wish to discuss the work of BAMI. This work provides an analogue of the work of A. C. and J. W. BAKER on the radical and Gelfand theory of $M_a(S)$ for commutative S in the field of non-commutative semigroups. I first need to remark that, in the non-commutative case, there is little hope of duplicating the theory of representations of locally compact groups and their algebras. The first obstruction is that, as I said above, we cannot represent $M_a(S)$ as an algebra of functions; the second is that although the convolution of measures extends naturally to semigroups the convolution of functions on groups inextricably involves the inversion and does not extend to semigroups in most cases. In particular, it would appear at this time that there is no analogue, even for a foundation semigroup of the left regular representation in $L^2(G,m)$ for the group G.

However, Bami found that he could obtain some useful information about representations for such semigroups. He found that in some cases it was not necessary to restrict attention to representations in Hilbert space and so he also considered representations in Banach spaces. I now describe Bami's theory which is to be found in [7]. The definitions are included for preciseness; they are the natural ones. Let F be a normed vector space over $\mathbb{C}$. A *representation* V of S by bounded operators on F is a homomorphism $x \mapsto V_x$ of S into $B(F)$, where $B(F)$ is the space of bounded operators on F under composition. We say V is *bounded* if there is $k \in R$ such that $\|V_x\xi\| \leq k\|\xi\|$ for all $x \in S$ and all $\xi \in F$; we then define $\|V\|$ as the infimum of all possible values for k. If A is a Banach algebra over $\mathbb{C}$ then a *bounded representation* of A by operators on F is a bounded homomorphism of A into the Banach algebra $B(F)$.

Theorem 3.1. *Let S be a foundation semigroup with identity* 1. *Let E be a reflexive Banach space. Suppose that R is a bounded and cyclic representation of $M_a(S)$*

by bounded operators on E^*. *Then there exists a unique continuous and bounded representation* V *of* S *by bounded operators on* E^* *with* $\|V\| = \|R\|, V_1 = 1$ *(the identity operator on* E^**) for which*

$$\langle R_\mu \xi, \eta \rangle = \int_S \langle V_x \xi, \eta \rangle d\mu(x) \quad (\mu \in M_a(S), \xi \in E^*, \eta \in E). \tag{2}$$

Further V *and* R *have the same closed invariant subspaces,* V *is faithful if* R *is, and in that case* $V_x \neq 0$ *for all* $x \in S$. *Conversely, it is clear that if* V *is a measurable and bounded representation of* S *by operators on* E^* *then* (2) *defines a bounded representation of* $M_a(S)$ *by operators on* E^*. □

[We say V is *faithful* if $x \neq y$ implies that $V_x \neq V_y$; we say that D is an *invariant* subspace for V if $\xi \in D$ and $x \in S$ imply that $V_x \xi \in D$.]

Bami was not able to discover if, starting with V on S, applying the converse part of 3.1 to obtain R and then applying the direct part of 3.1 to obtain $\tilde{V}$ say, one has $\tilde{V} = V$ under suitable conditions. This would ensure that in fact V was continuous. In the commutative case [2], [3], where representations just reduced to semicharacters it was possible to show that a semicharacter χ on S is continuous if and only if $\{x \in S \mid \chi(x) = 0\}$ is closed in S. (A semicharacter is a homomorphism of S into the multiplicative semigroup $\{\alpha \in C \mid |\alpha| \leq 1\}$ which is not identically zero.) In order to proceed any further in the non-commutative case, Bami found it necessary to turn his attention to representations on Hilbert space. To do this it was necessary to assume that S has an identity. An *involution* on S is a continuous mapping $x \mapsto x^*$ of S onto itself such that $(x^*)^* = x$ and $(xy)^* = y^* x^*$ for all $x, y \in S$. In that case, there is a natural involution on the algebra $M(S)$ defined by the rule

$$\int_S f(x) d\mu^*(x) = \overline{\int \overline{f(x^*)} d\mu(x)} \qquad \big(f \in C_{00}(S), \mu \in M(S)\big).$$

Further, $M_a(S)$ is closed under this involution. The notions of $*$*-representations* are the obvious ones. Of course the notion of involution generalizes the notion of inverse in a group G. We use the terminology of [9] for ideas concerning representations of Banach algebras.

Theorem 3.2. *Let* S *be a foundation semigroup with identity* **1** *and involution, and let* H *be a Hilbert space. Suppose that* R *is a* $*$*-representation of* $M_a(S)$ *by bounded operators on* H *such that for every non-zero* ξ *in* H *there is* μ *in* $M_a(S)$ *with* $R_\mu \xi \neq 0$. *Then there is a unique bounded and continuous* $*$*-representation* V

of S by bounded operators on H with $\|V\| \leq 1$ such that

$$\langle R_\mu \xi, \eta \rangle = \int_S \langle V_x \xi, \eta \rangle d\mu(x) \quad (\mu \in M_a(S); \xi, \eta \in H). \tag{3}$$

Further R and V have the same closed invariant subspaces, and if R is faithful then so is V and $V_x \neq 0$ for all x in S. Conversely, if V is a bounded and weakly measurable $*$-representation of S by bounded operators on H then formula (3) *defines a $*$-representation of $M_a(S)$, and even of $M(S)$, by such operators.* □

It follows immediately from (3) that if V is a bounded and weakly measurable $*$-representation of S is by bounded operators on H such that $V_1\xi \neq 0$ for all $\xi \in H$ then V is continuous if and only if $\{x \in S \mid \langle V_x\xi, \eta\rangle = 0\}$ is closed in S. (A representation of S by bounded operators is *weakly measurable* if it is a measurable function into the set of bounded operators on H with the weak operator topology.) As an application of the above representation theory, Bami shows in [7] that the following is true.

Theorem 3.3. *Let S be a foundation semigroup with identity and continuous involution. The following are equivalent:*
(i) *$M_a(S)$ is $*$-semisimple;*
(ii) *$M(S)$ is $*$-semisimple;*
(iii) *the set of bounded and continuous $*$-representations of S on Hilbert space separates the points of S.* □

In his second paper [8], Bami obtains results for positive definite functions on foundation semigroups with involution. These results both unify and generalize independent results previously known for discrete semigroups and for locally compact groups and for subsemigroups of such groups such as $\mathbb{R}_+ \cup \{0\}$. These should be described in the survey article on positive definite functions. The only other class of semigroups whose measure algebras seem to be accessible to study is a class of ordered semigroups. In 1966, Lardy studied the algebra $L^1(a, b)$, in which the semigroup is the interval $(a, b]$ with the operation maximum, see [40]. This work has recently been extended by Dhar and Vasudeva [14] to a semigroup which is a finite product of intervals. As mentioned earlier, we let I be the unit interval $[0, 1]$ with its usual topology and the operation maximum. We take S to be the product topological semigroup $I \times I$ (denoted by R in their papers). Of course the choice of a product of two, rather than an arbitrary finite number of copies of I, is no real restriction but it does serve to simplify the exposition. Such a semigroup is far from being a foundation semigroup, since $M_a(S)$ is trivial (it just consists of scalar multiples of the point mass at the zero $(1, 1)$ of S). However there is a subalgebra of $M(S)$ which to some extent is as nice as $M_a(S)$. Let m be the Lebesgue measure of $\mathbb{R}^2$ restricted to S. It is easy to show that if we regard elements of $L^1(S,m)$ as measures on S then $L^1(S,m)$ becomes a closed subalgebra of $M(S)$. So then $L^1(S,m)$ is a commutative Banach algebra. If elements of $L^1(S,m)$ are regarded as (equivalence classes of) functions the formula for the convolution product in $L^1(S,m)$ is given by

$$(f*g)(x,y) = f(x,y)\int_0^x\int_0^y g(u,v)dudv + g(x,y)\int_0^x\int_0^y f(u,v)dudv + \\ + \int_0^x\int_0^y f(x,v)g(u,y)dudv + \int_0^x\int_0^y f(u,y)g(x,v)dudv \\ (f,g\in L^1(S,m), x,y\in I).$$

It is known from earlier work of BAARTZ [1] that $M(S)$ is semisimple, so $L^1(S,m)$ must be semisimple. In [14] it is shown that the maximal ideal space of $L^1(S,m)$ can be identified, as a topological space, with the space $(0,1]\times(0,1]$, by the rule that for $(a,b)\in(0,1]\times(0,1]$, the corresponding homomorphism h takes the form

$$h(f) = \int_a^1\int_b^1 f(x,y)dm(x)dm(y) \quad (f\in L^1(S,m)).$$

Dhar and Vasudeva also show that $L^1(S,m)$ has an approximate identity. Most other work which has been done on this type of algebra is concerned with multipliers (see Section 6) and with vector-valued measures (see Section 9). The fact that it is necessary to restrict attention to semigroups embedded in a *finite* product of totally ordered sets if one requires a strong theory is justified to some extent in the parallel article by M. Mislove on semilattices. The most obvious outstanding problem arising from the topics in this section would appear to be that of finding applications of the representation theory initiated by Bami.

4. The algebra $\ell^1(S)$

To my surprise there have only been two papers which I can find which have been published recently on the topic of the algebra $\ell^1(S)$. One of these is by McLEAN and KUMMER [43] and contains a completely different proof of Theorem 3.3 (see Section 3) for the case in which S is a discrete semigroup. For a *commutative* semigroup S it is quite easy to determine the complex homomorphisms of the algebra $\ell^1(S)$. However, any attempt to describe the representations of $\ell^1(S)$ for non-commutative S seems to meet with considerable difficulty. There has, however, been a degree of success in the case in which S is an inverse semigroup. The one paper I discuss here is concerned with such a semigroup. In this paper [65] WORDINGHAM also provides applications of the theory to amenability which I will not discuss here. Let S be an inverse semigroup, i.e., S has an involution $*$ such that $sxx^*x = x$ for all x in S. Such a semigroup has a representation as a semigroup of partial maps (the Vagner-Preston representation). This representation is used to get a representation of $\ell^1(S)$ by bounded operators on $\ell^2(S)$. For x in S, $\bar{x}$ denotes the function on S which is the characteristic function of the singleton $\{x\}$. Then $\bar{x} \in \ell^1(S) \cap \ell^2(S)$. We define L_S to be the unique continuous linear map of $\ell^1(S) \times \ell^2(S) \to \ell^2(S)$ satisfying the formula

$$L_S(\bar{x}, \bar{y}) = \begin{cases} \overline{xy} & \text{if } x^*xy = y, \\ 0 & \text{otherwise.} \end{cases}$$

We then obtain a *-representation L of $\ell^1(S)$ by bounded linear mappings on $\ell^2(S)$ by the rule that

$$L_f(g) = L_S(f, g) \quad \text{for } f \in \ell^1(S), g \in \ell^2(S).$$

Wordingham constructs this representation and then shows that the representation is faithful.

5. Second duals and Arens regularity

This section discusses recent work on algebras closely related to the second dual of $M(S)$; other work on second duals is discussed in the section on weighted algebras. Let S be separately continuous. We denote by $LUC(S)$ the space of bounded, left uniformly continuous, complex-valued functions on S. By definition, f is in $LUC(S)$ if and only if $f \in C_b(S)$ and the map $x \mapsto {}_xf$ of S to $C_b(S)$ is continuous. Then $LUC(S)$ is a closed, translation-invariant subalgebra of $C_b(S)$. An 'Arens product' can be defined on the dual space $LUC(S)^*$ as follows. Let $\mu, \nu \in LUC(S)^*$ and let $f \in LUC(S)$. Define ${}_\nu f \in LUC(S)$ by

$${}_\nu f(x) = \nu({}_xf) \quad \text{for } \ x \in S,$$

and then define

$$(\mu * \nu)(f) = \mu({}_\nu f).$$

Under the multiplication $*$, $LUC(S)^*$ becomes a Banach algebra. In order to understand the problem to be discussed, let us consider the case in which S is discrete. Then $LUC(S) = \ell^\infty(S)$, so that $LUC(S)^*$ is the second dual of $\ell^1(S)$ and the multiplication is one of the two Arens products of $\ell^1(S)^{**}$. Recall that a Banach algebra is said to be *(Arens) regular* if the two Arens products on its second dual coincide. The natural problem is whether $\ell^1(S)$ is regular. If it is not one then one might ask how close it is to being regular. This reduces to determining the size of the set of elements of the second dual on which the two multiplications are equal. Returning to the original problem, let $f = LUC(S)$ and let $\mu, \nu \in LUC(S)^*$. Define a function f_μ on S by

$$f_\mu(x) = \mu(f_x) \quad \text{for } x \in S.$$

If $f_\mu \in LUC(S)$ we can define

$$(\mu_0 \nu)(f) = \nu(f_\mu).$$

We define Z (the *topological centre* of $LUC(S)^*$) to be

$$Z = \{\mu \in LUC(S)^* \mid f_\mu \in LUC(S) \text{ for all } f \in LUC(S) \text{ and } \nu \circ \mu = \mu * \nu \text{ for all } \nu \in LUC(S)^*\}.$$

It is shown by LAU [41] that $\mu \in Z$ if and only if the map $\nu \mapsto \mu * \nu$ of $LUC(S)^*$ to itself is continuous for the weak* topology. This is the reason for the term topological centre. Our problem is to determine this topological centre. Historically, these problems were first discussed by BUTCHER [10] and ZAPPA [67] who essentially consider the additive groups $\mathbb{Z}$ and $\mathbb{R}$. Butcher dealt with the case $\mathbb{Z}$, and showed that the centre of $\ell(\mathbb{Z})^{**}$ is $\ell^1(\mathbb{Z})$. (Since $\mathbb{Z}$ is commutative, the topological centre is the same as the algebraic centre.) At that time his ideas looked extremely promising since his proof in fact obtained the stronger fact that there is an element of $\ell^1(\mathbb{Z})^{**}$ which does not commute with more than half of the elements of $\ell^1(\mathbb{Z})^{**} \setminus \ell^1(\mathbb{Z})$, and another which does the same for the other half. His ideas were developed further by PARSONS [51]. However, the approach which led to the solution of the problem was that of Zappa. Her ideas were extended extensively by GROSSER and LOSERT [29] and then taken over by LAU [41]. Firstly we should remark that we can embed $M(S)$ naturally inside $LUC(S)^*$ by the rule

$$\mu(f) = \int_S f(x) d\mu(x) \quad (\mu \in M(S), f \in LUC(S))$$

It is then quite easy to show that $M(S) \subset Z$. Lau's theorem is then as follows.

Theorem 5.1. *Let S be either a locally compact group or else a discrete semigroup which is (left and right) topologically cancellative. Then $M(S) = Z$.* □

From this theorem a number of corollaries follow quickly. A related problem to the above is to take a locally compact group G and consider the algebra $L^1(G, m)$. One then defines Z to be the topological centre of $L^1(G, m)^{**}$, that is, the set of elements on which the two Arens products coincide. We denote by $*$ any one of the two Arens products on $L^1(G)^{**}$.

Theorem 5.2. [42] *Let G be a locally compact group. Let $\mu \in L^1(G,m)^{**}$. The following conditions on μ are equivalent:*
(i) $\mu \in L^1(G,m)$ *(under the natural embedding in its bidual);*
(ii) $\mu \in Z$*;*
(iii) *the maps $\nu \mapsto \mu * \nu$ and $\nu \mapsto \nu * \mu$ of $L^1(G)^{**}$ to itself are weak*-continuous (one of them is bound to be according to the choice of $*$).* □

This theorem was established by Işik, Pym and Ülger for G compact in [33].

6. Characterisation of multipliers

In this topic there re-emerges a dichotomy which arose earlier. The two classes of semigroups for which progress has been made in this field are foundation semigroups and (finite products of) totally ordered semigroups (see Section 3).

Let A be a Banach algebra. Then a (*right*) *multiplier* on A (or just *multiplier* if A is commutative) will be understood here, with one exception, to be a bounded linear mapping T of A into A satisfying the identity

$$T(xy) = xT(y) \quad (x, y \in A) \tag{4}$$

The set $\mathcal{M}(\mathcal{A})$ of right multipliers of A is a Banach algebra containing (a copy of) A with identity. For $x \in A$ we have the right multiplier T_x given by the formula

$$T_x(y) = yx \quad (y \in A)). \tag{5}$$

In many cases, e.g. if A has a bounded approximate identity, the equation (4) above is enough to ensure that T is a right multiplier. The aim of most research is to characterise multipliers of particular algebras. Very often this is done as follows. One constructs an algebra B which contains (a copy of) the given algebra A; this algebra is often an algebra of functions or measures under convolution. One then shows that equation (5) defines a multiplier if and only if x is in B. One will then just say that $\mathcal{M}(A) = B$, meaning that $\mathcal{M}(A) = \{T_x \mid x \in B\}$. We shall use the convention in the following discussion.

The theorem motivating the study of multipliers is that of Wendel which says that, for a locally compact group G, $\mathcal{M}(L^1(G)) = M(G)$, in the above sense. From this Wendel deduced that if $L^1(G,m)$ and $L^1(H,m)$ are isomorphic algebras for locally compact groups G and H then G is isomorphic to H as a topological group. Thus G is determined by $L^1(G,m)$.

Now if S is a semigroup with identity then $M(S)$ has an identity so that $\mathcal{M}(M(S)) = M(S)$. So a study of $\mathcal{M}(A)$ is only of interest for A a suitable subalgebra of $M(S)$ not containing that identity. If S does not have an identity then the situation is different. Define a *(right) multiplier* of S to be a continuous mapping ϕ of S to itself satisfying

$$\phi(st) = s\phi(t) \quad (s, t \in S).$$

We define $\Omega(S)$ to be the set of all such multipliers; then $\Omega(S)$ is a semigroup containing a copy of S in the same way that A is embedded in $\mathcal{M}(A)$. Then the formula

$$(T_\phi(\mu))(f) = \mu(f \circ \phi) \quad (f \in C_b(S), \mu \in M(S))$$

will define a multiplier T_ϕ of $M(S)$. From this it is clear that we cannot expect that $\mathcal{M}(M(S))$ can be embedded in $M(S)$ unless $\Omega(S) = S$ — which implies that S has an identity. Instead we can seek to embed $\mathcal{M}(M(S))$ in $M(\Omega(S))$ — for some suitable topology on $\Omega(S)$.

The first paper we shall mention is that of SLEIJPEN [60]. Although dating back to 1979 it appears to be the only result which attempts to generalize Wendel's Theorem to $M_a(S)$ where S is a foundation semigroup (see section 3). We take S to be a foundation semigroup with identity element 1, so that S must be a stip. Sleijpen's theorem needs a lot of terminology and notation to state. We define

$$S_0 = \{x \in S \mid U^{-1}x \cap xU^{-1} \text{ is a nbhd of 1 for all nbhds } U \text{ of } x\}$$

Then S_0 is a dense ideal of S; it is in fact the smallest dense ideal of S. Normally $S_0 \neq S$. We define $T = \Omega(S_0)$. Then for $U \subset S$ define

$$\tilde{U} = \{(x, y) \in T \times T \mid x(U \cap S_0) \cap y(U \cap S_0) \neq \phi\}.$$

Then $\{\tilde{U} \mid U$ is a neighbourhood of $1\}$ is a base for a uniform structure on T. We give T the uniformity and topology which is obtained in this way. Then T is a complete topological semigroup and S_0 (and hence S) is dense in T. In general $C_b(T)^*$ will not be a Banach algebra under convolution. However, if $\mu \in C_b(T)^*$ is such that

(a) the map $x \mapsto (\mu \circ f)(x) = \mu({}_x f)$ is continuous on T for all $f \in C_b(T)$, and
(b) for fixed $x \in S_0$ the map $f \mapsto \mu({}_x f)$ defines an element of $M(T)$,

then for each ν in $C_b(T)^*$ we can define $\nu * \mu \in C_b(T)^*$ by

$$(\nu * \mu)(f) = \nu(\mu \circ f). \quad (f \in C_b(T)).$$

We define $M_r(T)$ to be the set of all $\mu \in C_b(T)^*$ satisfying (a) and (b).

Theorem 6.1. (i) *$M_r(T)$ is a Banach algebra under $*$.*
(ii) *There is an isometric isomorphism $\mu \mapsto T_\mu$ from $M_r(T)$ onto $\mathcal{M}(M_a(S))$ where $T_\mu(\nu) = \nu * \mu$ for $\nu \in M_a(S)$ and $M_a(S)$ is embedded in $M(T)$ by the embedding of T into S in the natural way.* □

Warning: *Sleijpen calls 'right multipliers' what are usually understood as left multipliers.*

Sleijpen goes on to show that if S contains no sets of measurable cardinality then $M_r(T)$ can be represented as a certain set of measures on T.

The next case to be considered is where S is a product of totally ordered semigroups. I first discuss the paper of BAKER, PYM and VASUDEVA [4]. We suppose that for $1 \le j \le k, I_j$ is a totally ordered set which has the order topology, and is compact and connected. We give I_j the semigroup multiplication maximum. We put $B = I_1 \times I_2 \times \cdots \times I_k$. We regard B as having the product order; if $x \in B$ then x_j denotes its j-th component. For $a \in B$ we define the *box* B_a to be the set

$$B_a = \{x \in B \mid x \ge a\}.$$

We define the *inside* of B_a to be the set $B_a^0 = \{x \in B \mid x_j > a_j \text{ for } 1 \le j \le k\}$. In general B_a^0 is *not* the topological interior of B_a. Now let S be a closed subset of B, and let S have the topological and algebraic structure induced by B. We define $S^0 = \bigcup\{S \cap B_a^0 \mid a \in S\}$. For a proper subset J of $\{1, 2, \ldots, k\}$ we define the *face* F_a^J by

$$F_a^J = \{x \in B_a : x_j > a_j \text{ for } j \in J \text{ and } x_j = a_j \text{ for } j \notin J\}.$$

We denote by proj_J the projection of B_a onto F_a^J, so that $\mathrm{proj}_J(x) = y$ where $y_j = x_j$ for $j \in J$ and $y_j = a_j$ otherwise; this mapping induces a mapping of $M(B_a)$ onto $M(F_a^J)$ which is also denoted by proj_J. If $\mu \in M(B_a)$ then μ^J denotes the restriction of μ to F_a^J.

Now let A be a convolution measure algebra on S. So A is an L-subspace of $M(S)$ (see Section 3) and a closed subalgebra of $M(S)$ such that $\bigcup\{\mathrm{supp}\mu \mid \mu \in A\}$ is dense in S. Suppose also that $|\mu|(S \setminus S^0) = 0$ for all $\mu \in A$. If U is a Borel subset of S then A_U is the set of measures μ in A which satisfy $|\mu|(S \setminus U) = 0$. The first main theorem of [4] deals with a special case in which stronger conclusions are possible.

Theorem 6.2. *Suppose that S has an identity a. If $T \in \mathcal{M}(A)$ there is a unique element μ in $M(S)$ such that $T = T_\mu$. Further*
(i) *the restriction of μ to S^0 is in A*

(ii) *for any proper subset* J *of* $\{1, 2, \ldots k\}$ *we have*

$$\mu^J \in \bigcap \{\text{proj}_J M_U \mid U \textit{ is an open neighbourhood of } F_a^J\}.$$

Conversely, if $\mu \in M(B_a)$ *and* μ *satisfies* (i) *and* (ii) *then* μ *defines a multiplier* T_μ *of* A. *Further* $\|T_\mu\| = \|\mu\|$ *and if* T_μ *is positive then* μ *is positive.* □

If S does not have an identity the authors found need of an extra hypothesis. They assumed that there is a subsemigroup E of S such that, for all x in E, x is in the closure of $S \cap B_x^0$ and such that $\bigcup \{S \cap B_x^0 \mid x \in E\}$ is dense in S. This hypothesis is satisfied by many semigroups S.

Theorem 6.3. *Let* $T \in \mathcal{M}(\mathcal{A})$. *Then* $T = T_\mu$ *for some* $\mu \in M(B)$. *The restriction of* μ *to* S^0 *is in* A. □

Most of the other conclusions of theorem 6.2 may fail. If S has an infinite set of minimal elements then we cannot conclude that $\mu \in M(S)$ (essentially because A need not have an approximate identity). Further the uniqueness of μ and the identity $\|T\| = \|\mu\|$ may fail. If $k \geq 3$ we may have T positive but no positive μ such the $T = T_\mu$.

The above theorems contain a number of other published results as special cases, including work of DHAR and VASUDEVA [15], JOHNSON and LAHR [36] and TODD (Section 2 of [63]).

I now turn to certain algebras which appear as algebras of functions on a totally ordered semigroup. However, the multiplication in the algebras is obtained by embedding the space of functions in $M(S)$, so the results seem to be relevant here. I wish to discuss the work of BAKER, PYM and VASUDEVA [5]. Let X be a totally ordered space which is compact for the order topology, regarded as a semigroup for the operation maximum. Let μ be a positive continuous measure on X whose support is X. Let $p \geq 1$. We can embed $L^p(X, \mu)$ in $M(X)$ in the natural way since μ is bounded so that $L^p(X, \mu) \subseteq L^1(X, \mu)$. We consider the multiplication induced on $L^p(X, \mu)$ by convolution in $M(X)$. For $f, g \in L^p(X, \mu)$ we have

$$(f * g)(x) = g(x) \int^x f d\mu + f(x) \int^x g d\mu \quad (x \in X) \tag{6}$$

where $\int^x h d\mu$ denotes $\int_X \chi_{[0,x]} h d\mu$ and $[0, x] = \{y \in X \mid y \leq x\}$. One can then show that there is a constant K (probably > 1) such that $\|f * g\|_p \leq K \|f\|_p \|g\|_p$ ($f, g \in L^p(X, \mu)$). Therefore $*$ is jointly continuous for the norm $\|\cdot\|_p$. Now if $1 \leq r \leq p$ then $L^p \subset L^r$ and then if $f \in L^r, g \in L^p$ we have that $f * g$ exists and belongs to L^r. By a *multiplier* from $L^p(X, \mu)$ to $L^r(X, \mu)$ is meant a bounded linear mapping T such that

$$T(f * g) = (Tf) * g \quad (f, g \in L^p(X, \mu)).$$

Claim 6.4. (See the remark below) Under the above conditions, for $1 \leq r \leq p$ the following are equivalent.

(i) T is a multiplier from $L^p(X, \mu)$ to $L^r(X, \mu)$.

(ii) $Tf(x) = \frac{-1}{(\mu(0,x])^2} \int^x \alpha d\mu \int^x f d\mu + \frac{\alpha(x)}{\mu\left([0,x]\right)} \int^x f d\mu + \frac{f(x)}{\mu([0,x])} \int^x \alpha d\mu$,

μ-a.e., where α is a measurable function on X such that

(a) $x \mapsto \frac{1}{\mu([0,x])} \int^x \alpha d\mu$ is bounded on X, and

(b) $x \mapsto \frac{\alpha(x)}{\mu([0,x])} \int^x \alpha d\mu$ is an element of $L^r(X, \mu)$ for all f in $L^p(X, \mu)$.

(iii) $Tf(x) = kf(x) - f(x) \int_x t d\mu + t(x) \int^x f d\mu$,

where $k \in \mathbb{C}$, t is a measurable function on X,

$\int_x t d\mu$ denotes the integral $\int^\infty \chi_{X \setminus [0,x]} t d\mu$,

$x \mapsto \int_x t d\mu$ is a bounded function on X, and

$x \mapsto t(x) \int^x f d\mu$ is in $L^r(X, \mu)$ for all $f \in L^p(X, \mu)$.

If T is positive then we can obtain a much simpler formula for a multiplier T. In that case (iii) simplifies to $\mu = c\bar{0} + t, c \in \mathbb{R} + \cup\{0\}$. (Here $\bar{0}$ is the point mass at the identity of S which acts as an identity for convolution and t is as in (iii) above, with $t \geq 0$.)

Remark. As this survey was being drafted a mistake was found in the proof of 6.4; the erroneous argument had also been used in the proof of 6.6. It seems probable that the result in 6.4 for $r < p$ and in 6.6 for $r > p$ is incorrect and that the characterisation of the multipliers in these cases (which correspond closely) need modification.

In the next paper to be discussed, [38], the authors suggest a different definition for a multiplier. For $p \geq 1$ they consider the algebra $L^p(I, m)$ where m is Lebesgue measure on I and convolution is as in the previous example given by formula (6). But observe also that the same formula defines an action of $C(I)(= C_b(I))$ on $L^p(I, m)$ (by Hölder's inequality) i.e., if $f \in C(I), g \in L^p(I, m)$ then (6) defines $f * g \in L^p(I, m)$ and

$$\|f * g\|_p \leq 2\|f\|_\infty \ \|g\|_p.$$

In this way we can regard $L^p(I, m)$ as a left $C(I)$-module. Define q by the equation $\frac{1}{p} + \frac{1}{q} = 1$. Then, for f in $C(I)$ the adjoint of the action of f on $L^p(I, m)$ defines an action of f on $L^q(I, m)$; this is given by

$$(g \circ f)(s) = g(s) \int_0^s f(t) dt + \int_s^1 g(t) f(t) dt \text{ a. e. } m \quad (g \in L^q(I, m)). \tag{7}$$

This action makes $L^q(I, m)$ into a right $C(I)$ module. The two modules we have constructed are denoted by L_*^p and L_0^q respectively. The paper under discussion characterises

$$M_0^{r,p} = \mathrm{Hom}_{C(I)}(L_*^r, L_*^p), \text{ and } M_*^{r,p} = \mathrm{Hom}_{C(I)}(L_*^r, L_0^p),$$

where $\mathrm{Hom}_{C(I)}$ denotes the set of continuous module homomorphisms. The techniques involved are more abstract than those used in the other papers discussed in this section, making extensive use of the techniques of tensor products. The idea is to characterise $M_0^{r,p}$ and $M_*^{r,p}$ as dual spaces of certain spaces.

We define $A_0^{r,p} = L_*^r \hat{\otimes}_{C(I)} L_0^q$, and $A_*^{r,p} = L_*^r \hat{\otimes}_{C(I)} L_*^q$ (with $\frac{1}{p} + \frac{1}{q} = 1$), where $\hat{\otimes}$ denotes the projective tensor product of the two Banach spaces. It is shown that

$$(A_*^{r,p})^* = M_*^{r,p}, \quad (A_0^{r,p})^* = M_0^{r,p}.$$

Observe that if $\frac{1}{r} + \frac{1}{s} = 1$ then $A_*^{r,p} = A_*^{q,s}$. So in order to characterise $M_*^{r,p}$ it is enough to consider the case where $r \geq q$.

Theorem 6.5. *With the above terminology, for $r \geq q$*

$$M_*^{r,p} = \{\phi \mid \phi \text{ is a measurable function on } I \text{ and } \sup_{n \in \mathbb{N}} 2^{-n/s} \, \|P_n(\phi)\|_p < \infty\}$$

where $P_n(\phi) = \chi_{J_n}\phi$ with $J_n = [2^{-n}, 2^{1-n}]$ $(n = 1, 2, 3, \cdots)$, and each such ϕ induces a multiplier T_ϕ on L_*^r where $T_\phi(f) = f \circ \phi$ and $f \circ \phi$ is defined as in formula (7) above. □

Claim 6.6. (See the remark after 6.4.) With the above terminology

(i) for $r \geq p$, $M_0^{r,p} = \{\phi \mid \phi$ is a function on I with derivative ϕ' a. e. m and $\sup_{n \in \mathbb{N}} 2^{-n/s}\|P_n(\phi')\|_p < \infty\}$ where P_n is as in 6.5 and ϕ defines a multiplier T_ϕ on L_*^r by

$$(T_\phi(f))(x) = \phi'(x) \int_0^x f(y)dy + \phi(x)f(x) \text{ a. e. } m,$$

(ii) for $r < p$ we have $M_0^{r,p} = \{0\}$.

The ideas of the above paper were inspired, to some extent, by analogy to the theory of multipliers on spaces of functions on locally compact groups; the same is true of the next paper to discuss. This paper is by Kalra, Singh and Vasudeva [37]. They introduce the algebra $A_p(\mathbb{R}_+)$, where $\mathbb{R}_+$ is $(0, \infty)$ with operation maximum. The algebra $A_p(G)$ for G a locally compact group has been studied extensively. They aim to consider to what extent the theory of $A_p(G)$ is similar to that of $A_p(\mathbb{R}_+)$. Consider the space $L^1(\mathbb{R}_+, m)$ under order convolution similar to formula (6),

$$(f * g)(x) = g(x) \int_0^x f(t)dt + f(x) \int_0^x g(t)dt \text{a.e.} m, \tag{8}$$

where m is Lebesgue measure on $\mathbb{R}_+$. Then $L^1(\mathbb{R}_+, m)$ is a semi-simple commutative Banach algebra. We can identify the maximal ideal space of this algebra with $(0, \infty]$, where for $x \in (0, \infty]$ and $f \in L^1(\mathbb{R}_+, m)$ the Gelfand transform $\hat{f}$ is given by

$$\hat{f}(x) = \int_0^x f(t)dt.$$

For $1 \leq p \leq \infty$ we define

$$A_p(\mathbb{R}_+) = \{f \in L_1(\mathbb{R}_+, m) \mid \hat{f} \in L^p((0, \infty], m)\}$$

where the Lebesgue measure m is regarded as a measure on $(0, \infty]$. Then each $A_p(\mathbb{R}_+)$ is an ideal of $L_1(\mathbb{R}_+, m)$ and $A_\infty(\mathbb{R}_+) = L_1(\mathbb{R}_+, m)$. For the inherited product and the norm

$$|||f|||_p = ||f||_p + ||\hat{f}||_p,$$

$A_p(\mathbb{R}_+)$ becomes a semisimple Banach algebra. If $p < r$ then $A_p \mathbb{P} A_r$; if $1 \leq p < \infty$ then the maximal ideal space of A_p is $(0, \infty)$, identified by formula (8). The algebra Ap has no bounded approximate identity. The authors consider multipliers of the form T_ϕ on $A_p(\mathbb{R}_+)$ where ϕ is a function on $\mathbb{R}_+$ and T_ϕ is to satisfy the identity

$$T_\phi(\hat{f}) = \phi \hat{f} \quad \text{for } f \in A_p(\mathbb{R}_+).$$

It is shown that, for a function ϕ, T_ϕ is a multiplier if and only if ϕ is bounded and absolutely continuous on $[0, k]$ for all $k > 0$ and if $\phi' \hat{f} \in L_1(\mathbb{R}_+, m)$ for all $f \in A_p(\mathbb{R}_+)$ (where ϕ' is the derivative of ϕ). Clearly the last condition of the above is not very useful. The authors obtain both a necessary and a sufficient condition which are each more easily checked for T_ϕ to be a multiplier, but both conditions are clearly not necessary and sufficient.

Theorem 6.7. (i) *Let ϕ be bounded and absolutely continuous on $[0, k]$ for all $k > 0$. Suppose that $\phi' \in L^r((\delta, \infty), m)$ for some $\delta > 0$ and r with $\frac{1}{p} + \frac{1}{r} \geq 1$. Then T_ϕ is a bounded multiplier on $A_p(\mathbb{R}_+)$ in the above sense.*

(ii) *If T_ϕ is a bounded multiplier on $A_p(\mathbb{R}_+)$ then $\int_0^x |\phi'(t)|dt = 0(x^{1/p})$ as $x \to \infty$ and the function $x \mapsto x^{-r}\phi'(x)$ is in $L^1([1, \infty))$ for all $r > \frac{1}{p}$.* □

I would like to conclude this section with a short discussion of TODD's paper [63]. It had been shown previously by LAHR that if S is a commutative, cancellative semigroup (so that S can be embedded in a group G which is generated by S) then $\mathcal{M}(\ell^1(S)) \subseteq \ell^\infty(G)$. Todd was able to generalize this result considerably. Suppose that S is commutative. It can be shown that if a is an element of S such that
(i) $ax = ay$ implies $x = y$ $(x, y \in S)$, and
(ii) U open in S implies that aU is open in S

then there is a locally compact commutative semigroup T with identity 1, say, such that a is invertible in T (i.e., $1 \in aT$).

Theorem 6.8. (i) *With S and T as in the previous paragraph, there is an isometric homomorphism from $\mathcal{M}(M(S))$ into $M(T)$. In general the homomorphism is not onto $M(T)$.*

(ii) *With S and T as above, let μ be a positive measure on S for which $L^1(S, \mu)$ is a subalgebra of $M(S)$ and for which a is in the support of μ. Suppose also that there is a compact neighbourhood C of a such that $C^{-1}K$ is compact in S for all compact sets K in S. Then there is an isometric homomorphism of $\mathcal{M}(L^1(S, \mu))$ into $M(T)$.* □

In Section 4 of TODD's paper, this theorem is exploited (in the manner used by WENDEL) to show that, in certain cases, if $L^1(S, \mu)$ and $L^1(S_1, \mu_1)$ are isomorphic then there is an *algebraic* isomorphism between S and S_1. Observe that Wendel's Theorem on multipliers is a special case of Theorem 6.8 (ii) since the homomorphism is automatically onto because if S is a group G then $T = G$ and $L^1(G, m)$ is an ideal of $M(G)$.

7. Weighted convolution algebras

In this article we shall define a *weight function* on S to be a function, say w, mapping S into $\mathbb{R}_+$ such that
(a) w and $1/w$ are locally bounded, i.e., for each compact subset K of S there exist λ, μ in $\mathbb{R}_+$ such that

$$\lambda \leq w(x) \leq \mu \quad \text{for all } x \in K,$$

(b) w is Borel measurable, and
(c) $w(xy) \leq w(x)w(y)$ for all $x, y \in S$.

Conditions (a) and (b) of this definition have only been included for convenience. Many authors require w to be continuous, but most results seem to carry over to the more general case with a little care. Even on a simple semigroup like $(\mathbb{R}_+, +)$ there are many interesting discontinuous weight functions; most are semicontinuous, but that hypothesis seems to do little to improve the theory.

If w is a weight function then one can, with care, construct the Banach algebra to be denoted by $M(S, w)$. The reason for the need for care is that the elements of $M(S, w)$ need not be measures in the sense used in most texts since they are signed (do not take only non-negative real numbers and ∞ as their values) but not necessarily bounded. Technically, the definition given by some authors of $M(S, w)$ is not the one to produce a complete space, but in fact they knew what they *really* meant and if that had been correctly expressed then they would have obtained a complete space $M(S, w)$. This problem is analysed in depth in Rejali's papers [54], [55]; in those papers the problem is further complicated by adding the extra dimension that S need not be locally compact. One way to define $M(S, w)$ is as a space of linear functionals on $C_{00}(S)$; we shall use the alternative definition here. We define

$$M^{+}(S, w) = \{\mu \mid \mu \text{ is a positive measure on } S \text{ and } w \in L^{1}(S, \mu)\}. \tag{9}$$

(So the elements of $M^{+}(S, w)$ do not have to be bounded measures.) We can then define

$$M(S, w) = \{\mu_1 - \mu_2 + i(\mu_3 + \mu_4) \mid \mu_i \in M^{+}(S, w) \text{ for } 1 \leq i \leq 4\},$$

where linear combinations of elements of $M^{+}(S, w)$ are to be interpreted as follows. The linear combination is to be a set function defined on the compact subsets of S (where each μ_i takes only values in $\mathbb{R}$) defined in the obvious way, and any two linear combinations are regarded as equal if they coincide on the compact sets. In a natural way, $M(S.w)$ is a vector space over $\mathbb{C}$.

If $\mu \in M^{+}(S, w)$ then μ is $\sigma-$finite, so that any sensible meaning that is attached to integration with respect to an element of $M(S, w)$ can be used without ambiguity. The simplest is to define, for μ as in (9),

$$\int_S f d\mu = \int_S f d\mu_1 - \int_S f d\mu_2 + i(\int_S f d\mu_3 - \int_S f d\mu_4)$$

for, say, $f \in \bigcap_{i=1}^{n} L^{1}(S, \mu_i)$. We make $M(S, w)$ into a Banach space as follows. If $\mu \in M(S, w)$ there exists an element $w.\mu \in M(S)$ such that $\int f d(w.\mu) = \int f w d\mu$ for $f \in C_b(S)$ (in which case $f \in \bigcap_{i=1}^{n} L^{1}(S, \mu_i)$). The map $\mu \mapsto w.\mu$ is injective. For $\mu \in M(S, w)$ we define $\|\mu\|_w = \|w.\mu\|$, where the norm of $w.\mu$ is the total variation norm of $M(S)$. In this way we get an isometric isomorphism of $M(S, w)$ onto $M(S)$, *as Banach spaces*. Finally, we can make $M(S, w)$ into a Banach algebra by defining a convolution product by the rule

$$(\mu * \nu)(f) = \int\int f(xy)d\mu(x)d\nu(y) \ (\mu, \nu \in M(S,w), f \in C_{00}(S)).$$

If S is discrete then $M(S,w)$ is the same as the algebra

$$\ell^1(S,w) = \{f \mid f : S \to \mathbb{C} \text{ and } \|f\|_w = \sum_{x \in S} |f(x)|w(x) < \infty\}.$$

This algebra has been studied extensively, for example see DZINOTYIWEZI's book [22].

The main interest in both $\ell^1(S,w)$ and $M(S,w)$ has been to discover when it is Arens regular. Interest in this problem goes back to work of YOUNG [66], PYM [53], and CRAW and YOUNG [11]. As the title of his paper implied, Pym showed that 'most' unweighted (i.e., $w \equiv 1$) convolution algebras are irregular. In these three papers are attempts to establish necessary and sufficient conditions for the above algebras to be regular. The work left one or two rather prominent gaps in the theory. Recent work by REJALI [6], [54], [55] (including joint work with the author of this review) has cleared up most of these gaps. The biggest problem was that no-one could find an example of a semigroup S such that $\ell^1(S)$ is regular but $\ell^1(S,w)$ is not regular for some weight function w. Once we were able to show that no such semigroup S exists, a lot of the rest of the jigsaw dropped into place.

It is first necessary to introduce some notation and definitions. If w is a weight function on S then we define $\Omega: S \times S \to \mathbb{R}_+$ by the formula

$$\Omega(x,y) = \frac{w(xy)}{w(x)w(y)} \quad (x, y \in S).$$

Observe that $0 < \Omega(x,y) \le 1$ and that $\Omega \equiv 1$ if and only if w is multiplicative (equality in (c) of the definition of a weight function). If w has a subscript we shall endow Ω with the same subscript. It turns out that the regularity of $M(S,w)$ depends solely on Ω. We say that a bounded function $f: S \times S \to \mathbb{R}$ is *cluster* if for every pair $(x_n), (y_m)$ of sequences of distinct elements of S,

$$\lim_n \lim_m f(x_n, y_m) = \lim_m \lim_n f(x_n, y_m) \tag{10}$$

whenever both iterated limits in (10) exists. Observe that, for any $(x_n), (y_m)$ we can find subsequences of these sequences for which the corresponding iterated limits both exist. If f is cluster and both sides of (10) are zero [resp. positive] whenever they both exist we say f is *0-cluster* [resp. *positive cluster*].

Finally, we say that f is *uniformly 0-cluster* if $\lim_{n,m} f(x_n, y_m) = 0$ for all sequences (x_n) (y_m) of distinct elements of S (i.e., for each $\varepsilon > 0$ there exist finite subsets F, K of S such that $|f(x,y)| < \varepsilon$ whenever $x \notin F$ or $y \notin K$).

We shall need to consider infinite matrices of the form $(x_n y_m)$ where (x_n),(y_m) are two sequences in S and $x_n y_m$ is the entry in the nth column and the mth row of the matrix, for $n \in \mathbb{N}, m \in \mathbb{N}$. By a *submatrix* of $(x_n y_m)$ is meant an (infinite) matrix $(a_n b_m)$, say, where (a_n) is a subsequence of (x_n) and (b_n) a subsequence of (y_m); so $(a_n b_m)$ is obtained by deleting rows and columns of $(x_n y_m)$. A matrix $(x_n y_m)$ is a *type* C if either every row or every column (or both) of the matrix is constant. The fundamental theorem is as follows [6], [55].

Theorem 7.1. *Let w be a weight function on S. The following conditions are equivalent.*

(i) *$\ell^1(S, w)$ is regular.*

(ii) *for each pair $(x_n), (y_m)$ of sequences of distinct points of S there are subsequences $(a_n), (b_m)$ of $(x_n), (y_m)$ respectively such that either*

  (1) $\lim_n \lim_m \Omega(a_n, b_m) = 0 = \lim_m \lim_n \Omega(a_n, b_m)$ *or*

  (2) *the matrix $(a_n b_m)$ is of type C.*

(iii) *$M(S, w)$ is regular.*

(iv) *For every bounded, Borel measurable function h on S the map $(x, y) \mapsto h(xy)\Omega(x, y)$ is cluster.*

(v) *For each pair of sequences $(x_n), (y_m)$ in S and each subset A of S,*

$$\overline{\{\chi_A(x_n\, y_m)\Omega(x_n, y - m) \mid n < m\}} \cap \overline{\{\chi_A(x_n\, y_m)\Omega(x_n, y_m) \mid n > m\}} \neq \phi.$$

□

Corollary 7.2. *If $\ell^1(S)$ is regular then so is $M(S, w)$ for any weight function w on S.* □

Although Corollory 7.2 is an immediate consequence of Theorem 7.1, by regarding $\ell^1(S)$ as $\ell^1(S, 1)$, the method of proof requires the establishment of 7.2 before proving 7.1.

If S, as a discrete semigroup, is topologically cancellative, then one can show that $\ell^1(S, w)$ is regular if and only if Ω is 0-cluster, so that $\ell^1(S)$ is certainly irregular. It can also be deduced that (for arbitrary S), if $\ell^1(S, w)$ is regular and Ω is positive cluster then $\ell^1(S)$ is regular, and if Ω is 0-cluster then $\ell^1(S, w)$ is always regular. Another observation is that if we can find $\lambda, \mu \in \mathbb{R}_+$ such that $\lambda\Omega_1 \leq \Omega_2$ and $\mu\Omega_2 \leq \Omega_1$ (in particular if there are similar inequalities between w_1 and w_2) then $\ell^1(S, w_1)$ is regular if and only if $\ell^1(S, w_2)$ is.

In [23], DZINOTYIWEYI defines spaces of 'weighted (weakly) almost period functions' on S. REJALI [55] has introduced modified versions of these spaces which relate much more readily to the regularity of $\ell^1(S, w)$. Here it is necessary to assume that Ω is a separately continuous function on $S \times S$. We define

$$C_b(S,w) = \{f \mid f: S \to \mathbb{C} \text{ and } f/w \in C_b(S)\}$$

(so the elements of $C_b(S,w)$ need not be continuous unless w is). Then we define

$$WAP(S,w) = \{f \in C_b(S,w) \mid W(f) \text{ is relatively weakly compact in } C_b(S)\},$$
$$AP(S,w) = \{f \in C_b(S,w) \mid W(f) \text{ is relatively compact in } C_b(S)\},$$

where $W(f) = \{\frac{fx}{wx}\Omega_x \mid x \in S\}$ and $\Omega_x(y) = \Omega(y,x)$ for $x, y \in S$. Then $WAP(S,w), AP(S,w)$ are closed subspaces of $C_b(S,w)$, where $C_b(S,w)$ has the norm $\|f\|_w = \|\frac{f}{w}\|_\infty$. It is simple to show that if $f \in C_b(S,w)$ then $f \in WAP(S,w)$ if and only if the function $(x,y) \mapsto \frac{f(xy)}{w(x)w(y)}$ is cluster on $S \times S$. Rejali proves the following (see [55]).

Theorem 7.3. *Let Ω be separately continuous.*
(i) *$\ell^1(S,w)$ is regular if and only if $WAP(S_d,w) = \ell^\infty(S,w)$ where S_d is S with the discrete topology, and $\ell^\infty(S,w) = C_b(S_d,w)$.*
(ii) *$WAP(G,w) = C_b(G,w)$ if and only if G is compact or G is discrete and Ω is 0-cluster.*
(iii) *$AP(G,w)$ if and only if G is compact and Ω is jointly continuous or G is discrete and Ω is uniformly 0-cluster.* □

The only other work to report on weighted algebras is by Gharamani. In this work the weight functions are assumed to be continuous; I have not checked whether it applies to more general weight functions. We shall denote by $\ell^1(G,w)$ the set of those measurable functions f on G for which $\int_G |fw| dm < \infty$, which can be regarded as a subalgebra of $M(G,w)$. In [25] Gharamani considers a weight function w on $\mathbb{R}_+$ such that $w(0) = 1$ and $\lim_{t\to\infty} w(t)^{1/t} = 0$ (so that the algebra is radical). He shows that, for two such weight functions w_1, and w_2, the algebra $L^1(\mathbb{R}_+, w_1)$ is isomorphic to $L^1(\mathbb{R}_+, w_2)$ if and only if there exist a, b, m, M in $\mathbb{R}_+$ such that

$$m \le \frac{w_2(ax)}{w_1(ax)} b^x \le M \text{ for all } x \in \mathbb{R}_+.$$

In the other two papers [26], [27] Gharamani considers the full algebra $L^1(G,w)$. An element g of $L^1(G,w)$ is said to be *compact* if the multiplier $T_g: L^1(G,w) \to L^1(G,w)$ defined by right convolution multiplication by g is a compact operator. He shows that (a) if G is compact then every element is compact, (b) if G is not compact then no non-zero element of $L^1(G,m)$ is compact. Applications of this result are given in the second of the two papers. It is shown that a right multiplier of $L^1(G,w)$ is a compact (or weakly compact) operator if and only if it is induced by a compact element of $L^1(G,m)$. It follows that a necessary and sufficient condition for $L^1(G,m)$ to be an ideal in its bidual (in one of the Arens products— see Section 5) is that if G is compact. (The case $w \equiv 1$ is well-known.) However, as is pointed out in the MR review, if G is compact then $w \equiv 1$ is the only weight function on G since he has assumed that w is continuous.

8. Invariant measure semigroups

In his survey article [64] WILLIAMSON posed the question of finding out which semigroups S support a non-zero positive 'invariant' measure. Let us say that a measure μ on S is *invariant* if μ is positive, non-zero and $\mu(xB) = \mu(B)$ for all Borel sets B in S and all x in S such that xB is a Borel set. (There are lots of other ways to generalise the notion of invariant measure on a group; one can replace xB with $x^{-1}B$, and one can replace $=$ *by* $\geq$ or $\leq$ in both cases. These different cases have not received much attention except in those cases where it turns out that the situation reduces to the first case.) It would be useful to find semigroups which support an invariant measure since they would be foundation semigroups in the sense of Section 3. However, as we shall see, there are very few such semigroups. In this field it is not usually assumed that S is locally compact, but then it is assumed that μ is locally finite (i.e., for each $x \in S$ there is a neighbourhood U of x with $\mu(U) < \infty$).

The problem is to characterise those semigroups which support an invariant measure. As a first step to understanding the problem, let us take S to be discrete. It is then easy to show that an invariant measure must be (a scalar multiple of) the counting measure $\mu(A) = |A|$. This measure is invariant if S is (right) cancellative. For commutative S it follows that S supports an invariant measure μ if and only if S is embeddable in a group. However, a cancellative semigroup does not, in general, have to be embeddable in a group, so there is not such a nice answer if S is not commutative. Note also that if S is commutative then the invariant measure is essentially unique and equal to the restriction of Haar measure on the group to S.

We now turn to the general (non-discrete) case. The situation for commutative S was sorted out by RIGELHOFF [56] for a special case and PATERSON [52]. Rigelhoff considered the case in which the translation maps on S are open (U open implies xU open). He showed that (for commutative S), there is an invariant measure μ with support S if and only if S is homeomorphic and isomorphic to an open subsemigroup of a locally compact group G and then μ is the restriction to S of Haar measure on G. Paterson's theorem is as follows; it effectively kills off the problem for commutative S.

Theorem 8.1. *Let S be a commutative topological semigroup. If there is an invariant measure μ on S then there is a locally compact abelian group G and a* *continuous homomorphism q of S into G such that $q(S)$ has non-empty interior, and further there is a Haar measure m on G such that $\mu(C) = m(q(C))$ for every compact set C in S. The map q is injective if and only if S is cancellative; it is a homeomorphism onto $q(S)$ if and only if S has 'continuous inversion,' i.e., $\{s_\alpha\} \to s, \{s_\alpha t_\alpha\} \to st$ implies $\{t_\alpha\} \to t$.* □

Since the solution of the commutative case, Mirotin and Mukhin (separately and jointly) have published several papers tackling the non-commutative case. The results are far from providing a complete solution. I must confess to the fact that the papers are in Russian and I have had to rely on Mathematical Reviews for details of their contents.

In the first paper [44], S is assumed to be a locally compact left topological semigroup (i.e., the maps $y \mapsto xy$ are continuous on S for each $x \in S$) such that all (left and right) translation maps on S are open. Let μ be an invariant measure on S with support S. It is shown that we can write $S = \bigcup\{S_\alpha \mid \alpha \in I\}$ where each S_α is a component of S and is a minimal left ideal of S and is left cancellative with respect to $S(zx = zy$ implies $x = y$ for $x, y \in S_\alpha, z \in S)$. Also we have

$$\mu(B) = \sum_{\alpha\in I} \mu_\alpha(B) \quad \text{for each Borel set in } S,$$

where μ_α is a (left) invariant measure on S_α for all $\alpha \in I$. Further μ is unique to the extent that each μ_α is uniquely determined up to a non-negative constant.

A result parallel to this was obtained by Mukhin as follows.

Theorem 8.2. *Let S be a locally compact topological semigroup. Suppose that all translation maps are open and injective. Then S supports an invariant measure if and only if there is a uniformity $\mathcal{U}$ on S such that the topology obtained from $\mathcal{U}$ is Hausdorff and $\mathcal{U}$ has a base consisting of sets V for which $V(x) = \{y \in S \mid (x, y) \in V\}$ is open in the original topology on S for each $x \in S$ and such that $(y, z) \in V$ if and only if $(xy, xz) \in V$ for all $x \in S$. Further, we can write $S = \bigcup\{S_\alpha \mid \alpha \in I\}$ where each S_α is an open and closed ideal of S and any invariant measure on S_α is unique up to a positive constant.* □

In order to progress further it seems to be necessary to assume that S can be embedded (algebraically) in a group. The results of [46] and [50] provide the following theorem.

Theorem 8.3. *Let S be a topological semigroup in which all the translation maps (left and right) are open. If S has an invariant measure then S is locally compact. Suppose that S is locally compact and embeddable algebraically in a group. Then there is an invariant measure on S with support S if and only if S can be embedded homeomorphically and isomorphically as an open subsemigroup of a locally* *compact group, and then the invariant measure is the restriction to S of a left Haar measure on S.* □

The methods of proof of these results are all based on the proof of Weil that any topological group with an invariant measure is locally compact (see [30] for details). For a compact subset K of S we define a semimetric d_K on S by

$$d_K(x,y) = \mu(zxK\Delta zyK) \quad (x, y \in S)$$

where z is any element of S whose choice will not affect the value of d_K. The collection of all such semimetrics d_K can be used to obtain both a uniformity and a topology on S. If S is locally compact this topology will agree with the original topology of S. Also, if S is embeddable in a group G we can extend the topology to G, and then the completion of G becomes a locally compact group. However, if S does not have continuous inversion (see Theorem 8.1) the topology which we obtained from the semimetrics may not be Hausdorff. So in order to make G, and its completion, Hausdorff we need to take a quotient space of G. This means that our map of S into the locally compact group need not be injective. Finally, if S has open translations then S becomes an open subsemigroup of the group. If S has empty interior in the group we have an extra problem that the Haar measure of the group may vanish on S. The generalisation of Weil's construction is studied in [48], [49] as well as the papers mentioned above [52], [48], [47], [50].

Although most authors have concentrated on the condition that $\mu(xB) = \mu(B)$, it should be pointed out that MUKHERJEA and TSERPES [47], showed in 1979 that the condition $\mu(x^{-1}B) = \mu(B)$ can lead to much stronger conclusions about S and μ. Of course the latter condition is not satisfied for $S = \mathbb{R}_+$ and μ Lebesgue measure on $\mathbb{R}_+$.

9. Any other business

I have only found one paper which does not fit easily into one of the previous chapters. This paper concerns algebras of vector valued measures. Let A be a commutative unital Banach algebra with identity e. Then $M(S, A)$ denotes the vector space of countably-additive, regular, A-valued, Borel measures μ on S for which

$$\|\mu\| = \sup\left\{\sum_{n=1}^{\infty} \|\mu(E_n)\| \mid S = \bigcup_{n=1}^{\infty} E_n \text{ is a partition of } S \text{ into Borel sets}\right\}$$

is finite. Then $(M(S,A), \|\cdot\|)$ is a Banach space. We make $M(S,A)$ into a commutative unital Banach algebra by defining

$$\int f(x)d(\mu * v)(x) = \int\int f(xy)d\mu(x)dv(y) \text{ for } \mu, v \in M(S,A), f \in C_b(S).$$

In Dhar and Vasudeva [13] there is a preliminary study of this algebra for one particular choice of S. The algebra $M(I,A)$ had been studied earlier by Duchon [16]. The algebra studied here is $M(R,S)$ where R is the product semigroup $I \times I$.

We define $\hat{R}$ to be the set of all subsets of R of the form $J \times K$ where each J and K are of the form $[0,x]$ or $[0,x)$ for $x \in I$ and are non-empty. We topologise $\hat{R}$ as follows. The collection of intervals J can be given the order topology induced by set inclusion. We then give $\hat{R}$ the product of these two topologies.

Theorem 9.1. *Let h be a non-zero complex homomorphism of A. For $J \times K \in \hat{R}$ define $h_{J,K}$ on $M(R,A)$ by*

$$h_{J,K}(\mu) = \int\int_R \chi_{J\times K}(x)d\mu(x) \quad (\mu \in M(R,A)).$$

Then $h_{J,K}$ is a complex homomorphism on $M(R,A)$. Further each non-zero complex homomorphism of $M(R,A)$ can be uniquely expressed in the form $h_{J,K}$ for $J \times K \in \hat{R}$. The map $J \times K \to h_{J,K}$ is a homeomorphism of $\hat{R}$ onto the set of complex homomorphisms of $M(R,A)$ in the Gelfand topology. The algebra is semisimple if and only if A is semisimple. □

Other papers (apparently unpublished) by the same authors study the subalgebra $L^1(I,A)$ of $M(I,A)$ identifying the maximal ideal space (homeomorphically) with $(0,1] \times \hat{A}$ and the multipliers of $L^1(I,A)$ with $A \oplus L^1(I,A)$ where for $x \in A$ and $f \in L^1(I,A)$, $T_{(x,f)}(g) = xf + g * f$ $(g \in L^1(I,A))$. If A is also a Banach lattice (implicitly with $xy \geq 0$ whenever $x \geq 0$ and $y \geq 0$?) then the positive multipliers are of the form $T_{(x,f)}$ with $x \geq 0$ and $f \geq 0$ a.e.m. Also, in the more general case, $T_{(x,f)}$ is an isometry if and only if $f = 0$ and $\|x\| = 1$.

References

[1] Baartz, A. P., The measure algebra of a locally compact semigroup, Pacific J. Math. 21 (1967), 199–214,
MR35,4678

[2] Baker, A. C., and J. W. Baker, Algebras of measures on a locally compact semigroup II, J. London Math. Soc 2 (1970), 651–659,
MR43,402

[3] —, Algebras of measures on a locally compact semigroup III, J. London Math. Soc. 4 (1972), 685–695,
MR46,5928

[4] Baker, J. W., J. S. Pym, and H. L. Vasudeva, Multipliers on some measure algebras on compact semilattices,"Recent developments in the algebraic analytical and topological theory of semigroups", (Oberwolfach 1981), 8–30, Lecture Notes in Mathematics 998, Springer Verlag, Berlin etc., 1983,
MR86B43008

[5] —, Totally ordered spaces and their L^P algebras, Mathematika 29 (1982), 42–54, MR84E43004.

[6] Baker, J. W., and A. Rejali, On the Arens regularity of weighted convolution algebras, J. London Math. Soc., to appear

[7] Bami, M. Lashkarizadeh, Representations of foundation semigroups and their algebras, Canadian J. Math. 37 (1985), 29–47,
MR86G22005

[8] —, Bochner's theorem and the Hausdorff moment theorem on foundation topological semigroups, Canadian J. Math. 37 (1985), 785–809,
MR86M43006

[9] Bonsall, F., and J. Duncan, "Complete Normed Algebras", Springer-Verlag, Heidelberg, 1973

[10] Butcher R.J., "Dissertation", University of Sheffield, England, 1975

[11] Craw, I. G., and N. J. Young, Regularity of multiplication in weighted group and semigroup algebras, Quart. J. Math. 25 (1974), 351–358,
MR51,1282

[12] Dhar, R. K., and H. L. Vasudeva, $L^1[I, X]$ with order convolution, Proc. Amer. Math. Soc. 83,(1981), 499–505,
MR82J46070

[13] —, $M(R, X)$ with order convolution, Math. Nachr. 105 (1982), 271–279,
MR84B43003

[14] —, The L^1 algebra of a partially ordered semigroup, Rev. Roumaine Math. Pures Appl. 28 (1983), 811–821,
MR85E22006

[15] —, Characterisations of multipliers of $L^1(R)$, Rev. Roumaine Math. Pures Appl. 30 (1985), 325–332,
MR87C43003

[16] Duchon M., Structure theory for a class of convolution algebras of vector-valued measures, Math. Nachr. 60 (1974), 97–108,
MR50,7951

[17] Dzinotyiweyi, H. A. M., Continity of semigroup actions on normed linear spaces, Quart. J. Math. Oxford 31 (1980), 415–421,
MR82A22001

[18] —, Certain semigroups embeddable in topological groups, J. Austral. Math. Soc. 33 (1982), 30–39,
MR83K22010

[19] —, Non-separability of quotient spaces of function algebras on topological semigroups, Trans. Amer. Math. Soc. 272 (1982), 223–235,
MR84H43016

[20] —, Weak and norm continuity of semigroup actions on normed linear spaces, Quart. J. Math. Oxford 33 (1982), 85–90,
MR85C22004.

[21] —, Some aspects of abstract harmonic analysis, Semigroup Forum 29 (1984), 1–12,
MR85E22007

[22] —, "The analogue of the group algebra for topological semigroups", Pitman, Boston, Mass., London 1984,
MR85H43001

[23] —, Weighted function algebras on groups and semigroups, Bull. Austral. Math. Soc. 33 (1986), 307–318,
MR87H43005

[24] Dzinotyiweyi, H. A. M., and P. Milnes, Functions with separable orbits on foundation semigroups, J. Nigerian Math. Soc. 1 (1982), 31–38

[25] Gharamani, F., Isomorphisms between radical weighted convolution algebras, Proc. Amer. Math. Soc. 26 (1983), 343–351,
MR85H43002

[26] —, Weighted group algebra as an ideal in its second dual space, Proc. Amer. Math Soc. 90 (1984), 71–76,
MR85I43007

[27] —, Compact elements in weighted group algebras, Pac. J. Math. 113 (1984), 77–84,
MR85I43004

[28] Glicksberg, I., Convolution semigroups of measures, Pac. J. Math. 9 (1959), 51–67,
MR217405

[29] Grosser, M., and V. Losert, The norm strict bidual of a Banach algebra and the dual of $C_u(G)$, Manuscripta Math. 45 (1984), 127–146,
MR86B46073

[30] Halmos, P. R., "Measure Theory", Springer-Verlag, Heidelberg, 1974

[31] Hart, G., "Dissertation", Kansas State University, USA, 1970

[32] Hewitt, E., and K. A. Ross, "Abstract Harmonic Analysis, I",
Springer-Verlag, Heidelberg, 1970

[33] Işik, N., J. S. Pym, and A. Ülger, The second dual of the group algebra of a compact group, J. London Math. Soc. 35 (1987), 135–148

[34] Janssen, A., Integration separat stetiger Funktionen, Mathem. Scand. 48 (1981), 68–78,
MR82G2800

[35] Jean, Maw Ding, Algebra actions of semigroup algebras, Soochow J. Math. 10 (1984), 33–48,
MR86M43001

[36] Johnson, D. L. and C. D. Lahr, Multipliers of L^1 algebras with order convolution, Publ. Math. Debrecen 28, (1981)153–161,
MR82K43004

[37] Kalra, S., A. I. Singh and H. L. Vasudeva, The algebra $A_p((0,\infty))$, with order convolution and its multipliers, Preprint

[38] Kalra, S., W. Moran and H. L. Vasudeva, Multiplier spaces and their preduals on $[0,1]$ with order convolution, Preprint

[39] Kinzl, F., Einige Bemerkungen über absolut stetige Maße auf lokalkompakten Halbgruppen, Österreich. Akad. Wiss. math.-nat. Kl. Sitzungsber. II, 189 (1980), 361–370,
MR83D43301

[40] Lardy, L. J., $L^1(a,b)$ with order convolution, Studia Math. 28 (1966), 1–8,
MR36,631

[41] Lau, A. T. M., Continuity of Arens multiplication on the dual space of bounded uniformly continuous functions on locally compact groups and topological semigroups, Math. Proc. Cambridge Philos. Soc. 99 (1986), 273–283,
MR87I43001

[42] Lau, A. T. M, and V. Losert On the second conjugate of $L^1(G)$ of a locally compact group, J. London Math. Soc. 37 (1980), 464–470

[43] McLean, R. G. and H. Kummer, Representations of the Banach algebra (S), Semigroup Forum 37 (1988), 119–122

[44] Mirotin, A. R., Structure of invariant measures on locally compact semigroups with open shifts, Uspekhi Mat. Nauk. 37 (1982), 151–156,
MR83I43002

[45] —, Abelian semigroups with invariant measure, Teor. Funkciĭ Funkcional. Anal. i Priložen. 44 (1985), 67–68,
MR8722009

[46] Mirotin, A. R., and V. V. Mukhin, Weil topology in a group with a measure invariant on a subset, Sibirsk. Mat. Zh. 25 (1984), 132–136,
MR86B22007

[47] Mukherjea, A., and N. A. Tserpes, Invariant measures and the converse of Haar's theorem on semitopological semigroups, Pac. J. Math. 44 (1973), 251–262,
MR47,6932

[48] Mukhin, V. V., Invariant measures in semigroups with open shifts, Dokl. Akad. Nauk BSSR 27 (1983), 301–303,
MR85A28008

[49] —, Topologization of semigroups with invariant measure, Ukrain. Mat. Zh. 35 (1983), 103–106, 136,
MR84D43001

[50] —, Invariant measures and embedding of locally compact topological semigroups into topological groups, Dokl. Akad. Nauk SSSR, 278 (1984), 1063–1066, MR86A22003

[51] Parsons, D. J., The centre of second dual of a commutative semigroup algebra, Math. Proc. Cambridge Philos. Soc. 95 (1984), 71–92, MR85C43004

[52] Paterson, A. L. T., Invariant measure semigroups, Proc. London Math. Soc. 35 (1977), 313–332, MR58,28442

[53] Pym, J. S., Convolution measure algebras are not usually Arens-regular, Quart. J. Math. Oxford 25 (1974), 235–240, MR51,1276

[54] Rejali, A., The Arens regularity of weighted convolution algebras on semitopological semigroups, Preprint

[55] —, On Riesz type representation theorems, Preprint

[56] Rigelhoff, R., Invariant measures on locally compact semigroups, Proc. Math. Soc. 28 (1971), 173–176, MR43,3424

[57] Rudin, W., "Real and Complex Analysis", Tata-McGraw-Hill, 1974

[58] Sleijpen, G. L. J., Locally compact semigroups and continuous translations of measures, Proc. London Math. Soc. 37 (1978) 79–97, MR5817682a

[59] —, The action of a semigroup on a space of bounded Radon measures, Semigroup Forum 23 (1981), 137–152, MR83E22005

[60] —, L-multipliers for foundation semigroups with identity element, Proc. London Math. Soc. 39 (1979), 299–330, MR80K43003

[61] —, "Dissertation", Katholieke Universiteit Nijmegen, Holland, 1976

[62] —, The order structure of the space of measures with continuous translation, Ann. Inst. Fourier (Grenoble) 32 (1982), 67–110, MR83K43005

[63] Todd, D. G. Multipliers of certain convolution algebras over topological semigroups, Math. Proc. Cambridge Philos. Soc. 87 (1980), 51–59, MR81F43008

[64] Williamson, J. H., Harmonic analysis on semigroups, J. London Math. Soc. 42 (1967), 1–41

[65] Wordingham, J. R., The left * representation of an inverse semigroup, Proc. Amer. Math. Soc. 86 (1982), 55–58, MR83K43002

[66] Young, N. J., Semigroup algebras having regular multiplication, Studia Math. 47 (1973), 191–196,
MR48,9260

[67] Zappa A., The centre of the convolution algebra $Cu(G)^*$, Rend. Sem. Mat. Univ. Padova 52 (1974), 71–83

Positive definite and related functions on semigroups

Christian Berg

In this chapter we shall give a survey of the theory of positive definite and related functions on abelian semigroups with involution. In particular we shall take up some of the themes discussed in Berg, Christensen and Ressel [6] (referred to in the sequel as B–C–R) and follow their recent development.

We refer the readers to the following earlier survey articles about related subjects: Willliamson [48], Hofmann [24], Stewart [41], Berg [2].

1. Positive definite functions on abelian semigroups with involution

Behind the notion of a positive definite function on a group or a semigroup is the notion of a positive definite kernel which has roots back to positive definite matrices and the theory of integral equations.

We recall that for any set X a function $\Phi : X \times X \to \mathbb{C}$ is called a *positive definite kernel* on X if

$$\sum_{j,k=1}^{n} c_j \bar{c}_k \Phi(x_j, x_k) \geq 0 \tag{1}$$

for all $n \in \mathbb{N}$, $\{x_1, \ldots, x_n\} \subseteq X$ and $\{c_1, \ldots, c_n\} \subseteq \mathbb{C}$.

If the set X carries an algebraic operation $\circ$, it is possible to associate a kernel Φ to a function $\varphi : X \to \mathbb{C}$ in various ways, e.g., by defining $\Phi(x, y) = \varphi(x \circ y)$. The classical theories of Hankel and Toeplitz forms correspond to the kernels $\varphi(n+m)$ for $X = \mathbb{N}_0 = \{0, 1, 2, \ldots\}$ and $\varphi(n - m)$ for $X = \mathbb{Z}$.

A way of treating kernels on groups and semigroups in one theory is obtained by the notion of a semigroup $(S, \circ)$ with an *involution* $*$, i.e., a mapping $* : S \to S$ satisfying $(x \circ y)^* = y^* \circ x^*$ and $(x^*)^* = x$. For simplicity we shall always assume that semigroups have a neutral element, although many results have counterparts for semigroups without neutral element, see B–C–R, ch. 8.

A function $\varphi : S \to \mathbb{C}$ on a semigroup S with involution is called *positive definite* if the kernel $(s, t) \mapsto \varphi(s^* \circ t)$ is positive definite.

The set of positive definite functions on S is denoted by $\mathcal{P}(S)$, and the set of bounded functions in $\mathcal{P}(S)$ is denoted by $\mathcal{P}^b(S)$.

The above concept of positive definiteness on semigroups with involution occurs in the appendix by Sz.-Nagy [43] to the famous functional analysis monograph by Riesz and Sz.-Nagy.

We shall be concerned mainly with integral representations of positive definite functions on *abelian semigroups*. For an integral representation of positive definite functions on a Clifford semigroup see Paterson [34].

In the sequel $S = (S, +, *)$ denotes an abelian semigroup with neutral element 0 and involution $*$.

A *semicharacter* ρ on S is a complex–valued function ρ on S satisfying

$$\rho(0) = 1 \tag{2}$$
$$\rho(s + t^*) = \rho(s)\overline{\rho(t)} \quad \text{for } s, t \in S. \tag{3}$$

The set of semicharacters is denoted by S^*. We equip S^* with the topology inherited from $\mathbb{C}^S$, i.e., the topology of pointwise convergence. Therefore S^* is a completely regular space. Note that S^* is a topological semigroup under pointwise multiplication, the mapping $\rho \mapsto \overline{\rho}$ is an involution and the function $\rho \equiv 1$ is the neutral element. We call S^* the *dual semigroup* of S. The subsemigroup

$$\hat{S} = \{\rho \in S^* \mid |\rho(s)| \leq 1 \quad \text{for } s \in S\}$$

is called the *restricted dual semigroup*. It is compact.

The following fundamental result generalizing Bochner's theorem for discrete abelian groups is due to Lindahl and Maserick [29]. However, a footnote on p. 235 in Ionescu Tulcea [27] indicates that the result was known to him.

Theorem 1.1. *The set $\mathcal{P}_1^b(S)$ of bounded positive definite functions φ on S with $\varphi(0) = 1$ is a Bauer simplex, and its set of extreme points is $\hat{S}$. A bounded positive definite function φ has an integral representation*

$$\varphi(s) = \int_{\hat{S}} \rho(s)\, d\mu(\rho), \qquad s \in S,$$

where μ is a uniquely determined positive Radon measure on $\hat{S}$.

Lindahl and Maserick proved the theorem by considering the Banach algebra $\ell^1(S)$ of absolutely summable functions on S under the usual definition of convolution and involution. With each bounded function φ on S is associated a continuous linear functional L_φ on $\ell^1(S)$ defined by

$$L_\varphi(f) = \sum_{s \in S} f(s)\varphi(s).$$

Now $\varphi \mapsto L_\varphi$ is a homeomorphism of $\mathcal{P}_1^b(S)$ onto the set of positive linear functionals L on $\ell^1(S)$ satisfying $L(\varepsilon_0) = 1$, and a homeomorphism of $\widehat{S}$ onto the set of hermitian characters of $\ell^1(S)$. The results now follows because of a classical result stating that an extreme point of the set of normalized positive linear functionals is a character of $\ell^1(S)$.

Another possible proof is to establish directly that the set of extreme points of $\mathcal{P}_1^b(S)$ is equal to $\widehat{S}$. For the integral representation it is even enough to establish the inclusion ex $\mathcal{P}_1^b(S) \subseteq \widehat{S}$.

For semigroups with the identity involution this was done by Berg, Christensen and Ressel [5] without knowing the work of Lindahl and Maserick. The essential idea is that if $\varphi \in \mathcal{P}^b(S)$ and $s_1, \ldots, s_n \in S$, $c_1, \ldots, c_n \in \mathbf{C}$, then

$$\Phi(s) = \sum_{i,j=1}^{n} c_i \overline{c}_j \varphi(s + s_i + s_j^*) \tag{4}$$

belongs also to $\mathcal{P}^b(S)$.

A third possible proof is based on Hilbert space theory via a procedure similar to the GNS construction. This leads to a representation of S as operators on a reproducing kernel Hilbert space and the proof is finished by a version of the spectral theorem. For details see e.g., Schempp [38] and Berg and Maserick [7].

Positive definite functions on semigroups have been studied by Tagamlitzki and his pupils in Bulgaria. In lectures at the University of Sofia during the years 1965–1969, Tagamlitzki introduced the operators $W_{\xi,a}$ for functions f on an abelian group G by

$$W_{\xi,a}f(s) = 2f(s) + \xi f(s+a) + \overline{\xi} f(s-a), \tag{5}$$

where $a, s \in G$ and $\xi \in \mathbf{C}$.

He used these operators for proving Bochner's theorem for discrete groups as a direct consequence of Krein–Milman's theorem. A similar approach was published by Choquet [19]. This method was extended to semigroups with involution by T. Tonev in his Mag. Thesis [45], written under the supervision of Tagamlitzki. An account of this was later published in Semigroup Forum, see Tonev [46]. Shopova [40] proved that a function φ on an abelian group is positive definite if and only if

$$W_{\xi_1,a_1} \dots W_{\xi_n,a_n}\varphi(0) \geq 0$$

for all $a_1, \dots, a_n \in G$ and all $\xi_i \in \mathbf{C}$ with $|\xi_i| = 1$.

SHOPOVA's result was generalized in MASERICK [30], where the following result can be found:

Theorem 1.2. *Let S be an abelian semigroup with involution. For a function $\varphi : S \to \mathbf{C}$ the following conditions are equivalent:*
(i) $\varphi \in \mathcal{P}^b(S)$.
(ii) $W_{\xi_1,a_1} \dots W_{\xi_n,a_n}\varphi(0) \geq 0$ *for all* $n \geq 1$, $a_1, \dots, a_n \in S$, $\xi_1, \dots, \xi_n \in \{\pm 1, \pm i\}$.

Theorem 1.1 can be extended to α-bounded positive definite functions, where α is an *absolute value* on S, i.e., α is a non-negative function on S satisfying

$$\alpha(0) = 1 \ , \ \alpha(s^*) = \alpha(s) \ , \ \alpha(s+t) \leq \alpha(s)\alpha(t).$$

A function $\varphi \in \mathcal{P}(S)$ is called *α-bounded* if $|\varphi(s)| \leq C\ \alpha(s)$ for $s \in S$, where C is some non-negative constant depending on φ. The set of α-bounded positive definite functions on S is denoted by $\mathcal{P}^\alpha(S)$. If b denotes the absolute value $s \mapsto 1$ this notation is consistent with the one previously introduced. It is simple but crucial to observe that $\varphi \in \mathcal{P}^\alpha(S)$ satisfies $|\varphi(s)| \leq \varphi(0)\alpha(s)$. The α-boundedness condition has been considered by SZAFRANIEC [42].

Theorem 1.3. (BERG and MASERICK [7]) *The set $\mathcal{P}_1^\alpha(S)$ of α-bounded positive definite functions φ on S with $\varphi(0) = 1$ is a Bauer simplex, and its set of extreme points consists of the α-bounded semicharacters. A function $\varphi \in \mathcal{P}^\alpha(S)$ has an integral representation*

$$\varphi(s) = \int_{S^*} \rho(s)\, d\mu(\rho) \ , \quad s \in S,$$

where μ is a uniquely determined Radon measure supported by the compact set of α-bounded semicharacters.

Like in Theorem 1.1 one can directly establish the inclusion

$$\operatorname{ex} \mathcal{P}_1^\alpha(S) \subseteq \{\rho \in S^* \mid |\rho(s)| \leq \alpha(s),\ s \in S\},$$

and Theorem 1.3 follows.

In a recent paper by RESSEL [36] it is pointed out that many representation theorems of positive linear functionals as mixtures of multiplicative functionals can be deduced from Theorem 1.3. This reduction is based on the notion of a *convex semigroup*, i.e., a semigroup S which is also a convex subset of some vector space such that the semigroup operation – now written as multiplication – is compatible with the convex structure, i.e.,

$$x(\alpha y + \beta z) = \alpha(xy) + \beta(xz)\ ,\ \ (\alpha x + \beta y)^* = \alpha x^* + \beta y^*$$

for $x, y, z \in S$ and $\alpha, \beta \geq 0$, $\alpha + \beta = 1$.

The following result is a special case of Theorem 1 in Ressel [36].

Theorem 1.4. *Suppose $\varphi \in \mathcal{P}(S)$ is affine and has a representation*

$$\varphi(s) = \int_{S^*} \rho(s)\, d\mu(\rho)\ ,\quad s \in S$$

where μ is a Radon measure on S^*. Then μ is concentrated on the set of affine semicharacters.

As an application of the theory of convex semigroups Ressel proves Raikov's theorem on positive linear functionals on normed $*$-algebras, Riesz' representation theorem and the Plancherel–Godement representation for bitraces.

Convex semigroups have been considered from another point of view under the name of affine semigroups, see the chapter by Troallic [47] and the references therein.

It is tempting to try to obtain representation theorems for continuous positive definite functions on topological semigroups from the discrete version. As far as I know, this has turned out successfully only in few cases, of which I shall mention three.

1° For locally compact abelian groups this is done in Choquet [20], the idea being the following: Let φ be a continuous positive definite function on a locally compact abelian group G. Then there is a positive measure on the Bohr compactification $\hat{G}_d$ of $\hat{G}$ such that

$$\varphi(x) = \int_{\hat{G}_d} \overline{\rho(x)}\, d\mu(\rho) \quad \text{for } x \in G,$$

and Choquet succeeds in proving directly that the continuity of φ forces μ to be concentrated on the subset $\hat{G} \subseteq \hat{G}_d$ of continuous characters.

For another proof see Bucy and Maltese [17].

2° Let $S = [0, \infty[^k$ be considered as a semigroup under addition. If we consider S as a discrete semigroup, we can identify $\hat{S}$ with $[0, \infty]^k$ via the one-to-one correspondence $a \leftrightarrow \rho_a$, where $a = (a_1, \ldots, a_k) \in [0, \infty]^k$ and $\rho_a(s) = \exp(-\langle s, a\rangle)$ with $\langle s, a\rangle = \sum_{j=1}^k s_j a_j$. We see that ρ_a is continuous if and only if $a_j \neq \infty$ for $j = 1, \ldots, k$.

The functions $\varphi \in \mathcal{P}^b([0, \infty[^k)$ are given by

$$\varphi(s) = \int_{[0,\infty]^k} \exp(-\langle s, a\rangle) d\mu(a),$$

where μ is a positive Radon measure on $[0, \infty]^k$. If φ in addition is continuous then so is

$$s_j \mapsto \varphi(0, \ldots, 0, s_j, 0, \ldots, 0) = \int_{[0,\infty]} \exp(-s_j t) d\mu_j(t),$$

where μ_j is the j'th marginal distribution of μ, and it follows that $\mu_j(\{\infty\}) = 0$ for $j = 1, \ldots, k$, so μ is concentrated on $[0, \infty[^k$, and φ is the Laplace transform of μ. For details see B–C–R, pp. 113–115.

3° (See B–C–R, pp. 136–138.) Let X be a locally compact space and let $S = (\mathcal{K}, \cup)$ be the semigroup of compact subsets of X with union as semigroup operation and the identity involution.

Let $\mathcal{F}$ (respectively $\mathcal{G}$) denote the family of closed (respectively open) subsets of X. For $K \in \mathcal{K}$ and $G \in \mathcal{G}$ we denote by $\mathcal{V}(K, G) = \{L \in \mathcal{K} \mid K \subseteq L \subseteq G\}$, and these sets form the basis for a topology on $\mathcal{K}$ in which S is a Hausdorff topological semigroup. (It is erroneously stated in B–C–R p. 136 that the topology is not Hausdorff.) Functions $f : \mathcal{K} \to \mathbb{R}$ which are continuous in this topology are called *continuous from right*.

The dual semigroup S^* can be identified as a set with the family $\mathcal{S}$ of *faces* I in $\mathcal{K}$, i.e., non-empty subsets $I \subseteq \mathcal{K}$ satisfying

$$K, L \in I \Rightarrow K \cup L \in I \tag{6}$$

$$K \subseteq L, L \in I \Rightarrow K \in I. \tag{7}$$

The semigroup operation in $\mathcal{S}$ is intersection and then $I \mapsto 1_I$ is a semigroup isomorphism of $\mathcal{S}$ onto S^*. Defining

$$\tilde{K} = \{I \in \mathcal{S} \mid K \in I\} \quad \text{for } K \in \mathcal{K}$$

and equipping $\mathcal{S}$ with the coarsest topology in which the sets $\tilde{K}$ and $\mathcal{S} \setminus \tilde{K}$ are open (and hence clopen) for all $K \in \mathcal{K}$, then $I \mapsto 1_I$ is also a homeomorphism.

The integral representation of $\varphi \in \mathcal{P}(S) = \mathcal{P}^b(S)$ takes the form

$$\varphi(K) = \mu(\tilde{K}) \ , \ \ K \in \mathcal{K} \tag{8}$$

where μ is a positive Radon measure on $\mathcal{S}$.

The set of continuous semicharacters on $\mathcal{K}$ is identified with the set $\mathcal{S}_r$ of faces I which are clopen in $\mathcal{K}$, and this set can be identified with $\mathcal{F}$ via the mapping

$$F \mapsto I_F := \{K \in \mathcal{K} \mid K \cap F = \emptyset\} \tag{9}$$

of $\mathcal{F}$ onto $\mathcal{S}_r$, the inverse of which is

$$I \mapsto X \setminus \bigcup_{K \in I} K^\circ. \tag{10}$$

The topology on $\mathcal{F}$ transported from $\mathcal{S}_r$ with the subspace topology is too fine. Instead we use the coarser *Vietoris topology* on $\mathcal{F}$, a basis of which is the family of sets

$$\{F \in \mathcal{F} \mid F \cap K = \emptyset, \ F \cap G_1 \neq \emptyset, \ldots, F \cap G_n \neq \emptyset\},$$

where $K \in \mathcal{K}$, $G_1, \ldots, G_n \in \mathcal{G}$. With this topology $\mathcal{F}$ is compact. We now have the following result:

Proposition 1.5. *The mapping $c : \mathcal{S} \to \mathcal{F}$ defined by*

$$c(I) = X \setminus \bigcup_{K \in I} K^\circ$$

is continuous and maps $\mathcal{S}_r$ bijectively onto $\mathcal{F}$. For each continuous positive definite function φ on $\mathcal{K}$ there exists a unique positive Radon measure σ on $\mathcal{F}$ such that

$$\varphi(K) = \sigma(\{F \in \mathcal{F} \mid F \cap K = \emptyset\}) \ , \ \ K \in \mathcal{K},$$

and σ is the image measure under c of the unique measure μ on $\mathcal{S}$ such that (8) *holds.*

The idea of proof is the following:

Since φ is decreasing and continuous from right we have

$$\begin{aligned}\varphi(K) &= \sup\{\varphi(L) \mid L \in \mathcal{K}, K \subseteq L^\circ\} \\ &= \sup\{\mu(\tilde{L}) \mid L \in \mathcal{K}, K \subseteq L^\circ\} \\ &= \mu\Big(\bigcup_{K \subseteq L^\circ} \tilde{L}\Big) = \mu\Big(c^{-1}(\{F \in \mathcal{F} \mid F \cap K = \emptyset\})\Big) \\ &= \sigma(\{F \in \mathcal{F} \mid F \cap K = \emptyset\}).\end{aligned}$$

For other approaches to the result see Choquet [18], Matheron [32] and Talagrand [44]. For results about more general lattices of subsets of a topological space see Hofmann and Mislove [25] and references therein.

Several authors have studied continuous positive definite functions on locally compact subsemigroups of locally compact abelian groups, cf. Ionescu Tulcea [27] and Nussbaum [33], but locally compact abelian semigroups often have very few continuous semicharacters. As an example consider $S = [0, 1]$ as a compact semigroup under the maximum operation. Then S has only one continuous semicharacter namely the constant semicharacter. On the other hand the continuous positive definite functions on $[0, 1]$ are the non-negative, continuous and decreasing functions, so a topological version of Theorem 1.1 is not true without some restrictions. A paper by Lashkarizadeh Bami [28] contains a version of Theorem 1.3 for α-bounded continuous positive definite functions on foundation topological semigroups.

On semigroups S with the identical involution one studies *completely monotone* functions, i.e., functions $\varphi : S \to [0, \infty[$ satisfying

$$(I - E_{a_1}) \dots (I - E_{a_n})\varphi(s) \geq 0 \tag{11}$$

for all $n \geq 1$ and all $a_1, \dots, a_n, s \in S$, where E_a is the shift operator

$$E_a\varphi(s) = \varphi(s + a). \tag{12}$$

The set $\mathcal{M}(S)$ of completely monotone functions is a convex cone and an extremal subset of $\mathcal{P}^b(S)$. The set $\mathcal{M}^1(S)$ of functions $\varphi \in \mathcal{M}(S)$ satisfying $\varphi(0) = 1$ is a Bauer simplex with $\mathrm{ex}(\mathcal{M}^1(S)) = \hat{S}_+$, cf. Choquet [18]. If the semigroup is 2-divisible then $\hat{S} = \hat{S}_+$ and every bounded positive definite function is completely monotone. Here 2-divisibility means that every $s \in S$ is of the form $s = t + t$ for some $t \in S$. For integral representations of completely monotone functions see also Nussbaum [33], and for the case of S being a convex cone see Hoffmann-Jørgensen and Ressel [23] and Dettweiler [22].

2. Moment functions

As in the previous section S denotes an abelian semigroup with involution.

A function $\varphi : S \to \mathbf{C}$ is called a *Radon moment function* if it has the form

$$\varphi(s) = \int_{S^*} \rho(s)\, d\mu(\rho), \quad s \in S \tag{13}$$

for some positive Radon measure μ on S^*.

Here it is assumed that μ belongs to the set $E_+(S^*)$ of Radon measures on S^* for which

$$\int_{S^*} |\rho(s)| d\mu(\rho) < \infty \quad \text{for all } s \in S.$$

(In B–C–R the functions (13) were simply called moment functions. As we shall see below, it is necessary to consider functions (13), where μ belongs to a bigger class of measures than Radon measures, and we will reserve the name moment function for these more general functions.)

The set $\mathcal{H}_R(S)$ of Radon moment functions is a convex cone satisfying

$$\mathcal{P}^b(S) \subseteq \mathcal{H}_R(S) \subseteq \mathcal{P}(S). \tag{14}$$

It is possible to characterize the set of Radon moment functions having a representing measure μ with compact support as the set of exponentially bounded positive definite functions, see Berg and Maserick [7]. A positive definite function φ is *exponentially bounded*, if there exists an absolute value $\alpha : S \to [0, \infty[$ and a constant C such that $|\varphi(S)| \leq C\alpha(s)$ for $s \in S$.

The cone $\mathcal{P}(S)$ is clearly closed in the topology of pointwise convergence, but as pointed out to me by Bisgaard $\mathcal{H}_R(S)$ need not be closed. This is related to the fact that we only consider Radon measures on S^* in the representation (13).

For $s \in S$ we denote by χ_s the evaluation function of S^* into $\mathbf{C}$ given by

$$\chi_s(\rho) = \rho(s), \tag{15}$$

and we put

$$V(\mathbb{C}) = \operatorname{span}_{\mathbb{C}}\{\chi_s \mid s \in S\}\ ,\ V(\mathbb{R}) = \{\operatorname{Re} f \mid f \in V(\mathbb{C})\}. \tag{16}$$

Clearly $V(\mathbb{C}) = V(\mathbb{R}) + iV(\mathbb{R})$.

Bisgaard and Ressel [14] proposed to extend the concept of a Radon moment function in the following way:

Let $\mathcal{A}(S^*)$ denote the smallest σ-algebra on S^* for which the evaluations χ_s, $s \in S$ are measurable. Clearly $\mathcal{A}(S^*)$ is contained in the Borel σ-algebra $\mathcal{B}(S^*)$, so any (Radon) measure on $\mathcal{B}(S^*)$ induces by restriction a measure on $\mathcal{A}(S^*)$. Furthermore, it is easy to see that a finite Radon measure on $\mathcal{B}(S^*)$ is uniquely determined by its restriction to $\mathcal{A}(S^*)$.

By a *moment function* on S we understand any function $\varphi : S \to \mathbf{C}$ of the form (13), where μ belongs to the class $F_+(S^*)$ of measures on $\mathcal{A}(S^*)$ for which χ_s is integrable for all $s \in S$. The set of moment functions is a convex cone $\mathcal{H}(S)$. Clearly

$$\mathcal{H}_R(S) \subseteq \mathcal{H}(S) \subseteq \mathcal{P}(S). \tag{17}$$

Bisgaard has proved that $\mathcal{H}(S)$ is the closure of $\mathcal{H}_R(S)$ in the topology of pointwise convergence. (In the above paper by Bisgaard and Ressel the functions in $\mathcal{H}(S)$ are called quasi-moment functions).

If S is countable it is easy to see that S^* is a Polish space, $\mathcal{A}(S^*) = \mathcal{B}(S^*)$ and any finite measure on $\mathcal{B}(S^*)$ is automatically a Radon measure. For countable S we therefore have $\mathcal{H}_R(S) = \mathcal{H}(\mathcal{S})$.

Example. (Bisgaard) Let I be an arbitrary index set and let $S = \mathbb{N}_0^{(I)}$ be the set of families $s = (n_i)_{i \in I}$ of non-negative integers such that $n_i = 0$ for all but finitely many i. We consider S as a semigroup under addition and then the dual semigroup S^* can be identified with the product $\mathbb{R}^I$ via the mapping

$$\Phi \mapsto \rho_\Phi((n_i)) = \prod_{i \in I} \Phi(i)^{n_i} \quad , \quad \Phi \in \mathbb{R}^I .$$

Due to the convention $x^0 = 1$ for all $x \in \mathbf{R}$ only finitely many factors are different from 1 in this product.

If I is countable then $\mathcal{A}(S^*) = \mathcal{B}(S^*)$, but if I is uncountable then $\mathcal{A}(S^*)$ is strictly smaller than $\mathcal{B}(S^*)$.

Let σ be the normal distribution on $\mathbf{R}$ with moments $\sigma(n),\ n \geq 0$. Then

$$\varphi((n_i)) = \prod_{i \in I} \sigma(n_i) \quad , \quad (n_i) \in S \tag{18}$$

is a moment function on S represented by the product measure $\underline{\sigma} = \bigotimes_{i \in I} \sigma$. We shall show that $\varphi \notin \mathcal{H}_{\mathcal{R}}(S)$ if I is uncountable.

Suppose that τ is a Radon measure on $\mathbb{R}^I$ representing φ. For any finite subset $J \subseteq I$ let π_J be the corresponding projection $\pi_J : \mathbb{R}^I \to \mathbb{R}^J$. Then

$$\tau^{\pi_J} = \bigotimes_{i \in J} \sigma \tag{19}$$

because for $s = (n_i)$ with $n_i = 0$ for $i \notin J$ we have

$$\int \prod_{i \in J} x_i^{n_i} d\tau^{\pi_J} = \varphi(s) = \prod_{i \in J} \sigma(n_i),$$

showing that the two measures in (19) have the same moments, and by determinacy of the right-hand side of (19) the equality sign follows.

For any compact subset $C \subseteq \mathbb{R}^I$ and any finite subset $J \subseteq I$ we have

$$\begin{aligned} \tau(C) &\leq \tau\left(\pi_J(C) \times \mathbb{R}^{I \setminus J}\right) = \tau^{\pi_J}\left(\pi_J(C)\right) \\ &\leq \prod_{i \in J} \sigma\left(\pi_i(C)\right) \end{aligned}$$

and hence

$$-\log \tau(C) \geq \sum_{i \in I} -\log \sigma\left(\pi_i(C)\right).$$

The sum of uncountably many positive numbers is infinite. This shows that $\tau(C) = 0$ and hence $\tau = 0$ which is a contradiction.

We mention the following result from B–C–R.

Theorem 2.1. *Suppose S is finitely generated. We then have:*

(i) *S^* is locally compact with a countable basis.*

(ii) *$V(\mathbb{R})$ is an adapted space of continuous functions on S^*.*

(iii) *$\mathcal{H}_R(S)$ is closed in the topology of pointwise convergence.*

(iv) *Let $\varphi \in \mathcal{H}_R(S)$. The set of $\mu \in E_+(S^*)$ representing φ is a metrizable compact convex set in the weak topology.*

For general S we let Ω denote the set of finitely generated $*$-subsemigroups of S. For $X \in \Omega$ we define $\pi_X : S^* \to X^*$ by $\pi_X(\rho) = \rho \mid X$, so π_X is $\mathcal{A}(S^*) - \mathcal{A}(X^*)$ measurable. For $\mu \in F_+(S^*)$ the image measure $\mu^{\pi_X} \in F_+(X^*)$. A convenient topology on $F_+(S^*)$ is the coarsest for which the mappings $\mu \mapsto \mu^{\pi_X}$ of $F_+(S^*)$ into $F_+(X^*)$ are continuous for each $X \in \Omega$, where $F_+(X^*) = E_+(X^*)$ carries the weak topology. Using this topology on $F_+(S^*)$ property (iv) can be extended to general S:

(v) *Let $\varphi \in \mathcal{H}(S)$. The set of $\mu \in F_+(S^*)$ representing φ is a compact convex set.*

If S is finitely generated and S^* separates the points of S then $\mathbb{C}^S$ and $V(\mathbb{C})$ (cf. (16)) are in duality under the pairing

$$\langle\varphi, \sum_{j=1}^{n} a_j \chi_{s_j}\rangle = \sum_{j=1}^{n} a_j \varphi(s_j). \tag{20}$$

For a convex cone $C \subseteq V(\mathbb{C})$ we define the dual cone $C^\perp \subseteq \mathbb{C}^S$ by

$$C^\perp = \{\varphi \in \mathbb{C}^S \mid \langle\varphi, f\rangle \geq 0 \quad \text{for all} \;\; f \in C\}. \tag{21}$$

If we consider the convex cone Σ generated by the absolute squares of elements in $V(\mathbf{C})$, i.e.,

$$\Sigma = \left\{ \sum_{i=1}^{n} |f_i|^2 \mid f_i \in V(C) \right\} \tag{22}$$

then it is shown in B–C–R that

$$\mathcal{P}(S) = \Sigma^\perp \;\; , \;\; \mathcal{H}(S) = (V(\mathbf{R})_+)^\perp . \tag{23}$$

We recall that (23) is the basis for proving $\mathcal{P}(\mathbf{N}_0^k) \neq \mathcal{H}(\mathbb{N}_0^k)$ for $k \geq 2$ (Berg, Christensen and Jensen [4] and Schmüdgen [39]). For $k = 1$ the equality $\mathcal{P}(\mathbb{N}_0) = \mathcal{H}(\mathbb{N}_0)$ is the classical theorem of Hamburger.

For abelian groups G with the involution $s^* = -s$ one has $\mathcal{P}^b(G) = \mathcal{H}_R(G) = \mathcal{H}(G) = \mathcal{P}(G)$ by Bochner's theorem. If we consider the group $\mathbb{Z}^k$ with the identity involution the situation is analogous to the classical moment problem with $\mathcal{P}(\mathbf{Z}^k) \neq \mathcal{H}(\mathbb{Z}^k)$ for $k \geq 2$ but $\mathcal{P}(\mathbf{Z}) = \mathcal{H}(\mathbf{Z})$, and the elements in this space are the two-sided moment sequences

$$\varphi(n) = \int_{\mathbb{R}\setminus\{0\}} x^n \, d\mu(x) \;\; , \;\; n \in \mathbf{Z}. \tag{24}$$

The complex moment problem corresponds to the semigroup $S = \mathbf{N}_0^2$ with the involution $(n, m)^* = (m, n)$, the moment functions being

$$\varphi(n, m) = \int z^n \overline{z}^m \, d\mu(z) \tag{25}$$

for positive measures μ on $\mathbf{C}$. Like in the two dimensional moment problem $\mathcal{P}(S) \neq \mathcal{H}(S)$.

It is therefore very surprising that the following result holds:

Theorem 2.2. (BISGAARD [10]) *For the two-sided complex moment problem corresponding to $S = \mathbb{Z}^2$ with the involution $(n,m)^* = (m,n)$ every positive definite function is a moment function, i.e., for every $\varphi \in \mathcal{P}(S)$ there exists a positive measure μ on $\mathbb{C} \setminus \{0\}$ such that*

$$\varphi(n,m) = \int z^n \overline{z}^m \, d\mu(z) \ , \quad (n,m) \in \mathbb{Z}^2.$$

This theorem depends on a general result, and in order to explain this some terminology is convenient.

A semigroup with involution is called *Radon semiperfect* if $\mathcal{P}(S) = \mathcal{H}_R(S)$, and it is called *Radon perfect* if in addition every $\varphi \in \mathcal{H}_R(S)$ has a unique representing Radon measure.

In B–C–R we used the terminology perfect instead of Radon perfect, and it was proved that the class of Radon perfect semigroups is stable under the following operations

(i) finite products
(ii) homomorphic images
(iii) countable direct sums.

Semigroups S for which $\mathcal{P}(S) = \mathcal{P}^b(S)$ are Radon perfect. This is in particular true for abelian groups G with the involution $x^* = -x$, idempotent semigroups and finite semigroups.

In BISGAARD [10] the following was proved:

Theorem 2.3. *If S is Radon perfect and T is finitely generated and Radon semiperfect then $S \times T$ is Radon semiperfect.*

Idea of proof. Suppose φ is positive definite on $S \times T$. For each $t \in T$ there exists a unique complex Radon measure μ_t on S^* such that

$$\varphi(s,t) = \int \rho(s) \, d\mu_t(\rho) \ , \quad s \in S. \tag{26}$$

If t has the form $t = t_1 + t_1^*$, this is clear from the Radon perfectness because $s \mapsto \varphi(s, t_1 + t_1^*)$ is positive definite, so in this case μ_t is a positive measure. For general t (26) follows from the same "polarization" technique as used in B–C–R, p. 204 in the proof of the product stability (i).

The next step consists in proving that $t \mapsto \mu_t$ is a Radon measure-valued positive definite function, so we know in particular that $t \mapsto \mu_t(A)$ is positive definite for each $A \in \mathcal{B}(S^*)$. Since T is assumed Radon semiperfect there is a positive Radon measure σ_A on T^* such that

$$\mu_t(A) = \int \zeta(t) d\sigma_A(\zeta) \ , \ t \in T.$$

If T in addition is Radon perfect then σ_A is uniquely determined, and it is not difficult to prove that $A \mapsto \sigma_A$ is a Radon measure-valued Radon measure on $\mathcal{B}(S^*)$. However, T being only Radon semiperfect, σ_A is not uniquely determined, so for each $A \in \mathcal{B}(S^*)$ one has to make a choice of σ_A. By an elegant limit procedure Bisgaard achieves a choice of σ_A such that $A \mapsto \sigma_A$ is a Radon measure. The proof if finished following the lines of proof in B–C–R, p. 206: By the bimeasure theorem for Radon measures there exists a Radon measure κ on $S^* \times T^*$ such that $\kappa(A \times B) = \sigma_A(B)$ and finally one has

$$\varphi(s,t) = \int \rho(s)\zeta(t) d\kappa(\rho, \zeta).$$

To deduce Theorem 2.2 we let S denote the group $(\mathbb{Z}, +)$ with the involution $n^* = -n$ and let T denote the group $(\mathbb{Z}, +)$ with the identity involution. By Theorem 2.3 the group $G = \mathbb{Z}^2$ with the product involution $(n,m)^* = (-n,m)$ is Radon semiperfect, and so is the subgroup

$$H = \{(n,m) \in \mathbb{Z}^2 \mid n+m \ \text{even}\}$$

because if $\varphi \in \mathcal{P}(H)$ then $\tilde{\varphi} \in \mathcal{P}(G)$, where $\tilde{\varphi}$ is defined by $\tilde{\varphi} \mid H = \varphi$, $\tilde{\varphi} \mid (G \setminus H) = 0$.

Finally, the given semigroup $\mathbb{Z}^2$ with involution $(n,m)^* = (m,n)$ is $*$-isomorphic with H under the mapping $(n,m) \mapsto (n-m, n+m)$.

By replacing Radon measures with the bigger class $F_+(S^*)$ of measures on $\mathcal{A}(S^*)$ Bisgaard and Ressel [14] defined and exploited the following notions of perfectness and semiperfectness:

The semigroup S is called *semiperfect* if $\mathcal{P}(S) = \mathcal{H}(S)$, and it is called *perfect* if in addition every $\varphi \in \mathcal{H}(S)$ has a unique representing measure from $F_+(S^*)$. For countable semigroups S the two notions of semiperfectness coincide and similarly do the notions of perfectness.

A Radon semiperfect semigroup is clearly semiperfect, and a Radon perfect semigroup is perfect. In fact, for a Radon perfect semigroup S every $\mu \in F_+(S^*)$ is the restriction of a Radon measure, cf. Bisgaard [13].

Surprisingly enough the class of perfect semigroups is stable under the following operations

(i) finite products

(ii) homomorphic images

(iii) direct sums.

Also Theorem 2.3 can be generalized:

(iv) If S is perfect and T semiperfect then $S \times T$ is semiperfect.

The proof of these stability properties depends on a projective limit technique which will be explained below. On the other hand this is only possible because there are many $*$-subsemigroups $E \subseteq S$ for which every semicharacter can be extended to a semicharacter on S. Calling such $*$-subsemigroups *saturated*, Bisgaard [9] contains the following:

Theorem 2.4. *Let S be an abelian semigroup with involution. Any infinite $*$-subsemigroup of S can be extended to a saturated $*$-subsemigroup of the same cardinality.*

Let $\mathcal{D}_0(S)$ be the family of all saturated countable $*$-subsemigroups of S. This family is upwards filtering under inclusion because of Theorem 2.4, and

$$\mathcal{A}(S^*) = \bigcup_{E \in \mathcal{D}_0(S)} \pi_E^{-1}(\mathcal{B}(E^*)). \tag{27}$$

Since the restriction maps $\pi_E : S^* \to E^*$ are surjective, it is possible to construct the limit measure σ on $(S^*, \mathcal{A}(S^*))$ of any projective system $(\sigma_E)_{E \in \mathcal{D}_0(S)}$ of measures σ_E on $(E^*, \mathcal{B}(E^*))$. One simply has to define

$$\sigma(A) = \sigma_E(\pi_E(A)) \quad \text{for } A \in \mathcal{A}(S^*), \tag{28}$$

where $E \in \mathcal{D}_0(S)$ is chosen such that $A \in \pi_E^{-1}(\mathcal{B}(E^*))$. The right-hand side of (28) turns out to be independent of the choice of E.

The stability property (i) depends on a result about bimeasures like the corresponding result for Radon perfectness. In general a finite bimeasure λ on the product of two σ-algebras $\mathcal{A}$ and $\mathcal{B}$ is not induced by a measure on $\mathcal{A} \otimes \mathcal{B}$, but it holds in the following context:

Lemma 2.5. *Let S and T be semigroups with involution. Then every finite bimeasure on $\mathcal{A}(S^*) \times \mathcal{A}(T^*)$ is induced by a measure on $\mathcal{A}(S^*) \otimes \mathcal{A}(T^*)$.*

It is well-known that divisibility properties of semigroups are pertinent for perfectness, e.g., in establishing that the semigroup $\mathbf{Q}_+$ of non-negative rational numbers is perfect, cf. B–C–R. In Berg [1] it was proved that the dyadic numbers form a perfect semigroup and this was used to prove the perfectness of every countable 2-divisible semigroup S with the identical involution. This result was later extended by Sakakibara [37].

Let us call S *$*$-divisible* if every $s \in S$ can be written $s = nt + mt^*$ for some $t \in S$ and $n, m \in \mathbb{N}_0$ with $n + m \geq 2$.

BISGAARD and RESSEL [14] established the following fundamental result:

Theorem 2.6. *Every $*$-divisible semigroup with involution is perfect.*

It is an interesting problem to characterize (Radon) perfect and (Radon) semiperfect semigroups by algebraic conditions. No complete solution seems available, but BISGAARD [12] gives many partial results of which we shall only discuss a few.

Let I be an arbitrary index set and let $\mathbb{Q}^{(I)}$ be the $\mathbf{Q}$-vector space of functions $f : I \to \mathbf{Q}$ with finite support. Clearly $\dim_{\mathbb{Q}} \mathbb{Q}^{(I)} = \operatorname{card} I$. Considered as a semigroup under addition and having the identity involution we know from Theorem 2.6 that $\mathbb{Q}^{(I)}$ is perfect;it is Radon perfect if and only if I is countable. In fact, if I is uncountable it is easy to see that $\mathbb{Q}^{(I)}$ is not Radon perfect using the same method as in the previously discussed example of Bisgaard.

The vector space $W = \mathbb{Q}^{(I)}$, where $\operatorname{card} I$ is the smallest uncountable cardinal number, enters in the following characterization of Radon perfect groups with involution.

Theorem 2.7. (BISGAARD [12]) *Let G be an abelian group with involution.*

(i) *G is Radon perfect if and only if neither W nor $(\mathbb{Z}, \mathrm{id})$ is a $*$-homomorphic image of G.*

(ii) *G is Radon semiperfect if and only if neither W nor $(\mathbb{Z}^2, \mathrm{id})$ is a $*$-homomorphic image of G.*

3. Negative definite functions

We recall that a function $\psi : S \to \mathbb{C}$ is called *negative definite* if $\psi(s^*) = \overline{\psi(s)}$, $s \in S$ and

$$\sum_{j,k=1}^{n} \psi(s_j^* + s_k) c_j \overline{c_k} \leq 0$$

for all $n \geq 2, s_1, \ldots, s_n \in S$ and $c_1, \ldots, c_n \in \mathbf{C}$ with $\sum_{i=1}^{n} c_i = 0$. The set of negative definite functions on S is denoted by $\mathcal{N}(S)$. A fundamental relation between positive and negative definite functions is given in the theorem of SCHOENBERG:

$$\psi \in \mathcal{N}(S) \Leftrightarrow \exp(-t\psi) \in \mathcal{P}(S) \quad \text{for all } t > 0. \tag{29}$$

The set of functions $\psi \in \mathcal{N}(S)$ for which $\operatorname{Re}\psi$ is bounded below is denoted by $\mathcal{N}^{\ell}(S)$. A main point in B–C–R is an integral representation of the functions in $\mathcal{N}^{\ell}(S)$ depending on the existence of a Lévy function. The existence of Lévy functions was open at the time of writing but was subsequently established in BUCHWALTER [16].

We shall consider the set $\mathcal{L}$ of Lévy measures for S, i.e., the set of Radon measures λ on the locally compact space $\hat{S} \setminus \{1\}$ satisfying

$$\int (1 - \operatorname{Re}\rho(s))\,\rho(d\lambda(\rho) < \infty \quad \text{for all } s \in S. \tag{30}$$

Let

$$\mathcal{T} = \operatorname{span}_{\mathbb{C}}\{E_a \mid a \in S\}$$

denote the space of shift operators, cf. (12). For $R \in \mathcal{T}$ and $f \in \mathbb{C}^S$ we write $\langle R, f\rangle = Rf(0)$. With the natural operations $\mathcal{T}$ is a commutative complex $*$-algebra with unit I, the identity operator. The $*$-ideal in $\mathcal{T}$ consisting of those R such that $\langle R, 1\rangle = 0$ will be denoted by $\mathcal{T}_0$.

The result of BUCHWALTER can be formulated in the following way:

Theorem 3.1. *Let S be an abelian semigroup with involution. There exists a Lévy function, i.e., a function $L : S \times \hat{S} \to \mathbb{R}$ with the following properties:*

(i) *For every $\rho \in \hat{S}$ the function $L(\cdot, \rho)$ is $*$-additive on S, i.e., $L(s + t^*, \rho) = L(s, \rho) - L(t, \rho)$ for $s, t \in S$.*

(ii) *For every $s \in S$ the function $L(s, \cdot)$ is continuous on $\hat{S}$ and $L(s, \overline{\rho}) = -L(s, \rho)$ for $\rho \in \hat{S}$.*

(iii) *For every Lévy measure $\lambda \in \mathcal{L}$ and every $s \in S$*

$$\int |1 - \rho(s) + iL(s, \rho)|\,d\lambda(\rho) < \infty. \tag{31}$$

(iv) *For every $R \in \mathcal{T}_0$ satisfying $\langle R, \rho\rangle \geq 0$ for $\rho \in \hat{S}$*

$$\langle R, L(\cdot, \rho)\rangle = 0.$$

The idea of proof consists in the introduction of an abelian group $\tilde{G}$ associated with S in the following way. Let H_0 denote the $*$-subsemigroup of *almost hermitian* elements

$$H_0 = \{s \in S \mid \exists u = u^* \in S : u + s = u + s^*\}.$$

Let $\sim$ denote the equivalence relation on S defined by $s \sim t$ if $s + t^* \in H_0$ and let $[s] = \{t \in S \mid s \sim t\}$ for $s \in S$. The set $G := \{[s] \mid s \in S\}$ of equivalence classes is an abelian group under the addition of representatives. Let finally $\tilde{G}$ be the torsion free quotient group of G over the subgroup of elements of finite order. For $s \in S$ we denote by $\tilde{s}$ the image of s in $\tilde{G}$.

BUCHWALTER then chooses a family $(e_k)_{k\in\Lambda}$ of elements from S such that the system $(\tilde{e}_k)_{k\in\Lambda}$ is maximally free in $\tilde{G}$. For each $s \in S$ there exist an integer $n \geq 1$ and a family $(n_k) \in \mathbb{Z}^{(\Lambda)}$ such that

$$n\tilde{s} = \sum_{k\in\Lambda} n_k \tilde{e}_k, \tag{32}$$

where the sum is finite, and the family $(n_k/n)_{k\in\Lambda}$ is uniquely determined. The function $L : S \times \hat{S} \to \mathbb{R}$ is then well-defined by the formula

$$L(s,\rho) = \sum_{k\in\Lambda} \frac{n_k}{n} \operatorname{Im} \rho(e_k), \tag{33}$$

and satisfies the conditions of Theorem 3.1.

Using the shift operator Γ_a , $a \in S$ defined by

$$\Gamma_a f(s) = \frac{1}{2}(f(s+a) + f(s+a^*)) \tag{34}$$

we can formulate the integral representation of functions in $\mathcal{N}^\ell(S)$ given in B–C–R:

Theorem 3.2. *The following conditions are equivalent for a function $\psi : S \to \mathbb{C}$:*
(i) $\psi \in \mathcal{N}^\ell(S)$.
(ii) *ψ is hermitian and $(\Gamma_a - I)\psi \in \mathcal{P}^b(S)$ for each $a \in S$.*
(iii) *$\psi(0)$ is real and there exists a triple (ℓ, q, μ), where $\ell : S \to \mathbb{R}$ and $q : S \to [0,\infty[$ are functions satisfying*

$$\ell(s+t^*) = \ell(s) - \ell(t) \tag{35}$$
$$q(s+t) + q(s+t^*) = 2q(s) + 2q(t) \tag{36}$$

for $s, t \in S$, and $\mu \in \mathcal{L}$ is a Lévy measure (cf. (30)), such that

$$\psi(s) = \psi(0) + i\ell(s) + q(s) + \int_{\hat{S}\setminus\{1\}} (1 - \rho(s) + iL(s,\rho))d\mu(\rho) \tag{37}$$

for all $s \in S$.

If the conditions (i)–(iii) *are satisfied then the triple (ℓ, q, μ) is uniquely determined by ψ.*

The paper by BUCHWALTER [15] contains a comprehensive treatment of *quadratic forms* on S, i.e., functions $q : S \to \mathbb{C}$ satisfying

$$\left.\begin{aligned} q(ns) &= n^2 q(s), \\ q(s^*) &= \overline{q(s)} \quad \text{and} \\ (E_s - I)(E_{s^*} - I)q &= (E_s - I)(E_{s^*} - I)q(0) \quad \text{for } s \in S. \end{aligned}\right\} \tag{38}$$

For the case of an abelian group S with the involution $s^* = -s$ we have $\mathcal{N}(S) = \mathcal{N}^\ell(S)$ and the representation (37) is the Lévy-Khinchin formula.

If S is a topological semigroup with involution one looks for an integral representation of continuous functions in $\mathcal{N}^\ell(S)$ in the cases where an integral representation is known for continuous positive definite functions. Such a representation holds for the three cases studied in section 1.

It is possible to give an integral representation of functions in $\mathcal{N}(S)$ in case of a semiperfect semigroup S, and the representing measure is unique, if S is perfect (BISGAARD, private communication). The very simple case of $S = \mathbb{N}_0$ has been worked out in B–C–R, p. 188, but the result goes back to HORN [26]:

Theorem 3.3. *A function $\psi : \mathbb{N}_0 \to \mathbb{R}$ is negative definite if and only if it has a representation of the form*

$$\psi(n) = a + bn - cn^2 + \int_{\mathbf{R}\setminus\{1\}} (1 - x^n - n(1-x))d\mu(x) \quad , \quad n \in \mathbb{N}_0,$$

where $a, b \in \mathbb{R}$, $c \geq 0$ and μ is a positive Radon measure on $\mathbb{R} \setminus \{1\}$ satisfying

$$\int_{0<|x-1|<1} (1-x)^2 d\mu(x) < \infty \quad , \quad \int_{|x-1|\geq 1} |x|^n d\mu(x) < \infty \quad \text{for } n \geq 0.$$

Let now S carry the identity involution. A function $\psi : S \to \mathbb{R}$ is called *completely alternating* if

$$(I - E_{a_1}) \dots (I - E_{a_n})\psi(s) \leq 0 \tag{39}$$

for all $n \geq 1$ and all $a_1, \dots, a_n, s \in S$, cf. (11). The set $\mathcal{A}(S)$ of completely alternating functions is an extremal subcone of $\mathcal{N}^\ell(S)$. The Lévy measure of $\psi \in \mathcal{A}(S)$ is concentrated on $\hat{S}_+ \setminus \{1\}$. The theorem of SCHOENBERG, cf. (29), remains valid in the following form

$$\psi \in \mathcal{A}(S) \Leftrightarrow \exp(-t\psi) \in \mathcal{M}(S) \quad \text{for all} \quad t > 0.$$

4. Hoeffding's inequalities and Schur increasing functions

In this section S denotes an abelian semigroup with the identical involution.

Work of BICKEL and VAN ZWET has led CHRISTENSEN and RESSEL [21] to consider real-valued functions on S satisfying Hoeffding's inequality of order $n \geq 2$. To explain this notion we introduce the class $\mathrm{Mol}^1_+(S)$ of molecular probability measures on S, i.e., measures of the form $\mu = \sum_{i=1}^n c_i \varepsilon_{s_i}$ with $c_i \geq 0$, $\sum_{i=1}^n c_i = 1$, $s_i \in S$. The set $\mathrm{Mol}^1_+(S)$ is stable under convolution and for finitely many molecular probabilities $\mu_1, \ldots, \mu_n$ we denote their average by $\overline{\mu} := \frac{1}{n}(\mu_1 + \ldots + \mu_n)$.

A function $\psi : S \to \mathbb{R}$ satisfies *Hoeffding's inequality of order* $n \geq 2$ if

$$\int \psi \, d(\overline{\mu})^{*n} \leq \int \psi \, d(\mu_1 * \ldots * \mu_n) \tag{40}$$

for every finite set $\mu_1, \ldots, \mu_n \in \mathrm{Mol}^1_+(S)$.

The set of these functions is a closed convex cone denoted by $\mathcal{H}_n(S)$.

BICKEL and VAN ZWET [8] discovered that for a function ψ on the semigroup $(\mathbb{R}^k, +)$ Hoeffding's inequality of order 2 implies Hoeffding's inequality of every order and is equivalent to ψ being negative definite. These results were generalized to the general case in CHRISTENSEN and RESSEL [21] and in RESSEL [35]. Let us summarize the main results:

$$\mathcal{H}_2(S) = \mathcal{N}(S) \tag{41}$$

$$\mathcal{H}_n(S) \subseteq \mathcal{N}(S) \quad \text{for } n \geq 2 \tag{42}$$

$$\mathcal{H}_3(S) = \mathcal{CN}(S) \subseteq \mathcal{H}_n(S) \quad \text{for } n \geq 2, \tag{43}$$

where $\mathcal{CN}(S)$ is the convex cone of *completely negative definite functions*, i.e., the functions $\psi : S \to \mathbb{R}$ such that all shifts $E_a\psi$, $a \in S$ are negative definite.

If S is a 2-*divisible semigroup* then every negative definite function is automatically completely negative definite and hence

$$\mathcal{H}_n(S) = \mathcal{N}(S) = \mathcal{CN}(S) \quad \text{for } n \geq 2.$$

For the semigroup $S = \mathbb{N}_0$ we have $\mathcal{CN}(\mathbf{N}_0) \subset \mathcal{N}(\mathbb{N}_0)$ and the completely negative definite sequences can be characterized in terms of the integral representation in Theorem 3.3 by having a representing measure μ concentrated on $\mathbb{R}_+ \setminus \{1\}$.

For the additive semigroup $S = \mathbb{N}_0$ further information about the cones $\mathcal{H}_n(\mathbb{N}_0)$ is given in BISGAARD [11].

Theorem 4.1. *For every $m \geq 3$ there exists a sequence $\psi : \mathbb{N}_0 \to \mathbb{R}$ satisfying Hoeffding's inequality of every order $n > m$ but not satisfying Hoeffding's inequality* *of order* m, *i.e.*,

$$\bigcap_{n>m} \mathcal{H}_n(\mathbb{N}_0) \setminus \mathcal{H}_m(\mathbf{N}_0) \neq \emptyset. \tag{44}$$

In particular we have

$$\mathcal{H}_n(\mathbb{N}_0) \not\subseteq \mathcal{H}_m(\mathbb{N}_0) \quad \text{for } n > m \geq 3. \tag{45}$$

The proof does not exhibit a function belonging to the non-empty set in (44) but merely shows its existence by a highly non-trivial application of the Hahn-Banach theorem.

Using the idea of Schur majorization one is led to consider an inequality which is stronger than Hoeffding's inequality of order n.

For two vectors $x = (x_1, \dots, x_n)$ and $y = (y_1, \dots, y_n)$ with components belonging to a vector space we say that x is *majorized* by y, in symbols $x \prec y$, if there exits a doubly stochastic $n \times n$ matrix Ω such that $x = y\Omega$. For example $(\overline{\mu}, \dots, \overline{\mu}) \prec (\mu_1, \dots, \mu_n)$ if $\mu_i \in \mathrm{Mol}^1_+(S)$.

A function $\psi : S \to \mathbb{R}$ is called *Schur increasing of order* n if

$$\int \psi \, d(\nu_1 * \dots * \nu_n) \leq \int \psi \, d(\mu_1 * \dots * \mu_n) \tag{46}$$

for any sets $\nu = (\nu_1, \dots, \nu_n)$, $\mu = (\mu_1, \dots, \mu_n) \in \mathrm{Mol}^1_+(S)^n$ such that $\nu \prec \mu$.

The set of these functions is a closed convex cone denoted by $\mathcal{S}_n(S)$. Clearly $\mathcal{S}_n(S) \subseteq \mathcal{H}_n(S)$. It is easy to see that $\mathcal{S}_2(S) = \mathcal{N}(S)$ and that $\mathcal{S}_{n+1}(S) \subseteq \mathcal{S}_n(S)$. The functions in

$$\mathcal{S}(S) := \bigcap_{n=2}^{\infty} \mathcal{S}_n(S) \tag{47}$$

are called *Schur increasing*.

Theorem 4.2. *For a function* $\psi : S \to \mathbb{R}$ *the following conditions are equivalent:*
(i) ψ *belongs to* $\mathcal{CN}(S)$ *and is bounded below.*
(ii) ψ *is completely alternating, i.e.,*

$$(I - E_{a_1}) \cdots (I - E_{a_n})\psi \leq 0$$

for all $n \geq 1, a_1, \dots, a_n \in S$.

(iii) *There exists a non-negative additive function q and a Radon measure μ on $\hat{S}_+ \setminus \{1\}$ such that*

$$\psi(s) = \psi(0) + q(s) + \int_{\hat{S}_+\setminus\{1\}} (1 - \rho(s))\, d\mu(\rho) \;\; , \quad s \in S.$$

If ψ satisfies the three equivalent conditions then ψ is Schur increasing.

Proof. This follows from B–C–R, Theorems 4.6.7, 7.1.10 and 7.3.7. □

The inclusion

$$\mathcal{S}_3(S) \subseteq \mathcal{CN}(S) \tag{48}$$

can be strict as shown in Berg and Christensen [3] for $S = \mathbb{N}_0^2$.

For $S = \mathbb{N}_0$ or $S = \mathbb{Z}$ with the identity involution we have

$$\mathcal{S}(S) = \mathcal{CN}(S), \tag{49}$$

cf. Theorem 7.3.9 in B–C–R, which can be extended from Radon perfect semigroups to perfect semigroups, and probably to all semiperfect semigroups.

Theorem 4.3. *If S is perfect then $\mathcal{S}(S) = \mathcal{CN}(S)$.*

Proof. Suppose φ is completely positive definite, i.e., $E_a\varphi \in \mathcal{P}(S)$ for each $a \in S$. Let σ_a be the unique positive measure in $F_+(S^*)$ such that

$$E_a\varphi(s) = \int \rho(s)\, d\sigma_a(\rho) \;\; , \quad s \in S.$$

By perfectness it is easy to see that $\sigma_a = \rho(a)\,\sigma_0$ for each $a \in S$. The set $T_a := \{\rho \in S^* \mid \rho(a) < 0\}$ belongs to $\mathcal{A}(S^*)$ and the inequalities

$$0 \leq \sigma_a(T_a) = \int_{T_a} \rho(a)\, d\sigma_0(\rho) \leq 0$$

show that $\sigma_0(T_a) = 0$ for each $a \in S$.

If $(\nu_1, \ldots, \nu_n) \prec (\mu_1, \ldots, \mu_n)$ and if $\rho \in S^*$ is non-negative on a finite set $F \subseteq S$ carrying the measures $\nu_i, \mu_i,\; i = 1, \ldots, n$ then

$$\int \rho\, d(\nu_1 * \ldots * \nu_n) \geq \int \rho\, d(\mu_1 * \ldots * \mu_n),$$

cf. B–C–R, p. 243. Letting

$$P = \bigcap_{a \in F} T_a^c$$

we have $\sigma_0(P^c) = 0$ and hence

$$\begin{aligned}\int \varphi \, d(\nu_1 * \ldots * \nu_n) &= \int \left(\int_P \rho(s) \, d\sigma_0(\rho) \right) d(\nu_1 * \ldots * \nu_n)(s) \\ &\geq \int \varphi \, d(\mu_1 * \ldots * \mu_n),\end{aligned}$$

showing that $-\varphi$ is Schur increasing.

If φ is completely negative definite then $\exp(-t\varphi)$ is completely positive definite for each $t > 0$ by Schoenberg's theorem, hence $-\exp(-t\varphi) \in \mathcal{S}(S)$, and finally

$$\varphi = \lim_{t \to 0} \frac{1}{t}(1 - \exp(-t\varphi)) \in \mathcal{S}(S). \qquad \square$$

Corollary 4.4. *Suppose S is 2-divisible. Then*

$$\mathcal{H}_n(S) = \mathcal{S}_n(S) = \mathcal{N}(S) = \mathcal{CN}(\mathcal{S})$$

for all $n \geq 2$.

References

[1] Berg, C., Fonctions définies négatives et majoration de Schur, In: "Théorie du Potentiel". Proceedings, Orsay 1983. Eds. G. Mokobodzki and D. Pinchon. Lecture Notes in Mathematics 1096, 69–89. Springer, Berlin,1984

[2] Berg, C., The Multidimensional Moment Problem and Semigroups, In: "Moments in Mathematics". Proceedings of Symposia in Applied Mathematics 37, 110–124, Amer. Math. Soc., Providence, 1987

[3] Berg, C. and J. P. R. Christensen, Suites complètement définies positives, majoration de Schur et le problème des moments de Stieltjes en dimension k, C. R. Acad. Sci. Paris 297 (1983), Série I, 45–48

[4] Berg, C., J. P. R. Christensen and C.U. Jensen, A remark on the multidimensional moment problem, Math. Ann. 243 (1979), 163–169

[5] Berg, C., J. P. R. Christensen and P. Ressel, Positive definite functions on abelian semigroups, Math. Ann. 223 (1976), 253–272

[6] —, "Harmonic analysis on semigroups. Theory of positive definite and related functions", Graduate Texts in Mathematics 100. Springer, Berlin 1984

[7] Berg, C. and P. H. Maserick, Exponentially bounded positive definite functions, Illinois J. Math. 28 (1984), 162–179

[8] Bickel, P. J. and W. R. van Zwet, On a theorem of Hoeffding, In: "Asymptotic Theory of Statistical Tests and Estimation". Ed. I.M. Chakravarti, 307–324. Academic Press, New York, 1980

[9] Bisgaard, T. M.,Extension of characters on $*$-semigroups, Math. Ann. 282 (1988), 251–258

[10] Bisgaard, T. M., The two-sided complex moment problem, Ark. Mat. 27 (1989), 23–28

[11] Bisgaard, T. M., Hoeffding's inequalities: A counterexample, J. Theoretical Probability (to appear)

[12] Bisgaard, T. M., Characterization of perfect involution groups, Math. Scand. (to appear)

[13] Bisgaard, T. M., Method of moments on semigroups, (manuscript)

[14] Bisgaard, T. M. and P. Ressel, Unique disintegration of arbitrary positive definite functions on $*$-divisible semigroups, Math. Z. 200 (1989), 511-525

[15] Buchwalter, H., Formes quadratiques sur un semigroupe involutif, Math. Ann. 271 (1985), 619–639

[16] Buchwalter, H., Les fonctions de Lévy existent!, Math. Ann. 274 (1986), 31–34

[17] Bucy, R. S., and G. Maltese, Extreme positive definite functions and Choquet's representation theorem, J. Math. Analys. Appl. 12 (1966), 371–377

[18] Choquet, G., Theory of capacities, Ann. Inst. Fourier (Grenoble) 5 (1954), 131–295

[19] Choquet, G., Deux exemples classiques de représentations intégrale, Enseign. Math. 15 (1969), 63–75

[20] Choquet, G., Une démonstration du théorème de Bochner-Weil par discrétisation du groupe, Results in Mathematics 9 (1986), 1–9

[21] Christensen, J. P. R. and P. Ressel, A probabilistic characterization of negative definite and completely alternating functions, Z. Wahrsch. verw. Gebiete 57 (1981), 407–417

[22] Dettweiler, E., The Laplace transform of measures on the cone of a vector lattice, Math. Scand. 45 (1979), 311–333

[23] Hoffmann–Jørgensen, J. and P. Ressel, On completely monotone functions on $C_+(X)$, Math. Scand. 40 (1977), 79–93

[24] Hofmann, K. H., Topological semigroups. History, theory, applications, Jahresber. d. Deutsch. Math.-Verein. 78 (1976), 9–59

[25] Hofmann, K. H. and M. W. Mislove, Local compactness and continuous lattices. In: "Continuous Lattices". Proceedings, Bremen 1979. Eds. B. Banaschewski and R.-E. Hoffmann. Lecture Notes in Mathematics 871, 209–248. Springer, Berlin 1981

[26] Horn, R. A., Infinitely divisible positive definite sequences, Trans. Amer. Math. Soc. 136 (1969), 287–303

[27] Ionesco Tulcea, C. T., Sur certaines classes de fonctions de type positif, Ann. Ecole Norm. Sup. 74 (1957), 231–248

[28] Lashkarizadeh Bami, M.,Bochner's theorem and the Hausdorff moment theorem on foundation topological semigroups, Can. J. Math. 37 (1985), 785–809

[29] Lindahl, R.J. and P. H. Maserick, Positive-definite functions on involution semigroups, Duke Math. J. 38 (1971), 771–782

[30] Maserick, P. H., BV-functions, positive definite functions and moment problems, Trans. Amer. Math. Soc. 214 (1975), 137–152

[31] Maserick, P. H., A Lévy–Khinchin formula for semigroups with involutions, Math. Ann. 236 (1978), 209–216

[32] Matheron, G., " Random Sets and Integral Geometry", Wiley, New York, 1975

[33] Nussbaum, A. E., The Hausdorff-Bernstein-Widder theorem for semigroups in locally compact abelian groups, Duke Math. J. 22 (1955), 573–582

[34] Paterson, A. L. T., An integral representation of positive definite functions on a Clifford semigroup, Math. Ann. 234 (1978), 125–138

[35] Ressel, P., A general Hoeffding type inequality, Z. Wahrsch. verw. Gebiete 61 (1982), 223–235

[36] Ressel, P., Integral representations on convex semigroups, Math. Scand. 61 (1987), 93–111

[37] Sakakibara, N., The moment problem on abelian semigroups, (manuscript)

[38] Schempp, W., On functions of positive type on commutative monoids, Math. Z. 156 (1977), 115–121

[39] Schmüdgen, K., An example of a positive polynomial which is not a sum of squares of polynomials. A positive, but not strongly positive functional, Math. Nachr. 88 (1979), 385–390

[40] Shopova, D., On the condition of positivity, Annuaire Univ. Sofia, Fac. Math. Méc. 63 (1970), 53–59

[41] Stewart, J., Positive definite functions and generalizations, a historical survey, Rocky Mountain J. Math. 6 (1976), 409–434

[42] Szafraniec, F. H., Dilations on involution semigroups, Proc. Amer. Math. Soc. 66, (1977), 30–32

[43] Sz.-Nagy, B., "Prolongements des transformations de l'espace de Hilbert qui sortent de cet espace" Akademiai Kiado Budapest, 1955

[44] Talagrand, M., Quelques examples de représentation intégrale: Valuations, fonctions alternées d'ordre infini, Bull. Sci. Math. 100 (2), (1976), 321–329

[45] Tonev, T. W., Semigroups and positive definite functions, Mag. Thesis, University of Sofia, 1969

[46] Tonev, T. W., Positive-definite functions on discrete commutative semigroups, Semigroup Forum 17 (1979), 175–183

[47] Troallic, J.-P., Semigroupes affines semitopologiques compacts, this volume

[48] Williamson, J. H., Harmonic Analysis on Semigroups, J. London Math. Soc. 42 (1967), 1–41

Convolution semigroups and potential kernels on a commutative hypergroup

Herbert Heyer

The aim of this exposé is to survey recent advances in the potential theory of convolution semigroups on a commutative hypergroup. The problems to be discussed are related to the theory of space-homogeneous Markov processes on a commutative hypergroup. Their solutions are achieved by an application of harmonic analysis of commutative hypergroups. Some of the topics that will be covered in this paper arose from open questions posed in the author's previous survey [32]. The remarkable contributions of the last five years to the Schoenberg correspondence, the transience and embedding problems seem to make it worthwhile to once again look into the state of the art with curiosity. The reader will recognize that part of the theory presented here developed from the desire of extending the main theorems of the monograph [6] of C. Berg and G. Forst from Abelian locally compact groups to commutative hypergroups. In In the process, obstacles had to be overcome and at the same time new phenomena had to be envisaged. Both aspects are the motives for the author to collaborate with Walter R. Bloom on a book on probabilities and potentials on a hypergroup. The present survey can be considered as a synopsis for a central section of that book.

The author has profited from still unpublished work of M. Voit and H. Zeuner.

1. Preliminaries on commutative hypergroups

For a locally compact space K we introduce the system $\mathfrak{K}(K)$ of all compact subsets of K furnished with the Michael topology. By $\mathfrak{V}_x(K)$ we abbreviate the system of open neighborhoods of x in K. The symbols $\mathcal{C}^b(K)$, $\mathcal{C}^0(K)$ and $\mathcal{K}(K)$ denote the spaces of bounded continuous functions on K, continuous functions on K vanishing at infinity, and continuous functions on K having compact support. $\mathcal{C}^b(K)$ will carry the compact open topology $\mathcal{T}_{co}$. The inclusions $\mathcal{M}^1(K) \subset \mathcal{M}^{(1)}(K) \subset \mathcal{M}^b(K) \subset \mathcal{M}(K)$ concern the sets of probability measures, contraction measures (measures of norm ≤ 1), bounded measures and arbitrary (Radon) measures on K. In $\mathcal{M}(K)$ we consider the vague topology $\mathcal{T}_v$, in $\mathcal{M}^b(K)$ also the weak topology $\mathcal{T}_w$. For every $x \in K$, ϵ_x denotes the Dirac (point) measure in x. Given any measure $\mu \in \mathcal{M}_+(K)$ the abbreviation $L^p(K, \mu)$ for $p \in [1, \infty[$ refers to the Banach spaces of μ-integrable functions (function classes) on K.

Let K admit a topological involution $x \to x^-$.

Definition 1.1. K is called a *hypergroup* if there exists a convolution $*$ in $\mathcal{M}^b(K)$ such that $(\mathcal{M}^b(K), *, ^-)$ is an involutive Banach algebra and the following axioms hold:

(HG1) The mapping $(\mu, \nu) \to \mu * \nu$ from $\mathcal{M}^b(K) \times \mathcal{M}^b(K)$ into $\mathcal{M}^b(K)$ is bilinear, nonnegative and $\mathcal{T}_w$-continuous.

(HG2) For all $x, y \in K$ the measure $\epsilon_x * \epsilon_y \in \mathcal{M}^1(K)$ and has compact support.

(HG3) There exists a neutral element $e \in K$ satisfying $\epsilon_e * \epsilon_x = \epsilon_x * \epsilon_e = \epsilon_x$ for all $x \in K$, and $e \in \operatorname{supp}(\epsilon_x * \epsilon_y)$ if and only if $x = y^-$.

(HG4) The mapping $(x, y) \to \operatorname{supp}(\epsilon_x * \epsilon_y)$ from $K \times K$ into $\mathfrak{K}(K)$ is continuous.

In this paper we shall deal exclusively with *commutative* hypergroups K which by definition satisfy $\epsilon_x * \epsilon_y = \epsilon_y * \epsilon_x$ for all $x, y \in K$.

(1.2) The key notion in the theory of hypergroups is the generalized *translation operator* defined for $x \in K$ by

$$T^x f(y) := \int_K f(z) \epsilon_x * \epsilon_y(dz)$$

whenever f is an admissible function on K, $y \in K$. We also use the notation f_x for $T^x f$. Given a measure $\mu \in \mathcal{M}_+^b(K)$ and an admissible function f on K we introduce the function $\mu * f$ by

$$\mu * f(x) := \int f_x(y^-) \mu(dy)$$

for all $x \in K$. Applying the generalized translation operator we can introduce the space $\mathcal{C}^u(K)$ of uniformly continuous bounded functions on K and show that $\mathcal{K}(K) \subset \mathcal{C}^u(K)$. Properties of the *convolution operator* T^μ defined by

$$T^\mu f := \mu * f$$

for all f of appropriate spaces of functions have been studied and exposed in [40] and [24]. Occasionally we shall also consider T^μ for arbitrary measures $\mu \in \mathcal{M}_+(K)$ which involves convolvability of measures $\mu, \nu \in \mathcal{M}_+(K)$. It turns out that for every $f \in \mathcal{K}(K)$ the mapping $\mu \to T^\mu f$ from $\mathcal{M}_+(K)$ into the space $\mathcal{C}(K)$ of continuous functions on K is $\mathcal{T}_v - \mathcal{T}_{co}$-continuous ([12]). For any $\mu \in \mathcal{M}_+(K)$ we consider the set

$$D_+(\mu) := \{\nu \in \mathcal{M}_+(K) \colon \mu * \nu \quad \text{exists}\}.$$

Given $\mu \in \mathcal{M}_+(K)$ and $\nu \in D_+(\mu)$ the equality $\mu * \nu = 0$ implies $\nu = 0$.

(1.3) Since for commutative hypergroups K the Banach algebra $(\mathcal{M}^b(K), *, ^-)$ is commutative, *Gelfand's theorem* applies (as in the theory of Gelfand pairs) and an extensive harmonic analysis becomes available, the basic tool being the *translation invariant* (Haar) *measure* ω_K of K, whose existence has been established in [52]. Standard references for the harmonic analysis of commutative hypergroups are [18], [38] and [49], more recent sources are [29] and [54].

In the remaining part of this section we shall describe some of the fundamental notions and facts of harmonic analysis of commutative hypergroups and stress those properties known in the group case that become invalid for hypergroups.

(1.4) The *dual space* of K is defined as the space

$$K^\wedge := \{\chi \in \mathcal{C}^b(K) \colon \chi \not\equiv 0,\ \chi(x^-) = \overline{\chi(x)},\ T^x\chi(y) = \chi(x)\chi(y) \quad \text{for all} \quad x, y \in K\}$$

of characters of K, furnished with the topology $\mathcal{T}_{co}$. If in the definition of a character the boundedness and hermitian properties are dropped, we speak of a semicharacter of K.

(1.5) For any $\mu \in \mathcal{M}^b(K)$ the *Fourier transform* $\hat{\mu}$ on $K^\wedge$ is defined by

$$\hat{\mu}(\chi) := \int \overline{\chi(x)}\mu(dx)$$

for all $\chi \in K^\wedge$. Analoguously one introduces the *cotransform* $\check{\mu}$ of a measure $\mu \in \mathcal{M}^b(K^\wedge)$ as a function on K. The standard properties of the Fourier transform and cotransform extend to hypergroups. In particular we have the

(1.6) *Plancherel-Levitan Theorem* which states that the Fourier transform induces an isometric isomorphism from $L^2(K, \omega_K)$ onto $L^2(K^\wedge, \pi)$, where $\pi \in \mathcal{M}_+(K^\wedge)$

denotes the unique *Plancherel measure* associated with ω_K by the formula

$$\int_K |f|^2 \, d\omega_K = \int_{K^\wedge} |\hat{f}|^2 \, d\pi$$

valid for all $f \in L^1(K, \omega_K) \cap L^2(K, \omega_K)$. In general supp $\pi \subset K^\wedge$ but $\neq K^\wedge$. If supp $\pi = K^\wedge$ then the unit character $\mathbb{1}$ will be in supp π, and in this case K will be called a *Godement hypergroup*. In the complementary case K is said to be a *Kunze-Stein hypergroup*. There is an attempt to look at the structure of hypergroups in terms of *growth*. For a general approach see [23]. More information on hypergroups with polynomial or subexponential growth is contained in [53], [54] and [56]. Both types of hypergroups are Godement hypergroups.

(1.7) *Positive characters* have been established on every commutative hypergroup K. They can be constructed inside supp π, and they are isolated in supp π if and only if K is compact. As is shown in [56] such characters can be used in order to define modified convolutions on K. One just puts

$$\epsilon_x \circ \epsilon_y := \frac{1}{\chi_x(y)} \chi \cdot (\epsilon_x * \epsilon_y)$$

for $x, y \in K$.

(1.8) In general the dual space $K^\wedge$ of K is not a hypergroup. It does become a hypergroup if $K^\wedge$ carries a convolution $*$ defined (pointwise) by

$$\chi(x)\chi'(x) = \int_{K^\wedge} \tau(x) \epsilon_\chi * \epsilon_{\chi'}(d\tau)$$

for all $\chi, \chi' \in K^\wedge$ and $x \in K$ which satisfies the axioms of 1.1 with $\mathbb{1}$ as neutral element and complex conjugation as involution. If K has a *hypergroup dual* $K^\wedge$, K is said to be *strong*. In this case supp $\pi = \operatorname{supp} \omega_{K^\wedge}$ and $K \subset K^{\wedge\wedge}$. If $K \cong K^{\wedge\wedge}$ then by obvious reasons we call K a *Pontrjagin* hypergroup.

(1.9) The main objects to be discussed in this paper will be (continuous) *convolution semigroups* introduced as families $(\mu_t)_{t \geq 0}$ of measures in $\mathcal{M}_+^{(1)}(K)$ with the properties

(CS1) $\mu_t * \mu_s = \mu_{t+s}$ for all $t, s > 0$, and

(CS2) $\mathcal{T}_w - \lim\limits_{t \to 0} \mu_t = \epsilon_e$.

Convolution semigroups of particular interest are the *Poisson semigroups* $(\pi_t(\mu, m))_{t\geq 0}$ with defining measure $\mu \in \mathcal{M}_+^b(K)$ and parameter $m \geq \|\mu\|$ given by

$$\pi_t(\mu, m) := e^{-tm} \exp(t\mu)$$

for all $t > 0$. For $m = 1$ we obtain the Poisson semigroups $(\pi_t(\mu))_{t\geq 0}$ in $\mathcal{M}_+^{(1)}(K)$ with defining measure $\mu \in \mathcal{M}_+^{(1)}(K)$.

While general convolution semigroups in $\mathcal{M}^1(K)$ can be viewed as space-homogeneous Markov processes, Poisson semigroups admit an interpretation as random walks on K.

(1.10) Given a convolution semigroup $(\mu_t)_{t\geq 0}$ in $\mathcal{M}_+^{(1)}(K)$ there exists the corresponding *resolvent family* $(\rho_\lambda)_{\lambda>0}$ of measures in $\mathcal{M}_+^b(K)$ defined by

(R) $$\rho_\lambda(f) := \int_0^\infty e^{-\lambda t}\mu_t(f)\, dt$$

for all $f \in \mathcal{K}(K)$. In fact

(RF1) $\lambda\rho_\lambda \in \mathcal{M}_+^{(1)}(K)$ for all $\lambda > 0$, and

(RF2) $\rho_\lambda - \rho_{\lambda'} = (\lambda' - \lambda)\rho_\lambda * \rho_{\lambda'}$ for all $\lambda, \lambda' > 0$.

The converse that for every resolvent family $(\rho_\lambda)_{\lambda>0}$ in $\mathcal{M}_+^{(1)}(K)$ defined by the properties (RF1) and (RF2) there exists a unique convolution semigroup $(\mu_t)_{t\geq 0}$ in $\mathcal{M}_+^{(1)}(K)$ satisfying (R) can be proved under additional assumptions on K by a method to be described in Section 3. For a slightly more general approach to the *resolvent correspondence*, see [9].

2. A collection of examples

We shall list the majority of known examples, but will describe in more detail only those which have not yet been included in the survey literature. A basic reference for polynomial hypergroups is [41]. A broader variety of examples is contained in [24] and [32].

We start our collection by noting that

A. *Abelian* (locally compact) *groups* are trivial examples from the viewpoint of hypergroups. Nevertheless we shall often refer to these examples in order to indicate the origin of problems to be solved on hypergroups, their similarities to and discrepancies from the group case.

B. *Double coset hypergroups* $K := G/\!/H$ arise from Gelfand pairs (G, H) where this very origin yields the commutativity of K. It is known that the dual $K^\wedge$ of K can be identified with the set of (nonvanishing, hermitian) H-biinvariant spherical functions on G.

There will be various special types of double coset hypergroups listed below under a different classification. Here we only point out two subexamples:

(B1) The *duals* $K = (G/\!/H)^\wedge$ *of symmetric pairs* (G, H) *of compact type*. They are countably discrete hypergroups arising from strong compact hypergroups $G/\!/H$ (see [22]).

(B1.1) The *duals of compact connected Lie groups* L can be absorbed in the previous example by looking at the symmetric pairs (G, H) with $G := L \times L$ and H being the diagonal subgroup of G.

In particular the dual of SU(2) coincides with the dual of the double coset hypergroup SO(4)//SO(3).

(C) *Orbit hypergroups* $K := G_B$ for a locally compact group G and a relatively compact subgroup B of $\mathrm{Aut}(G)$ with $B \supset \mathrm{Int}(G)$. It is known that K is a Pontrjagin hypergroup (see [28]).

This property is particularly available if G is a compact group and B an arbitrary subgroup of $\mathrm{Aut}(G)$, or if G is a locally compact group $\in [FC]^- \cap \mathrm{SIN}$ and $B = \mathrm{Int}(G)$.

(C1) *Motion hypergroups* G_B with $G := \mathbb{R}^d$ and $B := \mathrm{SO}(d)$ for $d \geq 2$. In this case $G_B^\wedge \cong \mathbb{R}_+$.

(D) *Polynomial hypergroups* $(\mathbb{Z}_+, *(Q_n))$ are introduced via sequences $(Q_n)_{n\geq 0}$ of orthogonal polynomials on $\mathbb{R}$ given recursively by

$$\begin{cases} Q_0 & \equiv 1 \\ Q_1 & = 2\sqrt{a\gamma}\, id + \beta, \quad \text{and} \\ Q_1 Q_n & = a_n Q_{n+1} + b_n Q_n + c_n Q_{n-1} \end{cases}$$

for all $n \geq 1$. Here $(a_n)_{n\geq 0}$, $(b_n)_{n\geq 0}$ and $(c_n)_{n\geq 0}$ are sequences in $\mathbb{R}$ satisfying $a_n + b_n + c_n = 1$, $a_n, c_n > 0$ and $b_n \geq 0$ for all $n \geq 0$ such that the limits $\alpha := \lim_{n\to\infty} a_n$, $\beta := \lim_{n\to\infty} b_n$ and $\gamma := \lim_{n\to\infty} c_n$ exist and $\alpha, \gamma > 0$. If in the linearization

$$Q_n Q_m = \sum_{k=|n-m|}^{n+m} c(n,m,k) Q_k$$

the coefficients $c(n,m,k)$ are ≥ 0, then the formula

$$\epsilon_n * \epsilon_m := \sum_{k=|n-m|}^{n+m} c(n,m,k) \epsilon_k$$

defines a hypergroup structure on $\mathbb{Z}_+$. The mapping $x \to \chi_{(x)}$ with $\chi_{(x)}(n) := Q_n(x)$ for all $x \in \mathbb{R}$ establishes a homeomorphism between $\mathbb{R}$ and the set $\{\chi_{(x)}: x \in \mathbb{R}\}$ of all semicharacters of $\mathbb{Z}_+$. One has

$$\mathbb{Z}_+^{\wedge} = \{\chi_{(x)}: x \in \mathbb{R} \quad \text{such that} \quad \chi_{(x)} \quad \text{is bounded}\}$$

and that the measure orthogonalizing the sequence $(Q_n)_{n\geq 0}$ is the Plancherel measure π on $\mathbb{Z}_+^{\wedge}$. $\mathbb{Z}_+^{\wedge}$ is compact and hence supp π is compact, but in general $\neq \mathbb{Z}_+^{\wedge}$.

(D1) *Jacobi polynomial hypergroups* are defined by the sequences $(Q_n^{\alpha,\beta})_{n\geq 0}$ of normed Jacobi polynomials for $\alpha \geq \beta > -1$ and $\alpha + \beta + 1 \geq 0$. For $\beta \geq -\frac{1}{2}$ or $\alpha + \beta \geq 0$ they are all Pontrjagin hypergroups with dual hypergroup $\mathbb{Z}_+^{\wedge} \cong I := [-1, 1]$ and Plancherel (or orthogonalizing) measure

$$\pi(dx) = (1-x)^{\alpha}(1+x)^{\beta}\, dx\,.$$

Subexamples are

(D1.1) *Ultraspherical hypergroups* for $\alpha = \beta$.

(D1.1.1) *Spherical hypergroups* for $\alpha = \beta = \frac{d-3}{2}$, $d \geq 3$.

They include the hypergroups $\mathbb{Z}_+ \cong (\mathrm{SO}(d)/\!/\mathrm{SO}(d-1))^{\wedge}$.

(D1.1.1.1) *Chebychev hypergroups of first kind* for $d = 2$.

(D1.1.1.2) *Legendre hypergroups* for $d = 3$.

(D1.1.1.3) *Chebychev hypergroups of second kind* for $d = 4$.

The following three examples provide further polynomial hypergroups with $\mathbb{Z}_+^{\wedge} \cong I \cong \operatorname{supp} \pi$.

(D2) *q-ultraspherical hypergoups.*

(D3) *Associated Legendre hypergroups.*

(D4) *Generalized Chebychev hypergroups.*

For the
(D5) *Geronimus hypergroups* and

(D6) *Grinspun hypergroups*

we have at least $\mathbb{Z}_+^\wedge \cong I$.

(D7) *Cartier hypergroups* are defined for $a, b \in \mathbb{R}$, $a, b \geq 2$ by the sequence $(Q_n^{(a,b)})_{n\geq 0}$ of normalized Cartier polynomials given by

$$\begin{cases} Q_0^{(a,b)} \equiv 1, \\ Q_1^{(a,b)}(x) = c_1 x + c_2 \text{ with } c_1 := \frac{2}{a}\sqrt{\frac{a-1}{b-1}},\ c_2 := \frac{b-2}{a(b-1)}, \\ Q_1 Q_n = \frac{1}{a(b-1)} Q_{n-1} + \frac{b-2}{a(b-1)} Q_n + \frac{a-1}{a} Q_{n-1} \end{cases}$$

for all $n \geq 1$. The Haar measure of a Cartier hypergroup is given by

$$\begin{cases} \omega_{\mathbb{Z}_+}(\{0\} = 1 \\ \omega_{\mathbb{Z}_+}(\{n\}) = a(a-1)^{n-1}(b-1) \end{cases}$$

for all $n \geq 1$, $\mathbb{Z}_+^\wedge, \cong [-x_0, x_0]$ with $x_0 := \frac{1-c_2}{c_1} \geq 1$ is not a hypergroup and $\operatorname{supp} \pi \neq [-x_0, x_0]$ except for $a = b = 2$ in which case $x_0 = 1$.

(D7.1) The special case ($b = 2$, $a \in \mathbb{N}$, $a \geq 2$) of *Arnaud-Dunau hypergroups* is long known.

We note that the Cartier hypergroups arise from infinite distance transitive graphs as was shown in [55]. In fact, if a polynomial hypergroup $\mathbb{Z}_+$ is isomorphic to a double coset hypergroup, then $\mathbb{Z}_+$ is a Cartier hypergroup. The Arnaud-Dunau hypergroups arise from homogeneous trees.

Finally we note that there are hypergroup structures on $\mathbb{Z}_+$ which are not defined by polynomial sequences ([38], , 15.1D).

(E) *Two-variable Jacobi polynomial hypergroups* $(\mathbb{Z}_+^2, *(Q_{m,n}^\alpha))$ are introduced for $\alpha > 0$ via two-variable Jacobi polynomials of the form

$$Q_{m,n}^\alpha(z) := z^{|m-n|} Q_{m \wedge n}^{\alpha,|m-n|}(2|z|^2 - 1)$$

for all $z \in \mathbb{D} := \{z \in \mathbb{C} : |z| \leq 1\}$ (see [2]). The Haar measure of $\mathbb{Z}_+^2$ can be computed as

$$\omega_{\mathbb{Z}_+^2}(\{m,n\}) = \left(\int_{\mathbb{D}} |Q_{m,n}^\alpha(z)|^2 \pi(dz) \right)^{-1}$$

for all $(m,n) \in \mathbb{Z}_+^2$, where the orthogonalizing (Plancherel) measure π on $\mathbb{D}$ is given by

$$\pi(dx, dy) = \frac{\alpha + 1}{\pi} (1 - x^2 - y^2)^\alpha \, dx \, dy \, .$$

$\mathbb{Z}_+^2$ is a Pontrjagin hypergroup with $(\mathbb{Z}_+^2)^\wedge \cong \mathbb{D}$.

The following two related examples are the dual hypergroups of the types **(D1)** and **(E)** respectively.

(F) *Dual Jacobi polynomial hypergroups* $(I, *(Q_n^{\alpha,\beta}))$.

(F1) *Dual spherical hypergroups* for $\alpha = \beta = \frac{d-3}{2}$ of the form $\mathrm{SO}(d) /\!/ \mathrm{SO}(d-1)$, $d \geq 3$.

(G) *Dual two-variable Jacobi polynomial hypergroups* $(\mathbb{D}, *(Q_{n,m}^\alpha))$.

(G1) *Unitary hypergroups* for $\alpha = d - 2$ of the form $U(d) /\!/ U(d-1)$, $d \geq 3$.

(H) *Sturm-Liouville hypergroups of non-compact type* $(\mathbb{R}_+, *A)$ are defined in [59] as follows: Let A be a function in $\mathcal{C}(\mathbb{R}_+)$, > 0 and differentiable on $\mathbb{R}_+^\times$. We consider the Sturm-Liouville operator

$$L := -\frac{1}{A(x)} \frac{d}{dx} \left(A(x) \frac{d}{dx} \right)$$

on functions f which are twice differentiable on $\mathbb{R}_+^\times$. A *normalized* hypergroup $(\mathbb{R}_+, *)$ in the sense that

$$\begin{cases} \min \, \mathrm{supp}\,(\epsilon_x * \epsilon_y) = |x - y| & \text{and} \\ \max \, \mathrm{supp}\,(\epsilon_x * \epsilon_y) = x + y \end{cases}$$

for all $x, y \in \mathbb{R}_+$ is called a *Sturm-Liouville hypergroup* if for every even test function f on $\mathbb{R}$ restricted for $\mathbb{R}_+$ the function $u_f \in \mathcal{C}(\mathbb{R}_+ \times \mathbb{R}_+)$ given by

$$u_f(x, y) := \int f \, d(\epsilon_x * \epsilon_y)$$

for all $x, y \in \mathbb{R}_+$ is twice differentiable and satisfies the partial differential equation

$$\begin{cases} L_x u_f(x, y) = L_y u_f(x, y) \\ u_x(0, y) = 0 \end{cases}$$

for all $x, y \in \mathbb{R}_+^\times$. See also [51].

The Haar measure of the Sturm-Liouville hypergroup $\mathbb{R}_+$ is $A \cdot \lambda_{\mathbb{R}_+}$.

(H1) *Chébli-Trimèche hypergroups* as introduced in [13] are defined by the following additional conditions for A: $A(0) = 0$, $A(x) > 0$ for all $x > 0$, A is increasing with $\lim\limits_{x \to \infty} A(x) = \infty$, $\frac{A'}{A}$ is decreasing on $\mathbb{R}_+^\times$ to $2\rho \geq 0$, where ρ is the *index* of the hypergroup, and

$$\frac{A'}{A}(x) = \frac{a}{x} + B(x)$$

in a neighborhood of 0 with $a > 0$ and an odd $\mathcal{C}^\infty$-function B on $\mathbb{R}$. In this case

$$\operatorname{supp}(\epsilon_x * \epsilon_y) = [|x - y|, x + y]$$

for all $x, y \in \mathbb{R}_+$.

For the Plancherel measure π of $\mathbb{R}_+$ we have $\operatorname{supp} \pi = [\rho^2, \infty[$. It follows that in general $\mathbb{R}_+^\wedge$ is not a hypergroup. It is, however, a hypergroup if $\mathbb{R}_+$ is of subexponential growth which is equivalent to $\rho = 0$ and to $\operatorname{supp} \pi = \mathbb{R}_+^\wedge$ ([56]). In the case $\rho > 0$ (of exponential growth) $\mathbb{R}_+^\wedge$ is not a hypergroup.

(H1.1) *Bessel-Kingman hypergroups* are Chébli-Trimèche hypergroups with

$$A(x) = x^{2\alpha+1}, \ \alpha > -\frac{1}{2} \, .$$

They are of exponential growth and have the Pontrjagin property.

(H1.1.1) *Motion hypergroups* for $\alpha := \frac{d}{2} - 1$, $d \geq 2$. They coincide with the double coset hypergroups $M(d)/\!/\mathrm{SO}(d)$ arising from the motion group $M(d)$ of $\mathbb{R}^d$. Compare with **(C1)**.

(H1.2) *Jacobi hypergroups* defined by

$$A(x) := (\operatorname{sh} x)^{2\alpha+1}(\operatorname{ch} x)^{2\beta+1}, \quad \alpha \geq \beta \geq -\frac{1}{2}$$

satisfy $\rho > 0$ and are therefore Kunze-Stein hypergroups.

(H1.2.1) *Hyperbolic hypergroups* for $\beta = -\frac{1}{2}$.

(H1.2.1.1) The *Naimark hypergroup* for ($\beta = -\frac{1}{2}$ and) $\alpha = \frac{1}{2}$ can be viewed as the double coset hypergroup $\mathrm{SL}(2, \mathbb{C}) /\!/ \mathrm{SU}(2)$.

(H1.2.2) *Rank 1 hypergroups* are Jacobi hypergroups with $\alpha = \frac{p-1}{2}$ and $\beta := \frac{q-1}{2}$, $p \geq q \geq 0$.

They are double coset hypergroups arising from Riemannian symmetric pairs (G, H) of non-compact type and rank 1. In particular

(H1.2.2.1) *Hyperbolic hypergroups of rank 1* of the form $\mathrm{SO}(d, 1) /\!/ \mathrm{SO}(d)$, $d \geq 2$ are obtained by the choice

$$A(x) = (\operatorname{sh} x)^p, \quad p := d - 1 .$$

Here $\mathrm{SO}(d, 1)$ denotes the Lorentz group of dimension d.

So far we have listed Sturm-Liouville hypergroups $\mathbb{R}_+$ with interval support $[|x - y|, x + y]$ of the convolution $\epsilon_x * \epsilon_y$ for $x, y \in \mathbb{R}_+$. We add two more examples with two-point support $\{|x - y|, x + y\}$ of $\epsilon_x * \epsilon_y$.

(H2) The *symmetric hypergroup* $\mathbb{R}_+ \cong \mathbb{R} \rtimes \mathbb{Z}_2 /\!/ \mathbb{Z}_2$ defined by the convolution

$$\epsilon_x * \epsilon_y = \frac{1}{2}(\epsilon_{|x-y|} + \epsilon_{x+y})$$

for all $x, y \in \mathbb{R}_+$ is a Sturm-Liouville hypergroup with

$$A(x) := 1$$

for all $x \in \mathbb{R}_+$.

(H3) The cos*h-hypergroup* $\mathbb{R}_+$ introduced via the convolution

$$\epsilon_x * \epsilon_y = \frac{\operatorname{ch}(x - y)}{2 \operatorname{ch} x \operatorname{ch} y} \epsilon_{|x-y|} + \frac{\operatorname{ch}(x + y)}{2 \operatorname{ch} x \operatorname{ch} y} \epsilon_{x+y}$$

for all $x, y \in \mathbb{R}_+$ is a Sturm-Liouville hypergroup with

$$A(x) := (\operatorname{ch} x)^2$$

for all $x \in \mathbb{R}_+$. It has been shown in [58] that $\mathbb{R}_+^\wedge = \mathbb{R}_+ \cup i[0,1]$ and $\pi = \frac{2}{\pi} \lambda_{\mathbb{R}_+}$, hence that supp $\pi = \mathbb{R}_+ \underset{\neq}{\subseteq} \mathbb{R}_+^\wedge$.

(J) *Sturm-Liouville hypergroups of compact type* $(J, *A)$ with $J := [0,1]$ have been studied in analogy to those of noncompact type in [1]. Let A be a function on J such that A is increasing on $\left[0, \frac{1}{2}\right]$, $\frac{A'}{A}$ is decreasing, $A(1-x) = A(x)$ for all $x \in J$, and such that there exists an $\alpha > 0$ and a $\mathcal{C}^\infty$-function B on J with $B'(0) = 0$, $B(x) > 0$ for all $x \in J$ satisfying

$$A(x) = (\sin \pi x)^\alpha B(x)$$

for all $x \in J$. For all $x, y \in J$ there exists a measure $\epsilon_x * \epsilon_y \in \mathcal{M}^1(J)$ with

$$\operatorname{supp}(\epsilon_x * \epsilon_y) = [|x-y|, 1 - |1-x-y|]$$

such that the function u_f on $J \times J$ defined by

$$u_f(x,y) := \int f \, d(\epsilon_x * \epsilon_y)$$

satisfies (with the Sturm-Liouville operator L on $\mathring{J}$) the equations

$$\begin{cases} u_f(x,0) = f(x) \quad \text{and} \\ L_x u_f(x,y) = L_y u_f(x,y) \quad \text{for all} \quad x, y \in \mathring{J} \end{cases}$$

and every $\mathcal{C}^\infty$-function f on J with

$$f^{(2n+1)}(0) = f^{(2n+1)}(1) = 0, \quad n \geq 1 .$$

The hypergroup J has Haar measure $A \cdot \lambda_J$.

(J1) The *Jacobi double coset hypergroups* listed in Theorem 6.1b of [59] provide subexamples.

We note that in [59] all Sturm-Liouville structures on $\mathbb{R}_+$ and J which arise from double coset hypergroups have been determined.

In contrast to (J) the

(K) *symmetric hypergroup* $J = \mathbb{T} \otimes \mathbb{Z}_2 // \mathbb{Z}_2$ defined by the convolution

$$\epsilon_x * \epsilon_y = \frac{1}{2}(\epsilon_{|x-y|} + \epsilon_{1-|1-x-y|})$$

for all $x, y \in J$ has a two-point support. In fact, $\mathbb{T} \otimes \mathbb{Z}_2 // \mathbb{Z}_2$ provides the only hypergroup structure on J with this property.

3. The Schoenberg correspondence

Positive and negative definite functions on a hypergroup K have been introduced in analogy to groups. Since we will need similar properties for functions on the dual space $K^\wedge$ of K, a more general approach has to be chosen. We report on results of M. Voit proved in [55]. Let

$$\mathcal{M}_{\mathbb{1}}(K^\wedge) := \{\mu \in \mathcal{M}^b(K^\wedge) : \mu = c\epsilon_{\mathbb{1}} + g \cdot \pi,\ c \in \mathbb{C},\ g \in \mathcal{K}(K^\wedge)\}.$$

A function $\varphi \in L^1_{\text{loc}}(K^\wedge, \pi)$ is said to be *strictly positive definite* if for all $\mu \in \mathcal{M}_{\mathbb{1}}(K^\wedge)$ such that $\check{\mu} \geq 0$ one has

$$\int_{K^\wedge} \varphi \, d\mu \geq 0.$$

The set of all strictly positive definite functions on $K^\wedge$ will be abbreviated by $\mathcal{SP}(K^\wedge)$.

The definition of strict positive-definiteness modifies the notion of strong positive-definiteness introduced for continuous function on $K^\wedge$ in [42] which in the case of a hypergroup dual $K^\wedge$ of K is equivalent to strong positive-definiteness proposed in [9]. If K is strong, then for continuous functions on $K^\wedge$ all definitions coincide.

Theorem 3.1. (Bochner)
(i) *For any $\mu \in \mathcal{M}^b_+(K)$ one has $\hat{\mu} \in \mathcal{SP}(K^\wedge) \cap \mathcal{C}(K^\wedge)$.*
(ii) *For $\varphi \in \mathcal{SP}(K^\wedge)$ there exists a measure $\mu \in \mathcal{M}^b_+(K)$ such that $\hat{\mu} = \varphi$ locally π-a.e. If, in addition, $\varphi \in \mathcal{C}(K^\wedge)$, then*

$$\text{Res}_{\text{supp}\,\pi}\, \hat{\mu} = \text{Res}_{\text{supp}\,\pi}\, \varphi. \qquad \square$$

The usual properties established for strictly positive definite functions on $K^\wedge$ include the sequential closedness with respect to the pointwise convergence of the set $\mathcal{SP}(K^\wedge)$.

Theorem 3.2. (Lévy) *Let $(\mu_n)_{n\geq 1}$ be a sequence in $\mathcal{M}_+^b(K)$ such that*

$$\begin{cases} \sup\limits_{n\geq 1} \|\mu_n\| < \infty & \textit{and} \\ \lim\limits_{n\to\infty} \hat{\mu}_n = \varphi & \textit{pointwise on} \quad \operatorname{supp} \pi \,. \end{cases}$$

There exists a unique measure $\mu \in \mathcal{M}_+^b(K)$ satisfying

$$\begin{cases} \hat{\mu} = \varphi & \textit{locally} \quad \pi \textit{ a.e. and} \\ \mathcal{T}_v - \lim\limits_{n\to\infty} \mu_n = \mu \,. \end{cases}$$

If, in addition, K is a Godement hypergroup and φ is continuous in $1\!\!1$, then

$$\begin{cases} \hat{\mu} = \varphi & \textit{(everywhere) and} \\ \mathcal{T}_w - \lim\limits_{n\to\infty} \mu_n = \mu \,. \end{cases}$$ □

Earlier versions of this *continuity theorem* can be found in [8] and [44].

A function $\psi \in \mathcal{C}(K^\wedge)$ is called *strictly negative definite* if $\psi(1\!\!1) \geq 0$, ψ is hermitian and if for all $\mu \in \mathcal{M}_{1\!\!1}(K^\wedge)$ such that $\check{\mu} \geq 0$ and $\check{\mu}(e) = 0$ one has

$$\int_{K^\wedge} \psi \, d\mu \leq 0 \,.$$

Let $\mathcal{SN}(K^\wedge)$ denote the set of all strictly negative definite functions on $K^\wedge$.

If K is strong, then any strictly negative definite function ψ on $K^\wedge$ is negative definite and $\operatorname{Re}\psi \geq \psi(1\!\!1) \geq 0$. On the other hand it can be shown that strongly negative definite functions on $K^\wedge$ in the sense of [44] are strictly negative definite.

As for negative definite functions on $K^\wedge$ one shows that $\mathcal{SN}(K^\wedge)$ is sequentially $\mathcal{T}_{\text{co}}$-closed, and for $\varphi \in \mathcal{SP}(K^\wedge) \cap \mathcal{C}(K^\wedge)$ one gets $\varphi(1\!\!1) - \varphi \in \mathcal{SN}(K^\wedge)$.

Theorem 3.3. (Schoenberg)

(i) *For every convolution semigroup $(\mu_t)_{t\geq 0}$ in $\mathcal{M}_+^b(K)$ there exists exactly one function $\psi \in \mathcal{SN}(K^\wedge)$ satisfying*

$$\hat{\mu}_t = \exp(-t\psi)$$

for all $t > 0$.

(ii) *If conversely $\psi \in \mathcal{SN}(K^\wedge)$, then there exists exactly one convolution semigroup $(\mu_t)_{t\geq 0}$ in $\mathcal{M}_+^{(1)}(K)$ such that*

$$\operatorname{Res}_{\operatorname{supp} \pi} \hat{\mu}_t = \operatorname{Res}_{\operatorname{supp} \pi} \exp(-t\psi)$$

holds for all $t > 0$. *Moreover one has*

$$\|\mu_t\| = \hat{\mu}_t(\mathbb{1}) \le \exp(-t\psi(\mathbb{1}))$$

for all $t > 0$. □

The one-to-one correspondence between convolution semigroups $(\mu_t)_{t\ge 0}$ in $\mathcal{M}_+^{(1)}(K)$ and strictly negative definite functions $\psi \in \mathcal{SN}(K^\wedge)$ stated in the theorem is called the *Schoenberg correspondence*. Given a convolution semigroup $(\mu_t)_{t\ge 0}$ in $\mathcal{M}_+^{(1)}(K)$ the associated function $\psi \in \mathcal{SN}(K^\wedge)$ is said to be the *exponent* of $(\mu_t)_{t\ge 0}$.

The *proof* of (ii) of the theorem is based on the fact that for $\psi \in \mathcal{SN}(K^\wedge)$ and $\lambda > 0$ the fraction $(\lambda + \psi)^{-1}$ is a well-defined continuous function on supp $\pi \cup \{\mathbb{1}\}$ and any of its (measurable) extensions to $K^\wedge$ belongs to $\mathcal{SN}(K^\wedge)$.

The Schoenberg correspondence for hypergroups has been the theme of various publications, e.g. [44] and [9]. In [57] a restricted version of the Schoenberg duality has been proved.

Theorem 3.4. *Any convolution semigroup* $(\mu_t)_{t\ge 0}$ *in* $\mathcal{M}_+^{(1)}(K)$ *with bounded exponent* $\psi \in \mathcal{SN}(K^\wedge)$ *is a Poisson semigroup* $(\pi_t(\mu, m))_{m\ge 1}$ *for some defining measure* $\mu \in \mathcal{M}_+^b(K)$ *and parameter* $m \ge \|\mu\|$, *and*

$$\psi = m - \hat{\mu}\,.$$ □

Corollary 3.5. *On a discrete hypergroup* K *any convolution semigroup in* $\mathcal{M}_+^{(1)}(K)$ *is a Poisson semigroup.* □

This follows from the fact that on the compact dual $K^\wedge$ of K the exponent of any convolution semigroup in $\mathcal{M}_+^{(1)}(K)$ is necessarily bounded.

A statement similar to that of the corollary is contained in [21].

4. Transience of convolution semigroups

Let $(\mu_t)_{t\ge 0}$ be a convolution semigroup in $\mathcal{M}_+^{(1)}(K)$ with corresponding resolvent family $(\rho_\lambda)_{\lambda>0}$. For every $f \in \mathcal{K}_+(K)$ we have

$$\lim_{\lambda\to 0} \rho_\lambda(f) = \int_0^\infty \mu_t(f)\,dt \le \infty\,.$$

If this limit is $< \infty$ for all $f \in \mathcal{K}_+(K)$, then it defines a measure $\kappa \in \mathcal{M}_+(K)$. In this case $(\mu_t)_{t\geq 0}$ is said to be *transient*, and the measure κ is called the *potential kernel* of $(\mu_t)_{t\geq 0}$. $(\mu_t)_{t\geq 0}$ is said to be *recurrent* if it is not transient.

We note that the Poisson semigroup with defining measure $\mu \in \mathcal{M}_+^{(1)}(K)$ is transient if and only if

$$\sum_{n\geq 0} \mu^n < \infty \quad \text{(in the sense of } \mathcal{T}_v),$$

and in this case the corresponding kernel is *elementary* of the form

$$\kappa = \sum_{n\geq 0} \mu^n .$$

We recall that a measure $\mu \in \mathcal{M}_+(K)$ is called *shift-bounded* if $\mu * f \in \mathcal{C}^b(K)$ for all $f \in \mathcal{K}(K)$, and it is said to *vanish at infinity* if $\mu * f \in \mathcal{C}^0(K)$ for all $f \in \mathcal{K}(K)$.

Theorem 4.1. *The potential kernel of a transient convolution semigroup in* $\mathcal{M}_+^{(1)}(K)$ *is shift-bounded.* □

A *proof* of this theorem has been given in [23], where yet another approach to positive definite functions and measures is taken.

In order to exhibit necessary and sufficient conditions for a convolution semigroup to be transient we first look at Poisson semigroups.

Theorem 4.2. *For any Poisson semigroup* $(\mu_t)_{t\geq 0}$ *in* $\mathcal{M}_+^{(1)}(K)$ *with defining measure* μ *we consider the following condition:*

(CF) *There exists a neighborhood* $V \in \mathfrak{V}_{1\!\!1}(K^\wedge)$ *such that*

$$\overline{\lim_{t\uparrow 1}} \int_V \operatorname{Re} \frac{1}{1 - t\hat{\mu}}\, d\omega_{K^\wedge} < \infty .$$

(i) *If* $(\mu_t)_{t\geq 0}$ *is transient, then* (CF) *holds.*
(ii) *If* K *is a strong hypergroup and* (CF) *holds, then* $(\mu_t)_{t\geq 0}$ *is transient.* □

We note that condition (CF) implies that for the exponent $\psi = 1 - \hat{\mu}$ of $(\mu_t)_{t\geq 0}$

$$\operatorname{Re} \frac{1}{\psi} \in L^1_{\text{loc}}(K^\wedge, \pi) .$$

The *proof* of the *Chung-Fuchs criterion* for hypergroups is due to L. GALLARDO and O. GEBUHRER who in [22] also show that for discrete hypergroups a Poisson semigroup in $\mathcal{M}^1(K)$ is transient if and only if (CF) holds. In addition these authors provide the following:

4.3. Application based on previous work on the central limit theorem for compact symmetric pairs (G, H) [14] and on ideas developed in order to characterize recurrent Lie groups [4].

A convolution semigroup $(\mu_t)_{t\geq 0}$ in $\mathcal{M}_+^{(1)}(K)$ will be called *adapted* if the closed subhypergroup of K generated by the set $\bigcup_{t>0} \operatorname{supp}(\mu_t)$ equals K.

A hypergroup K is said to be *transient* if every adapted Poisson semigroup in $\mathcal{M}_+^{(1)}(K)$ is transient.

In the spirit of this definition the hypergroup $(G /\!/ H)^\wedge$ of Example B1 is transient if and only if the dimension of the tangent space attached to the homogeneous space G/H at the point eH is ≥ 3. In particular the dual of SU(2) listed in B1.1 is transient. See [27] for further information.

Theorem 4.4. *For a hypergroup K we consider the following condition*

(M) *For every $f \in \mathcal{K}_+(K)$ there exists a $g \in \mathcal{K}(K^\wedge)$ such that $f \leq \check{g}$.*

(i) *Let $(\mu_t)_{t\geq 0}$ be a convolution semigroup in $\mathcal{M}_+^{(1)}(K)$ with exponent $\psi \in \mathcal{SN}(K^\wedge)$. If $(\mu_t)_{t\geq 0}$ is transient, then* $\operatorname{Re} \frac{1}{\psi} \in L^1_{\text{loc}}(K^\wedge, \pi)$.

(ii) *Let condition* (M) *be satisfied and let ψ be a function in $\mathcal{SN}(K^\wedge)$ with associated convolution semigroup $(\mu_t)_{t\geq 0}$ in $\mathcal{M}_+^{(1)}(K)$. If $\frac{1}{\psi} \in L^1_{\text{loc}}(K^\wedge, \pi)$, then $(\mu_t)_{t\geq 0}$ is transient.* □

The *proofs* of this and the next theorem are due to W. R. BLOOM and the author [9], [12]. The reformulation of the statements in terms of strictly negative definite functions has been suggested by M. VOIT.

We note that discrete hypergroups, compact hypergroups, strong hypergroups, and dual hypergroups (of commutative hypergroups) fulfill condition (M).

4.5. Examples of transient convolution semigroups are the adapted ones on polynomial hypergroups of type D having the Kunze-Stein property, on polynomial hypergroups of types D1, D4 and D5 all of them subject to the restriction $\alpha > 0$.

Positive definite measures in $\mathcal{M}(K)$ are defined for hypergroups in complete analogy to groups. The set of all such measures forms a $\mathcal{T}_v$-closed convex cone $\mathcal{M}_p(K)$ in $\mathcal{M}(K)$ which is stable with respect to involution and conjugation. Clearly $\varphi \in \mathcal{C}(K)$ is the ω_K-density of a measure in $\mathcal{M}_p(K)$ if and only if φ is a (bounded) positive definite function on K. We shall need the fact that the set $\mathcal{M}_p^b(K) := \{\mu \in \mathcal{M}_p(K) : \mu * f * f^\sim \in \mathcal{C}^b(K)$ for all $f \in \mathcal{K}(K)\}$ is $\mathcal{T}_v$-closed in $\mathcal{M}^b(K)$.

Theorem 4.6. *Let K be a Pontrjagin hypergroup and let $(\mu_t)_{t\geq 0}$ be a convolution semigroup in $\mathcal{M}_+^{(1)}(K)$ with exponent $\psi \in \mathcal{SN}(K^\wedge)$. If $\frac{1}{\psi} \in L^1_{\text{loc}}(K^\wedge, \omega_{K^\wedge})$, then*
(1) $\frac{1}{\psi}\omega_{K^\wedge} \in \mathcal{M}_p(K^\wedge)$,
(2) $(\mu_t)_{t\geq 0}$ *is transient, and*
(3) *the potential kernel κ of $(\mu_t)_{t\geq 0}$ is given by*

$$\kappa = \left(\frac{1}{\psi} \cdot \omega_{K^\wedge}\right)^\vee .$$

(4) *If, in addition, $(\mu_t)_{t\geq 0}$ is symmetric then $\kappa \in \mathcal{M}_p(K)$, and the measures κ and $\frac{1}{\psi} \cdot \omega_{K^\wedge}$ are shift-bounded.* □

In the example A of an Abelian group K Theorem 4.4 can be strengthened to the statement that a convolution semigroup $(\mu_t)_{t\geq 0}$ in $\mathcal{M}^1(K)$ with exponent ψ is transient if and only if Re $\frac{1}{\psi} \in L^1_{\text{loc}}(K^\wedge, \omega_{K^\wedge})$. The crucial step that local integrability of Re $\frac{1}{\psi}$ implies transience of $(\mu_t)_{t\geq 0}$ has been proved for $K := \mathbb{R}$ only in 1969 by D. S. Ornstein [45]. Later, in 1971 S. C. Port and C. J. Stone presented an elaborate probabilistic demonstration in their paper [47] based on previous work [46], for arbitrary Abelian locally compact groups K with a countable basis of their topology. The first analytic treatment, mainly potential theoretic, for compactly generated Abelian groups, is due to M. Itô [37]. For symmetric convolution semigroups an analytic proof is contained in [6]. In this special case the full analogy to groups can be established for arbitrary hypergroups, by a method to be introduced in the next section.

In contrast to the group case the renewal behavior of convolution semigroups on a hypergroup K can be described solely in terms of the structure of K.

Theorem 4.7. *Let K be a Kunze-Stein hypergroup and let χ_0 be the unique positive character in* supp π. *Then any adapted convolution semigroup $(\mu_t)_{t\geq 0}$ in $\mathcal{M}_+^{(1)}(K)$*
(i) *is transient,*
(ii) *its potential kernel κ vanishes at infinity,*
(iii) $\chi_0 \cdot \kappa \in \mathcal{M}^b(K)$ *with*

$$\|\chi_0 \cdot \kappa\| = \frac{-1}{ln\|\chi_o \cdot \mu_t\|}, \quad \textit{and}$$

(iv) $\kappa * f \in L^2(K, \omega_K)$ *for all* $f \in L^2(K, \omega_K)$. □

For the special case of Poisson semigroups more information is available.

Theorem 4.8. *Let K and χ_0 be given as in Theorem 4.7. Then for any measure $\mu \in \mathcal{M}^1(K)$ such that the closed subhypergroup of K generated by* supp μ *coincides with K*

(i) *the Poisson semigroup $(\mu_t)_{t\geq 0}$ with defining measure μ is transient,*

(ii) *for the spectral radius $\rho(\mu)$ of the convolution operator T^μ on $L^2(K, \omega_K)$ one has*

$$\rho(\mu) = \hat{\mu}(\chi_0) < 1 ,$$

(iii) *the potential kernel κ of $(\mu_t)_{t\geq 0}$ vanishes at infinity,*

(iv) *$\chi_0 \cdot \kappa \in \mathcal{M}^b(K)$ with*

$$\|\chi_0 \cdot \kappa\| = \frac{1}{1 - \|\chi_0 \cdot \mu\|}, \quad \textit{and}$$

(v) *$\kappa * f \in L^2(K, \omega_K)$ for all $f \in L^2(K, \omega_K)$, hence*

(vi) *the Poisson equation*

$$\varphi - \varphi * \mu = f$$

has a unique solution $\varphi := \kappa * f \in L^2(K, \omega_K)$ for any $f \in L^2(K, \omega_K)$. □

Theorems 4.7 and 4.8 are essentially due to O. Gebuhrer [26] (see also [23]). New proofs have been supplied by M. Voit in [56].

In studying

4.9. Examples of transient hypergroups it appears useful to know that a hypergroup K is in fact transient if and only if every Poisson semigroup in $\mathcal{M}^1(K)$ with a symmetric defining measure with compact support is transient ([4]).

4.9.1. A *polynomial hypergroup* $\mathbb{Z}_+$ (of type D) is transient if and only if the elementary Poisson semigroup with defining measure ϵ_1 is transient, and this is true if and only if

$$\int_{-1}^{1} \frac{1}{1-x}\, \pi(dx) < \infty .$$

We obtain once again that the Arnaud-Dunau hypergroups (of type D7.1) for $a > 2$ are transient ([20]).

The dual polynomial hypergroups I (of type F) are clearly recurrent. In fact, any convolution semigroup $(\mu_t)_{t\geq 0}$ in $\mathcal{M}_+^{(1)}(I)$ with exponent ψ is transient if and only if $\psi(\mathbb{1}) > 0$, and this excludes $(\mu_t)_{t\geq 0}$ to be in $\mathcal{M}^1(I)$.

4.9.2. A *Chébli-Trimèche hypergroup* $\mathbb{R}_+$ (of type H1) is transient if and only if

$$\int_0^1 \frac{1}{x}\,\pi(dx) < \infty\,.$$

In the case $\rho > 0$ (Kunze-Stein) every adapted convolution semigroup in $\mathcal{M}_+^{(1)}(\mathbb{R}_+)$ is transient, so $\mathbb{R}_+$ is a transient hypergroup. In the case $\rho = 0$ (Godement) the above integrability condition becomes a nontrivial restriction. If we specialize to type H1.1.1 hypergroups $\mathbb{R}_+$, then transience is available if and only if $d \geq 3$ (and recurrence for $d \leq 2$), a fact that is consistent with the behavior of the Brownian semigroup on $\mathbb{R}^d$ for $d \geq 1$.

There is a dream result in the case that K is a compactly generated Abelian group which can be assumed to be of the form $\mathbb{R}^d \times G_1$, with $d \geq 0$ and a discrete group G_1. Then K is transient if and only if $d + \operatorname{rank}(G_1) \geq 3$. See Chapter 3 of the monograph [48] by D. Revuz and further references therein. We also mention that it has been proved in [3] that a (not necessarily Abelian) connected Lie group is transient if and only if it is of polynomial growth of order ≥ 3.

5. Perfect kernels

J. Deny's theory of convolution kernels on an Abelian locally compact group initiated in [15] and [16] has been exposed systematically in Chapter III of the book [6] by C. Berg and G. Forst. Recently W. R. Bloom and the author have extended the main results on potential theoretic principles, excessive measures and fundamental families for strong hypergroups. Here we restrict ourselves to describing the characterization of potential kernels in terms of fundamental families [12].

We start with a result proved in [34].

Theorem 5.1. *Any transient convolution semigroup in $\mathcal{M}_+^{(1)}(K)$ vanishes at infinity in the sense that*

$$\mathcal{T}_v - \lim_{t\to\infty} \mu_t = 0\,. \qquad \square$$

From now on we assume that K is a strong hypergroup.

Let κ be any measure in $\mathcal{M}_+(K)$. A *fundamental family* associated with κ is a net $(\sigma_V)_{V\in\mathfrak{V}}$ of measures in $\mathcal{M}_+^{(1)}(K)$, where $\mathfrak{V}$ is a base of the system $\mathfrak{V}_e^c(K)$ of all compact neighborhoods of e in K, satisfying the following conditions

(a) $\sigma_V \in D_+(\kappa)$, $\sigma_V * \kappa \leq \kappa$ but $\sigma_V * \kappa \neq \kappa$.

(b) $\sigma_V * \kappa = \kappa$ on V^c.

(c) $\mathcal{T}_v - \lim_{n\to\infty} \sigma_V^n * \kappa = 0$.

$\kappa \in \mathcal{M}_+(K)$ is said to be a *perfect kernel* if there exists a fundamental family associated with κ.

As an *example* of a fundamental family we mention the constant net $(\mu_L)_{L\in\mathfrak{V}_e^c(K)}$ with $\mu_L := \mu \in \mathcal{M}_+^{(1)}(K)$ for all $L \in \mathfrak{V}_e^c(K)$. The elementary kernel

$$\kappa := \sum_{n\geq 0} \mu^n$$

is a perfect kernel with associated fundamental family $(\mu_L)_{L\in\mathfrak{V}_e^c(K)}$.

Let $\kappa \in \mathcal{M}_+(K)$ be a perfect kernel with associated fundamental family $(\sigma_V)_{V\in\mathfrak{V}}$. Then for every $V \in \mathfrak{V}$ we have

(1) $\kappa - \sigma_V * \kappa$ is a measure in $\mathcal{M}_+(K)\backslash\{0\}$ with compact support $\subset V$,

(2) there exists a unique $a_V \in \mathbb{R}_+^\times$ such that

$$\eta_V := a_V(\kappa - \sigma_V * \kappa) = a_V \kappa * (\epsilon_e - \sigma_V) \in \mathcal{M}^1(K).$$

Let $V \in \mathfrak{V}$ and $N \geq 1$ be fixed. The sequence

$$\left(\frac{1}{a_V}\sum_{n=0}^{N} \sigma_V^n\right)_{N\geq 1}$$

is $\mathcal{T}_v$-convergent, and with

$$\kappa_V := \frac{1}{a_V}\sum_{n\geq 0} \sigma_V^n$$

we obtain for $N \to \infty$ from condition (c) that

(3) $\kappa_V * \eta_V = \kappa$.

Since η_V has compact support, $\eta_V * f \in \mathcal{K}_+(K)$ whenever $f \in \mathcal{K}_+(K)$, and hence

$$\kappa * f = \kappa * (\eta_V * f) \in \mathcal{C}^b(K).$$

This shows

Theorem 5.2. *Any perfect kernel $\kappa \in \mathcal{M}_+(K)$ is shift-bounded.* □

Theorem 5.3. *For any measure $\kappa \in \mathcal{M}_+(K)$ the following statements are equivalent:*

(i) *κ is a perfect kernel.*

(ii) *κ is the potential kernel of a unique transient convolution semigroup.* □

Clearly the transient convolution semigroup corresponding to the elementary kernel of the above example, coincides with the Poisson semigroup with defining measure μ.

Application 5.4. Every transient convolution semigroup in $\mathcal{M}_+^{(1)}(K)$ is the $\mathcal{T}_v$-limit of a net of Poisson semigroups.

More precisely: Let $(\mu_t)_{t\geq 0}$ be a transient convolution semigroup in $\mathcal{M}_+^{(1)}(K)$ with potential kernel κ arising (by the theorem) from a fundamental family $(\sigma_V)_{V\in\mathfrak{V}}$ and a norming net $(a_V)_{V\in\mathfrak{V}}$ (as in the above discussion). Then

$$\mu_t = \mathcal{T}_w - \lim_{V\in\mathfrak{V}} \mu_t^V ,$$

where

$$\mu_t^V := \pi_t(a_V\sigma_V,\, a_V)$$

for all $t > 0$.

Examples of perfect kernels are at present only known for the groups $\mathbb{R}^d$, $d \geq 1$. In particular the Newton kernel and the Riesz kernel of order $\alpha \in]0,2]$ are perfect. The associated fundamental families are given in[6].

6. Invariant Dirichlet forms

In this section we shall describe the method of invariant Dirichlet forms over a hypergroup K and their application to the transience of convolution semigroups. For Abelian groups and symmetric Dirichlet forms the theory was initiated by J. Deny much prior to his Stresa lectures [17]. A first discussion of non-symmetric Dirichlet forms is due to C. Berg and G. Forst in [5].

Let $(\mu_t)_{t\geq 0}$ be a convolution semigroup in $\mathcal{M}_+^{(1)}(K)$. On any of the Banach spaces $E = \mathcal{C}^0(K)$, $\mathcal{C}^u(K)$ and $L^2(K,\omega_K)$ $(\mu_t)_{t\geq 0}$ induces a strongly continuous contraction semigroup $(P_t)_{t\geq 0}$ of positive operators on E by $P_t := T^{\mu_t}$ for all $t \geq 0$. Clearly, $(P_t)_{t\geq 0}$ is *translation invariant* in the sense that $P_tE \subset E$ and $T^xP_t = P_tT^x$ holds for all $x \in K\,(t \geq 0)$. Moreover if $(\mu_t)_{t\geq 0}$ is a convolution semigroup in $\mathcal{M}^1(K)$, then $(P_t)_{t\geq 0}$ is *Markovian* in the sense that

$$\sup\{P_tf\colon f \in E,\ 0 \leq f \leq 1\} = 1$$

for all $t \geq 0$. Finally, if $(\mu_t)_{t\geq 0}$ is symmetric and $E = L^2(K,\omega_K)$, then $(P_t)_{t\geq 0}$ is *selfadjoint* in the sense that P_t is a selfadjoint operator for every $t \geq 0$. The converse of these statements are contained in the following

Theorem 6.1. (Feller correspondence) *There is a one-to-one correspondence between convolution semigroups $(\mu_t)_{t\geq 0}$ in $\mathcal{M}_+^{(1)}(K)$ and translation invariant, strongly continuous semigroups $(P_t)_{t\geq 0}$ of positive contraction operators in E which is given by $P_t = T^{\mu_t}$ for all $t \geq 0$. In particular $(\mu_t)_{t\geq 0}$ is in $\mathcal{M}^1(K)$ if and only if $(P_t)_{t\geq 0}$ is Markovian, and for $E = L^2(K, \omega_K)$, $(\mu_t)_{t\geq 0}$ is symmetric if and only if $(P_t)_{t\geq 0}$ is selfadjoint.* □

The *generator* of a convolution semigroup $(\mu_t)_{t\geq 0}$ in $\mathcal{M}_+^{(1)}(K)$ is introduced as the infinitesimal generator A with domain $D(A)$ of the contraction semigroup $(P_t)_{t\geq 0}$ on E which corresponds to $(\mu_t)_{t\geq 0}$ by the theorem.

Let now K be a strong hypergroup. For $i = 0, u, 2$ A_i denotes the generator of $(\mu_t)_{t\geq 0}$ considered as an operator with domain $D(A_i)$ on $\mathcal{C}^0(K)$, $\mathcal{C}^u(K)$ and $L^2(K, \omega_K)$ respectively. As an application of Theorem 5.3 one obtains that $\mathcal{K}(K) \cap D_0 \cap D_2$ is a dense subspace of $\mathcal{K}(K)$, $\mathcal{C}^u(K)$ and $L^2(K, \omega_K)$ ([35]).

We quote two facts about generators:

(1) The implications

$$\operatorname{supp}(A_i f) \subset \operatorname{supp} f$$

for all $f \in D_i$ are equivalent whenever i runs through $\{0, u, 2\}$.

(2) Let $\psi \in \mathcal{SN}(K^\wedge)$ be the exponent of $(\mu_t)_{t\geq 0}$. Then

$$(A_2 f)^\wedge = -\hat{f}\psi$$

for all $f \in D_2 = \{g \in L^2(K, \omega_K) \colon \hat{g}\psi \in L^2(K^\wedge, \omega_{K^\wedge})\}$.

Let (β, V) be a positive closed form on $L^2(K, \omega_K)$ with corresponding strongly continuous resolvent $(R_\lambda)_{\lambda>0}$ and strongly continuous contraction semigroup $(P_t)_{t\geq 0}$. (β, V) is said to be *translation invariant* if $T^x R_\lambda = R_\lambda T^x$ for all $x \in K$ $(\lambda > 0)$. A translation invariant positive closed form (β, V) on $L^2(K, \omega_K)$ such that the unit contraction T_I satisfies $T_I V \subset V$, is called a (translation invariant) *Dirichlet form* on $L^2(K, \omega_K)$ if one of the subsequent equivalent conditions hold:

(a) T_I operates with respect to β, i.e.,

$$\operatorname{Re}\, \beta(f + T_I f,\ f - T_I f) \geq 0$$

for all $f \in V$.

(b) λR_λ is sub-Markovian for all $\lambda > 0$.

Theorem 6.2. (Deny correspondence) *There is a one-to-one correspondence between translation invariant Dirichlet forms (β, V) on $L^2(K, \omega_K)$ and functions $\psi \in \mathcal{SN}(K^\wedge)$ satisfying the condition*

$$|\operatorname{Im} \psi| \leq c \operatorname{Re} \psi$$

valid for some $c > 0$, which is given by

$$\beta(f,g) = \int_{K^\wedge}^{\cdot} \hat{f}\ \bar{\hat{g}}\ \psi\, d\omega_{K^\wedge}$$

for all $f, g \in V := \{h \in L^2(K, \omega_K) \colon \int_{K^\wedge} |\hat{h}|^2 \mathrm{Re}\,\psi\, d\omega_{K^\wedge} < \infty\}$. □

A *proof* of this theorem based on the one available in the group case has been given by W. R. Bloom and the author in [10].

Let (β, V) be a translation invariant Dirichlet form on $L^2(K, \omega_K)$ with corresponding strongly negative definite function $\psi \in \mathcal{SN}(K^\wedge)$. Then the strongly negative function corresponding to the hermitian part (α, V) of (β, V) is $\mathrm{Re}\ \psi$. This remark leads us to the *discussion of symmetric translation invariant Dirichlet forms* (β, V) on $L^2(K, \omega_K)$ which by definition arise from selfadjoint contraction semigroups $(P_t)_{t \geq 0}$ and hence by the Feller correspondence 6.1 from symmetric convolution semigroups $(\mu_t)_{t \geq 0}$ in $\mathcal{M}_+^{(1)}(K)$.

For any $t > 0$ and $f \in L^2(K, \omega_K)$ we look at

$$\beta_t(f) := \frac{1}{t} \langle f - \mu_t * f, f \rangle .$$

Applying the Plancherel-Levitan Theorem 1.6 and the Schoenberg correspondence 3.3 we obtain

$$\beta_t(f) = \frac{1}{t} \int_{K^\wedge} |\hat{f}|\,(1 - e^{-t\psi})\, d\pi \geq 0 .$$

Clearly the generator A_2 of $(\mu_t)_{t \geq 0}$ satisfies $-A_2 \geq 0$ and $D(-A) \subset L^2(K, \omega_K)$. Since $(\mu_t)_{t \geq 0}$ was assumed to be symmetric, $-A_2$ is selfadjoint. We define

$$V := D(\sqrt{-A_2}) \subset L^2(K, \omega_K)$$

and

$$\beta(f) := \| \sqrt{-A_2}\, f \|_2^2$$

for all $f \in V$. It is evident that $\beta_t(f) \leq \beta(f)$ for all $f \in V$, $t > 0$ and that

$$\lim_{t \to 0} \beta_t(f) = \sup_{t > 0} \beta_t(f) = \beta(f)$$

for all $f \in V$. Consequently

$$\beta(f) = \lim_{t \to 0} \int_{K^\wedge} |\hat{f}|^2 \frac{1}{t} (1 - e^{-t\psi})\, d\pi$$

$$= \int_{K^\wedge} |\hat{f}|^2 \psi \, d\pi \,,$$

which means that (β, V) is a symmetric translation invariant Dirichlet form on $L^2(K, \omega_K)$.

One notes that the previous discussion remains valid for any commutative hypergroup K. Hence, for symmetric convolution semigroups in $\mathcal{M}_+^{(1)}(K)$ we obtain the transience criterion 4.4 without any further restriction on K.

Theorem 6.3. *Let K be an arbitrary commutative hypergroup and $(\mu_t)_{t \geq 0}$ a symmetric convolution semigroup in $\mathcal{M}_+^{(1)}(K)$ with exponent $\psi \in \mathcal{SN}(K^\wedge)$. The following statements are equivalent:*
(i) $(\mu_t)_{t \geq 0}$ *is transient.*
(ii) $\frac{1}{\psi} \in L^1_{\mathrm{loc}}(K^\wedge, \pi)$. □

We sketch the *proof* of this result which is due to L. Gallardo and O. Gebuhrer [24], in order to indicate the use of Deny's method. The implication (i) $\Longrightarrow$ (ii) is clear. From (ii) follows that β defined above is a norm on V. Introducing $\hat{\beta}$ on $\hat{V} = \{\hat{v} : v \in V\}$ by $\hat{\beta}(\hat{v}) := \beta(v)$ for all $v \in V$ we obtain a norm $\sqrt{\hat{\beta}}$ on $\hat{V}$. Let $\mathcal{H}_{\hat{V}}$ denote the Hilbert completion of $\hat{V}$ with respect to $\sqrt{\hat{\beta}}$. Then $\mathcal{H}_{\hat{V}} \cong L^2(K^\wedge, \psi \cdot \pi) \subset L^1_{\mathrm{loc}}(K^\wedge, \pi)$, and the *energy*

$$I(f) := \int_0^\infty \mu_t(f * f^-) \, dt$$

is finite for any $f \in L^2(K, \omega_K)$ such that $\hat{f}$ is a bounded π-measurable function on $K^\wedge$ with compact support. But this implies that for all $f \in \mathcal{K}_+(K)$

$$\kappa(f) = \int_0^\infty \mu_t(f) \, dt < \infty \,.$$

As an immediate *application* we get that a Poisson semigroups with symmetric defining measure $\mu \in \mathcal{M}^1(K)$ is transient if and only if

$$\frac{1}{1 - \hat{\mu}} \in L^1_{\mathrm{loc}}(K^\wedge, \pi) \,.$$

This criterion extends to all Poisson semigroups in $\mathcal{M}^1(K)$ provided K is hermitian. In particular it applies to all one-dimensional hypergroups $\mathbb{R}_+$ and I and yields some of the transience results in the examples of Section 4.

Translation invariant Dirichlet forms on the strong hypergroups of type F and type G have been calculated in terms of (the canonical form of) the corresponding strongly negative definite functions in [10].

7. Convolution semigroups of local type

The purpose of this section is to report on recent work by the author [35] on convolution semigroups admitting local generators. For hermitian hypergroups this study can be related to their canonical form in the sense of LÉVY and KHINTCHINE.

We assume K to be a strong hypergroup and $(\mu_t)_{t\geq 0}$ a convolution semigroup in $\mathcal{M}_+^{(1)}(K)$ with corresponding resolvent family $(\rho_\lambda)_{\lambda>0}$ and exponent ψ. For $i = 0, u, 2$ let A_i denote the generator of $(\mu_t)_{t\geq 0}$ with domain D_i. A slight extension of Proposition 3.3 of [44] is contained in the following

Theorem 7.1. *There exists a measure $\eta \in \mathcal{M}_+(K^\times)$ satisfying*

$$\lim_{t\to 0} \frac{1}{t} \int f \, d\mu_t = \int f \, d\eta$$

for every $f \in \mathcal{C}^b(K)$ such that supp $f \subset K^\times := K\backslash\{e\}$. *The measure η is uniquely determined by the equality*

$$(1 - \check{\sigma}) \cdot \eta = \mathrm{Res}_{K^\times}\eta_\sigma$$

valid for all symmetric measures $\sigma \in \mathcal{M}^1(K^\wedge)$ with compact support, where η_σ denotes the unique measure in $\mathcal{M}_+^b(K)$ satisfying $\hat{\eta}_\sigma = \psi * \sigma - \psi$. □

η is called the *Lévy measure* of $(\mu_t)_{t\geq 0}$.

Theorem 7.2. *For any measure $\eta \in \mathcal{M}_+(K^\times)$ the following statements are equivalent:*

(i) $\eta = \mathcal{T}_v - \lim\limits_{t\to 0} \frac{1}{t}\, \mathrm{Res}_{K^\times}\mu_t$.

(ii) $\eta = \mathcal{T}_v - \lim\limits_{\lambda\to\infty} \mathrm{Res}_{K^\times}\lambda^2\rho_\lambda$.

(iii) $\eta(f) = A_0 f^-(e)$ *for all* $f \in \mathcal{K}(K)$ *with* $f^- \in D_0$, supp $f \subset K^\times$.

(iv) $\eta(f) = A_u f^-(e)$ *for all* $f \in \mathcal{C}^u(K)$ *with* $f^- \in D_u$, supp $f \subset K^\times$.

(v) $\eta(f^- * \bar{g}\ ) = \langle A_2 f, g\rangle$ *for all* $f, g \in \mathcal{K}(K)$, $f \in D_2$ *with* supp $(f^- * \bar{g}\ ) \subset K^\times$. □

Here f^- denotes the function $x \to f(x^-)$ on K.

A convolution semigroup $(\mu_t)_{t\geq 0}$ in $\mathcal{M}_+^{(1)}(K)$ is said to be of *local type* if any of the conditions

$$\operatorname{supp}(A_i f) \subset \operatorname{supp} f ,$$

valid for every $f \in D_i$, $i = 0, u, 2$, holds.

Theorem 7.3. *The following statements are equivalent:*
(i) $(\mu_t)_{t\geq 0}$ *is of local type.*
(ii) $\lim_{t\to 0} \frac{1}{t}\, \mu_t(U^c) = 0$ *for all* $U \in \mathfrak{V}_e(K)$.
(iii) $\eta \equiv 0$.

If, in addition K *satisfies property (F) of* **[43]** *and* η *is symmetric, then we have also equivalence to*
(iv) *There exists a triplet* (c, ℓ, q) *consisting of a number* $c \geq 0$, *a homomorphism* $\ell\colon K^{\wedge} \to \mathbb{R}$, *and a nonnegative quadratic form* $q\colon K^{\wedge} \to \mathbb{R}$ *(both in the sense of* **[43]***) such that*

$$\psi + i\ell + q . \qquad \square$$

The *proof* of this theorem relies on the fact that for given relatively compact open subsets U, W of a hypergroup K with $\bar{U} \subset V$ there exists a function $f \in D_0 \cap D_2$ satisfying $0 \leq f \leq 1$, $f = 1$ on U, and $f = 0$ on V^c.
We add a

Remark 7.4. relating locality to transience. Let $(\sigma_V)_{V\in\mathfrak{V}}$ denote a fundamental family associated with the potential kernel κ of a transient convolution semigroup $(\mu_t)_{t\geq 0}$ in $\mathcal{M}_+^{(1)}(K)$. Then

$$A_0 f = \lim_{V\in\mathfrak{V}} a_V(\sigma_V - \epsilon_e) * f$$

for all $f \in D_0$, and

$$\eta = \mathcal{T}_v - \lim_{V\in\mathfrak{V}} a_V \operatorname{Res}_{K^\times}\sigma_V ,$$

where the $a_V > 0$ are chosen such that $a_V(\kappa - \sigma_V * \kappa) \in \mathcal{M}^1(K)$ for all $V \in \mathfrak{V}$. $(\mu_t)_{t\geq 0}$ is of local type if and only if there exists a fundamental family $(\sigma_V)_{V\in\mathfrak{V}}$ associated with κ such that $\sigma_V \subset V$ for all $V \in \mathfrak{V}$. $\square$

Examples 7.5.
7.5.1. Convolution semigroups of local type in hypergroups of *type A* are the Brownian semigroup on $\mathbb{R}^d$ which for $d \geq 3$ is transient, and the heat semigroup on $\mathbb{R}^{d+1}$ which is transient for all $d \geq 1$. The symmetric stable semigroup of order $\alpha \in]0, 2[$ on $\mathbb{R}^d$, however, is not of local type. See [6], 17.16 and 18.23

for details. For more general Abelian locally compact groups the exponents of convolution semigroups of local type can be calculated in terms of their canonical representation with vanishing Lévy measure. See [30], Chapter VI.

7.5.2. On *type C1* hypergroups $\mathbb{R}_+$ convolution semigroups of local type have been exhibited by calculating the quadratic forms on $\mathbb{R}_+$, in [19].

More generally this has been done for

7.5.3. hypergroups of *type H.1.1* in [39].

On discrete hypergroups there are no convolution semigroups of local type, by Corollary 3.5.

7.5.4. for *type F* hypergroups I quadratic forms on $\mathbb{Z}_+$ have been calculated in [44] and hence convolution semigroups of local type are available on I. In particular for $\alpha \geq \beta \geq -\frac{1}{2}$ the quadratic forms are of the form

$$g(n) = a\,\frac{n(n + \alpha + \beta + 1)}{\alpha + \beta + 2}$$

for all $n \in \mathbb{Z}_+$ $(a \geq 0)$.

Quadratic forms have also been determined for hypergroups of type D2, D3, D4 and D7.1 in [43].

7.5.5. In the case of *type B* hypergroups the lack of a canonical representation sets a limit to the previous method of determining convolution semigroups of local type. In special situations results are available. See [31].

7.5.6. It can be deduced from [2] that a convolution semigroup $(\mu_t)_{t \geq 0}$ on the *type E* hypergroup $\mathbb{D}$ with symmetric Lévy measure is local if and only if the exponent ψ of $(\mu_t)_{t \geq 0}$ is of the form $\psi = c + q$ with

$$q(m, n) := a(m - n)^2 + b\left(m + n + \frac{2mn}{\alpha + 1}\right)$$

whenever $(m, n) \in \mathbb{Z}_+^2$.

8. Embedding infinitely divisible probability measures into convolution semigroups

This is a profound problem of structural probability theory. It has been thoroughly studied on groups. See [30], Chapter III. The advances achieved up to 1985 are contained in [33], and more recent developments are described in [36]. Despite of extensive investigations and farreaching results the problem has remained widely open for arbitrary locally compact groups. On the other hand, for Abelian groups appealing results are available since 1963. Some of these results can be established within the framework of a commutative hypergroup K, as was indicated by L. GALLARDO and O. GEBUHRER in [21] and fully established by M. VOIT in [55].

In order to report on this recent work we need to introduce *algebraic convolution semigroups* $(\mu_t)_{t>0}$ in $\mathcal{M}^1(K)$ defined solely by the homomorphism property $\mu_t * \mu_s = \mu_{t+s}$ for all $t, s > 0$. An algebraic convolution semigroup $(\mu_t)_{t>0}$ in $\mathcal{M}^1(K)$ is said to be *continuous* if the homomorphism $t \to \mu_t$ from $\mathbb{R}_+$ into $\mathcal{M}^1(K)$ is $\mathcal{T}_w$-continuous. In this case $\mu_0 := \mathcal{T}_w - \lim_{t\to 0} \mu_t$ exists, is an idempotent in $\mathcal{M}^1(K)$ and hence of the form $\mu_0 = \omega_H$ for some compact subhypergroup H of K. If we wish to be more precise we will speak in this case of an *H-continuous* convolution semigroup. The convolution semigroups introduces in 1.9 coincide with the $\{e\}$-continuous ones.

The following fact prepares the subsequent theorem. Let $(\mu_t)_{t>0}$ be an algebraic convolution semigroup in $\mathcal{M}^1(K)$. If $\mathcal{T}_w - \lim_{t\to 0} \mu_t := \mu_0$ exists, then $\mu_0 * \mu_s = \mu_s$ for all $s \geq 0$, μ_0 is of the form ω_H for some compact subhypergroup H of K, and $(\mu_t)_{t>0}$ is H-continuous.

Theorem 8.1. *Let K be a hypergroup satisfying one of the following independent conditions:*

(a) *K is a Godement hypergroup.*

(b) *The identity character $\mathbb{1}$ of K is not isolated in $K^\wedge$, and there exists an open neighborhood V of $\mathbb{1}$ such that each $\chi \in V\backslash\{\mathbb{1}\}$ belongs to $\mathcal{C}^0(K)$.*

Then any symmetric algebraic convolution semigroup $(\mu_t)_{t\geq 0}$ in $\mathcal{M}^1(K)$ is H-continuous for some compact subhypergroup H of K. □

The result which is proved in [11] compares with a similar one, for hermitian hypergroups without further assumptions, given in [55].

A measure $\mu \in \mathcal{M}^1(K)$ is said to be *H-embeddable* if there exists an H-continuous convolution semigroup $(\mu_t)_{t\geq 0}$ in $\mathcal{M}^1(K)$ such that $\mu_1 = \mu$.

The set of all H-embeddable measures will be abbreviated by $\mathcal{E}_H(K)$. We also put

$$\mathcal{E}(K) := \cup \{\mathcal{E}_H(K) : H \text{ a compact subhypergroup of } K\}.$$

Clearly, every measure $\mu \in \mathcal{E}(K)$ is *infinitely divisible* in the sense that for every $n \geq 1$ there exists an n-th root $\mu_n \in \mathcal{M}^1(K)$ such that $\mu_n^n = \mu$. Thus, denoting the set of all infinitely divisible measures in $\mathcal{M}^1(K)$ by $\mathcal{I}(K)$, we surely have $\mathcal{E}(K) \subset \mathcal{I}(K)$. The inverse inclusion is in general not even true for arbitrary Abelian groups. The *embedding property* $\mathcal{I}(K) \subset \mathcal{E}(K)$ is true, however, for the groups $\mathbb{Z}$, $\mathbb{T}$, $\mathbb{R}$ and their finite products, also for certain classes of non-Abelian Lie groups, and for commutative hypergroups, under additional hypotheses.

Theorem 8.2. (Continuous embedding) *Let K be a commutative hypergroup*
(i) *If $K^\wedge$ is pathwise connected, then*

$$\mathcal{I}(K) \subset \mathcal{E}_{\{e\}}(K).$$

(ii) *If K is hermitian, then in general*

$$\mathcal{I}(K) \subset \mathcal{E}(K),$$

and if additionally $K^\wedge$ contains no proper compact subhypergroup, then again

$$\mathcal{I}(K) \subset \mathcal{E}_{\{e\}}(K).$$

The embedding convolution semigroups in both (i) and (ii) are necessarily unique. □

It turns out that the *proof* of this theorem can be carried out in analogy to the Euclidean groups $\mathbb{R}^d$, $d \geq 1$, but the arguments are based on the theory presented in Section 3.

We note that statement (i) of the theorem becomes false if one drops the hypothesis of pathwise connectedness. In fact, consider the solenoidal group $\mathbb{Q}_d^\wedge$ which is connected but not pathwise connected. For every $q \in \mathbb{Q}_d$, $q \neq 0$ the measure $\epsilon_q \in \mathcal{I}(K)$ but $\notin \mathcal{E}(K)$.

There is an extensive theory of H-continuous convolution semigroups in $\mathcal{M}_+^{(1)}(K)$ developed in [9] which for strong hypergroups yields analogs of the resolvent and Schoenberg correspondences. On the attempt presented in [7] of studying embeddability for root compact hypergroups we reported in [32]. In connection with the embedding problem various problems of factorization of measures can be discussed. See the recent monograph [50]. There are however, only rather restricted results available. From [55] we quote this

Theorem 8.3. (Closedness of $\mathcal{I}(K)$) *Let K be a commutative hypergroup satisfying one of the following hypotheses:*
(a) *K is hermitian.*
(b) *K is second countable and has a pathwise connected dual.*

Then $\mathcal{I}(K)$ is $\mathcal{T}_w$-closed in $\mathcal{M}^1(K)$. □

Even for Abelian locally compact groups K the assertion of the theorem does not remain true without additional assumptions: If for example K admits a proper dense subgroup

$$H := \{x \in K\colon \text{ For every } n \geq 1 \text{ there is an } x_n \in K\colon x_n^n = x\},$$

then $\mathcal{I}(K)$ is not $\mathcal{T}_w$-closed in $\mathcal{M}^1(K)$.

Examples 8.4. of hypergroups admitting the embedding property are all one-dimensional hypergroups $\mathbb{R}_+$ and I, since they are hermitian, but also the polynomial hypergroups $\mathbb{Z}_+$, so in particular all hypergroups of types D1 to D7.

References

[1] Achour, A., and K. Trimèche, Opérateurs de translation généralisée associés à un opérateur différentiel singulier sur un intervalle borné, C. R. Acad. Sc. Paris, 288 (19 février 1979), Série A, 399–402

[2] Annabi, H., and K. Trimèche, Convolution généralisée sur le disque unité, C. R. Acad. Sc. Paris, 278 (2 janvier 1974), Série A, 21–24

[3] Baldi, P., Caractérisation des groupes de Lie connexes récurrents, Ann. L'Inst. Henri Poincaré, Section B, 17 (1981), 281–308

[4] Baldi, P., N. Lahoué, and J. Peyrière, Sur la classification des groupes récurrents, C. R. Acad. Sc. Paris, 285 (19 décembre 1977), Série A, 1103–1104

[5] Berg, C., and G. Forst, Non-symmetric translation invariant Dirichlet forms, Inventiones math. 21 (1973), 199–212

[6] —, "Potential Theory on Locally Compact Abelian Groups," Springer, New York etc., 1975

[7] Bloom, W. R., Infinitely divisible measures on hypergroups, in: Probability Measures on Groups VI, Springer Lecture Notes in Math. 928 (1982), 1–15

[8] Bloom, W. R., and H. Heyer, The Fourier transform for probability measures on hypergroups, Rendiconti di Matematica (2) Vol. 2, Serie VII (1982), 315–334

[9] —, Convolution semigroups and resolvent families of measures on hypergroups, Math. Z. 188 (1985), 449–474

[10] —, Non-symmetric translation invariant Dirichlet forms on hypergroups, Bull. Austral. Math. Soc. 36 (1987), 61–72

[11] —, Continuity of convolution semigroups on hypergroups, J. Theoretical Prob. 1 (1988), 271–286

[12] —, Characterisation of potential kernels of transient convolution semigroups on a commutative hypergroup, in: Probability Measures on Groups IX, Springer Lecture Notes in Math. 1379 (1989), 21–35

[13] Chébli, H., Opérateurs de translation généralisée et semi-groupes de convolution, in: Théorie du Potentiel et Analyse Harmonique, Springer Lectures Notes in Math. 404 1974, 33–59

[14] Clerc, J. L., and B. Roynette, Un théorème central limite, in: Analyse Harmonique sur les Groupes de Lie, Springer Lectures Notes in Math. 739 (1979), 122–131

[15] Deny, J., Familles fondamentales. Noyaux associés, Ann. L'Inst. Fourier (Grenoble), 3 (1951), 73–101

[16] Deny, J., Noyaux de convolution de Hunt et noyaux associés à une famille fondamentale, Ann. L'Inst. Fourier (Grenoble) 12 (1962), 643–667

[17] Deny, J., "Méthodes hilbertiennes en théorie du potentiel, Potential Theory" (C. I. M. E. 1 Ciclo, Stresa), Rome: Ed. Cremonese, 1970

[18] Dunkl, C. F., The measure algebra of a locally compact hypergroup, Trans. Amer. Math. Soc. 179 (1973), 331–348

[19] Faraut, J., and K. Harzallah, Distances hilbertiennes invariantes sur un espace homogène, Ann. L'Inst. Fourier (Grenoble) 24 (1974), 171–217

[20] Gallardo, L., Exemples d'hypergroupes transients, in: Probability Measures on Groups VIII, Springer Lecture Notes in Math. 1210 (1986), 68–76

[21] Gallardo, L., and O. Gebuhrer, Lois infiniment divisibles sur les hypergroupes commutatifs, discrets, dénombrables, in: Probability Measures on Groups VII, Springer Lecture Notes in Math. 1064 (1984), 116–130

[22] —, Un critère usuel de transience pour certains hypergroupes commutatifs. Application au dual d'un espace symétrique, in: Probabilités sur les Structures Géométriques, pp. 41–56. L'Université P. Sabatier, Toulouse, 1985

[23] —, Marches aléatoires sur les hypergroupes localement compacts et analyse harmonique commutative. Publ. de l'IRMA, Strasbourg, 1985

[24] —, Marches aléatoires et hypergroupes, Expo. Math. 5 (1987), 41–73

[25] Gebuhrer, O., "Analyse harmonique sur les espaces de Gelfand-Levitan et applications à la théorie des semigroupes de convolution", Publ. de l'IRMA, Strasbourg 1989

[26] —, Quelques propriétés du noyau potentiel d'une marche aléatoire sur les hypergroupes de type Kunze-Stein, in: Probability Measures on Groups VIII, Springer Lecture Notes in Math. 1210, (1986), 77–83

[27] Guivarc'h, Y., M. Keane, and B. Roynette, "Marches Aléatoires sur les Groupes de Lie," Springer Lecture Notes in Math. 624 (1977)

[28] Hartmann, L., R. W. Henrichs, and R. Lasser, Duals of orbit spaces in groups with relatively compact inner automorphism groups are hypergroups, Mh. Math. 88 (1979), 229–238

[29] Hauenschild, W., E. Kaniuth, and A. Kumar, Harmonic analysis on central hypergroups and induced representations, Pac. J. Math. 110 (1984), 83–112

[30] Heyer, H., "Probability Measures on Locally Compact Groups," Springer, 1977

[31] —, Convolution semigroups of probability measures on Gelfand pairs, Expo. Math. 1 (1983), 3–45

[32] —, Probability theory on hypergroups: a survey, in: Probability Measures on Groups VII, Springer Lecture Notes in Math. 1064 (1984), 481–550

[33] —, Recent contributions to the embedding problem for probability measures on a locally compact group, J. Multivar. Analysis 19, (1986), 119–131

[34] —, Convolution semigroups of measures on Sturm-Liouville structures, in: Analisi Armonica, Symposia Mathematica Vol. 39, pp. 131–162. Ist. Mat. F. Severi, Roma, 1986

[35] —, Convolution semigroups of local type on a commutative hypergroup, Hokkaido Math. J. 18 (1989),321–337

[36] —, Das Einbettungsproblem der Wahrscheinlichkeitstheorie: Leopold Schmetterers Beiträge zur strukturellen Wahrscheinlichkeitstheorie und neuere Entwicklungen, to appear

[37] Ito, M., Transient Markov convolution semigroups and associated negative definite functions, Nagoya Math. J. 92 (1983), 153–161. Remarks on that paper, Nagoya Math. J. 102 (1986), 181–184

[38] Jewett, R. I., Spaces with an abstract convolution of measures, Advances in Math. 18 (1975), 1–101

[39] Kingman, J. F. C., Random walks with spherical symmetry, Acta Math. 109 (1963), 11–53

[40] Lasser, R., Fourier-Stieltjes transforms on hypergroups, Analysis 2 (1982), 281–303

[41] —, Orthogonal polynomials and hypergroups, Rendiconti di Matematica, Serie VII, 3 (1983), 185–209

[42] —, Bochner theorems for hypergroups and their applications to orthogonal polynomial expansions, J. of Approx. Theory 37 (1983), 311–325

[43] —, On the Lévy-Hinčin formula for commutative hypergroups, in: Probability Measures on Groups VII, Springer Lecture Notes in Math. 1064 (1984), 298–308

[44] —, Convolution semigroups on hypergroups, Pac. J. Math. 127 (1987), 353–371

[45] Ornstein, D. S., Random Walks I, Trans. Amer. Math. Soc. 138 (1969), 1–43

[46] Port, S. C., and C. J. Stone, Potential theory of random walks on Abelian groups, Acta Math. 122 (1969), 19–114

[47] —, Infinitely divisible processes and their potential theory I, II, Ann. de l' Inst. Fourier (Grenoble) 21.2 (1971), 157–275; 21.4 (1971), 179–265

[48] Revuz, D., "Markov Chains," North-Holland Publ. Co., 1975

[49] Ross, K. A., Hypergroups and centers of measure algebras. Symposia Mathematica, 22 (1977), 189–203

[50] Rusza, I. Z., and G. J. Székely, "Algebraic Probability Theory," John Wiley & Sons, 1988

[51] Schwartz, A. L., Classification of one-dimensional hypergroups, Proc. Amer. Math. Soc. 103 (1988), 1073–1081

[52] Spector, R., Mesures invariantes sur les hypergroupes, Trans. Amer. Math. Soc. 239 (1978), 147–165

[53] Vogel, M., Spectral synthesis on algebras of orthogonal polynomial series, Math. Z. 194 (1987), 99–116

[54] —, Harmonic analysis and spectral synthesis in central hypergroups, Math. Ann. 281 (1988), 369–385

[55] Voit, M., "Positive Charaktere und ihr Beitrag zur Wahrscheinlichkeitstheorie auf kommutativen Hypergruppen", Dissertation, Technische Universität, München, 1987

[56] —, Positive characters on commutative hypergroups and some applications, Math. Z. 198 (1988), 405–421

[57] —, Negative definite functions on commutative hypergroups, in: Probability Measures on Groups IX, Springer Lecture Notes in Math. 1379 (1989), 376–388

[58] Zeuner, H., Properties of the $cosh$ hypergroup, in: Probability Measures on Groups IX, Springer Lecture Notes in Math. 1379 (1989), 425–434

[59] Zeuner, H., One-dimensional hypergroups, to appear in Advances of Math., 1989

Amenability of semigroups

Anthony To-Ming Lau

1. Introduction

The purpose of this article is to report on the recent development of certain aspects of amenability of semigroups since the fundamental paper of DAY (1957, [9]) and his survey article (1969, [11]).

The subject of amenability began in 1904 when LEBESGUE asked if the uniqueness of the Lebesgue integral is still preserved if the Monotone Convergence Theorem (which is equivalent to countable additivity) is replaced by finite addivity. BANACH (1923, [3]) showed that there exists a mean on the bounded real-valued functions on the integers which is invariant under all translations, i.e. the group of integers under addition is amenable. This contrasted the result of HAUSDORFF (1914, [43]) who showed that there does not exist any mean on the bounded real-valued functions on the sphere in three dimensions, which is invariant under all rotations. VON NEUMANN (1929, [98]) introduced and studied the class of discrete amenable groups. The term "amenable" was due to DAY (1950, [8]). Since then the theory has grown in many directions to include amenability of discrete and topological semigroups, locally compact groups, and Banach algebras (in particular C^*-algebras and von Neumann algebras).

Readers who are interested in the theory and applications on amenability of locally compact groups should read the classic of GREENLEAF (1969, [39]) and the very recent books (with up-to-date references) of PIER (1984, [88]) and PATERSON (1988, [87]). An excellent introductory exposition of the subject may also be found in Chapter 18 of HEWITT and ROSS (1963, [44]). Readers interested in amenability of Banach algebras and C^*-algebras should consult the original Memoir of B. E. JOHNSON (1972, [51]), the article of KHELEMSKIĬ (1984, [54]) and a recent book of PIER (1988, [89]).

We also include, at the end of each section, a list of related problems that we believe to be open.

2. Some notations

Throughout this paper, S will denote a *semitopological semigroup*, i.e. S is a semigroup with a Hausdorff topology such that for each $a \in S$, the mappings $s \to as$ and $s \to sa$ from S into S are continuous. Let $\ell^\infty(S)$ denote the C^*-algebra of bounded complex-valued functions on S with the supremum norm and pointwise multiplication. For each $a \in S$ and $f \in \ell^\infty(S)$, let $\ell_a f$ and $r_a f$ denote, respectively, the left and right translate of f by a, i.e. $(\ell_a f)(s) = f(as)$ and $(r_a f)(s) = f(sa)$, $s \in S$. Let X be a closed subspace of $\ell^\infty(S)$ containing constants and invariant under translations. Then a linear functional $m \in X^*$ is called a *mean* if $\|m\| = m(1) = 1$; m is called a *left invariant mean*, denoted by LIM, if $m(\ell_a f) = m(f)$ for all $a \in S$, $f \in X$.

Let $C(S)$ denote the space of all bounded continuous complex-valued functions on S and let $LUC(S)$ denote the space of *left uniformly continuous* complex-valued functions on S, i.e. all $f \in C(S)$ such that the mapping $a \to \ell_a f$ from S into $C(S)$ iscontinuous when $C(S)$ has the sup norm topology. Then $LUC(S)$ is a C^*-subalgebra of $C(S)$ invariant under translations and contains the constant functions. Further informations on $LUC(S)$ and its dual Banach algebra may be found in Mitchell (1970, [81]), Namioka (1967, [83]) and Berglund, Junghenn and Milnes (1978, [77]).

S is called *left reversible* if S has the finite intersection property for closed right ideals (i.e. $\overline{aS} \cap \overline{bS} \neq \emptyset$ for any $a, b \in S$). If S is a discrete left reversible semigroup, we may define an equivalence relation on S by sRt for $s, t \in S$ if there exists $x \in S$ with $sx = tx$. The set S/R of equivalence classes is a right cancellative semigroup under the induced multiplication called the *right cancellative quotient semigroup* of S.

For any set A, let $|A|$ denote the cardinality of A and χ_A the characteristic function on A. If A is a subset of a semigroup S, $x \in S$, let $x^{-1}A = \{s \in S;\ xs \in A\}$ and $Ax^{-1} = \{s \in S;\ sx \in A\}$.

Let X be a topological space and S be a semitopological semigroup, by an *action [anti-action]* of S on X we shall mean a map $\psi : S \times X \to X$ denoted by $(s, x) \to s \cdot x, s \in S,\ x \in X$, such that

(i) $s_1 \cdot (s_2 \cdot x) = (s_1 s_2) \cdot x$, $[s_1 \cdot (s_2 x) = (s_2 s_1) \cdot x,\ s_1, s_2 \in S,\ x \in X]$ for all $s_1, s_2 \in S,\ x \in X$.

(ii) For each $s \in S$, the map $x \to s \cdot x$, $x \in X$, from X into X is continuous.

(iii) For each $x \in X$, the map $s \to s \cdot x$ from S into X is continuous.

The action $S \times X \to X$ is *jointly continuous* if ψ is continuous when $S \times X$ has the product topology.

If X is a locally compact Hausdorff space, let $C_0(X)$ denote all continuous complex-valued functions on X vanishing at infinity and let $M(X)$ denote the space of complex regular Borel measures on X with total variation norm. Also let

$P(X)$ denote the probability measures in $M(X)$ and let $P_c(X)$ denote all measures in $P(X)$ with compact support.

If A is a subset of a topological vector space (E, τ), let $co\ A$ denote the convex hull of A and $\overline{co}\ A$ or $\overline{co}^{\tau}\ A$ denote its closed convex hull.

3. Amenable semitopological semigroups

A semitopological semigroup S is called *left amenable* if $LUC(S)$ has a LIM (see Namioka (1967, [83])).

Remark 3.1. (a) If S is commutative, then S is left amenable (see von Neumann (1929, [98]) and Day (1942, [7])).

(b) If S is compact and has jointly continuous multiplication, then S is left amenable if and only if S has unique minimal right ideal. (This follows from Namioka (1967, [83], Lemma 1.3) and Rosen (1956, [92])).

(c) If S is amenable and S' is a continuous homomorphic image of S, then S' is also left amenable (see proof of Proposition 6.1 in Lau (1970, [58]).

(d) If S and T are left amenable semitopological semigroups, then $S \times T$ is also left amenable (see proof of Proposition 6.4 in Lau (1970, [58])).

(e) If $S = \bigcup\{S_\alpha;\ \alpha \in I\}$ where each S_α is a left amenable subsemigroup of S with the induced topology and for each $\alpha, \beta \in I$, there exist $\gamma \in I$ such that $S_\alpha \cup S_\beta \subseteq S_\gamma$, then S is also left amenable (see proof of Proposition 6.2 in Lau (1970), [58]).

(f) If $\{S_\alpha : \alpha \in I\}$ are semitopological semigroups with identities e_α, let $S = \prod^w S_\alpha$ denote the *weak direct product* of $\{S_\alpha; \alpha \in I\}$ i.e. all functions f on I such that $f(\alpha) \in S_\alpha$ for all α, and $f(\alpha) = e_\alpha$ for all but finite number of α. Then S with the pointwise topology and pointwise multiplication is a semitopological semigroup. Furthermore, S is left amenable if and only if each S_α is left amenable (see Lau (1970, [58]), proof of Proposition 6.5). □

For $f \in LUC(S)$, let $R(f)$ denote the pointwise closure of $co\{r_a f;\ a \in S\}$. Also let K_S denote the linear span of $\{f - \ell_a f;\ f \in LUC(S),\ a \in S\}$.

Theorem 3.2. *The following are equivalent:*

(a) *S is left amenable.*

(b) (Localization property) *For each $f \in LUC(S)$, there exists a mean m on $LUC(S)$ such that $m(\ell_a f) = m(f)$ for all $a \in S$.*

(c) *For each $f \in LUC(S)$, $a \in S$, there exists a mean m on $LUC(S)$, such that $m(\ell_s(f - \ell_a f)) = 0$ for all $s \in S$.*

(d) (Stationary property) *For each* $f \in LUC(S)$, $R(f)$ *contains a constant function. (In this case, for each constant function* $\alpha 1$ *in* $R(f)$, *there is a* LIM m *on* $LUC(S)$ *such that* $m(f) = \alpha$*).*

(e) *For each* $f \in LUC(S)$, $a \in S$, $R(f - \ell_a f)$ *contains zero.*

(f) *For each* $h = \Sigma\{f_k - \ell_{a_k} f_k;\ k = 1, \ldots, n\}$, *where* $f_1, f_2, \ldots, f_n \in LUC(S)$ *and* $a_1, \ldots, a_n \in S$, *then* $\sup Re\, h(s) \geq 0$.

(g) $\inf\{\|1 - h\|,\ h \in K_S\} = 1$.

(h) K_S *is not uniformly dense in* $LUC(S)$.

(i) (Fixed point property) *For any jointly continuous affine action of* S *on a convex compact subset* Y *of a separated locally convex linear topological space,* Y *contains a common fixed point for* S.

(j) (Hahn-Banach extension property) *For any linear anti-action of* S *on a real topological vector space* E, *if* p *is a continuous sublinear map on* E *such that* $p(s \cdot x) \leq p(x)$ *for all* $s \in S$, $x \in E$, *and if* φ *is an invariant linear functional on an invariant subspace* F *of* E *such that* $\varphi \leq p$, *then there exists an invariant extension* $\tilde{\varphi}$ *of* φ *to* E *such that* $\tilde{\varphi} \leq p$.

(k) (Geometric form of Hahn-Banach extension property) *For any linear anti-action of* S *on a topological vector space* E, *if* U *is an invariant open convex subset of* E *containing an invariant element, and* M *is an invariant subspace of* E *which does not meet* U, *then there exists a closed invariant hyperplane* H *of* E *such that* H *contains* M *and* H *does not meet* U.

(ℓ) (Hahn-Banach separation property) *For any linear anti-action of* S *on a Hausdorff topological vector space* E *with a base of the neighbourhoods of the origin consisting of invariant open convex sets, then any two points in* $E_f = \{x \in E;\ sx = x \text{ for all } s \in S\}$ *can be separated by a continuous invariant linear functional on* E.

(m) *For any linear anti-action of* S *on a separated real topological vector space* E, *if* L *is an invariant subspace of* E *and* K *is a convex subset of* E *such that* $K - x_0$ *is invariant for some* x_0 *contained in* $L \cap \operatorname{Int} K$, *then for each invariant linear functional* φ *on* L *such that* $\varphi(x) \leq \alpha$ *for all* $x \in L \cap K$ *and some fixed real number* α, *then there exists an invariant linear extension* $\tilde{\varphi}$ *of* φ *to* E *such that* $\tilde{\varphi}(x) \leq \alpha$ *for all* $x \in K$.

(n) (Invariant ideal property) *For any positive linear anti-action of* S *on a partially ordered topological vector space* E *(i.e.,* $s \cdot x \geq 0$ *whenever* $x \geq 0$, $s \in S$*) with a topological order unit* e *such that* $se = e$ *for all* $s \in S$, *if* F *is an invariant subspace of* E *containing* e, *and* φ *is an invariant monotonic linear functional on* F, *then there exists an invariant monotonic linear functional* $\tilde{\varphi}$ *on* E *extending* φ.

(o) *For any positive linear anti-action of* S *on a partially ordered topological vector space* E *with a topological order unit* e, E *contains a maximal proper ideal which is invariant under* S.

(p) (Invariant subspace property) S *satisfies property* $P(n)$ *for each positive integer* n*:*

$P(n)$ Let E be a separated locally convex space and $S \times E \to E$ be a linear action of S on E which is jointly continuous on compact convex subsets of E. Let X be a subset of E such that there exists closed invariant subspace H of E with codimension n and $x + H \cap X$ is compact convex for each $x \in E$. If $\mathcal{L}_n(X)$, the collection of n-dimensional subspaces of X, is non-empty and $s(L) \in \mathcal{L}_n(X)$ for each $s \in S, L \in \mathcal{L}_n(X)$, then there exists $L_0 \in \mathcal{L}_n(X)$ such that $s(L_0) = L_0$ for each $s \in S$.

(q) *S satisfies $P(n)$ for some positive integer n.*

(r) *For any linear anti-action of S on a normed linear space X such that $\|s \cdot x\| \leq \|x\|$ for all $s \in S$, $x \in X$, then* $\operatorname{dist}(0, co\, O(x)) = d(x, K_X)$ *for all $x \in X$ where $O(x) = \{s \cdot x;\ s \in S\}$ and K_X is the linear span of $\{x - s \cdot x;\ s \in S,\ x \in X\}$.* □

Remark 3.3. (i) The equivalence of (a) and (d) of Theorem 3.2 is due to MITCHELL (1965, [78]) for discrete semigroups. The equivalence (a) $\Leftrightarrow$ (b) $\Leftrightarrow$ (c) $\Leftrightarrow$ (d) $\Leftrightarrow$ (e) as stated were proved by GRANIRER and LAU (1971, [38]).

(ii) Theorem 3.2 (a) $\Leftrightarrow$ (f) is due to Dixmier (1950, [15]) and (a) $\Leftrightarrow$ (g) $\Leftrightarrow$ (h) is due to DAY (see HEWITT and ROSS (1963, [44], p. 235) and DAY (1969, [11]) for discrete semigroups. Proofs for the topological case are similar).

(iii) The equivalence (a) $\Leftrightarrow$ (i) of the theorem above is MITCHELL's generalization (1970, [81]) of DAY's fixed point theorem for discrete left amenable semigroups (1961, [10]).

(iv) The equivalence (a) $\Leftrightarrow$ (j) of the theorem is due to SILVERMAN (1956, [94]) for discrete semigroups. The equivalences (a) $\Leftrightarrow$ (j) $\Leftrightarrow$ (k) $\Leftrightarrow$ (ℓ) $\Leftrightarrow$ (n) $\Leftrightarrow$ (o) can be found in LAU (1974, [62]) (see also FORREST (1989, [25])). Furthermore, condition (m) is an analogue of a property found by KY FAN (1977, [23]) for almost periodic actions. The equivalence (a) $\Leftrightarrow$ (m) can be proved by method similar to that for Theorem 1 in LAU (1977, [65]).

(v) KY FAN (1965, [23]) proved that any discrete left amenable semigroup has property $P(n)$ for each positive integer n. The proofs for the equivalence of (a) $\Leftrightarrow$ (p) $\Leftrightarrow$ (q) are contained in LAU (1983, [68]) and LAU, PATERSON and WONG (1988, [72]) (see LAU and WONG (1987, [73]) for correction of the proof (a) $\Rightarrow$ (p), and LAU and WONG (1988, [74]) for similar results established).

(vi) GLICKSBERG (1963, [28]) proved that any discrete left amenable semigroup has property (r). Also, as shown by GRANIRER (1969, [35]), in order to prove left amenability, it is enough to assume (r) when $X = \ell^\infty(S)$ and S acts on X by left translation. The proof for the topological case of (a) $\Leftrightarrow$ (r) is a simple modification of GRANIRER's argument in (1969, [35]) (see also RIEMERSMA (1971, [91]) and FORREST (1989, [25])). □

Let E be a Banach space, and $A \subseteq E^*$. An element $\varphi_0 \in A$ is said to be a $w^* - G_\delta$ *point* of A if there are $x_n \in E$ and scalars γ_n, $n = 1, 2, 3\ldots$ such that $\{\varphi_0\} = \bigcap_{n=1}^{\infty} \{\varphi \in A;\ \varphi(x_n) = \alpha_n\}$. A discrete semigroup S is said to have $w^* - G_\delta$ *sequential property* if whenever $\{T_s : s \in S\}$ is a representation of S as w^*-continuous linear operators on a dual Banach space E^*, and K is a bounded convex S invariant subset of E^*, then the w^*-G_δ points of F_K (the fixed point set in the weak *-closure of K) is in the weak*-*sequential* closure of K.

Theorem 3.4. GRANIRER (1972, [36] and 1985, [37]) *Let S be a discrete countable semigroup. If S is left amenable, then S has the $w^* - G_\delta$ sequential property. If S is a group and S has the $w^* - G_\delta$ sequential property, then S is left amenable.* □

Let $\Delta(S)$ be the spectrum of $LUC(S)$. Then S is *n-extremely left amenable* (n-ELA) if there exists a subset $H_0 = \{\gamma_1, \ldots, \gamma_n\}$ of $\Delta(S)$, $|H_0| = n$, which is minimal with respect to the property: $L_a H_0 = H_0$ for all $a \in S$, where $L_a = \ell_a^*$. In this case $m = \frac{1}{n} \sum_{i=1}^{n} \gamma_i$, is a LIM on $LUC(S)$. Conversely if $LUC(S)$ has a LIM in $co\ \Delta(S)$, then S is n-ELA for some n (see MITCHELL (1966, [79]), GRANIRER (1965, [32], 1967, [34], 1969, [35]), SORENSON (1966, [95]) and LAU (1970, [58,59])).

The following is proved by MITCHELL (1970, [81]) for $n = 1$ and LAU (1970, [59]) for all n.

Theorem 3.5. *For any semitopological semigroup S and fixed positive integer n:*

(a) *If $LUC(S)$ is n-ELA, then for any jointly continuous action of S on a compact Hausdorff space X,*
   2(*) there exists a nonempty subset $F \subseteq X$ such that $|F| \leq n$, $|F|$ divides n and $aF = F$ for all $a \in S$.

(b) *If S satisfies (*) when S acts on $\Delta(S)$, then $LUC(S)$ is m-ELA for some $m \leq n$, m divides n.* □

For discrete semigroups, SORENSON (1966, [95]) obtained the following generalization of a result of MITCHELL (1966, [79]) and GRANIRER (1965, [32]) for $n = 1$:

Theorem 3.6. *For any discrete semigroup S, S is n-ELA if and only if S is left reversible and the right cancellative quotient semigroup S/R is a group of order n.* □

Remark 3.7. If S is a connected n-ELA semitopological semigroup, then S is ELA (LAU (1970, [58])). Also if S is a subsemigroup of a locally compact group G and S is n-ELA, then S is a finite group of G of order n (GRANIRER and LAU (1971, [38])). □

Problem 1. Let S be a left amenable semitopological semigroup. Does this imply $C(S)$ has a LIM? (The answer to this question is affirmative when S is a locally compact group, GREENLEAF (1969, [39]), or a measurable non-locally null subsemigroup of a locally compact group, LAU (1973, [61]) or certain classes of locally compact semigroups, PATERSON (1978, [85])).

Problem 2. Let S be a semitopological semigroup such that for each $f \in C(S)$ there exists a mean m (depending on f) on $C(S)$ such that $m(\ell_s f) = m(f),\ s \in S$. Does $C(S)$ have a LIM?

This problem is equivalent to:

Problem 2′. If for each $f \in C(S)$, $R(f)$ contains a constant function, does $C(S)$ have a LIM?

Problem 3. Each of the following properties implies left amenability of a discrete semigroup. Is the converse true? (The answer is yes when S is a group, (see LAU (1982, [66]) and LAU and LOSERT (1986, [70]).)

(C) When X is a weak*-closed left translation invariant subspace of $\ell^\infty(S)$ and X is complemented, then X admits a closed left translation invariant complement.

(C*) Each weak*-closed left translation invariant subalgebra of $\ell^\infty(S)$ closed under conjugation admits a closed left translation invariant complement.

Problem 4. ((GRANIRER (1972, [37], p. 29)) Does the $w^* - G_\delta$ sequential property characterize the class of countable discrete left amenable semigroups? (This is the case for countable groups.)

Problem 5. Let $\{S_\alpha : \alpha \in I\}$ be semitopological semigroups such that S_α is n_α-ELA for some $n_\alpha \leq n$, $n_\alpha > 1$ for finitely many α's, and $n =$ product $\{n_\alpha; \alpha \in I\}$. Is $S = \prod S_\alpha$ (the full direct product) n-ELA? (Note that this is the case when each S_α is discrete. See LAU (1970, [58]). Also DAY (1957, [9]) has given an example to show that the full direct product of left amenable semigroups need not be left amenable.)

Problem 6. If for some fixed n and for each $f \in LUC(S)$, the pointwise closure of $\{\frac{1}{n}\sum_{i=1}^{n} r_{a_i}(f);\ a_1, \ldots, a_n \in S\}$ contains a constant function, is $LUC(S)$ m-ELA for some $m \leq n$? (Note that this is true for $n = 1$, see GRANIRER and LAU (1971, [38])).

Problem 7. If S is n-ELA for some n, is every extreme point of $LIM(S)$ of the form $\frac{1}{n}\sum_{i=1}^{n}\varphi_i,\ \varphi_i \in \Delta(S)$? (This is the case when S is discrete (LAU (1970, [58]); see also GOYA (1972, [29])).

Problem 8. If S is n-ELA for some n, is the supp φ for each extreme point φ of $LIM(S)$ S-minimal in $\Delta(S)$? (This is the case when S is discrete by a result of SORENSON (1966, [95]); see also LAU (1970, [58], p. 75, Corollary)).

Problem 9. If S is left amenable, and supp φ is S-minimal for each extreme point φ of $LIM(S)$, is S n-ELA? (This is the case when S is a commutative countable discrete semigroup as shown in NASSAR (1984, [84]). See also FAIRCHILD (1972, [21])).

Problem 10. Let S be an n-extremely amenable topological group. Is S finite? (see GRANIRER (1967, [34]) and GRANIRER and LAU (1971, [38])).

Problem 11. If S is n-ELA, is $Ext(LIM(S))$ weak*-closed? (This is the case when S is discrete, see NASSAR (1984, [84])).

Problem 12. (NASSAR (1984, [84])) If $Ext(LIM(S))$ is weak*-closed, is S n-ELA? (See TALAGRAND (1979, [97])).

4. Locally compact semigroups

Throughout this section S will denote a locally compact semitopological semigroup. Then $M(S)$ is a Banach algebra with total variation norm and multiplication defined by

$$\int f d\mu * \nu = \iint f(xy)d\mu(x)d\nu(y) = \iint f(xy)d\nu(y)d\mu(x)$$

for $f \in C_0(S)$, $\mu, \nu \in M(S)$. (See GLICKSBERG (1961, [27]), JOHNSON (1969, [50]), WONG (1978, [102]) and KHARAGHANI (1988, [53])). Furthermore $(P(S), *)$ is a semigroup.

Consider on $M(S)$ the following properties:

(LSU_ρ) There exists a net (μ_n) in $P_c(S)$ such that for each compact set K contained in S, $\|\nu * \mu_n - \mu_n\|$ tends to zero uniformly over all ν in P_c which are supported in K.

(LSU_δ) There exists a net (μ_n) of elements of $P_c(S)$ such that for each compact set K in S, $\|\delta_s * \mu_n - \mu_n\|$ tends to zero uniformly for s in K.

(LS_ρ) There exists a net (μ_n) in $P_c(S)$ such that $\|\nu * \mu_n - \mu_n\|$ tends to zero for each ν in $P_c(S)$.

(LS_δ) There exists a net (μ_n) in $P_c(S)$ such that $\|\delta_s * \mu_n - \mu_n\|$ tends to zero for each $s \in S$.

Condition (LSU_c), defined by WONG (1976, [100]), can easily be seen to be equivalent to (LSU_δ) (denoted by (LSU)). All other conditions are defined by DAY (1976, [13], 1982, [14]). Furthermore the following implications hold:

$$(LSU_\rho) \Longleftrightarrow (LSU_\delta) \Longrightarrow (LS_\rho) \Longrightarrow (LS_\delta) \Longrightarrow (S \text{ is left amenable}).$$

However nothing more seems to be known unless S is a locally compact group. In this case, all the conditions are equivalent (see WONG (1976, [100]), DAY (1982, [14]) and JUNGHENN (1981, [52])). Also (LSU) is the analogue of REITER's condition P_1 and (LS_ρ) is the analogue of DAY's condition for strong invariance for $L_1(S)$ (see GREENLEAF (1969, [39])). Furthermore, as shown by DAY in (1982, [14], p. 86) if S is a locally compact group, and $\varphi \in M(S)$ is defined by $\varphi(E) = \int_E f(x)d\lambda(x)$, where λ is a fixed left Haar measure on S, and f is continuous, bounded nonnegative with compact support, and $\int_S f d\lambda = 1$, then for each net (μ_n) which satisfies (LS_δ), the set $(\varphi * \mu_n)$ is equicontinuous and satisfies (LSU). However, nothing else is known for the general case.

Since $M(S)^* = C_0(S)^{**}$ is the second dual of a commutative C^*-algebra, it is again a commutative C^*-algebra. For each $F \in M(S)^*$, $\varphi \in P(S)$, define $F \cdot \varphi$ (resp. $\varphi \cdot F$) $\in M(S)^*$ by $(F \cdot \varphi)(\mu) = F(\varphi * \mu)$ (resp. $(\varphi \cdot F)(\mu) = F(\mu * \varphi)$), $\mu \in M(S)$. A positive linear functional of norm one on $M(S)^*$ is called a *topological left invariant mean* if $m(F \cdot \varphi) = m(F)$ for each $\varphi \in P(S)$, $F \in M(S)^*$.

Let $I_0(S) = \{\psi \in M(S);\ \psi(S) = 0\}$.

Theorem 4.1. *The following conditions on a locally compact semitopological semigroup S are equivalent:*
(a) *S has property (LS_ρ).*
(b) *$M(S)^*$ has a topological left invariant mean.*
(c) *The Banach algebra $M(S)$ is left amenable, i.e. whenever X is a two-sided Banach A-module such that $\varphi \cdot x = x$ for each $\varphi \in P(S)$, $x \in X$, then any bounded derivation from A into X^* is inner.*
(d) *For each $\psi \in I_0(S)$ and $\epsilon > 0$ there exists $\varphi \in P(S)$; $\|\psi \cdot \varphi\| < \epsilon$.*
(e) *$|\psi(S)| = \inf\{\|\psi \cdot \varphi\|;\ \varphi \in P(S)\}$ for each $\psi \in M(S)$.*

(f) *$\mathcal{N}(S)$ is closed under addition, where $\mathcal{N}(S)$ denotes all $F \in M(S)^*$ such that* $\inf\{\|\varphi \cdot F\|; \varphi \in P(S)\} = 0$.
(f) $\mathrm{dist}(I_1, I_2) = 0$ *for any two right ideals I_1, I_2 of the semigroup $P(S)$.* □

Remark 4.2. In Theorem 4.1, the equivalences (a) ⇔ (c) ⇔ (d) are proved in LAU (1983, [67]); the equivalences (a) ⇔ (b) ⇔ (e) are due to WONG (1973, [99]); and the equivalences (b) ⇔ (f) ⇔ (g) are due to WONG and RIAZI (1983, [103]).□

Theorem 4.3. (LAU (1983, [67])) *$I_0(S)$ has a bounded approximate identity if and only if S has property (LS_ρ) and $M(S)$ has a bounded approximate identity.*□

Problem 13. (DAY (1982, [14])) Does $(LS_\rho) \Rightarrow (LSU_\delta)$?

Problem 14. (DAY (1982, [14])) Does $(LS_\delta) \Rightarrow (LS_\rho)$?

Since $BM(S)$ can be embedded isometrically into $M(S)^*$ by $f \to \tau(f)$, where $\tau(f)(\mu) = \int f(x) d\mu(x)$, $\mu \in M(S)$, and $BM(S)$ denotes the space of bounded complex-valued Borel measurable functions on S with the supremum norm, the following implications hold:

$$(LS_\delta) \Rightarrow \text{“}BM(S) \quad \text{has} \quad LIM\text{”} \Rightarrow \text{“}C(S) \text{ has } LIM.\text{”}$$

Problem 15. Does “$C(S)$ has LIM” ⇒ “$BM(S)$ has a LIM”?

Problem 16. Does “$BM(S)$ has LIM” ⇒ “(LS_δ)”?

It was proved in LAU (1973, [61]) that if S is a non-locally null measurable subsemigroup of a locally compact group, then $BM(S)$ has a LIM if and only if S is left amenable. PATERSON (1978, [85]) proved that if $P(S)$ contains an absolutely continuous measure, then $(LS) \Leftrightarrow$ “$C(S)$ has LIM” (see also DZINTOYIWEYI (1978, [18]))).

5. Følner type conditions and Følner numbers

Throughout this section, S denotes a discrete semigroup. FØLNER (1955, [24]) introduced the following necessary and sufficient condition for S to be left amenable.

(FC) (Følner Condition) For each finite subset F of S and $\epsilon > 0$, there exists a finite subset A of S such that $|sA \setminus A| < \epsilon |A|$ for each $s \in F$.

In his thesis, FREY (1960, [26]) showed that every discrete left amenable semigroup satisfies FC; however the converse is false since every finite semigroup satisfies FC, though not every finite semigroup is left amenable (an elegant proof of FREY's result was given by NAMIOKA (1964, [82])). If S is left cancellative, then FC is equivalent to:

(SFC) (Strong Følner Condition) Given any finite subset F of S and any $\epsilon > 0$, there exists a finite subset A of S such that $|A \setminus sA| \leq \epsilon|A|$ for all $s \in F$.

As was shown by ARGABRIGHT and WILDE (1967, [1]), SFC implies left amenability LA of S. They showed that if S is left amenable and S/R (the right cancellative quotient semigroup) is left cancellative, then S satisfies the SFC. Finally KLAWE (1977, [56]) proved that the converse also holds:

Theorem 5.1. *A semigroup S satisfies SFC if and only if S is left amenable and S/R is left cancellative.* □

Example 5.2. (KLAWE (1977, [56])) Let $S = U \times_\rho T$ (where U is the free abelian semigroup generated by the elements $\{u_i \mid i = 0, 1, 2, \ldots\}$, T is the infinite cyclic semigroup with generator $\{a\}$, $\rho_a(u_i) = u_{i-1}$, $i \geq 1$, and $\rho_a(\mu_0) = u_0$) is left amenable, but does not satisfy the SFC. The semigroup S is also right cancellative but not left cancellative. Hence this example also settles the SORENSON's conjecture (see GRANIRER (1967, [34])).

When S is a discrete group, FØLNER (1955, [24]) also proved that the following condition is equivalent to amenability of S:

(WFC) (Weak Følner Condition). There exists a number k, $0 < k < 1$, such that for any elements $s_1, \ldots, s_n \in S$ (not necessarily distinct), there is a finite subset A of S satisfying

$$\frac{1}{n}\sum_{i=1}^{n} |A \setminus s_i A| \leq k|A|.$$

NAMIOKA (1964, [82]) observed that WFC does not imply LA. He introduced two sufficient conditions stronger than WFC:

($WNFC$) (Weak Namioka-Følner Condition). There exists a number k, $0 < k < 1$, such that for any elements $s_1, \ldots, s_n$, $s'_1, \ldots, s'_n$ of S, there is a finite subset A of S satisfying

$$\frac{1}{n}\sum_{i=1}^{n} |s_i A \cap s'_i A| \geq k|A|.$$

$(SNFC)$. There exists a number $k, 0 < k < \frac{1}{2}$, such that for any elements $s_1, \dots, s_n$ of S, there is a finite subset A of S satisfying

$$\frac{1}{n}\sum_{i=1}^{n} |A \setminus s_i A| \leq k|A|.$$

NAMIOKA (1964, [82]) proved that $SNFC$ implies $WNFC$ and $WNFC$ implies LA. In fact he proved that if $SNFC$ holds for k, then $WNFC$ holds for $1-2k$. YANG (1987, [105]) observed that if $WNFC$ holds for k, then $SNFC(WFC)$ holds for $1-k$ and showed that KLAWE's example does not satisfy $WNFC$, thus settling a problem raised by NAMIOKA (1964, [82]). The following diagram summarizes the known implications about the Følner-type conditions.

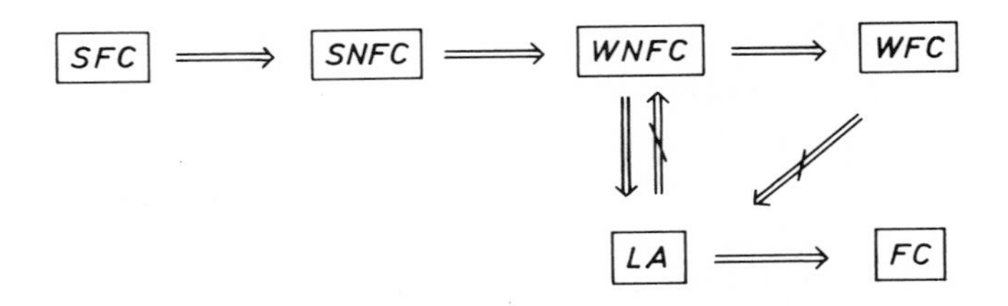

WONG (1988, [100]) defined $\varphi(S)$, *the* FØLNER *number* of a semigroup S, to be the infinimum of all numbers $k \leq 1$ such that WFC holds. Furthermore, as shown by YANG (1987, [105]) the following hold:

Theorem 5.3. *The following implications hold:*
(a) *S satisfies WFC* $\iff$ $\varphi(S) < 1$
(b) *S satisfies $SNFC$* $\iff$ $\varphi(S) < \frac{1}{2}$
(c) *S satisfies SFC* $\iff$ $\varphi(S) = 0$
(d) *If S does not satisfy SFC, then $\varphi(S) \geq \frac{1}{6}$.*
(e) *If S is cancellative, then $\varphi(S) = 0 \Leftrightarrow \varphi(s) < 1 \Leftrightarrow S$ is left amenable.* □

Problem 17. (YANG (1987, [105])). If S is a left reversible semigroup, then $\varphi(S) \leq \varphi(S/R)$. Does equality hold?

Problem 18. Can any of the following arrows be reversed?

$$SFC \Rightarrow SNFC \Rightarrow WNFC \Rightarrow LA + WFC.$$

(Note that these conditions are equivalent when S is finite, left cancellative, abelian, or certain semidirect products. See YANG (1987, [105]).)

Problem 19. (Yang (1987, [105])) Is there any left amenable semigroup S such that $0 < \varphi(s) < 1$? (If not, then all the conditions $SFC, SNFC, WNFC$ and $LA+WFC$ are equivalent.)

6. The set of invariant means

Let S be a semitopological semigroup and let $ML(S)$ be the set of left invariant means on $LUC(S)$. It is natural to study the cardinality and structure of $ML(S)$ when S is left amenable. The first results in this direction were given by Day (1957, [9]) who proved that for discrete solvable groups, infinite amenable non-torsion groups, and infinite locally finite groups, all have more than one left invariant mean. Luthar (1959, [75]) showed that a discrete commutative semigroup has a unique left invariant mean if and only if it has a finite ideal. Continuing the search for conditions that $\langle ML(S)\rangle$, the linear span of $ML(S)$, be finite dimensional, Granirer (1963, [30] and [31]) (for the countable and left cancellative case) and Klawe (1977, [55]) (for the general case), proved the following definitive result:

Theorem 6.1. (Granirer-Klawe) *For any discrete left amenable semigroup S, $\dim\langle ML(S)\rangle = n$ if and only if S contains exactly n minimal finite left ideals.* □

If S is left amenable, then $ML(S)$ is a weak*-compact non-empty-convex subset of $LUC(S)^*$ and hence must have extreme points. Chou (1971, [6]) proved that if S is a countable discrete group, then $ML(S)$ has no (weak*) exposed points. Granirer (1972, [36]) proved that if S is a countable discrete left amenable semigroup, then $ML(S)$ has exposed points if and only if S has finite left ideals. Finally Yang (1986, [104]) was able to drop the countability condition of Chou and Granirer and proved:

Theorem 6.2. (Chou-Granirer-Yang) *For any discrete left amenable semigroup, $ML(S)$ has exposed points if and only if S has finite left ideals. The number of exposed points of $M(S)$ is exactly the number of finite left ideals of S.* □

A subset A in a discrete semigroup S is *left thick* if for any finite subset σ of S, there exists $s \in S$ such that $\sigma s \subseteq A$. Left thick subsets in semigroups were introduced by T. Mitchell (1965, [78]). They play a vital role in determining card $(ML(S))$. In fact, if $\{A_\gamma : \gamma \in \Gamma\}$ is a collection of pairwise disjoint left thick subsets of a left amenable semigroup S, then by Mitchell's theorem (1965, [78]), we can choose $\mu_\gamma \in ML(S)$ with $\mu_\gamma(\chi_{A_\gamma}) = 1$. This set $\{\mu_\gamma : \gamma \in \Gamma\}$ of left invariant means on S is linearly independent (see Klawe (1977, [55])).

We say that $A \subseteq S$ is *strongly left thick* if for each $B \subset S$, with $|B| < |A|$, the set $A \setminus B$ is left thick in S. Strongly left thick subsets are defined by KLAWE (1980, [57]) in order to determine a lower bound for the dimension of $\langle ML(S)\rangle$. She proved that a subset $A \subseteq S$ is strongly left thick if and only if there exists a collection $\{D_\gamma : \gamma \in \Gamma\}$ of pairwise disjoint subsets of A, which are left thick, such that $|\Gamma| = |A|$. Note that unless the semigroup is infinite and right cancellative, left thick subsets need not be strongly left thick.

For a discrete semigroup S, KLAWE (1980, [57]) defined the cardinal

$$\kappa(S) = \min\{|B| : B \subset S, \quad \mu(B) = 1 \quad \text{forall } \mu \in M(S)\}$$

and gave a proof that $|ML(S)| = 2^{2^{\kappa(S)}}$ when $\langle ML(S)\rangle$ is infinite dimensional. However, there is a gap in her proof as pointed out by PATERSON (1985, [86]). PATERSON defined the cardinal

$$\rho(S) = \min\{|\bigcup_{i=1}^{n} s_i S_i| : n \geq 1, \quad \{S_1, \ldots, S_n\} \text{ is a partition of } S, s_1, \ldots, s_n \in S\}$$

and proved:

Theorem 6.3. *If $\langle ML(S)\rangle$ is infinite dimensional, then $|ML(S)| = 2^{2^{\rho(S)}}$.* □

Finally YANG (1988, [106]) defined the cardinal

$$\tau(S) = \sup\{|A| : A \subset S, \ A \text{ is strongly left thick}\}$$

and proved that KLAWE's result is correct:

Theorem 6.4. *If S is a left amenable semigroup such that $ML(S)$ is infinite dimensional, then $|ML(S)| = 2^{2^{\kappa(S)}} = 2^{2^{\tau(S)}}$. In this case $\kappa(S) = \tau(S) = \rho(S)$.* □

The general problem of determining $|ML(S)|$ when S is a semitopological semigroup is still open. However the following are known (see also LAU (1986, [69])):

Theorem 6.5. (LAU and PATERSON (1986, [71])) *If S is an amenable locally compact non-compact group, then $|ML(S)| = 2^{2^{d(S)}}$ where $d(S)$ is the smallest possible cardinality of a covering of S by compact sets.* □

In a recent paper, DZINOTYIEWEYI (1988, [20], Corollary 3.4), showed that the conclusion of Theorem 6.4 holds when S is a non-compact left amenable topological semigroup with an identity such that $S = \bigcup\{\text{supp}(\mu);\ \mu \in M_a(S)\}$ (where $M_a(S)$ is the collection of all $\mu \in M(S)$ such that the maps $x \to |\mu|(x^{-1}K)$ and

$x \to |\mu|(Kx^{-1})$ of S into $\mathbb{R}$ are continuous for all compact $K \subseteq S$), and the sets $C^{-1}D$ and DC^{-1} are compact whenever C and D are compact subsets of S (see also DZINOTYIWEYI (1985, [19])).

GRANIRER (1965, [34]) proved that if S is a topological semigroup containing exactly n compact left ideal groups, then $\dim\langle ML(C(S)\rangle = n$. However he showed that the converse is false even when S is an abelian topological group and $n = 1$.

It follows from Theorem 6.4 that if S is an amenable locally compact group then either $\dim\langle ML(C(S)\rangle = 1$ or $\dim\langle ML(C(S)\rangle = \infty$. Moreover, $\dim\langle ML(C(S)\rangle = 1$ if and only if S is compact. The following is proved by GRANIRER (1965, [34]).

Theorem 6.6. *Let S be any separable (not necessarily closed) subgroup of a locally convex linear topological space. Then either* $\dim\langle ML(C(S)\rangle = 1$ *or* $\dim\langle ML(C(S)\rangle = \infty$. *Also,* $\dim\langle ML(C(S)\rangle = 1$ *if and only if* $S = \{0\}$. □

Problem 20. Find necessary and sufficient conditions for $\langle ML(S)\rangle$ of a semitopological semigroup S to be finite dimensional.

Problem 21. Let S be a semitopological semigroup such that $\dim\langle ML(S)\rangle = \infty$. Find $|ML(S)|$.

Problem 22. When will $ML(S)$ have weak* exposed points?

(Note that this problem is completely answered when S is discrete (Theorem 6.2), and S is a locally compact group (see YANG (1988, [106])).

7. Amenability of $\ell^1(S)$

A Banach algebra A is called *amenable* if $H^1(A, X^*) = 0$ for all Banach A-bimodules X, i.e. if all bounded module derivations from A into dual modules of Banach A-bimodules are inner. This notion of amenability was introduced by B. E. JOHNSON (1972, [50]), who proved that if G is a locally compact group, then G is amenable if and only if the Banach algebra $L^1(G)$ is amenable. Also if $\ell^1(S)$ is amenable, then S is amenable but the converse if false (even when S is commutative (see DUNCAN and NAMIOKA (1978, [16])). In particular left and right amenability of $\ell^1(S)$ of a discrete semigroup as defined in Theorem 4.1 do not imply amenability. Recently, DUNCAN and PATERSON (1989, [17]) proved that if $\ell^1(S)$ is amenable then S is regular and the number of idempotents is finite. In particular, for the case of an inverse semigroup, $\ell^1(S)$ is amenable if and only if S has a finite number of idempotents and every subgroup of S is amenable (see DUNCAN and NAMIOKA (1978, [16])). Also when S is commutative, $\ell^1(S)$ is amenable if and only if S is a finite semilattice of groups (GRØNBÆK (1989, [40])).

Problem 23. When is $M(S)$ of a locally compact semi-topological semigroup amenable?

This problem seems to be unknown even when S is a locally compact group. Note that if $M(S)$ is amenable, then $C(S)$ has a left invariant mean and a right invariant mean by Theorem 4.1.

Let A be a commutative Banach algebra. A Banach A-bimodule X is said to be commutative if $u \cdot x = x \cdot u$ for every $u \in A,\ x \in X$. A is called *weakly amenable* if every continuous derivative of A into a commutative Banach A-module is zero. The notion of weak amenability is introduced recently by BADE, CURTIS and DALE (1987, [2]).

Problem 24. When is $M(S)$ of a commutative locally compact semi-topological semigroup weakly amenable?

It is shown by N. GRØENBÆK (1989, [40]) that if S is a commutative discrete semigroup which is a union of groups, then $\ell^1(S)$ is weakly amenable. He also shows that if $S = [0,1]$ with the binary operation $st = \min\{s+t, 1\},\ s,t \in S$, then $\ell^1(S)$ is not weakly amenable.

8. Left reversibility and almost periodic functions

Let $A(S)$ denote the space of all $f \in C(S)$ such that $\mathcal{LO}(\{) = \{\ell_\int\{; \int \in \mathcal{S}\}$ is relatively compact in the norm topology of $C(S)$ and $W(S)$ denote the space of all $f \in C(S)$ such that $\mathcal{LO}(\{)$ is relatively compact in the weak topology of $C(S)$. Functions in $A(S)$ (resp. $W(S)$) are called *almost periodic* (resp. *weakly almost periodic*). If S is left reversible, then $A(S)$ has a LIM (LAU 1973, [60]). If S is normal and $C(S)$ has a LIM, then S is left reversible (see HOLMES and LAU (1972, [47])). At the 1984 conference in Richmond, Virginia, T. MITCHELL showed that the discrete semigroups S_1 and S_2 are not left reversible but $A(S_i),\ i = 1,2$ has a LIM (since $\bar{S}_i^a$, the almost periodic compactification of S_i, is a compact group, by the Swelling Lemma (see HOFMANN and MOSTERT (1966, [46]) and RUPPERT (1984, [93]; 4.19 (iv)), where S_1 is the semigroup generated by $\{a,b,c,1\}$, where 1 is the identity, with the relation $ab = 1$ and $ac = 1$, and S_2 is the semigroup generated by $\{a,b,c,d,1\}$ with the relations $ac = 1$ and $bd = 1$.

More recently, HSU (1985, [48]) proved that if S is a discrete left reversible semigroup, then $W(S)$ has a LIM. Hence for discrete semigroups, the following is known (see LAU (1976, [64])):

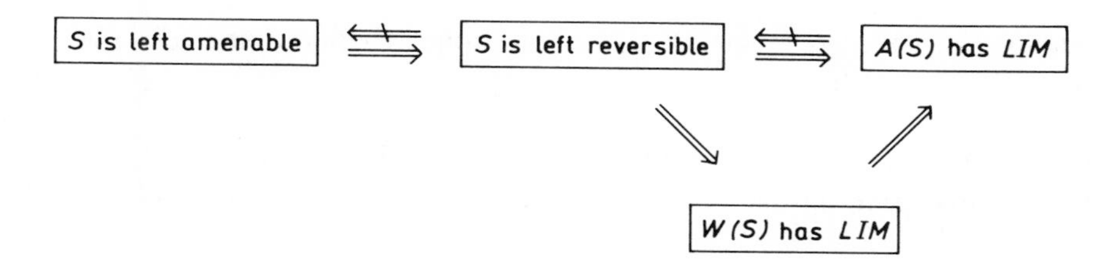

Note that if S is a semitopological semigroup, then left amenability of S does not imply left reversibility, as shown in HOLMES and LAU (1972, [47]). The same example there shows that for *arbitrary* semitopological semigroups, "$W(S)$ has a LIM" does not imply left reversibility. Hence in this case we have

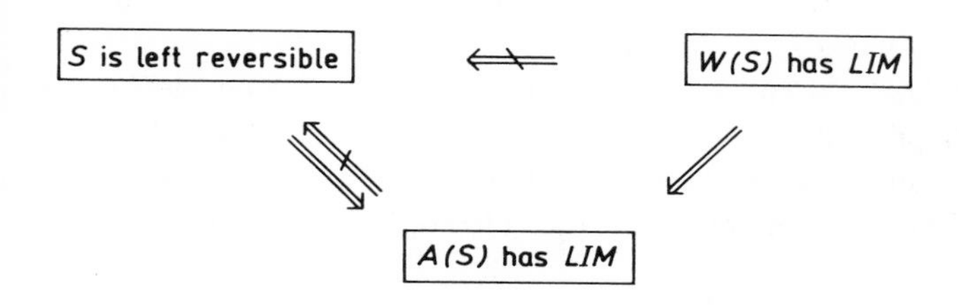

Problem 25. Let S be a semitopological semigroup. Does "$A(S)$ has LIM" imply "$W(S)$ has LIM"?

Problem 26. Let S be a left reversible semitopological semigroup. Does $W(S)$ have a LIM? This problem is equivalent to: If S is left reversible, is S^w, the weakly almost periodic compactification of S, left reversible?

Problem 27. Let S be a *discrete* semigroup. Does "$W(S)$ have a LIM" imply left reversibility?

Note that this is not true for arbitrary semitopological semigroups. Also, with MITCHELL's examples, the answer (for discrete semigroups) to either Problem 24 or Problem 26 must be negative. This problem is equivalent to Problem 3 in LAU (1976, [64]).

Problem 28. Let S be a subsemigroup of an amenable group. If $A(S)$ has a LIM, is S left reversible?

Note that the space $A(S)$ of a subsemigroup S of an amenable group need not have a LIM. Indeed, M. HOCHSTER (1969, [45]) showed that it is possible to embed T, the free semigroup on two generators $\{a, b\}$, into a solvable group. But, as readily checked, 1_{aT} and 1_{bT} are both almost periodic and $aT \cap bT = \varphi$. Hence $A(T)$ cannot have a LIM.

Problem 29. If S is a discrete and left reversible, does there exist a closed left translation invariant subspace Z of $\ell^\infty(S)$, such that Z properly contains $W(S)$ and Z has a LIM?

HUANG (1988, [49]) has recently shown that this is true for a large class of discrete groups. CHING CHOU (oral communication) has informed us that this is even true for all groups.

Problem 30. Let S be a discrete semigroup. Consider on S the following fixed point property:

(G^w). Whenever S is a representation of S as weakly continuous and Q-non-expansive mappings on a weakly compact convex set X of a separated locally convex space (E, Q), then X contains a common fixed point for S.

Does"$W(S)$ has LIM" imply (G^w)?

Note that (G^w) implies "$W(S)$ has LIM" (see LAU (1976, [64])) and HSU (1985, [48]) proved that if S is left reversible, then (G^w) holds. In particular, HSU's result answers Problems 2, 7 and 8 raised in LAU (1976, [64]). The remaining five problems posed in LAU (1976, [64]) are still open.

References

[1] Argabright, L. N., and C. O. Wilde, Semigroups satisfying a strong Følner condition, Proc. Amer. Math. Soc. 18 (1967), 587–591

[2] Bade, W. G., P. C. Curtis, Jr., and H. G. Dale, Amenability and weak amenability for Beurling and Lipschitz algebras, Proc. Lond. Math. 57 (1987), 359–377

[3] Banach, S., Sur le problème de la mesure, Fund. Math. 4 (1923), 7–33

[4] Berglund, J. F., H. D. Junghenn, H.D., and P. Milnes, "Compact right topological semigroups and generalizations of almost periodicity", Lecture Notes in Mathematics 663 (1978), Springer-Verlag, Berlin etc

[5] Chou, C., On the size of the set of left invariant means on a semigroup, Proc. Amer. Math. Soc. 23 (1969), 199–205

[6] —, On a geometric property of the set of invariant means of a group, Proc. Amer. Math. Soc. 30 (1971), 296–302

[7] Day, M. M., Ergodic theorems for abelian semigroups, Trans. Amer. Math. Soc. 51 (1942), 583–608

[8] —, Amenable groups, Bull. Amer. Math. Soc. 56 (1950), 46

[9] —, Amenable semigroups, Illinois J. Math. 1 (1957), 509–544

[10] —, Fixed point theorems for compact convex sets, Illinois J. Math. 5 (1961), 585–590

[11] —, Semigroups and amenability, in: Semigroups (K.W. Folley, ed.), Academic Press, New York-San Francisco-London, 1969, pp. 5–53.

[12] —, "Normed linear spaces", 3rd ed., Springer-Verlag, Berlin-Heidelberg-New York, 1973

[13] —, Lumpy subsets in left amenable locally compact semi-groups, Pac. J. Math. 62 (1976), 87–93

[14] —, Left thick to left lumpy - a guided tour, Pac. J. Math. 101 (1982), 71–92

[15] Dixmier, J., Les moyennes invariantes dans les semi-groupes et leurs applications, Acta Sci. Math. Szeged 12 (1950), 213–227

[16] Duncan, J. and I. Namioka, I., Amenability of inverse semigroups and their semigroup algebras, Proc. Roy. Soc. Edinburgh, Sect. A 80 (1978), 309–321

[17] Duncan, J. and A. Paterson, Amenability for discrete convolution semigroup algebras, Preprint (1989)

[18] Dzinotyiweyi, H. A. M., Algebras of functions and invariant means on semigroups, Preprint (1978)

[19] —, The cardinality of the set of invariant means on a locally compact topological semigroup, Compositio Math. 54 (1985), 41–49

[20] —, The cardinality of the set of left invariant means on topological semigroups, Bull. Austr. Math. Soc. 37 (1988), 247–262

[21] Fairchild, L. R., Extreme invariant means without minimal support, Trans. Amer. Math. Soc. 172 (1972), 83–93

[22] Fan, K., Invariant subspaces for a semigroup of linear operators, Indag. Math. 27 (1965), 447–551

[23] —, Extension of invariant linear functionals, Proc. Amer. Math. Soc. 66 (1977), 23–29

[24] Følner, E., On groups with full Banach mean value, Math. Scand. 3 1955, 243–254

[25] Forrest, B., Invariant means, right ideals and the structure of semitopological semigroups, Semigroup Forum (1989), to appear

[26] Frey, A. H., "Studies in amenable semigroups", Thesis, University of Washington, Seattle, Washington, 1960

[27] Glicksberg, I., Weak compactness and separate continuity, Pac. J. Math. 11 (1961), 205–216

[28] —, On convex hulls of translates, Pac. J. Math. 13 (1963), 153–164

[29] Goya, E., "Extremely amenable semigroups", Dissertation, Michigan State University, 1972

[30] Granirer, E., On amenable semigroups with a finite-dimensional set of invariant means, I, Illinois J. Math. 7 (1963), 32–48

[31] —, On amenable semigroups with a finite-dimensional set of invariant means, II, Illinois J. Math. 7 (1963), 49–58

[32] —, Extremely amenable semigroups, Math. Scand. 17 (1965), 177–197

[33] —, On the invariant mean on topological semigroups and on topological groups, Pac. J. Math. 15 (1965), 107–140

[34] —, Extremely amenable semigroups II, Math. Scan. 20 (1967), 93–113

[35] —, Functional analytic properties of extremely amenable semigroups, Trans. Amer. Math. Soc. 137 (1969), 53–75

[36] —, "Exposed points of convex sets and weak sequential convergence", Memoirs of the Amer. Math. Soc. 123 (1972)

[37] —, Geometric and topological properties of certain w^*-compact convex sets which arise from the study of invariant means, Can. J. Math. 37 (1985), 107–121

[38] Granirer, E. E., and A. T. Lau, Invariant means on locally compact groups, Illinois J. Math. 15 (1971), 249–257

[39] Greenleaf, F., "Invariant means on topological groups", Van Nostrand, New York, 1969

[40] Grønbæk, N., Amenability of weighted discrete convolution algebras on cancellative semigroups, Proc. Royal Soc. Edinburgh, (1989), to appear

[41] —, Amenability of discrete convolution algebras, "the commutative case", Preprint (1989)

[42] —, A characterization of weakly amenable Banach algebras, Studia Math., (1989), to appear

[43] Hausdorff, F., "Grundzüge der Mengenlehre", Leipzig, 1914

[44] Hewitt E., and K. Ross, "Abstract harmonic analysis, I", Springer-Verlag, Berlin etc., 1963

[45] Hochster, M., Subsemigroups of amenable groups, Proc. Amer. Math. Soc. 21 (1969), 363–364

[46] Hofmann, K. H., and P. S. Mostert, "Elements of compact semigroups", C. E. Merrill, Columbus, Ohio (1966)

[47] Holmes, R. D., and A. T. Lau, Non-expansive actions of topological semigroups and fixed points, J. London Math. Soc. 5 (1972), 330–336

[48] Hsu, R., "Topics on weakly almost periodic functions", Dissertation, SUNY at Buffalo, 1985

[49] Huang, Y., Topics on invariant means and weakly almost periodic functions on semigroups, Dissertation, State Univ. of New York, Buffalo, 1988

[50] Johnson, B. E., Separate continuity and measurability, Proc. Amer. Math. Soc. 20 (1969), 420–422

[51] —, "Cohomology in Banach algebras", Memoirs Amer. Math. Soc. 127 (1972)

[52] Junghenn, H. D., Topological left amenability of semidirect products, Canad. Math. Bull. 24 (1981), 79–85

[53] Kharaghani, H., The convolution of Radon measures, Proc. Amer. Math. Soc. 103 (1988), 1189-1191

[54] Khelemskiĭ, A. Ya, Flat Banach modules and amenable Banach algebras, Trans. Moscow Math. Soc. 47 (1984) 179–218; Amer. Math. Soc. Transl. 47 (1985), 199–227

[55] Klawe, M., On the dimension of left invariant means and left thick subsets, Trans. Amer. Math. Soc. 231 (1977), 507–518

[56] —, Semidirect product of semigroups in relation to amenability, cancellation properties and strong Følner conditions, Pac. J. Math. 73 (1977), 91–106

[57] —, Dimensions of the sets of invariant means of semigroups, Illinois J. Math. 24 (1980), 233–243

[58] Lau, A. T., Topological semigroups with invariant means in the convex hull of multiplicative means, Trans. Amer. Math. Soc. 148 (1970), 69–84

[59] —, Functional analytic properties of topological semigroups and n-extreme amenability, Trans. Amer. Math. Soc. 152 (1970), 431–439

[60] —, Invariant means on almost periodic functions and fixed point properties, Rocky Mountain J. Math. 3 (1973), 69–76

[61] —, Invariant means on subsemigroups of locally compact groups, Rocky Mountain J. Math. 3 (1973), 77–81

[62] —, Amenability and invariant subspaces, J. Austral. Math. Soc. 18 (1974), 200–204

[63] —, Semigroups of operators on dual Banach spaces, Proc. Amer. Math. Soc. 54 (1976), 393–396

[64] —, Some fixed point theorems and their applications to W^*-algebras, in: "Fixed Point Theory and its Applications", S. Swaninathan Ed., Academic Press, New York, 1976, pp. 121–129

[65] —, Extension of invariant linear functions: a sequel to Fan's paper, Proc. Amer. Math. Soc. 63 (1977), 259–262

[66] —, Invariantly complemented subspaces of $L_\infty(G)$ and amenable locally compact groups, Illinois J. Math. 26 (1982), 226–235

[67] —, Analysis on a class of Banach algebras with applications to harmonic analysis on locally compact groups and semigroups, Fund. Math. 118 (1983), 161–175

[68] —, Finite-dimensional invariant subspaces for a semigroup of linear operators, J. Math. Anal. Appl. 97 (1983), 374–379

[69] —, Continuity of Arens multiplication on the dual space of bounded uniformly continuous functions on locally compact groups and topological semigroups, Math. Proc. Cambridge Philos. Soc. 99 (1986), 273–283

[70] Lau, A.T., and V. Losert, Weak* closed complemented invariant subspaces of $L_\infty(G)$ and amenable locally compact groups, Pac. J. Math. 123 (1986), 149–159

[71] Lau, A. T., and A. L. T. Paterson, The exact cardinality of the set of topological left invariant means on an amenable locally compact group, Proc. Amer. Math. Soc. 98 (1986), 75–80

[72] Lau, A. T., A. L. T. Paterson, and J. C. S. Wong, Invariant subspace theorems for amenable groups, Proc. Edinburgh Math. Soc., to appear

[73] Lau, A. T., and J. C. S. Wong, Finite-dimensional invariant subspaces for measurable semigroups of linear operators, J. Math. Anal. Appl. 127 (1987), 548–558

[74] Lau, A. T., and J. C. S. Wong, Invariant subspaces for algebras of linear operatoars and amenable locally compact groups, Proc. Amer. Math. Soc. 102 (1988), 581–586

[75] Luthar, I. S., Uniqueness of the invariant mean on an abelian semigroup, Illinois J. Math. 3 (1959), 28–44

[76] —, Uniqueness of the invariant mean on abelian topological semigroups, Trans. Amer. Math. Soc. 104 (1962), 403–411

[77] Milnes, P., Left mean ergodicity, fixed points and invariant means, J. Math. Anal. Appl. 65 (1978), 32–43

[78] Mitchell, T., Constant functions and left invariant means on semigroups, Trans. Amer. Math. Soc. 119 (1965), 244-261

[79] —, Fixed points and multiplicative left invariant means, Trans. Amer. Math. Soc. 122 (1966), 195–202

[80] —, Fixed points of reversible semigroups of non-expansive mappings, Kodai Math. Sem. Rep. 22 (1970), 322–323

[81] —, Topological semigroups and fixed points, Illinois J. Math. 14 (1970), 630–641

[82] Namioka, I., Følner's conditions for amenable semigroups, Math. Scand. 15 (1964), 18–28
[83] —, On certain actions of semi-groups on L-spaces, Studia Math. 29 (1967), 63–77
[84] Nassar, M., "Ergodic measures and recurrent points in the Stone-Čech Compactification of amenable semigroups", Dissertation, SUNY at Buffalo, 1984
[85] Paterson, A. L. T., Amenability and locally compact semigroups, Math. Scand. 42 (1978), 271—288
[86] —, The cardinality of the set of left invariant means on a left amenable semigroups, Illinois J. Math. 29 (1985), 567–583
[87] —, "Amenability", Amer. Math. Soc. Mathematical Surveys and Monographs, 29, 1988
[88] Pier, J. P., "Amenable locally compact groups", John Wiley and Sons, New York, 1984
[89] —, "Amenable Banach algebras", Pitman Research Notes in Mathematics 172, 1988
[90] Riazi, A. and J. C. S. Wong, Characterizations of amenable locally compact semigroups, Pac. J. Math. 108 (1983), 479–496
[91] Riemersma, M., On a theorem of Glicksberg and fixed point properties of semigroups, Indag. Math. 33 (1971), 340–345
[92] Rosen, W. G., On invariant means over compact semigroups, Proc. Amer. Math. Soc. 7 (1956), 1076–1082
[93] Ruppert, W. A. F., "Compact semitopological semigroups: an intrinsic theory", Lecture Notes in Mathematics 1079, 1984, Springer-Verlag, Berlin etc
[94] Silverman, R. J., Means on semigroups and the Hahn-Banach extension property, Trans. Amer. Math. Soc. 83 (1956), 22–237
[95] Sorenson, J. R., "Existence of measures that are invariant under a semigroup of transformations", Thesis, Purdue Univ., 1966
[96] Takahashi, W., Fixed point theorem for amenable semigroup of non-expansive mappings, Kodai Math. Sem. Rep. 21 (1969), 383–386
[97] Talagrand, M., Geométrie des simplexes de moyennes invariantes, J. of Funct. Anal. 34 (1979), 304–337
[98] von Neumann, J., Zur allgemeinen Theorie des Maßes, Fund. Math. 13 (1929), 73–116
[99] Wong, J. C. S., An ergodic property of locally compact semigroups, Pac. J. Math. 48 (1973), 615–619
[100] —, A characterization of topological left thick subsets in locally compact left amenable semigroups, Pac. J. Math. 62 (1976), 295–303
[101] —, Convolution and separable continuity, Pac. J. Math. 75 (1978), 601–611
[102] —, On Følner conditions and Følner numbers for semigroups, Preprint, 1988
[103] Wong, J. C. S., and A. Riazi, Characterizations of amenable locally compact semigroups, Pac. J. Math. 108 (1983), 479–496
[104] Yang, Z. Exposed points of left invariant means, Pac. J. Math. 125 (1986), 487–494
[105] —, On Følner numbers and Følner-type conditions for amenable semigroups, Illinois J. Math. 31 (1987), 496–517
[106] —, On the set of invariant means, J. London Math. Soc. 37 (1988), 317–330

Part IV

Applications: Systems theory, number theory, probability, topology

Applications of semigroups to geometric control theory

Ivan Kupka

Introduction

Quite a few problems in control theory boil down to the study of the action of special classes of semigroups to be specified below. Actually, in the general case, one must study the action of *pseudo-semigroups*. But we shall content ourselves with semigroups and illustrate our statement above by discussing some of the main problems in control theory.

1. Control semigroups and their actions

Let U be a set and let $F(U)$ denote the set of all mappings $u: J \to U$, where J is a closed compact interval (depending on u). Then $F(U)$ is a semigroup under *the concatenation operation*: Let $u: [\alpha, \beta] \longrightarrow U$ and $v: [\gamma, \delta] \to U$ be given. Then

$$u + v: [\alpha, \beta + \delta - \gamma] \to U, \quad (u + v)(t) = \begin{cases} u(t), & \text{if } \alpha \leq t < \beta \\ v(t + \gamma - \beta), & \text{if } \beta \leq t \leq \beta + \delta - \gamma \end{cases}$$

is the concatenation of u and v. The real additive group $\mathbb{R}$ acts on $F(U)$ by time translation: Let $\tau \in \mathbb{R}$ and let $u: [\alpha, \beta] \to \mathbb{R}$ be an element of $F(U)$. Then $\tau \cdot u: [\alpha - \tau, \beta - \tau] \to U$ is given by $(\tau \cdot u)(t) = u(t + \tau)$. These time translations are semigroup automorphisms and the orbits of this action are congruence classes of a congruence on $F(U)$. Since we will have to integrate, we need more structure on U. We shall assume that U has been provided with a σ-algebra of subsets.

Definition 0. A *control semigroup* $S(U)$ is a subsemigroup of $F(U)$ such that
(i) all elements u in $S(U)$ are measurable, and
(ii) if u belongs to $S(U)$ and K is any closed subinterval of the domain of u, the restriction $u|K$ of u to K also belongs to $S(U)$.
Since in the following we shall restrict ourselves to autonomous system we shall assume moreover that
(iii) $S(U)$ is stable under time translations. □

Since the time translations induce automorphisms of the semigroup $S(U)$, the orbits of the action of $\mathbb{R}$ are also congruence classes of $S(U)$ and the orbit space for this action is a quotient semigroup $QS(U)$ of $S(U)$. The most popular example of a control semigroup is $PC(U)$, the set of all piecewise constant mappings. Assuming that U is a subset of a Banach space B and that the σ algebra chosen on U is just the Borel algebra, then the set $L^1(U)$ of all $u: J \to U$ such that u belongs to $L^1[J; B]$ is another such example. A third is $L^\infty(U)$, the set of all $u: J \to U$ that are measurable and bounded.

Now we have to define the actions of those semigroups which are of interest for control theory.

Definition 1. Given a control semigroup, a *control system having* $S(U)$ *as set of admissible controls* will be a smooth (C^∞ or C^ω) paracompact connected manifold X, called *state space*, endowed with a family $\{F_u | u \in U\}$ of smooth (C^∞ or C^ω) vector fields parametrized by U and satisfying the following conditions:

(i) The correspondence $u \mapsto F_u$ is measurable (for the canonical topology on the space of vector fields on a manifold).

(ii) For each $\widehat{u}: [\alpha, \beta] \to U$ in $S(U)$, and each x_0 in X, there exists a unique absolutely continuous curve $\widehat{x}: [\alpha, \beta] \to X$ such that $\widehat{x}(\alpha) = x_0$ and

$$\frac{d\widehat{x}}{dt} = f_{\widehat{u(t)}}(\widehat{x}(t))$$

for almost all t in $[\alpha, \beta]$. □

Such a system defines an action of $S(U)$ on X:

$$(\widehat{u}, x_0) \mapsto \widehat{u}\cdot x_0: S(U) \times X \to X, \quad \widehat{u}\cdot x_0 = \widehat{x}(\beta)$$

for $\widehat{u}: [\alpha, \beta] \to U$ in $S(U)$ and for x_0 in X. Under these very general assumptions it does not follow that the mapping, $x_0 \mapsto \widehat{u}\cdot x_0: X \to X$ is a diffeomorphism. In fact it may not even be continuous. But usually the mappings $\rho(\widehat{u}): X \to X$, $\rho(\widehat{u})x_0 = \widehat{u}\cdot x_0$ will be diffeomorphisms, and $\rho: S(U) \to \mathrm{Diff}(M)$ will be a representation of $S(U)$.

An quite general example of this kind arises when $S(U) = PC(U)$ and all the vector fields F_u, $u \in U$, are complete. Then $\rho\big(PC(U)\big)$ will be the subsemigroup of Diff(X) generated by the one-parameter semigroups $\{\exp tF_u \mid t \geq 0\}$, $u \in U$.

One last remark: If $\widehat{u}$ is in $S(U)$ and $\tau \in \mathbb{R}$, then the response of the system to the time translate $\tau\cdot\widehat{u}$ is just the time translate $\tau\cdot\widehat{x}$ reponse $\widehat{x}$ to $\widehat{u}$. Hence the action of $S(U)$on X can be factored through the quotient $QS(U)$ of $S(U)$.

Now we go to the *accessibility problem* in control theory.

2. Accessibility

The control semigroup we are going to consider here will be $PC(U)$ unless otherwise indicated. Further $(X, \{F_u \mid u \in U\})$ will be a control system having $PC(U)$ as admissible control semigroup.

Definition 2. (i) The orbit $PC(U){\cdot}x$ of a point x in X under the action of $PC(U)$ is called *the accessibility set of* x and is denoted customarily by $A(x)$.

(ii) The system is called *controllable* (or better *transitive*) if $A(x) = X$ for all $x \in X$, in other words, if the action of $PC(U)$ is transitive on X. □

The main problems of accessibility theory are these:

(i) What can one say about the orbits $P(C){\cdot}x$?

(ii) Give conditions for a system to be transitive or give conditions for a system to be non-transitive.

The only general fact known about question (i) is the following.

Theorem 0. (Orbit theorem) (i) *Assume the following reversibility hypothesis: For any* $x, y \in X$, *either* $A(x) \cap A(y) = \emptyset$ *or* $A(x) = A(y)$. *Then the orbits* $A(x)$ *are immersed manifolds, that is, each* $A(x)$ *carries a smooth manifold structure such that the canonical injection* $i: A(x) \hookrightarrow X$ *is an injective smooth immersion.*

(ii) *If* $t \mapsto \big(x(t), u(t)\big): J \to X \times U$ *is any trajectory of the system, then* x *is entirely contained in one accessibility set* A. □

Remarks 1. a) The reversibility hypothesis is equivalent to the following: For any $x, y \in X$ the condition $x \in A(y)$ implies the condition $y \in A(x)$.

b) The reversibility hypothesis is satisfied if, for example, the subsemigroup $\rho\big(PC(U)\big)$ of Diff(X) defined by the action of $PC(U)$ on X is a group. This happens, for instance, if U carries an involution $u \mapsto u^*$ such that $F_{u^*} = -F_u$. □

It is difficult in general to determine the tangent space to $A(x)$ in the orbit theorem. If $\mathcal{L}$ is the Lie algebra generated by the vector fields $\{F_u \mid u \in U\}$, then for any $x \in X$, we have $\mathcal{L}(x) \subset T_xA(x)$, but in the C^∞ case, $\mathcal{L}(x)$ may be much smaller that $T_xA(x)$. On the other hand, in the analytic (C^ω) case, $\mathcal{L}(x) = T_xA(x)$ for all $x \in X$. We also get the following corollary in both the $C^\infty$and the C^ω case.

Corollary. *If* $\mathcal{L}(x) = T_xX$ *for all* $x \in X$, *and if the system is reversible, then* $A(x) = X$ *for all* $x \in X$. □

3. Lie saturates

Let us turn to question (ii). The above corollary gives a sufficient condition for *reversible* systems to be transitive, i.e., $\mathcal{L}(x) = T_xX$ for all $x \in X$. In 1976, V. JURDJEVIC and the author introduced an object which was to play, in the controllability problem, the same role with respect to a general system, as $\mathcal{L}$ plays with respect to reversible systems in the corollary above. The object we defined was that of *Lie saturates*.

For left (or right) invariant systems on Lie groups, to be defined below, this concept is a particular case of the notion of *Lie wedge*, introduced completely independently of the Lie saturate by K. H. HOFMANN and J. D. LAWSON.

Definition 3. (i) Given any system $F = \{F_u \mid u \in U\}$ on a manifold X, *the saturate* of the system, Sat(F), is the set of all smooth vector fields G on X such that, for any $x_0 \in X$, there exists a trajectory $x : [0,T] \to X$ of G starting at x_0 and contained in the closure $\overline{A}(x_0)$ of the accessibility set $A(x_0)$ of x_0. (ii) The Lie saturate $\mathcal{LS}(F)$ of the system F is the intersection Sat$(F) \cap$ Lie(F) of the saturate of F and its Lie algebra.

The interest of this concept lies in the following easy theorem which is entirely analogous to the corollary above:

Theorem 1. *Let F be a system such that for any $x \in X$, one has $\mathcal{LS}(F)(x) = T_xX$. Then F is transitive.* □

An nice application of this theorem is the following result due basically to C. LOBRY.

Proposition 0. *Let $F = F_0 + \sum_{i=1}^{m} u^i F_i$ be an affine control system such that*
(i) Lie$(F)(x) = T_xX$ *for all $x \in X$, and*
(ii) *the recurrent points of F_0 are everywhere dense.*
Then F is transitive.

Proof. Since $\pm F_j = \lim_{u\to\pm\infty} \frac{F_0+uF_j}{|u|}$, all the vector fields $\{\pm F_j, 1 \leq j \leq m\}$ belong to $\mathcal{LS}(F)$. Let us show that $\pm F_0$ also belongs to $\mathcal{LS}(F)$. To do this we shall show that for any $x \in X$ and any $t > 0$, the element $\exp -tF_0(x)$ belongs to $\overline{A(x)}$. Now (i) implies that the interior int $A(x)$ of $A(x)$ is dense in $A(x)$. Since the set $R(F_0)$ of recurrent points of F_0 is also dense, then $R(F_0) \cap A(x)$ is dense in $A(x)$. Take any open neighborhood M of $\exp -tF_0(x)$. Then one can find a point r in $R(F_0) \cap A(x) \cap \exp -tF_0(M)$. Since $\exp -tF_0(r) \in M$ and $r \in R(F_0)$, there exists a $T > 0$ such that $\exp(TF_0)(r)$ belongs to M. Now r and $\exp(TF_0)(r)$ belong to $A(x)$. Hence $M \cap A(x)$ is not empty for any neighborhood M of $\exp -tF_0(x)$

Hence we have proved that $\pm F_j, 0 \leq j \leq m$, all belong to $\mathcal{LS}(F)$. This shows that $\mathcal{LS}(F) \supset \text{Lie}(F)$ and, consequently, for any $x \in X, \mathcal{LS}(F)(x) = T_x X$. All we have to do now is to apply the theorem. □

4. Systems generated by group actions

Let G be a Lie group, Lie(G) its Lie algebra. Any subset U of Lie(G) gives birth to a system having control space U on any manifold X on which G acts smoothly.

Definition 4. Let $\mu: G \times X \to X$ be a smooth action of G on the manifold X. Any vector $u \in \text{Lie}(G)$ generates a smooth vector field F_u on X as follows: For $x \in X, F_u(x) = d_G\mu(1_G, x)u$, where d_G denotes the derivative along the first factor.

The system induced by U on X is just $\{F_u \mid u \in U\}$. It is called *a system subordinated to the action of G on X*. □

In the special case when $G = X$ and μ is the left translation action of G on itself, if we identify Lie(G) with the set of right-invariant vector-fields on G as is usually done, then the system induced by U is just U itself. The accessibility set $A(1_G, U)$ of the neutral element 1_G for that system is a subsemigroup of G. Given any action of G on a manifold X, the accessibility set $A(x, F_u, u \in U)$ of the point $x \in X$ under the system $F^U = \{F_u, u \in U\}$ induced by U on X is just $A(1_G, U)x$,

Now how about the Lie saturates of such systems? Again, given U, a subset of Lie (G), and any action $\mu : G \times X \to X$ of G on X, the Lie saturate $\mathcal{LS}(F^U)$ of the induced system F^U is also subordinated to the action μ and induced by a subset $\mathcal{LS}(U, \mu)$ of Lie(G), depending on μ. In particular if μ is the left translation action of G on itself, the corresponding subset $\mathcal{LS}(U)$ of Lie(G) is a Lie wedge in the sense of Hilgert, Hofmann, and Lawson [7]. We collect all the pertinent information in the following proposition.

Proposition 1. (i) *The Lie saturate $\mathcal{LS}(U)$ induced by a subset $U \subset$ Lie(G) on G by the left translation action, is a Lie wedge. The semigroup it generates contains $A(1_G, U)$ and is contained in the closure $\overline{A(1_G, U)}$.*

(ii) *If $\mu : G \times X \to X$ is any smooth action of G on the manifold X, and if $\mathcal{LS}(U, \mu) \subset$ Lie(G) induces the Lie saturate of the system F^U generated by U on X, then $\mathcal{LS}(U, \mu) \supset \mathcal{LS}(U)$. They are not equal, in general.*

(iii) *If U is transitive on G, then it will generate a transitive system F^U on any manifold X on which G acts transitively.* □

5. Accessibility for subordinated system

The theory of accessibility for systems associated with group actions has progressed quite far, at least for the action of a group on itself by translation. At the present time there two kinds of results. One kind, due to HILGERT, HOFMANN and LAWSON uses purely "algebraic" methods in contrast with the recurrence properties used in the other kind of results due to V. JURDJEVIC and the author. The first method is very suitable for nilpotent and solvable groups which are almost flat and hence have no recurrence. The second method applies very well to semi-simply groups which contain lots of "winding".

For the sake of completeness, before discussing the results mentioned just now, let us state the following trivial proposition:

Proposition. *Assume G is a compact connected Lie group. For a subset U in* Lie(G) *to generate a transitive system on G, it is necessary and sufficient that U generates* Lie(G) *as a Lie algebra.* □

A) *Nilpotent groups.* We have the following result of HILGERT, HOFMANN and LAWSON [7].

Theorem 2. *Let G be a connected nilpotent* Lie *group, U a wedge in* Lie(G) *generating it as a* Lie *algebra. Then U is transitive on G if and only if either of the following conditions is satisfied:*
(i) *The commutator* CLie(G) *of* Lie(G) *meets the interior of W.*
(ii) *The inverse image $\exp_G^{-1}(1_G)$ of G under the exponential mapping meets the interior of the extended wedge $W +$* CLie(G). □

The proof of this result is based on the following characterization:

let G be as above and let S be a maximal open proper semigroup in G. Then Lie wedge generating S is a halfspace of Lie(G) bounded by a hyperplane which is a Lie subalgebra of Lie(G).

B) *Solvable groups and semi-direct products.* The preceeding result was greatly generalized by J. LAWSON in [11]. This paper contains a wealth of results about maximal subsemigroups of Lie groups but we shall mention here only the ones that are directly relevant to accessibility.

Theorem 3. *Let G be a connected simply connected* Lie *group such that $G/$*Rad *G is compact* (Rad$G$ *is the maximal solvable normal subgroup of G). Then the maximal semigroups with non-empty interior are the semigroups generated by half-spaces in* Lie (G) *whose boundary is a subalgebra.* □

This implies the following corollary:

Corollary 1. *Let G be as in* Theorem 3. *Then a subset U in* Lie (G) *will generate a transitive system on G if and only if it is not contained in a half space of* Lie (G) *bounded by a* Lie *subalgebra.* □

From this we get easily another corollary. The result was proved previously by BONNARD et al. [3].

Corollary 2. *Let G be a semi-direct product of a compact group K and a vector group V on which K acts linearly without fixed points* (*except* 0 *of course*). *Then the system induced by a subset U in* Lie (G) *on G is transitive if and only if U generates* Lie (G) *as a* Lie *algebra.* □

C) *The semi-simple case.* The case when G is a semi-simple group was discussed in a series of papers starting with the paper [9] which laid the foundations and introduced the methods for the subsequent papers [1], [2], [3], [5], [11]. To state these results, we need a few definitions from [10].

Definition 5. An element B in a real semisimple algebra L is called *strongly regular* if
(i) all the non-zero eigenvalues (possibly complex) of $\operatorname{ad} B$ are simple, and
(ii) zero is an eigenvalue of $\operatorname{ad} B$ having a multiplicity equal to the rank of L. □

We order the complex numbers lexicographically, as follows: let $z_1, z_2 \in \mathbb{C} : z_1 \leq z_2$ if either $\Re z_2 > \Re z_1$ or $\Re z_2 = \Re z_1$ and $\Im z_2 \geq \Im z_1$. Here $\Re z$, $\Im z$ denote the real and imaginary part of the complex number z, respectively.

Definition 6. Let S be a subset of $\mathbb{C}$. An element $s \in S$ is called *maximal in S* if for any $z \in S$ which is ≥ 0 for the ordering, $s + z$ does not belong to S.

Definition 7. Let S be a subset of $\mathbb{C}$. An element σ in S is called *simple* if
(i) $\sigma \geq 0$,
(ii) there does not exist z_1, z_2 in S, $z_1 > 0, z_2 > 0$, such that $\sigma = z_1 + z_2$. □

Let U be a subset of Lie(G) where G is a real semi-simple group with a finite center. We want to give condition on U for the left invariant system induced by U on G to be controllable (or transitive) on G. The following basic assumptions on U were introduced in [10]:

AS1 $U \cap (-U)$ contains a strongly regular element B.

Denote by $S(B)$ the set of all eigenvalues of B and by $S_0(B)$ the subset of the non-zero ones. If $L_{\mathbb{C}}$ is the complexification of Lie(G) and $\operatorname{ad}_{\mathbb{C}} B$ that of $\operatorname{ad} B$, then $L_{\mathbb{C}}$ is a direct sum $L_0 \oplus \sum_{\alpha \in S_0(B)} L_\alpha$, where L_0 is the kernel of $\operatorname{ad}_{\mathbb{C}} B$ corresponding to the eigenvalues α. The dimension of L_0 is equal to the rank of

$L_{\mathbb{C}}$. The vector spaces L_α, $\alpha \in S_0(B)$, are one-dimensional (over $\mathbb{C}$!). If A is an element of $L_{\mathbb{C}}$ (or L) we denote by $A(0)$, $A(\alpha)$, $\alpha \in S_0(B)$, its components in L_0, L_α, so that $A = A(0) + \sum_{\alpha \in S_0(B)} A(\alpha)$.

AS2 If σ is a maximal element in $S(B)$ there exist A_1, A_2 in U such that $A_1(\sigma)$, $A_2(-\sigma)$ are non zero.

AS3 If σ is a maximal element in $S(B)$ and its real part $s = \Re\sigma$ belongs to $S_0(B)$, then there exists A_+, A_- in U such that $\mathrm{Kil}\big(A_+(s), A_-(-s)\big) < 0$ where $\mathrm{Kil}\colon L_{\mathbb{C}} \times L_{\mathbb{C}} \to \mathbb{C}$ denotes the Killing form of $L_{\mathbb{C}}$.

The assumptions **AS1,2,3** insure that there exist elements X in Lie(G) such that the whole line $\mathbb{R}{\cdot}X$ is contained in $LS(U)$ and are powerful enough to allow us to generate many other reversible elements Y in $LS(U)$ (that is, elements such that $\mathbb{R}{\cdot}Y \subset LS(U)$) by performing the following operations:

– operating with $e^{t\,\mathrm{ad}\,X}$, $t \in \mathbb{R}$, on $LS(U)$,

– taking positive combinations and limits.

Hence the assumtions **AS1, 2, 3** will insure that $LS(U)$ contains many reversible elements; but if $LS(U)$ is too small, they will not guarantee that U is transitive. One needs one more assumption **AS4** to warrant this. The assumption **AS4** is generic. Examples were given in the paper [10].

On the other hand it is clear from the orbit theorem that a necessary condition for U to be transitive is that the Lie algebra $L(U)$ generated by U is just Lie(G). Hence the minimal candidate for assumption **AS4** would be that $L(U)$ is Lie(G). The assumptions made in [10] obviously imply this but they are stronger. Indeed the following result was stated in the preprint [2], at least in the case where U is the set $\{A + uB \mid u \in \mathbb{R}\}$, $A, B \in$ Lie (G), and B satisfies **AS1**.

Theorem 4. *Assume that U is a subset of* Lie (G) *satisfying the assumptions* **AS1, 2, 3**. *Then U is transitive if and only if the Lie algebra generated by U is* Lie (G) *itself.*

Let us sketch a proof of this theorem at least in the case where Lie (G) is simple. For any $\alpha \in S_0(B)$, denote by $\overline{\alpha}$ its complex conjugate and by $L(\alpha)$ the real "eigenspace" $(L_\alpha \oplus L_{\overline{\alpha}}) \cap L$. Let also $L(0)$ be $L_0 \cap L$. It follows from Proposition 11 and 12 of [9], that $L(U)$ is the algebra generated by the $L(\alpha), \alpha \in S_0(B)$, such that $A(\alpha) \neq 0$ for some $A \in U$, and by $U \cap L(0)$. Then one considers the subalgebra I generated by

(i) the $L(\sigma)$ and $L(-\sigma)$ such that σ is maximal in $S_0(B)$,

(ii) the $L(s)$ and $L(-s)$ where $s = \Re\sigma$, σ is maximal in $S_0(B)$, $s \in S_0(B)$,

(iii) all the $L(\alpha), \alpha \in S_0(B)$ such that $(\alpha) \subset LS(U)$ and there exists a maximal σ in $S_0(B)$ such that either $\alpha + \sigma$ or $\alpha - \sigma$ belongs to $S_0(B)$.

Then one proves that I is in fact an ideal in $L(U)$. Hence if we assume that $L(U)$ = Lie (G), since Lie (G) is simple and I is an ideal, I = Lie (G) (since I contains $L(\sigma)$, σ maximal, I is not 0). The fact that I is an ideal is not hard to see. □

D) *Final remarks.* The preceding theorems give a fairly good picture of transitivity for systems subordinated to a group action in the nilpotent and solvable case on one hand and the semisimple case on the other. One of the main problems, now, is to blend these results together to get results about general Lie groups. I believe this can be done.

References

[1] El Assoudi R., and J. P. Gauthier, Controllability of right invariant systems on real semisimple groups of type F_n, G_2, B_n, C_n, Math. for Control, Signals and Systems, to appear

[2] —, "Controllability of right invariant systems on semisimple Lie groups", Preprint

[3] Gauthier, J. P., and G. Bornard, Controllabilité des systèmes bilinéaires, SIAM J. Control and Opt. 20 (1982), 377–384

[4] Gauthier, J. P., I. Kupka, and G. Sallet, Controllability of right invariant systems on real semisimple Lie groups, Systems and Control Letters 5 (1984), 187–190

[5] Hilgert, J., "Maximal semigroups and controllability in products of Lie groups", Archiv d. Math. 49 (1987), 189–195

[6] Hilgert, J., K. H. Hofmann, and J. D. Lawson, Controllability of systems on a nilpotent Lie group, Beiträge Algebra Geometrie 30 (1985), 185–190

[7] Hofmann, K. H., and J. D. Lawson, On Sophus Lie's fundamental theorem, I, Indag. Math. 45 (1983), 453–466

[8] Hofmann, K. H., and J. D. Lawson, On Sophus Lie's fundamental theorem, II, Indag. Math. 46 (1984), 255–265

[9] Jurdjevic, V., and I. Kupka, Controllability of right invariant systems on semisimple Lie groups and their homogeneous spaces, Annales de l' Institut Fourier 31 (1981), 151–179

[10] Lawson, J. D., Maximal subsemigroups of Lie groups that are total, Proc. Edinb. Math. Soc. 30 (1987), 479–501

[11] Leite, S., and P. Crouch, Controllability on classical Lie groups, Math. for Control, Signals and Systems 1 (1988), 31–42

The semigroup $\beta\mathbb{N}$ and its applications to number theory

Neil Hindman

1. Introduction

We take $\beta\mathbb{N}$, the Stone-Čech compactification of the discrete set N of positive integers, to be the set of ultrafilters on N, the principal ultrafilters – that is, those with non-empty intersection – being identified with the points of N. The topology on $\beta\mathbb{N}$ is given by taking a basis for the open sets (and also for the closed sets), sets of the form $\overline{A} = \{p \in \beta\mathbb{N}: A \in p\}$ where $A \subseteq N$.

The branch of number theory with which we are concerned is called "Ramsey Theory" after the famous theorem of F. P. Ramsey [31]. A typical problem in Ramsey Theory can be thought of as follows: One has a set X and a collection $\mathcal{G}$ of "good" subsets of X. One asks whether, whenever X is divided into finitely many pieces, one piece must contain a member of $\mathcal{G}$. Viewed in this way one can see why we prefer the construction of $\beta\mathbb{N}$ which have the points being ultrafilters on $\mathbb{N}$; an ultrafilter on $\mathbb{N}$ is a filter on $\mathbb{N}$ with the additional property that whenever $\mathbb{N}$ is divided into finitely many pieces some one of those pieces is a member of the ultrafilter.

In this paper we will be interested in the applications of the algebraic structure of $\beta\mathbb{N}$ to Ramsey Theory. The connection originated in an idea of Fred Galvin who asked around 1970 whether there could exist $p \in \beta\mathbb{N}$ such that whenever $A \in p$ one has $\{x \in \mathbb{N}: A - x \in p\} \in p$. (Where $A - x = \{x \in \mathbb{N}: y + x \in A\}$. That is, A is translated to the left by x, with anything passing 0 falling off the end.) Galvin called such ultrafilters "almost translation invariant". He was interested in the existence of such an ultrafilter because he had a simple construction showing that any member of such an ultrafilter must contain $FS(B)$ for some finite set B. (Here $FS(B) = \{\Sigma F: F$ is a finite non-empty subset of $B\}$.) As a consequence, Galvin knew he would have a proof of what was then known as the Graham-Rothschild conjecture: If $\mathbb{N}$ is divided into finitely many pieces, some one of these pieces will contain $FS(B)$ for some infinite $B \subseteq N$.

I worked on this problem and initially succeeded only in showing that the continuum hypothesis implied the converse of Galvin's construction. That is, under CH the truth of the Graham-Rothschild conjecture implied the existence of an almost translation invariant ultrafilter. Soon thereafter, using a complicated combinatorial agrument, I showed directly [18] that the Graham-Rothschild conjecture, now known as the Finite Sum Theorem, is true.

GALVIN's almost translation invariant ultrafilters thus became figments of the continuum hypothesis, and even then the existence relied on two very complicated proofs. He was interested in establishing their existence in ZFC and one day in 1975 he asked STEVEN GLASER if such ultrafilters existed. When GLAZER quickly answered "yes", GALVIN tried to explain that GLAZER was missing something – it couldn't be that easy. It was!

It so happened that GLAZER was in the intersection of two sets of mathematicians; he may in fact have been the only member of this intersection. The first of these sets consisted of those who knew that, given a discrete semigroup $(S, +)$ one could extend the operation to the Stone-Čech compactification βS so that for each $p \in \beta S$, the function λ_p defined by $\lambda_p(q) = p + q$ is continuous (so that βS is "left-topological") and for each $x \in S$, the function ρ_x defined by $\rho_x(y) = y + x$ is continuous. This extension was implicitly established by DAY [9] in 1957 using methods of ARENS [1]. The first explicit statement seems to have been made by CIVIN and YOOD [7]. These mathematicians were also familiar with the following result of ELLIS [13]: If $(X, +)$ is any compact left-topological semigroup, it has an idempotent.

The second set involved the intersection of those mathematicians who thought of the Stone-Čech compactification of a discrete space S as the set of ultrafilters on S. (Most of the mathematicians in the first group thought of βS as a subspace of the dual of the space of bounded real or complex valued functions on S.) Accordingly, GLAZER wanted to know given p and q in $\beta\mathbb{N}$, precisely what ultrafilter was $p+q$. That is, for which $A \subseteq N$ is $A \in p+q$? The answer was that $A \in p+q$ if and only if $\{x \in \mathbb{N}: A - x \in p\} \in q$. From this characterization it follows immediately that an idempotent in $(\beta\mathbb{N}, +)$ is almost translation invariant.

Combined with GALVIN's original construction (which can be found in Lemma 2.3 below), this observation yielded a short and simple proof of the Finite Sum Theorem. Since then, there have been numerous other applications of the algebraic structure of $\beta\mathbb{N}$ to Ramsey Theory. It is the other applications and some related algebraic developments which we survey in this paper. (PAPAZYAN [29] has recently obtained a direct proof of the existence of almost translation invariant ultrafilters.)

Section 2 will consist of some results utilizing the extension of both addition and multiplication and their interactions. Section 3 will consider solutions to certain equations in $\beta\mathbb{N}$ and their relationship to Ramsey Theory and the algebraic structure of $\beta\mathbb{N}$. In Section 4 we present a new and easy proof of van der Waerden's Theorem. Section 5 presents some results about the algebraic structure of $(\beta\mathbb{N}, +)$ which were motivated by combinatorial considerations. Many of the results of Sections 2 and 4 are consequences of my collaboration with VITALY BERGELSON, to whom I am indebted.

Given a set A and a cardinal κ we write $[A]^{\kappa} = \{B \subseteq A\colon |B| = \kappa\}$ and $[A]^{<\kappa} = \{B \subseteq A\colon |B| < \kappa\}$. We write ω for the first infinite cardinal (which is the first infinite ordinal) and take $\omega = \{0, 1, 2, \ldots\} = N \cup \{0\}$.

All hypothesized topological spaces are presumed to be Hausdorff.

2. Addition and multiplication in $\beta\mathbb{N}$

In the title of this paper "Semigroup" should really be "Semigroups". Some of the most important applications of the algebraic structureof $\beta\mathbb{N}$ have utilized the fact that ordinary addition and multiplication both extend to $\beta\mathbb{N}$ and interact. Exactly analogously with addition, given $A \subseteq \mathbb{N}$ and $p, q \in \beta\mathbb{N}$, we have that $A \in p \cdot q$ if and only if $\{x \in \mathbb{N}\colon A/x \in p\} \in q$, where $A/x = \{y \in \mathbb{N}\colon y \cdot x \in A\}$.

The theorem of ELLIS mentioned in the introduction is a basic tool, so we state it here.

Theorem 2.1 (ELLIS **[13]**). *Let $(X, +)$ be a compact Hausdorff left-topological semigroup. Then X has an idempotent.*

As we mentioned, Theorem 2.1 provides what we believe is the simplest and prettiest proof of the Finite Sum Theorem. (Here, "simplest" refers to ease of presentation. The original combinatorial proof is apparently simplest in terms of the logical structure needed to carry out the proof – see [6].) There are, however, other proofs. By way of contrast the simple result presented in Corollary 2.5 is now ten years old and has to date no other proof.

Definition 2.2. $\Gamma = \{p \in \beta\mathbb{N}$: for all $A \in p$, there exists $B \in [A]^{\omega}$ with $FS(B) \subseteq A\}$.

Lemma 2.3. (a) $\Gamma = c\ell\{p \in \beta\mathbb{N}\colon p + p = p\}$. (b) Γ *is a right ideal of* $(\beta\mathbb{N}, \cdot)$.

Proof. (a) One has immediately from it definition that Γ is closed. To see that $\{p \in \beta\mathbb{N}\colon p + p = p\} \subseteq \Gamma$, let $p \in \beta\mathbb{N}$ with $p + p = p$. (We now present GALVIN's original argument.) Let $A \in p$ and let $C_1 = A$. Since $C_1 \in p = p + p$, one has $\{x \in N\colon C_1 - x \in p\} \in p$ so that $\{x \in C_1\colon C_1 - x \in p\} \in p$. Pick $x_1 \in C_1$ with $C_1 - x_1 \in p$ and let $C_2 = C_1 \cap (C_1 - x_1)$. Inductively given $C_n \in p$, pick $x_n \in C_n$ with $x_n > x_{n-1}$ and $C_n - x_n \in p$ and let $C_{n+1} = C_n \cap (C_n - x_n)$. Let $B = \{x_n\colon n \in N\}$. Then $FS(B) \subseteq A$. (To see, for example that $x_6 + x_4 + x_3 + x_1 \in A$, note that $x_6 \in C_6 \subseteq C_5 \subseteq C_4 - x_4$. Thus $x_6 + x_4 \in C_4 \subseteq C_3 - x_3$ so that $x_6 + x_4 + x_3 \in C_3 \subseteq C_2 \subseteq C_1 - x_1$. Therefore $x_6 + x_4 + x_3 + x_1 \in C_1 = A$.)

To see that $\Gamma \subseteq c\ell\{p \in \beta\mathbb{N}\colon p + p = p\}$, let $q \in \Gamma$ and let $A \in q$. Pick $B \in [A]^{\omega}$ with $FS(B) \subseteq A$. We show there is some $p \in \beta\mathbb{N}$ with $FS(B) \in p$ and $p + p = p$, so that $\overline{A} \cap \{p \in \beta\mathbb{N}\colon p + p = p\} \neq \emptyset$. Let $T = \{p \in \beta\mathbb{N}$: for each $F \in [B]^{<\omega}$,

$FS(B\backslash F) \in p\}$. Since $\{FS(B\backslash F)\colon F \in [B]^{<\omega}\}$ has the finite intersection property one has $T \neq \emptyset$. By its definition T is closed. We claim T is a subsemigroup under addition. To this end, let $p, r \in T$ and let $F \in [B]^{<\omega}$. We show $FS(B\backslash F) \subseteq \{x \in \mathbb{N}\colon FS(B\backslash F) - x \in p\}$, so that $FS(B\backslash F) \in p + r$ as required. To this end, let $x \in FS(B\backslash F)$ and pick $G \in [B\backslash F]^{<\omega}$ with $x = \Sigma G$. Then $FS(B\backslash(F \cup G)) \in p$. Let $y \in FS(B\backslash(F \cup G))$ and pick $H \in [B\backslash(F \cup G)]^{<\omega}$ with $y = \Sigma H$. Then $y + x = \Sigma(H \cup G)$ and $H \cup G \in [B\backslash F]^{<\omega}$. Thus $FS(B\backslash(F \cup G)) \subseteq FS(B\backslash F) - x$ so $FS(B\backslash F) - x \in p$ as required. Since T is a compact semigroup, we have by Theorem 2.1 that T has an idempotent p. But then $FS(B) \in p$ which is what we were trying to show.

(b). We have by part (a) that $\Gamma \neq \emptyset$. Let $p \in \Gamma$ and let $q \in \beta\mathbb{N}$. To see that $p \cdot q \in \Gamma$, let $A \in p \cdot q$. Then $\{x \in \mathbb{N}\colon A/x \in p\} \in q$ so is non-empty. Pick $x \in \mathbb{N}$ with $A/x \in p$ and pick $B \in [A/x]^{\omega}$ with $FS(B) \subseteq A/x$. Then $Bx \in [A]^{\omega}$ and $FS(Bx) \subseteq A$. □

Note in part (b) above that we established an additively defined set is a multiplicative right ideal. Note further that the validity of the distributive law on $\mathbb{N}$ is essential to the proof. Be cautioned, however, that the distributive law does *not* hold in $\beta\mathbb{N}$ [12, Section 6]. In fact it is not known whether any $p, q, r \in \beta\mathbb{N}\backslash\mathbb{N}$ satisfy either of $p \cdot (q + r) = p \cdot q + p \cdot r$ or $(p + q) \cdot r = p \cdot r + q \cdot r$.

Theorem 2.4. *There exists $p \in \Gamma$ with $p \cdot p = p$.*

Proof. By Lemma 2.3(b), Γ is a compact subsemigroup of $(\beta\mathbb{N}, \cdot)$ so that Theorem 2.1 applies. □

In the following $FP(B)$ is defined analogously to $FS(B)$. That is, $FP(B) = \{\Pi F\colon F \in [B]^{<\omega}\}$.

Corollary 2.5. *Let $m \in \mathbb{N}$ and let $N = \cup_{i=1}^{m} A_i$. There exists $i \in \{1, 2, \ldots, m\}$ and there exist $B, C \in [A_i]^{\omega}$ with $FS(B) \subseteq A_i$ and $FP(C) \subseteq A_i$.*

Proof. Pick p as guaranteed by Theorem 2.4 and pick $i \in \{1, 2, \ldots, m\}$ with $A_i \in p$. Since $p \in \Gamma$, pick $B \in [A_i]^{\omega}$ with $FS(B) \subseteq A_i$. Since $p \cdot p = p$, repeat GALVIN's argument from the proof of Lemma 2.3 to get $C \in [A_i]^{\omega}$ with $FP(C) \subseteq A_i$. □

Since Corollary 2.5 first appeared [19], several other results utilizing the interaction between $(\beta\mathbb{N}, +)$ and $(\beta\mathbb{N}, \cdot)$ have appeared in [4], [11], and [25]. To illustrate the arguments, we present in the remainder of this section a portion of those results which are easy to state and prove.

Definition 2.6. (a) For $A \subseteq N$, $\overline{d}(A) = \limsup_{n\to\infty} |A \cap \{1,2,\ldots,n\}|/n$.
(b) $\Delta = \{p \in \beta B$: for each $A \in p$, $\overline{d}(A) > 0\}$.
(c) $\Delta_1 = \{p \in \beta\mathbb{N}$: for each $A \in p$ there exists $k \in \mathbb{N}$ such that $\cup_{t=1}^{k} A - t$ contains arbitrarily long blocks of $N\}$.
(d) Given $m, v, c \in \mathbb{N}$ and $\vec{x} \in \mathbb{N}^m$, $s(m,v,c,\vec{x}) = \{cx_t + \Sigma_{i=t+1}^{m} \lambda_i x_i : t \in \{1,2,\ldots,m\}$ and $\lambda_i \in \{-v, -v+1, \ldots, v-1, v\}\}$.
(e) $U = \{p \in \beta\mathbb{N}$:for each $A \in p$ and each $(m,v,c) \in \mathbb{N}^3$ there exists $\vec{x} \in \mathbb{N}^m$ with $S(m,v,c,\vec{x}) \subseteq A\}$.

The importance of the set U to Ramsey Theory comes from the relationship between the sets $S(m,v,c,\vec{x})$ and partition regular systems of homogeneous linear equations. A system of homogeneous linear equations with integer coefficients is said to be *partition regular* provided for each finite partition F of $\mathbb{N}$ there is a solution to the system with all entries coming from one cell of F. (We will say in this situation that one cell "contains a solution" to the system.) A result of Deuber [10, Satz 2.5] establishes that a set contains solutions to all partition regular systems of homogeneous linear equations if and only if it contains for each $(m,v,c) \in \mathbb{N}^3$ some $S(m,v,c,\vec{x})$. Thus $p \in U$ if and only if each member of p contains solutions to all partition regular systems of homogeneous linear equations. (The famous theorem of Rado [30] characterizes those systems which are partition regular.)

Note that, from the way they are defined (by "for each $A \in p\ldots$") one has immediately each of Δ, Δ_1, and U is topologically closed, hence compact. (If $A \in p$ and A fails the defining requirement, then since that requirement does not refer to p, one has $\overline{A}$ is a neighborhood of p missing the specified set.) We need that each of Δ, Δ_1, and U has certain algebraic properties. We present a detailed proof for the algebraic assertions about Δ_1 to illustrate the arguments, giving references for the other assertions.

Lemma 2.7. (a) *Δ is a regular right ideal of $(\beta\mathbb{N}, +)$ and of $(\beta\mathbb{N}, \cdot)$.*
(b) *Δ_1 is a left ideal of $(\beta\mathbb{N}, +)$ and a right ideal of $(\beta\mathbb{N}, \cdot)$.*
(c) *U is a subsemigroup of $(\beta\mathbb{N}, +)$ and a two-sided ideal of $(\beta\mathbb{N}, \cdot)$.*

Proof. (a) This is [12], Theorem 7.4 (or see [20], Lemma 10.8).

(b) That $\Delta_1 \neq \emptyset$ follows from [23], Lemma 3.4. To see that Δ_1 is a left ideal of $(\beta\mathbb{N}, +)$, let $p \in \Delta_1$, $q \in \beta\mathbb{N}$, and $A \in q + p$. Let $B = \{x \in \mathbb{N}: A - x \in q\}$. Then $B \in p$ so pick $k \in \mathbb{N}$ such that $\bigcup_{t=1}^{k} B - t$ contains arbitrarily long blocks of $\mathbb{N}$. (To see that $\bigcup_{t=1}^{k} A - t$ contains abitrarily long blocks of $\mathbb{N}$, let $m \in \mathbb{N}$ and pick $y \in \mathbb{N}$ with $\{y+1, y+2, \ldots, y+m\} \subseteq \bigcup_{t=1}^{k} B - t$. For each $i \in \{1,2,\ldots,m\}$ pick $t(i) \in \{1,2,\ldots,k\}$ with $y+i+t(i) \in \mathbb{N}$. Pick $x \in \bigcap_{i=1}^{m} A-(y+i+t(i))$, which is non-empty since it is a member of q. Then $\{x+y+1, x+y+2, \ldots, x+y+m\} \subseteq \bigcup_{t=1}^{k} A-t$.

To see that Δ_1 is a right ideal of $(\beta\mathbb{N}, \cdot)$, let $p \in \Delta_1$, $q \in \beta\mathbb{N}$, and $A \in p \cdot q$. Then $\{x \in \mathbb{N}: A/x \in p\} \in q$ so pick x with $A/x \in p$. Pick $k \in \mathbb{N}$ such that $\bigcup_{t=1}^{k}(A/x) - t$ contains arbitrarily long blocks of $\mathbb{N}$. Let $\ell = x \cdot (k+1)$. To see that $\bigcup_{t=1}^{\ell} A - t$ contains arbitrarily long blocks of $\mathbb{N}$, let $m \in \mathbb{N}$ and pick $y \in \mathbb{N}$ with $\{y+1, y+2, \ldots, y+m\} \subseteq \bigcup_{t=1}^{k}(A/x) - t$. Let $z = x \cdot y$. To see that $\{z+1, z+2, \ldots, z+m\} \subseteq \bigcup_{t=1}^{\ell} A - t$, let $i \in \{1, 2, \ldots, m\}$ and let a be the least integer with $a \cdot x \geq i$. Then $a \in \{1, 2, \ldots, m\}$ so pick $s \in \{1, 2, \ldots, k\}$ with $y + a + s \in A/x$, so that $y \cdot x + a \cdot x + s \cdot x \in A$. That is, $z + i + (a \cdot x + s \cdot x - i) \in A$. Since $1 \leq a \cdot x + s \cdot x - i < s \cdot x + x \leq (k+1) \cdot x = \ell$, we have $z + i \in \bigcup_{t=1}^{\ell} A - t$.

(c) These conclusions follow from [11], Lemma 2 and [3], Lemma 2.2. □

Notice in the following proof that one alternately uses the additive and multiplicative structures of $\beta\mathbb{N}$ to get that smaller and smaller sets are non-empty.

Theorem 2.8. *There exists $p \in \beta\mathbb{N}$ such that $p \in \Gamma \cap \Delta \cap \Delta_1 \cap U$, $p = p \cdot p$, and p is in a minimal right ideal of $(\beta\mathbb{N}, \cdot)$.*

Proof. Since Δ is a right ideal of $(\beta\mathbb{N}, +)$ and Δ_1 is a left ideal of $(\beta\mathbb{N}, +)$ we have $\Delta \cap \Delta_1 \cap U \neq \emptyset$ and is hence a compact subsemigroup of $(\beta\mathbb{N}, +)$. Pick by Theorem 2.1 $q \in \Delta \cap \Delta_2 \cap U$ with $q + q = q$. By Lemma 2.3(a), $q \in \Gamma$ and hence $\Gamma \cap \Delta \cap \Delta_1 \cap U \neq \emptyset$. Since $\Gamma \cap \Delta \cap \Delta_1 \cap U$ is a right ideal of $(\beta\mathbb{N}, \cdot)$ it contains by a routine Zorn's Lemma application a minimal right ideal R of $(\beta\mathbb{N}, \cdot)$, which is closed and hence contains a multiplicative idempotent p. □

Corollary 2.9. *Let $m \in \mathbb{N}$ and let $\mathbb{N} = \bigcup_{i=1}^{m} A_i$, there exists $i \in \{1, 2, \ldots, m\}$ and there exist $B, C \in [A_i]^\omega$ and $k \in \mathbb{N}$ such that*

(a) $FS(B) \subseteq A_i$

(b) $FP(C) \subseteq A_i$

(c) $\overline{d}(A_i) > 0$,

(d) $\bigcup_{t=1}^{k} A_i - t$ *contains arbitrarily long blocks of* $\mathbb{N}$

(e) A_i *contains no solutions to all partition regular systems of homogeneous linear equations with integer coefficients.*

Proof. Pick p as guaranteed by Theorem 2.8 and pick $i \in \{1, 2, \ldots, m\}$ with $A_i \in p$. The conclusion follows respectively from the facts: (a) $p \in \Gamma$, (b) $p \cdot p = p$, (c) $p \in \Delta$, (d) $p \in \Delta_1$, and (e) $p \in U$. □

For a much longer list of conclusions, see [3], Corollary 3.16. The fact from Theorem 2.8 that p lies in a minimal right ideal was not utilized in Corollary 2.9. We postpone such utilization to Section 4.

3. Solving equations in $\beta\mathbb{N}$

For some time after the GALVIN-GLAZER proof of the Finite Sum Theorem, it remained open that one might have $p \in \beta\mathbb{N}$ with $p + p = p$ and $p \cdot p = p$. Applying GALVIN's argument to such an ultrafilter one would have: If $m \in \mathbb{N}$ and $\mathbb{N} = \bigcup_{i=1}^{m} A_i$, then there exist $i \in \mathbb{N}$ and $B \in [A_i]^\omega$ with $FS(B) \cup FP(B) \subseteq A_i$. (We have already seen in Corolary 2.5 how the existence of a multiplicative idempotent which is close to the set of additive idempotents yields a somewhat weaker combinatorial conclusion.)

Theorem 3.1. *There exists A_1, A_2 such that $\mathbb{N} = A_1 \cup A_2$ but there do not exist $i \in \{1, 2\}$ and $B \in [A_i]^\omega$ with $FS(B) \cup FP(B) \subseteq A_i$. Consequently no $p \in \beta\mathbb{N}$ has $p + p = p = p \cdot p$.*

Proof. [21]. □

It was then established [22] that the equation $p + q = p \cdot q$ has no solutions with $p, q \in \beta\mathbb{N}\backslash\mathbb{N}$. A question of VAN DOUWEN's [12, Question 8.7], initially raised in 1979, remains open: Is there a solution in $\beta\mathbb{N}\backslash\mathbb{N}$to the equation $p + q = r \cdot s$? To put it another way, letting $N^* = \beta\mathbb{N}\backslash\mathbb{N}$, is $(N^* + N^*) \cap (N^* \cdot N^*) \neq \emptyset$? We do have a result establishing that for many $r \in \mathbb{N}^*$, there do not exist solutions to $p + q = r \cdot s$.

Theorem 3.2. *Let $r \in \mathbb{N}^*$ and assume that for infinitely many $n \in \mathbb{N}$, $\mathbb{N}n \in r$. Then there do not exist $p, q, s \in \mathbb{N}^*$ with $p + q = r \cdot s$.*

Proof. [22], Theorem 5.3. □

We want to apply Theorem 3.2 to obtain some algebraic corollaries. Before we do this we need to state a portion of the basic structure theorem for left topological semigroups, due to RUPPERT.

Theorem 3.3. *Let $(X, *)$ be a compact left topological semigroup. Then X has a smallest two-sided ideal K and $K = \cup\{R\colon R$ is a minimal right ideal of $X\} = \cup\{L\colon L$ is a minimal left ideal of $X\}$. Further, if R is a minimal right ideal and L is a minimal left ideal, then $R \cap L$ is a group.*

Proof. [32], Satz 3. □

Theorem 3.4. *Let M be the smallest ideal of $(\beta\mathbb{N}, +)$ and let K be the smallest ideal of $(\beta\mathbb{N}, \cdot)$. Then $M \cap K = \emptyset$.*

Proof. Suppose we have $p \in M \cap K$. Pick R_1 and R_2, minimal right ideals of $(\beta\mathbb{N}, +)$ and $(\beta\mathbb{N}, \cdot)$, respectively with $p \in R_1 \cap R_2$. Then $p + \mathbb{N}^* \subseteq R$ and $p + \mathbb{N}^*$ is a right ideal of $\beta\mathbb{N}$ (since one easily has $\mathbb{N}^* + \beta\mathbb{N} \subseteq \mathbb{N}^*$). Thus $p + \mathbb{N}^* = R$, so there is some $q \in \mathbb{N}^*$ with $p + q = p$. Similarly, there is some $s \in \mathbb{N}^*$ with $p \cdot s = p$. But $\bigcap_{n=1}^{\infty} \overline{\mathbb{N}n}$ is an ideal of $(\beta\mathbb{N}, \cdot)$ so $K \subseteq \bigcap_{n=1}^{\infty} \overline{\mathbb{N}n}$ and hence $p \in \bigcap_{n=1}^{\omega} \overline{\mathbb{N}n}$. But then, by Theorem 3.2, $p + q \neq p \cdot s$. □

It is an easy fact that $(\beta\mathbb{N}, +)$, ρ_p is never continuous for $p \in \mathbb{N}^*$. It is a somewhat more difficult fact that the restriction of ρ_p to $\mathbb{N}^*$ is never continuous for $p \in \mathbb{N}^*$. (See [12].) We utilize Theorem 3.2 to establish that another simple function is not continuous.

Theorem 3.5. *Let $p \in \mathbb{N}^*$ and define $f: \mathbb{N}^* \to \mathbb{N}^*$ by $f(p) = p + p$. Then f is not continuous.*

Proof. We show $f[\mathbb{N}^*]$ is not closed in $\mathbb{N}^*$. By Lemma 2.3, $\Gamma = c\ell\{p \in \beta\mathbb{N}: p+p = p\}$ and Γ is a right ideal of $(\beta\mathbb{N}, \cdot)$. Pick a minimal right ideal R of $(\beta\mathbb{N}, \cdot)$ with $R \subseteq \Gamma$ and pick by Theorem 2.1 $q \in R$ with $q \cdot q = q$. As above we have $q \in \bigcap_{n=1}^{\infty} \overline{\mathbb{N}n}$ so that by Theorem 3.2, $q \cdot q \neq p + p$ for any $p \in \mathbb{N}^*$. That is, $q \cdot q \notin f[\mathbb{N}^*]$. But $q \cdot q = q \in \Gamma = c\ell\{p \in \beta\mathbb{N}: p + p = p\} \subseteq c\ell f[\mathbb{N}^*]$. □

4. Van der Waerden's Theorem

I have often talked about the applications of the algebraic structure of $\beta\mathbb{N}$ to Ramsey Theory and it often happens that I would be asked, "Can you prove van der Waerden's Theorem?". (This theorem, one of the oldest and most famous in Ramsey Theory, asserts that whenever $m \in \mathbb{N}$ and $\mathbb{N} = \bigcup_{i=1}^{m} A_i$, some A_i contains arbitrarily long arithmetic progressions.) It was always annoying that the answer was "no".

It should be pointed out for example that van der Waerden's Theorem is a consequence of Corollary 2.9. However, the fact that $U \neq \emptyset$ was utilized and this required at some state that Rado's Theorem (which implies van der Waerden's Theorem) be involved. At a more basic level, if we let $AP = \{p \in \beta\mathbb{N}$: for each $A \in p$, A contains arbitrarily long arithmetic progressions$\}$, then van der Waerden's Theorem is equivalent to the assertion that $AP \neq \emptyset$. We had however, no way of concluding that $AP \neq \emptyset$ without invoking van der Waerden's Theorem or something stronger.

I am happy to report that this situation has now changed. Furstenberg and Katznelson [15] have recently devised an easy proof of van der Waerden's Theorem using the topological algebraic structure of enveloping semigroups. Vitaly Bergelson and I [4] were able to utilize their ideas to obtain an even simpler proof using the topological algebraic structure of $\beta\mathbb{N}$. The four of us have now simplified the proof further [2]. We present this proof here in its entirety.

We fix $\ell \in \mathbb{N}$ and set out to show that each member of the smallest ideal of $\beta\mathbb{N}$ has all of its members containing length ℓ arithmetic progressions.

Definition 4.1. (a) $S = (\beta\mathbb{N})^\ell$ with the product topology and coordinatewise addition.
(b) $E^* = \{(a, a+d, a+2d, \ldots, a+(\ell-1)d) : a \in \mathbb{N}$ and $d \in \omega\}$
(c) $I^* = \{(a, a+d, a+2d, \ldots, a+(\ell-1)d) : a, d \in N\}$
(d) $E = c\ell_S E^*$
(e) $I = c\ell_S I^*$.
Note that the only difference in the definition of E^* and I^* is that, for E^* the number d is allowed to be 0.

Lemma 4.2. *S is a compact left topological semigroup.*
Further, if $\vec{x} = (x_1, x_2, \ldots, x_\ell) \in \mathbb{N}^\ell$, then $\rho_{\vec{x}} : S \to S$ is continuous.

Proof. One has immediately that S is a compact semigroup.
Given $\vec{p} = (p_1, p_2, \ldots, p_\ell) \in S$, the continuity of each λ_{p_i} in $\beta\mathbb{N}$ yields the continuity of $\lambda_{\vec{p}}$ in S. Likewise given $\vec{x} = (x_1, x_2, \ldots, x_\ell) \in \mathbb{N}^\ell$, the continuity of each ρ_{x_i} in $\beta\mathbb{N}$ yields the continuity of $\rho_{\vec{x}}$ in S. □

Lemma 4.3. *E is a compact subsemigroup of S and I is a two-sided ideal of E.*

Proof. E is trivially compact. We let $\vec{p}, \vec{q} \in E$ and show that $\vec{p} + \vec{q} \in E$. We show further that if either $\vec{p}$ or $\vec{q}$ is in I, then $\vec{p} + \vec{q} \in I$. To this end, let U be a neighborhood of $\vec{p} + \vec{q}$. By the continuity of $\lambda_{\vec{p}}$, pick a neighborhood V of $\vec{q}$ with $\vec{p} + V \subseteq U$. Since $\vec{q} \in E = c\ell E^*$, pick $a \in \mathbb{N}$ and $d \in \omega$ with $(a, a+d, \ldots, a+(\ell-1)d) \in V$. If $\vec{q} \in I$, choose $d \neq 0$. Let $\vec{x} = (a, a+d, \ldots, a+(\ell-1)d)$. Then $\vec{p} + \vec{x} \in U$ so by the continuity of $\rho_{\vec{x}}$ pick a neighborhood W of $\vec{p}$ with $W + \vec{x} \subseteq U$. Since $\vec{p} \in c\ell E^*$, pick $b \in \mathbb{N}$ and $e \in \omega$ (with $e \neq 0$ if $\vec{p} \in I$) such that $(b, b+e, \ldots, b+(\ell-1)e) \in W$ and let $\vec{Y} = (b, b+e, \ldots, b+(\ell-1)e)$. Then $\vec{y} + \vec{x} \in U \cap E^*$. Further, if $d \neq 0$ or $e \neq 0$, in particular if $\vec{p} \in I$ of $\vec{q} \in I$, $\vec{y} + \vec{x} \in U \cap I^*$. □

Theorem 4.4. *Let $p \in K(\beta\mathbb{N})$, the smallest ideal of $(\beta\mathbb{N}, +)$, and let $\vec{p} = (p, p, \ldots, p) \in S$. Then $\vec{p} \in I$. Consequently, each member of p contains a length ℓ arithmetic progression.*

Proof. We first show $\vec{p} \in E$. Indeed, let U be a neighborhood of $\vec{p}$ and pick $A_1, A_2, \ldots, A_\ell \in p$ with $\overline{A}_1 \times \overline{A}_2 \times \ldots \times \overline{A}_\ell \subseteq U$. Pick $a \in \bigcap_{i=1}^{\ell} A_i$. Then $(a, a, \ldots, a) \in U \cap E^*$.

Since $p \in K(\beta\mathbb{N})$ pick by Theorem 3.3 a minimal right ideal R of $\beta\mathbb{N}$ with $p \in R$. Now $\vec{p}+E$ is a right ideal of E so pick a minimal right ideal R^* of E with $R^* \subseteq p+E$. Now minimal right ideals in a compact left topological semigroup are closed so pick by Theorem 2.1 an idempotent $\vec{q}=(q_1,q_2,\ldots,q_\ell)$ in R^*. Then $\vec{q} \in \vec{p}+E$ so pick $\vec{s}=(s_1,s_2,\ldots,s_\ell)$ in E with $\overline{q}=\vec{p}+\vec{s}$.

We show that $\vec{p}=\vec{q}+\vec{p}$. To this end, let $i \in \{1,2,\ldots,\ell\}$. Then $q_i = p+s_i$ so $q_i \in R$ so $q_i+\beta\mathbb{N} \subseteq R$. Since R is minimal, $q_i+\beta\mathbb{N}=R$ so $p \in q_i+\beta\mathbb{N}$. Pick $t_i \in \beta N$ with $p=q_i+t_i$. Then $q_i+p=q_i+q_i+t_i=q_i+t_i=p$. Thus $\vec{p}=\vec{q}+\vec{p}$ as desired. Therefore $\vec{p} \in \vec{q}+E=R^*$ so $\vec{p} \in K(E) \subseteq I$ since $K(E)$ is the smallest ideal of E.

Finally, let $A \in p$. Then $\overline{A}\times\overline{A}\times\ldots\times\overline{A}$ is a neighborhood of $\vec{p}$ and $\vec{p} \in c\ell\ I^*$ so pick $a,d \in \mathbb{N}$ with $(a,a+d,\ldots,a+(\ell-1)d) \in \overline{A}\times\overline{A}\times\ldots\times\overline{A}$. Then $\{a,a+d,\ldots,a+(\ell-1)d\} \subseteq A$. □

Corollary 4.5 (VAN DER WAERDEN). *Let $m \in \mathbb{N}$ and let $N=\bigcup_{i=1}^m A_i$. Then there exists i such that A_i contains arbitrarily long arithmetic progressions.*

Proof. Pick a member p of the smallest ideal of $(\beta N,+)$ and pick i with $A_i \in p$. Apply Theorem 4.4 (once for each length ℓ). □

Definition 4.6. Let $(S,+)$ be a discrete semigroup. A subset A of S is called *central* provided there is an idempotent p in the smallest ideal of βS with $A \in p$.

Using refinements of the arguments used in the proof of van der Waerden's Theorem above, BERGELSON and I [4] proved the following theorem (which is originally due to FURSTENBERG and WEISS in [14]). In fact we proved a generalization which yields both Theorem 4.7 and the new Theorem 4.8.

Theorem 4.7. *Let $\ell \in \mathbb{N}$ and for each $i \in \{1,2,\ldots,\ell\}$, let $\langle y_{i,n}\rangle_{n=1}^\infty$ be a sequence in $\mathbb{N}$ and let A be a central set in $(N,+)$. There exist a sequence $\langle a_n\rangle_{n=1}^\infty$ in $\mathbb{N}$ and a sequence $\langle H_n\rangle_{n=1}^\infty$ of pairwise disjoint finite non-empty subsets of $\mathbb{N}$ such that whenever F is a finite non-empty subset of $\mathbb{N}$ one has*

$$\sum_{n\in F}(a_n+\sum_{m\in H_n} y_{1,m}),\ \sum_{n\in F}(a_n+\sum_{m\in H_n} y_{2,m}),\ldots,\sum_{n\in F}(a_n+\sum_{m\in H_n} y_{\ell,m})) \in A^{\ell+1}\ .$$

Theorem 4.8. *Let $\ell \in \mathbb{N}$ and for each $i \in \{1,2,\ldots,\ell\}$, let $\langle y_{i,n}\rangle_{n=1}^\infty$ be a sequence in $\mathbb{N}\backslash\{1\}$ and let A be a central set in $(\mathbb{N},\cdot)$. There exist a sequence $\langle a_n\rangle_{n=1}^\infty$ in $\mathbb{N}\backslash\{1\}$ and a sequence $\langle H_n\rangle_{n=1}^\infty$ of pairwise disjoint finite non-empty subsets of $\mathbb{N}$ such that, whenever f is a finite non-empty subset of $\mathbb{N}$, one has*

$$(\prod_{n\in F}(a_n\cdot\prod_{m\in H_n} y_{1,m}),\ \prod_{n\in F}(a_n\cdot\prod_{m\in H_n} y_{2,m}),\ldots,\prod_{n\in F}(a_n\cdot\prod_{m\in H_n} y_{\ell,m})) \in A^{\ell+1}\ .$$

Once again, the intersection of addition and multiplication in $\beta\mathbb{N}$ becomes useful. As we saw, the ultrafilter p chosen in Theorem 2.8 was in a minimal right ideal of $(\beta\mathbb{N}, \cdot)$. Thus every member of p is multiplicatively central. Thus the conclusion of Theorem 4.8 can be added to the conclusions of Corollary 2.9. What we did not show in Theorem 2.8 was that in fact p can be chosen so that each of its members are additively central, so that the conclusion of Theorem 4.7 can be added to the conclusions of Corollary 2.9. (This is done by showing that $c\ell\{p: p$ is a minimal idempotent in $(\beta\mathbb{N}, +)\}$ is a right ideal of $(\beta\mathbb{N}, \cdot)$ and bringing in this set at an appropriate point in the proof of Theorem 2.8.)

Bergelson and I have recently used the methods of this section to obtain a new proof of the HALES-JEWETT Theorem [17] and some generalizations.

5. Algebraic structure of $(\beta\mathbb{N},+)$

We present here several results about the algebraic structure of $(\beta\mathbb{N}, +)$ which were motivated, at least in large part, by questions in Ramsey Theory. (Often the chain of motivation is twisted and lengthy so we will not take pain to detail the motivations. For a description of one such chain, see [8].)

Definition 5.1. (a) $T = \bigcap_{n=1}^{\infty} \overline{Nn}$
(b) $I = \bigcap_{n=1}^{\infty} \overline{N2^n}$.

Observe that trivially $T \subseteq I$. PYM and I established the following [27]. Here c is the cardinality of the continuum.

Theorem 5.2. *Let p be an idempotent of $\beta\mathbb{N}$. Then $(p + \beta\mathbb{N} + p) \cap T$ contains a copy of the free semigroup on 2^c generators. If p is in the smallest ideal of βN, then $(p + \beta\mathbb{N} + p) \cap T$ contains a copy of the free group on 2^c generators.*

This theorem left open the question of whether $\beta\mathbb{N}$ has copies of the free group on 2^c generators which miss the smallest ideal. This question was answered nicely by LISAN [28].

Theorem 5.3. *$\beta\mathbb{N}$ contains c pairwise disjoint copies of I which miss the smallest ideal of $\beta\mathbb{N}$. In particular, there are 2^c pairwise disjoint copies of the fre group on 2^c generators outside of the smalles ideal.*

The plentiful existence of these copies of I lent added interest to a question of VAN DOUWEN [12], namely whether $\beta\mathbb{N}\backslash\mathbb{N}$ contains a topological and algebraic copy of all of $\beta\mathbb{N}$. (It is an old and well known fact that any infinite closed subset of $\beta\mathbb{N}$ contains a topological copy of all of $\beta\mathbb{N}$.) LISAN and I [26] showed that such copies cannot be found where he fond his copies of I nor can they be found inside the smallest ideal.

Theorem 5.4. *Assume* $\varphi: \beta\mathbb{N} \to \beta n\backslash\mathbb{N}$ *is an isomorphism and a homeomorphism to its range and let* $p = \varphi(1)$. *Then* p *is not in the smallest ideal of* $\beta\mathbb{N}$. *Also no member of* p *is very thin— that is, if* $A \in p$, *then there exist* $F, G \in [A]^{<\omega}$ *with* $\Sigma F = \Sigma G$ *but* $F \neq G$.

Definition 5.5. (a) Given $A \subseteq \mathbb{N}$, $d^*(A) = \sup\{a\colon$ for each $n \in \mathbb{N}$ there exists $x \in \mathbb{N}$ with $|\{x+1,\ x+2, \ldots, x+n\} \cap A|/n \geq a\}$.
(b) $\Delta^* = \{p \in \beta\mathbb{N}\colon d^*(A) > 0$ for each $A \in p\}$.

It is easy to see in the fashion of Section 2 that Δ^* is a closed two sided ideal of $(\beta\mathbb{N}, +)$. It is also easy to see that $d^*(A) = 1$ if and only if A contains arbitrarily long blocks of $\mathbb{N}$. It is a fact from [24] that if $d^*(A) > 0$ and $\epsilon > 0$, then there exists $k \in \mathbb{N}$ with $d^*(\bigcup_{t=1}^{k} A - t) > 1 - \epsilon$. Consequently we have the following characterizations.

(1) $p \in \Delta_1$ if and only if for each $A \in p$ there exists $k \in \mathbb{N}$ with $d^*(\bigcup_{t=1}^{k} A - t) = 1$,
(2) $p \in \Delta^*$ if and only if for each $A \in p$ and each $\epsilon > 0$ there exists $k \in \mathbb{N}$ with $d^*(\bigcup_{t=1}^{k} A - t) > 1 - \epsilon$.

We also note that Δ_1 is known [23] to be the closure of the smallest ideal of $\beta\mathbb{N}$ (and to be a closed two sided ideal.) Accordingly, Davenport and I were led to suspect that Δ^* is not much bigger than Δ_1, possibly even the next biggest closed two sided ideal of $\beta\mathbb{N}$. (This assertion we knew to have several nice Ramsey Theory consequences, including at least one now known to be false.) The true state of affairs is as far from this assertion as possible.

Theorem 5.6. Δ_1 , *is the intersection of closed two sided ideals lying strictly between it and* Δ^*. *That is, given* $p \in \Delta^*\backslash\Delta_1$ *there is a closed two sided ideal* J *with* $\Delta_1 \subsetneq J \subsetneq \Delta^*$ *and* $p \notin J$.

There is a standard way of comparing the structure of two ultrafilters known as the Rudin-Keisler order. One says $p \leq q$ if there is some $f\colon N \to N$ such that for each $A \in q$, $f[A] \in p$. (Do not confuse this order with the ordering of idempotents used in section 4.) Note that $p \leq q$ says that p is obtained from q by identifying or collapsing some points of $\mathbb{N}$ so that the structure of q is at least as rich as that of p. Blass and I [5] studied the relationship between the Rudin-Keisler order and the operation + in$\beta\mathbb{N}$. Since a member $p+q$ has translates which are in p, it is not surprising that often $p \leq p+q$. It is in fact somewhat surprising that this inequality may fail. What is more surprising is that often $q \leq p+1$. A corollary to our main result from [5] establishes that the smallest ideal of $(\beta\mathbb{N}, +)$ contains elements of arbitrarily complex set-theoretic structure.

Theorem 5.7. *The smallest ideal of* $(\beta\mathbb{N}, +)$ *contains no minimal elements and is cofinal upward in the Rudin-Keisler order.*

References

[1] Arens, R., The adjoint of a bilinear operator, Proc. Amer. Math. Soc. $\underline{2}$ (1951), 839-848

[2] Bergelson, V., H. Furstenberg, N. Hindman, and Y. Katznelson, A new proof of van der Waerden's Theorem, manuscript

[3] Bergelson, V., and N. Hindman, A combinatorially large cell of a partition of $\mathbb{N}$, J. Comb. Theory (Series A) 48 (1988), 39–52

[4] —, Nonmetrizable topological dynamics and Ramsey Theory, manuscript

[5] Blass, A., and N. Hindman, Sums of ultrafilters and the Rudin-Keisler and Rudin-Frolik orders, manuscript

[6] Blass, A., J. Hirst, and S. Simpson, Logical analysis of some theorems of combinatorics and topological dynamics, in "Logic and Combinatorics", S. Simpson ed., Contemporary Mathematics 65 (1987), 124–156

[7] Civin, P., and B. Yood, The second conjugate space of a Banach algebra as an algebra, Pac. J. Math. 11 (1961), 847–870

[8] Davenport, D., and N. Hindman, Subprincipal closed ideals in $\beta\mathbb{N}$, Semigroup Forum 36 (1987), 223–245

[9] Day, M., Amenable semigroups, Illinois J. Math. 1 (1959), 509–544

[10] Deuber, W., Partitionen und lineare Gleichungssysteme, Math. Z. 133 (1973), 109–123

[11] Deuber, W., and N. Hindman, Partitions and sums of (m, p, c)-sets, J. Comb. Theory (Series A) 45 (1987), 300–302

[12] van Douwen, E., The Čech-Stone Compactification of a discrete groupoid, Topology and its Appl., to appear

[13] Ellis, R., "Lectures on topological dynamics", Benjamin, New York, 1969

[14] Furstenberg, H., "Recurrence in ergodic theory and combinatorial number theory", Princeton University Press, Princeton, 1981

[15] Furstenberg, H., and Y. Katznelson, Idempotents and coloring theorems, manuscript

[16] Graham, R., B. Rothschild and J. Spencer, "Ramsey Theory", Wiley, New York, 1980

[17] Hales, A., and R. Jewett, Regularity and positional games, Trans. Amer. Math. Soc. 106 (1963), 222–229

[18] Hindman, N., Finite sums from sequences within cells of a partition of $\mathbb{N}$, J. Comb. Theory (Series A) 17 (1974), 1–11

[19] —, Partitions and sums of products of integers, Trans. Amer. Math. Soc. 247 (1979), 227–245

[20] —, Ultrafilters and combinatorial number theory, in "Number Theory Carbondale 1979", M. Nathanson ed., Lecture Notes in Math. 751 (1979), 119–184

[21] —, Partitions and sums and products—two counterexamples, J. Comb. Theory (Series A) 29 (1980), 113–120

[22] —, Sums equal to products in $\beta\mathbb{N}$, Semigroup Forum 21 (1980), 221–255

[23] —, Minimal ideals and cancellation in $\beta\mathbb{N}$, Semigroup Forum 25 (1982), 291–310

[24] —, On density, translates, and pairwise sums of integers, J. Comb. Theory (Series A) 33 (1982), 147–157

[25] —, Ramsey's Theorem for sums, products, and arithmetic progressions, J. Comb. Theory (Series A) 38 (1985), 82–83

[26] Hindman, N., and A. Lisan, Does $\mathbb{N}^*$ contain a topological and algebraic copy of $\beta\mathbb{N}$?, manuscript

[27] Hindman, N., and J. Pym, Free groups and semigroups in $\beta\mathbb{N}$, Semigroup Forum 30 (1984), 177–193

[28] Lisan, A., Free groups in $\beta\mathbb{N}$ which miss the minimal ideal, Semigroup Forum 37 (1988), 233–239

[29] Papazyan, T., The existence of almost translation invariant ultrafilters on any semigroup, manuscript

[30] Rado, R., Studien zur Kombinatorik, Math. Z. 36 (1933), 424–480

[31] Ramsey, F., On a problem of formal logic, Proc. London Math. Soc. (2) 30 (1930), 264–286

[32] Ruppert, W., Rechtstopologische Halbgruppen, J. Reine Angew. Math 261 (1973), 123–133

Probability measures on semigroups of nonnegative matrices

R. W. R. Darling and A. Mukherjea

1. Products of random finite-dimensional matrices

In this chapter our aim is to show how certain semigroup results can play an important role in certain problems in probability. For example, consider the following problem:

Suppose that there are d sites $\{1, 2, \dots, d\}$. Consider the d-vector

$$a_0 = (a_0(1), a_0(2), \dots, a_0(d)) ,$$

where the ith entry is the amount of some commodity at site i at time 0. Suppose that the corresponding vector a_n at time n is given by

$$a_n = a_{n-1} A_n ,$$

where $A_1, A_2, \dots$ are a sequence of d by d random matrices, which are independent and identically distributed (or i. i. d., in short). Then the problem is to describe the asymptotic behavior of the sequence a_n) as n goes to infinity. This clearly leads to the study of the product $M_n = A_1 A_2 \cdots A_n$ of d by d random matrices as n goes to infinity.

Problems of this type can be considerd in the infinite dimensional as well as finite dimensional context. In this section, we restrict our attention to the finite dimensional case. The infinite dimensional case will be considered in the next section.

The study of products of random matrices was apparently started by BELLMAN [1]. He showed that if the sequence (A_n) is independent and has strictly positive entries, then under certain conditions, the limit of

$$\frac{1}{n} E[\log(M_n)_{ij}]$$

exists, where the subscript ij refers to the matrix entry on the ith row and jth column. Bellman's lead was followed by Furstenberg and Kesten in their important paper [11], where they proved a number of strong law and central limit theorem type results for products of random nonnegative matrices under more general conditions. These results involved stationary distributions and led to a general analysis of related problems in semisimple Lie groups. See FURSTENBERG [10]. See also GRENANDER [12], GUIVARC'H [14], GUIVARC'H and RAUGI [15], TUTUBALIN [24], and the references in these papers. A different type of random matrix product has been considered by BERGER [2]. Recently KERSTEN and SPITZER [19] (see also BOUGEROL [3] and MUKHERJEA [22]) considered the following problem:

Let $A_1, A_2, \ldots$ be a sequence of i. i. d. nonnegative d by d matrices. Suppose that they satisfy:
(KS_1) $\Pr(A_1 \in J) = 0$; (KS_2) For some positive integer m, $\Pr(A_1 A_2 \cdots A_m \in I) > 0$,
where J is the set of all d by d nonnegative matrices which have at least one zero row or one zero column or both, and I is the set of all d by strictly positive matrices. The problem then is to find when the sequence $M_n = A_1 A_2 \cdots A_n$ converges in distribution, as n tends to infinity, to a distribution not concentrated at the zero matrix.

In this section we will restrict our attention to the context of the above problem and show how certain semigroup ideas can be very useful in a proper understanding and solution of this problem.

Suppose that μ is the distribution of A_i, that is, μ is a probability measure on the Borel set of the topological semigroup of d by d nonnegative matrices (with respect to matrix multiplication and usual topology derived from the Euclidean d^2-space). Let S be the closed subsemigroup generated by the support $S(\mu)$ of μ, that is,

$$S = \text{closure}\left[\bigcup_{n=1}^{\infty} S(\mu)^n\right] .$$

Then, μ^m (the mth convolution power of μ) is the distribution of $A_1 A_2 \cdots A_m$, and the Kesten-Spitzer conditions (KS_1) and (KS_2) are translated as

$$(KS_1)\ \mu(J) = 0\ ; \quad (KS_2)\ \mu^m(I) > 0\ .$$

Notice that J^c, the complement of J, is a semigroup and I, a subset of J^c, is an ideal of J^c, that is, for any x in J^c, $Ix \subset I$ and $xI \subset I$. Because of this ideal property and condition (KS_1), it follows that

$$\mu^{k+1}(I) = \int_{J^c} \mu^k(Ix^{-1})\mu(dx)$$

which is at least $\mu^k(I)$, since for x in J^c, $I \subset Ix^{-1}$, where $Ix^{-1} = \{y: yx \in I\}$. Also, if (Y_n) is a sequence of i. i. d. d by d random nonnegative matrices with distribution μ^m, then by (KS_2),

$$\sum_{n=1}^{\infty} \Pr(Y_n \in I) = \infty$$

so that by the Borel-Cantelli Lemma, $\Pr(Y_n \in I$ infinitelyoften $= 1$, and this means (since by (KS_1), $\mu^k(J) = 0$ for all k, and since I is an ideal) that $\Pr(Y_1 Y_2 \cdots Y_n \in I$ eventually) $= 1$ so that $\mu^{mn}(I)$ goes to 1 as n goes to infinity. It follows that $\mu^n(I)$ goes to 1 as n goes to infinity. Before we get into any more details for general d, let us observe what happens in the simplest case when $d = 1$.

Let us then consider a sequence (B_n) of i. i. d. nonnegative random variables with distribution σ, a probability measure on $[0, \infty)$. There are two cases to consider.

Suppose first that $S(\sigma)$ contains a point in $[0, 1)$. Then if (σ^n) is a tight sequence of probability measures, the sequence (σ^n) will converge weakly to the unit mass at 0. [Let us remark that this result holds in general; namely that when $S(\sigma)$ generates a locally compact Hausdorff topological semigroup with a zero, then (σ^n) converges weakly to the unit mass at 0 iff the sequence (σ^n) is tight.] Now suppose that $S(\sigma) \subset [1, \infty)$. In this case, if the sequence (σ^n) is tight, then it can be shown that the sequence

$$\frac{1}{n} \sum_{k=1}^{n} \sigma^k$$

converges weakly to an idempotent probability measure ν on $[1, \infty)$. See Proposition 4.3 in [22]. Since the support of ν has to be a compact group in this case (see [22], where it is proven that the support of an idempotent probability measure on a locally compact Hausdorff topological semigroup is a completely simple subsemigroup with a compact group factor), it is clear that ν has to be the unit mass at 1. Since $\sigma\nu = \nu$ (convolution being continuous with respect to weak topology), it follows that σ also has to be the unit mass at 1. Let us remark that if we assume that $\Pr(B_i = 0) = 0$, then the above assertions follow also by applying KOLMOGOROV's Three Series Theorem to the sequence $\sum_{i=1}^{n} \log B_i$, which is defined almost surely. Thus, nothing interesting happens when $d = 1$. However, more interesting results hold for $d > 1$.

Theorem 1.1. *Let μ be a probability measure on the $d \times d$ nonnegative matrices, and let S be the closed semigroup generated by $S(\mu)$, with respect to matrix multiplication and the usual topology. Then the sequence $(\frac{1}{n} \sum_{k=1}^{n} \mu^k)$ (denoted by (μ_n), for short) converges weakly to a probability measure ν such that $S(\nu)$ intersects J^c*

iff S is compact and $m(S)$ intersects J^c, where $m(S)$ is the set of all matrices in S with minimal rank.

Proof. Notice that the weak limit of (μ_n) is an idempotent probability measure ν such that $\mu\nu = \nu\mu = \nu$ so that $S(\mu)S(\nu)$ as well as $S(\nu)S(\mu)$ is contained in $S(\nu)$ an ideal of S. Since $S(\nu)$ is a completely simple subsemigroup of S, it follows from [5] that $S(\nu) = m(S)$. The "if" part follows. For the "only if" part, suppose that the weak limit ν of (μ_n) exists. As before, $S(\nu)$ is a completely simple ideal of S, and equals $m(S)$. For x in $S(\nu) \cap J^c$, $xS(\nu)x$ is a compact group (since $S(\nu)$ has a compact group factor) of nonnegative matrices, and therefore, finite by [8]. Also, $x \in xS(\nu)x$ so that $xS(\nu)x \subset J^c$. Let e be the identity of $xS(\nu)x$. Then e has a strictly positive diagonal (being an idempotent nonnegative matrix with no zero rows or columns). Also, eSe and $eS(\nu)e$ are both equal to $xS(\nu)x$ so that there exists a positive M such that for any s in S and for any i and j,

$$e_{ii}s_{ij}e_{jj} \leq (ese)_{ij} \leq M$$

(since $xS(\nu)x$ is finite). It follows that S is compact. □

We observe that the result in Theorem 1.1 is somewhat surprising in the sense that the weak convergence of (μ_n) in this theorem needed S to be compact. This is, of course, not so when the support of μ_n is contained in J. Here is an example.

Example 1.2. Consider the following multiplicative matrix semigroups:

$$K_1 = \left\{ \begin{pmatrix} 1 & m \\ 0 & n \end{pmatrix} : m, n \quad \text{nonnegative integers} \right\}$$

and

$$K_2 = \left\{ \begin{pmatrix} 1 & m \\ 0 & n \end{pmatrix} : n \quad \text{is a nonnegative integer} \right\}$$

Let $S_0 = K_1 \cup K_2$. Consider any probability measure m such that its support contains the matrices

$$\begin{pmatrix} 1 & 1 \\ 0 & 1 \end{pmatrix}, \begin{pmatrix} 1 & 1 \\ 0 & 0 \end{pmatrix}, \begin{pmatrix} 1 & 0 \\ 0 & 0 \end{pmatrix}.$$

It can be verified that μ^n converges weakly to a probability measure ν, where $S(\nu) = K_2$. Note that K_2 is not compact. This leads to the following open question.

Problem 1.3. Is there anything specific we can say, as in Theorem 1.1, when we do not require the condition that $S(\nu)$ intersect J^c?

Let us remark that using Theorem 4.3 in [21], we can, of course, say that whenever the sequence (μ^n) is tight, the sequence

$$\frac{1}{n}\sum_{k=1}^{n}\mu^k$$

converges weakly to ν, where $S(\nu) = m(S)$. In other words, we can still identify $S(\nu)$, but when can we say something about S? One obvious question in this context, is, of course, how to decide when the sequence (μ^n) is tight. Notice that if there exists a strictly positive vector $x = (x_1, x_2, \ldots, x_d)$, where each x_i is positive, such that

$$sX(\mu) = x \quad \text{or} \quad S(\mu)x^T = x^T \ ,$$

then it is clear that either $xS = X$ or $Sx^T = x^T$ and in either case, it follows that S is compact since for any s in S and $1 \leq i, j \leq d$, we have: in case $xS = x$,

$$x_{ii}s_{ij} \leq (xs)_{ij} = x_{ij} \quad \text{or} \quad s_{ij} \leq x_{ij}/x_{ii} \ ,$$

meaning that S is bounded. (The same happens in the other case.) Consequently, the sequence μ^n is tight. Such a condition is not only too strong, but also does not seem to work when the matrices are not nonnegative. However, in the case when $d = 2$, Kesten and Spitzer have shown in [19] that under their conditions (KS_1) and (KS_2), the sequence μ^n converges weakly to a probability measure ν different from the unit mass at 0 iff there exists a strictly positive vector x such that either $xS(\mu) = x$ or $S(\mu)x^T = x^T$. It was shown by Mukherjea [22] that this result is false for $d > 2$. In this context, the following result has been proven in [22].

Theorem 1.4. *Let μ be a probability measure on $d \times d$ real matrices with the usual topology. Let S be the closed semigroup (with respect to matrix multiplication) generated by $S(\mu)$. Suppose that for every open subset G of S containing all the matrices in S with rank one,*

$$\lim_{n\to\infty} \mu^n(G) = 1 \ .$$

Suppose also that there exists a non-zero vector x such that $xS(\mu) = x$ or $S(\mu)x^T = x^T$. Then the sequence (μ^n) converges weakly to a probability measure ν and every matrix in the support of ν has rank one. □

The condition in this theorem is met, for example, when for some positive integer m, the set of all rank one matrices in S has positive μ^m-measure and 0 is not in S, the reason being that then the rank one matrices form an ideal of S. In the proof of the above theorem, two results played key roles. The first one is the following fact: Suppose that x, y and z are d by d real matrices such that $x = x^2$, x has no zero rows, rank y = rank z = 1, $xy = x$ and $xz = x$. Then, y and z are also idempotent and $yz = y$.

The second result is the observation that a compact group of d by d real matrices each with rank one (with respect to multiplication and the usual topology) is either a singleton or has exactly two elements. (This is because every d by d rank one real matrix x such that the set of all powers of x is bounded and does not have the zero matrix as its cluster point must be either an idempotent or else the negative of an idempotent.) Clearly these results use the rank one propety crucially and as such the condition in the theorem seems to be rather restrictive, and we need a better theorem for general d by d real matrices. This is still an open question. KESTEN and SPITZER looked into this problem for nonnegative matrices in [19], but under their conditions (KS_1) and (KS_2). Under these conditions, they have essentially proven that the sequence (μ^n) is tight and 0 does not belong to S iff S is compact with no zero element. This clearly shows that the conditions of Kesten and Spitzer are too strong for a noncompact S to support a probability measure μ such that the sequence (μ^n) is tight and $S(\mu)$ generates S as a closed semigroup. This leads to the following problem.

Problem 1.5. Suppose that μ is a probability measure on d by d real matrices with the usual topology. Let S be the closed multiplicative semigroup generated by $S(\mu)$. When can we say that the sequence (μ^n) is tight? What structures does the tightness of this sequence impose on S?

BOUGEROL in [3] has also considered the above problem under various conditions. However, the problem remains still open and we need more satisfactory answers. We will now end this section with two examples and a fairly complete answer to the above problem in the case of $d = 2$.

Example 1.6. (An application of Theorem (μ_n)) Let μ be a probability measure such that

$$S(\mu) = \left\{ \begin{pmatrix} s & rs \\ (1-s)/r & 1-s \end{pmatrix}, \begin{pmatrix} t & r(1-v) \\ (1-t)/r & v \end{pmatrix} \right\}$$

where r, s, t, v are real numbers and r is non-zero. Then if x represents the vector $(1-r)$, it can be verified that for each y in $S(\mu)$, $xy = x$. It follows that the sequence (μ^n) converges weakly to a probability measure ν such that $S(\nu)$ contains only matrices with rank one.

Example 1.7. Consider real numbers p, q, r, s such that $p > 0$, $s > 0$, $0 < q < 1$, $0 < r < 1$, and $ps = (1 - q)(1 - r)$. Suppose that the support of μ is given by

$$S(\mu) = \left\{ \begin{pmatrix} 1 & p \\ 0 & q \end{pmatrix}, \begin{pmatrix} r & 0 \\ s & 1 \end{pmatrix} \right\}.$$

Notice that (KS_1) and (KS_2) hold. Also if x represents the vector $(1, p/(1 - q))$, then $xy = x$ for each y in $S(\mu)$. Here the closed semigroup S generated by $S(\mu)$ is compact, the sequence (μ^n) converges weakly to a probability measureν, $S(\nu)$ consists of only rank one matrices in S, and $\nu(I) = 1$, where I is the set of strictly positive matrices in S. However, $S(\nu)$ contains the matices

$$\begin{pmatrix} 1 & b \\ 0 & 0 \end{pmatrix} = \lim_{n\to\infty} \begin{pmatrix} 1 & p \\ 0 & q \end{pmatrix}^n$$

where $b = p/(1 - q)$, and

$$\begin{pmatrix} 0 & 0 \\ c & 1 \end{pmatrix} = \lim_{n\to\infty} \begin{pmatrix} r & 0 \\ s & 1 \end{pmatrix}^n$$

where $c = s/(1 - r)$. The point of this example is that here $S(\nu)$ intersects J.

Let us now state the last theorem in this section. This theorem was given in [22].

Theorem 1.8. *Let μ be a probability measure on 2 by 2 real matrices with usual topology and matrix multiplication. Let S be the closed semigroup generated by $S(\mu)$. Let $m(S)$ denote the set of all matrices in S with minimal rank, and* rank m(S) *denote the rank of the matrices in $m(S)$. Then the following assertions hold:*

(1) *Suppose that (μ^n) converges weakly to ν. Then*
  2(i) *$S(\nu) = m(S)$;*
  2(ii) *ν is the unit mass at 0 iff 0 is in S;*
  2(iii) rank m(S) = 2 *iff S is a compact group of invertible matrices and ν is the Haar measure on S;*
  2(iv) rank m(S) = 1 *and $m(S)$ contains no subgroup with exactly two elements iff there exists a nonzero x in S such that x is an idempotent, x has rank one and either$xS = x$ or $Sx^T = x^T$.*

(2) *If* rank m(S) = 2 *and S is non-compact, then the sequence (μ^n) converges vaguely to the zero measure, that is, $\mu^n(K)$ goes to zero for any compact set K as n goes to infinity.*

(3) *If* rank m(S) = 2 *and S is compact, then S is a compact group. Furthermore, then Kawada-Ito's theorem holds so that (μ^n) converges weakly to ν and ν is not the unit mass at 0 iff there does not exist a compact normal subgroup H of S such that $S(\mu)$ is contained in Hx with x not in H.*

(4) *Suppose that* rank m(S) = 1 *and that there exists a non-zero vector x such that $sX(\mu) = x$ or $S(\mu)x^T = x^T$. Then the sequence (μ^n) converges weakly to a probablity measure ν iff for every open set G containing $m(S)$,* $\lim_{n\to\infty} \mu^n(G) = 1$.

2. Infinite-dimensional random matrices

Let V be a countably infinite set. We shall be concerned here with certain probability measures on the semigroup

$$M \equiv \{f \in \{0,1,2,\ldots\}^{V\times V} : \sum_x f(x,y) < \infty, \quad \text{each } y\} \tag{2.1}$$

of V by V matrices with non-negative integer entries, such that each column has only finitely many non-zero entries; the composition $(f,g) \to fg$ is matrix multiplication, so $fg(x,y) = \sum_z f(x,z)g(z,y)$. For the applications we have in mind, there are two natural metrizable topologies to use on M:

(2.2) Pointwise convergence of the columns, called the ϑ-topology; i.e. $\{f_n\}$ converges to f in (M,ϑ) whenever, for each y, there exists $N \equiv N(y)$ such that the "column" $f_n(\cdot,y) = f(\cdot,y)$ for all $n \geq N$.

This could be rephrased as: given y, for all sufficiently large n,

$$\sum_{x\in A} f_n(x,y) = \sum_{x\in A} f(x,y) \quad \forall \quad A \subset V \; ;$$

thus (2.2) is stronger than

(2.3) Convergence of column support, called the ϑ_0-topology; i.e. $\{f_n\}$ converges to f in (M,ϑ_0) whenever, for each y, there exists $N \equiv N(y)$ such that for all $n \geq N$,

$$\left\{A \subset V : \sum_{x\in A} f_n(x,y) = 0\right\} = \left\{A \subset V : \sum_{x\in A} f(x,y) = 0\right\} .$$

With a little work, the reader may convince himself that, for each of these topologies, the map $(f,g) \to fg$ is jointly continuous, making M a topological semigroup; M is not locally compact, however.

The rationale for the ϑ_0-topology will now be explained. Our study of the semigroup M has been motivated by applications to stochastic automata; see WOLFRAM's book [25] and DRIFFEATH's [13] tutorial paper. In these applications, such as the voter model and contact process examples below, the semigroup M has a continuous "Boolean" right action on the space $X \equiv \{0, 1\}^V$, where X has the topology of pointwise convergence, as follows; for $\eta \in X$ and $f \in M$, $\eta \cdot f \in X$ is given by

$$\eta \cdot f(y) \equiv 1_{\{\Sigma_x \eta(x) f(x,y) \geq 1\}}, y \in V \tag{2.4}$$

It is straightforward to verify that this is indeed an action, i.e. $(\eta \cdot f) \cdot g = \eta \cdot (fg)$, and that $(\eta, f) \to \eta \cdot f$ is jointly continuous. If β is a probability measure on the Borel sets of M, then the triple (X, M, β) with the structure above is an example of a *binary stochastic automaton*; for more details, see DARLING and MUKHERJEA [7].

Let $P(M, \vartheta)$, $P(M, \vartheta_0)$ denote the sets of probability measures on the Borel sets of (M, ϑ), (M, ϑ_0), and X, respectively, endowed with the topology of weak convergence; the bounded continuous function from (M, ϑ) to R are denoted $C(M, \vartheta)$, and similarly for $C(M, \vartheta_0)$ and $C(X)$. In addition to the usual convolution operator $*$ on $P(M, \vartheta)$, viz. $\beta * \beta$, given by

$$\int_M \psi(h) \beta * \beta'(dh) = \int_{M \times M} \psi(fg) \beta(df) \beta'(dg), \psi \in C(M, \vartheta) \ , \tag{2.5}$$

there is a convolution operation $P(X) \times P(M, \vartheta) \to P(X)$, namely $(\mu, \beta) \to \mu^* \beta$, given by

$$\int_X \phi(\zeta) \mu^* \beta(d\zeta) \equiv \int_{X \times M} \phi(\eta \cdot f) \mu(d\eta) \beta(df), \phi \in C(X) \ . \tag{2.6}$$

It is readily verified that $(\mu^* \beta) * \beta' = \mu^* (\beta * \beta')$, and that $(\mu, \beta) \to \mu^* \beta$ is jointly continuous for the weak topologies on $P(M, \vartheta)$ and $P(X)$. We call $\mu \in P(X)$ *invariant* for the stochastic automaton (X, M, β) if $\mu^* \beta = \mu$.

In applications to stochastic automata, we are frequently interested in the asymptotic behaviour of $\{\mu^* \beta^n\}$, for a specific $\beta \in P(M, \vartheta)$, and for all $\mu \in P(X)$; here β^n is the nth convolution iterate of β. It is possible for $\{\mu^* \beta^n\}$ to converge weakly for all μ, even when $\{\beta^n\}$ does not converge weakly in $P(M, \vartheta)$; this is because the ϑ-topology is too strong in this context. The appropriate topology is the ϑ_0-topology, as the next lemma shows:

Lemma 2.1. *Let $\{\beta^n\}$ be a sequence in $P(M, \vartheta)$. Then (i)$\Leftrightarrow$(ii):*
(i) *$\{\beta^n\}$ converges weakly in $P(M, \vartheta_0)$ to some $\lambda \in P(M, \vartheta)$.*
(ii) *For all $\mu \in P(X)$, $\{\mu^* \beta^n\}$ converges weakly in $P(X)$ to $\mu^* \lambda$ for some $\lambda \in P(M, \vartheta)$.*

Proof. Suppose that $\{f_n\}$ converges to f in (M, ϑ_0). Then for all $\eta \in X$, $\eta \cdot f_n$ converges pointwise to $\eta \cdot f$, so for all $\phi \in C(X)$,

$$\psi(f) \equiv \int_X \phi(\eta \cdot f)\mu(d\eta) = \lim_{n\to\infty} \int_X \phi(\eta \cdot f_n)\mu(d\eta) = \lim_{n\to\infty} \psi(f_n) \ .$$

This shows that the map ψ defined on the line above is in $C(M, \vartheta_0)$.

Let us assume that (i) holds. Then for all $\mu \in P(X)$ and all $\phi \in C(X)$,

$$\begin{aligned}\int_X \phi(\zeta)\mu^* \beta_n(d\zeta) &= \int_M \{\int_X \phi(\eta \cdot f)\mu(d\eta)\}\beta_n(df) = \int_M \psi(f)\beta_n(df) \\ \to \int_M \psi(f)\lambda(df) &= \int_M \{\int_X \phi(\eta \cdot ff)\mu(d\eta)\}\lambda(df) = \int_X \phi(\zeta)\mu^* \lambda(d\zeta)\end{aligned}$$

as $n \to \infty$, by (i). This proves (ii).

Conversely suppose (ii) holds. Fix $A \subset V$, let $\eta = 1_A$, and let $\mu = \delta_{\{\eta\}}$. It follows from (ii) that for all k and all $\{x_1, \ldots, x_k\} \subset V$,

$$\int_M 1_C(\eta \cdot f)\beta_n(df) \to \int_M 1_C(\eta \cdot f)\lambda(df),$$

where $C = \{\zeta \in X : \zeta(x_i) = 0, \ 1 \leq i \leq k\}$, which is both open and closed in X. In other words,

$$\beta_n\big(\{f: \sum_{z\in A} f(z, x_i) = 0, \ 1 \leq i \leq k\}\big) \to \lambda\big(\{f: \sum_{z\in A} f(z, x_i) = 0, \ 1 \leq i \leq k\}\big) \ .$$

Sets of the form $\big(\{f: \sum_{z\in A} f(z, x_i) = 0, \ 1 \leq i \leq k\}\big)$ form a convergence determining class in (M, ϑ), and so (i) holds. □

Example 2.2. Stochastic flows on a countable set. Suppose β is a probability measure on M such that $\beta(M_1) = 1$, where

$$M_1 \equiv \{f \in \{0,1\}^{V\times V} : \sum_x f(x, y) = 1, \quad \text{each} \quad y\} \ ; \tag{2.7}$$

note that M_1 is in one-to-one correspondence with the set Γ of mappings from V to V, where F in Γ corresponds to the f inM_1 such that $f(x, y) = 1_{\{F(y)=x\}}$. No confusion will arise if we regard β as a probability measure on the Borel sets of Γ. The set M_1 is itself a subsemigroup of M, on which the two topologies above coincide, and correspond to the pointwise topology on Γ. The composition operation on M, which is matrix multiplication, corresponds to composition of functions on Γ; thus if $g(x,y) = 1_{\{G(y)=x\}}$,

$$fg(x,y) = \sum_z f(x,z)g(z,y) = \sum_z 1_{\{F(z)=x\}}1_{\{G(y)=z\}} = 1_{\{F\circ G(y)=x\}} \ .$$

The nth convolution power β^n is the law of the composition $F_n \circ F_{n-1} \circ \cdots \circ F_1$ of independent, identically distributed ("i. i. d.") random mappings $F_1, F_2, \ldots$ from V to V, with law β. The asymptotics of $\{\beta^n\}$ are considered in Darling and Mukherjea [6], and summarized in Section 3 below.

2.2A. Specific case. The following example of the construction of a probability measure β is taken from DARLING and MUKHERJEA [7]. The idea here is to construct a renewal process on the positive integers, and create a random mapping by having all the sites between two renewals mapped one unit in the same direction. Suppose $V = \{0,1,2,\ldots\}$, and let $\{a_k,\ k=1,2,\ldots\}$ be a collection of strictly positive numbers such that $\Sigma_{k\geq 1}\, a_k = 1$, $\Sigma_{k\geq 1}\, ka_k \equiv \mu < \infty$, and $a_1 < \mu/2$. Let $\{\xi_1, \xi_2, \ldots\}$ be i. i. d. positive integer valued random variables such that $\Pr(\xi_1 = k) = a_k$. Define a renewal process $\{T_k,\ k=0,1,2,\ldots\}$ by

$$T_0 = 0, T_k = T_{k-1} + \xi_k \quad \text{for} \quad k \geq 2 \ .$$

The conditions on $\{a_k, k=1,2,\ldots\}$ ensure that there exists an integer $b \geq 2$ such that $\Sigma_{k\geq b}\, ka_k)/\mu \geq 1/2 + 3\epsilon$ for some $\epsilon > 0$. Finally define a random mapping $G\colon V \to V$ by

$$\begin{aligned}
&G(x) = x, \quad \text{if} \quad x = 0 \quad \text{and} \quad \xi_1 \geq b,\\
&G(x) = x-1, \quad \text{if} \quad x \geq 1 \quad \text{and} \quad x \in [T_{m-1}, T_m) \quad \text{such that} \quad \xi_m \geq b,\\
&G(x) = x+1, x \in [T_{m-1}, T_m) \quad \text{such that} \quad \xi_m < b \ .
\end{aligned}$$

The probability measure β on V^V is now defined by:

$$\beta(\{f \in V^V\colon f(x_i) = y_i,\ i=1,2,\ldots,s\}) = \Pr(G(x_i),\ i=1,2,\ldots,s) \ . \qquad (2.8)$$

In this random transformation from V to V, all the sites from T_{m-1} to $T_m - 1$ are mapped in the same direction, so events at nearby sites are highly dependent. We shall have more to say about the convolution iterates of β later on.

2.2B. Application to the discrete-time voter model. The original voter model is a continuous-time stochastic process, which is described fully in LIGGETT [20]. The discrete-time version, corresponding to the probability measure β on $\Gamma \equiv V^V$, is a stochastic process $\{\eta_k,\ k=0,1,\ldots\}$ on $X \equiv \{0,1\}^V$ which is constructed using a sequence of independent, identically distributed random mappings $F_1, F_2, \ldots$ from

V to V, with law β; given η_0 in X, defined for $k = 1, 2, \ldots$,

$$\eta_k(y) \equiv \eta_{k-1}(F_k(y)), \quad y \in V\ . \tag{2.9}$$

The idea is simple: at time instant k, each site y looks to a randomly chosen neighbour $F_k(y)$, and notes that neighbour's value ("opinion"); then each site simultaneously adopts the value of its selected neighbour. Notice that we do not require that the choices $F(y)$ and $F(z)$ be independent to different sites y and z. To put this in matrix language, let M_1 act on X on the right according to (2.4) above:

$$\eta \cdot f(y) \equiv 1_{\{\Sigma x \eta(x) f(x,y) \geq 1\}} = \eta(F(y))\ , \tag{2.10}$$

for $\eta \in X$, $f \in M$, and $F: V \to V$ given by $1_{\{F(y)=x\}} = f(x, y)$; regarding β as a probability measure on M_1, we see that

$$\eta_k = \eta_0 \cdot A_1 A_2 \ \cdots \ A_k\ , \tag{2.11}$$

where $A_1, A_2, \ldots$ are i. i. d. random matrices in M_1 with law β. The implications of the behaviour of $\{\beta^n\}$ for the voter model will be discussed below.

Example 2.3. Discrete-time contact processes. Suppose β is a probability measure on M as in (2.1). As in the previous example, we construct, for η_0 in X, an X-valued Markov chain $\{\eta_k\}$ by

$$\eta_k = \eta_0 \cdot A_1 A_2 \ \cdots \ A_k\ ,$$

where $A_1, A_2 \ \cdots$ are i.i.d random matrices in M with law β, and the action of M on X is given by (2.4). To see that this corresponds to a discrete-time contact process (cf. Durrett and Schonmann [9]), whatever the choice of the probability measure β, observe that the updating rule is as follows. For the configuration η, a site x is *alive* if $\eta(x) = 1$, and *lifeless* if $\eta(x) = 0$. The choice of a random element F of M means that each site y has a random set $\{x \in V: F(x, y) = 1\}$ of *contacts* (which could include y itself, or could be empty); if y is lifeless, it becomes alive at the next step if any of the contacts are alive; on the other hand if site y is alive, it dies if none of its contacts are alive. We do not require that the *contact sets* $\{x \in V: F(x, y) = 1\}$ and $\{x \in V: F(x, z) = 1\}$ be independent of $y \neq z$. This Markov chain could be regarded as a simple model for propagation of a plant species, for example.

2.3A Oriented bond percolation. This is a specific case of Example 2.3 where $V = Z$, and the law β is as follows: fix $p \in (0, 1)$, and let the random matrix A_1, whose law is β, satisfy

(2.12) The columns $\{A_1(\cdot, y), y \in Z\}$ are independent, and

(2.13) $\{A_1(y, y), A_1(y + 1, y)\}$ are i. i. d. Bernoulli (p) random variables, while $A_1(x, y) = 0$ for $x \notin \{y, y + 1\}$.

For this model the time index becomes the second spatial coordinate. If $\{(x, 0): x \in B\} \subset Z^2$, represents the set of *wet* sites at the level 0, and $\eta_0 = 1_B$, then $\{(x, k): \eta_k(x) = 1\}$ represents $\Sigma_{x\in B}(A_1, A_2 \cdots A_k)(x, y) \geq 1$. The problem is to find the smallest value of p, called p_c, such that, with positive probability, η_k is non-zero for every k : i.e. wetness percolates upwards without bound. According to computer simulations presented in DURRETT [8], $p_c \cong 0.64$; the author also derives rigorous upper and lower bounds, using graphical arguments.

2.3B. Oriented site percolation. This is identical to the previous example, except that (2.13) is replaced by

$$\begin{aligned} \Pr(A_1(y, y) = A_1(y + 1, y) = p = \\ \Pr(A_1(y, y) = A_1(y + 1, y) = 0), \end{aligned} \tag{2.14}$$

while

$$A_1(x, y) = 0 \quad \text{for} \quad x \notin \{y, y + 1\} .$$

See Durrett [8] for a further discussion.

2.3C. A contact process on Z^d. This model has been analyzed in DARLING and MUKHERJEA [7]; it is (like the previous two examples) a special case of a Stochastic Growth Model in the sense of DURRETT and SCHONMANN [9]. Let $V = Z^d$, let α and δ be non-negative real numbers such that $0 < \delta + 2d\alpha \leq 1$, and let β be the law of a random $Z^d \times Z^d$ matrix A such that

(2.15) The columns $\{A(\cdot, y), y \in Z^d\}$ are independent; and

(2.16) For each y in Z^d, the column $A(\cdot, y)$ has $2d + 2$ possible values: $A(\cdot, y)$ is identically zero with probability δ, $A(\cdot, y) = 1_{\{y\}}(\cdot)$ with probability $1-\delta-2d\alpha$, and for each x adjacent to y, $A(\cdot, y) = 1_{\{x,y\}}(\cdot)$ with probability α; here "x adjacent to y" means that x and y differ by 1 in exactly one entry, and is abbreviated to $x \leftrightarrow y$.

To understand the effect of this prescription, suppose $\eta_0 = \eta$, and $\eta_1 \equiv \eta$. A_1 is the first update; then for each $y \in Z^d$,

$$\Pr(\eta_1(y) = 1|\eta) = (1 - \delta)1_{\{\eta(y)=1\}} + (\alpha\Sigma_{z\leftrightarrow y}1_{\{\eta(z)=1\}}1_{\{\eta(y)=0\}} .$$

Because of assumption (2.15), the random variables $\{\eta_1(y), y \in Z^d\}$ are conditionally independent given η. Thus the probability that a nonliving site will come to life is proportional to the number of living neighbours; this is a characteristic feature of the original continuous time contact process introduced by HARRIS [16]. Here the problem is to identify the set of $(\alpha, \delta) \in (0, 1) \times (0, 1)$ such that this stochastic automaton has an invariant measure other than Dirac measure at 0, and to describe this invariant measure; efforts in this direction are reported in DARLING and MUKHERJEA [7].

2.3D. Generalization of the last three models. The last three models have a lot in common, and this suggests that the following context may be a fruitful one in which to work. Let $(V, +)$ be a countable Abelian group, and for $u \in V$ and $f \in M$, define f_u in M by: $f_u(x, y) = f(x + u, y + u)$. Suppose that β is a probability measure on M which is translation-invariant, in the sense that $\beta(\{f: f_u \in C\}) = \beta(C)$ for all Borel sets C in M. A further simplifying assumption is that if A_1 is a random matrix whose law is β, then the columns $\{A_1(\cdot, y), y \in V\}$ are independent. The problem is now to classify the asymptotic behaviour of $\{\beta^n\}$ in $P(M, \vartheta_0)$.

3. Some partial results

To gain some insight into stochastic flows on a countable set (Example 2.2), DARLING and MUKHERJEA [6] examined the structure of Γ ($\equiv$ the set of mappings from V to V with the pointwise topology) as a topological semigroup. Here are some of the results. Here $\text{Im}(f)$ denotes the image in V of $f \in \Gamma$, and for any subsemigroup S of Γ,

(3.1) $S^* \equiv \{f \in S \mid |\text{Im}(f)| < \infty \text{ and } |\text{Im}(f)| \text{ is minimal}\}$;

(3.2) $I \equiv I(S) \equiv \{f \in S \mid \text{Im}(f \circ g) = \text{Im}(f) \ \forall \ g \in S\}$.

Proposition 3.1. *(i) If $S^* \neq \emptyset$, then S^* is a completely simple subsemigroup of S (i.e. contains no proper ideals, and contains a primitive idempotent), and is the completely simple minimal ideal of S. Also the group factor of S^* is compact.*

(ii) Let E denote the set of idempotents of S. If S is completely simple, then for all $f \in S$ and all $e \in E$, $\text{Im}(f) = f(\text{Im}(e))$, and f is one-to-one on $\text{Im}(e)$. (A complete characterization of the completely simple subsemigroups of Γ is given in the paper cited.)

(iii) If S contains a completely simple minimal ideal K, then $K = I(S)$ (see (3.2); so $I = S^*$ if $S^* \neq \emptyset$).

Many of the theorems about convolution of probability measures on locally compact semigroups (see MUKHERJEA and TSERPES [23]) generalize to the case of a complete separable metric semigroup. Suppose β is a probability measure on M with support $S(\beta)$, and let S denote the closure of the semigroup generated by $S(\beta)$, so S is a subsemigroup of M.

Proposition 3.2. *Suppose the sequence $\{\beta^n\}$ is tight; then*

(i) *S has a completely simple minimal ideal K with a compact group factor such that for every open $W \supseteq K$, $\lim_n \beta^n(W) = 1$.*

(ii) $(1/n)\sum_{1\leq\kappa\leq n}\beta^k$ *converges weakly to a probability measure* λ *on* M *whose support is* K, *such that* $\lambda^2 = \lambda = \beta * \lambda = \lambda * \beta$.

Conversely, suppose S *has a completely simple minimal ideal* K *with a compact group factor, such that* $\beta^n(K) > 0$ *for some* n; *then* $\lim_n \beta^n(K) = 1$, *and* $\{\beta^n\}$ *is tight.*

Proof. See Darling and Mukherjea [6], Section 5.

For a probability measure β on Γ, there is an easily checked condition for tightness of $\{\beta^n\}$. The *one point motion* induced by β is the Markov chain on V with transition probability $p(x, y) \equiv \beta(\{f\colon f(x) = y\})$.

Lemma 3.3. *For a probability measure* β *on* Γ, $\{\beta^n\}$ *is tight if and only if the one point motion induced by* β *has a transition matrix* P *such that* P^n *converges elementwise to a stochastic matrix* Q *as* $n \to \infty$.

Proof. See Darling and Mukherjea [7].

Theorem 3.4. (Positive recurrent stochastic flows) *Suppose* β *is a probability measure on* Γ, *and* $\{\beta^n\}$ *is tight; then* $(1/n)\sum_{1\leq\kappa\leq n}\beta^k$ *converges weakly to a probability measure* λ, *such that* $\lambda^2 = \lambda = \beta * \lambda = \lambda * \beta$. *The support of* λ *is* $I(S)$ *(see (3.2)), where* S *is the closure of the semigroup generated by* S_β; *note that* $I(S) = S^*$ *if* $S^* \neq \emptyset$.

With further conditions, weak convergence of $(1/n)\sum_{1\leq k\leq n}\beta^k$ can be improved, as the following result from Darling and Mukherjea [7] shows. For any $k \geq 2$, the *k-point motion* is the Markov chain on V^k with transition probability

$$p((x_1, \ldots, x_k), (y_1, \ldots, y_k)) = \beta(\{f\colon f(x_1) = (y_1, \ldots, f(x_k) = y_k\}) \ .$$

Theorem 3.5. *Suppose that* β *is a probability measure on* Γ, *such that the induced one-point motion is irreducible and positive recurrent. Then* $\{\beta^n\}$ *converges weakly in* $P(\Gamma)$ *(or equivalently, for any probability measure* $\mu \in P(X)$ *the sequence* $\{\mu * \beta^n\}$ *converges weakly in* $P(X)$; *see Lemma 2.1), provided either (i) or (ii) holds:*

(i) $S(\beta)$, *the support of* β, *contains an element* $e = e \circ e$ *in* Γ *such that, for all* $f \in S \equiv cl(\cup_{n\geq 1} S(\beta)^n)$, $e(V) = e(f(V))$.

(ii) *For every* $k \geq 1$, *the* k*-point motion is aperiodic.*

Remark. This is a generalization of a theorem of Holley and Liggett [18]; note that we require *no* assumption about the random transformation (whose law is β) acting independently at different sites, nor is there any restriction of μ.

This may be applied to Example 2.2A above:

Proposition 3.6. *The one-point motion associated with β in (2.8) is irreducible and positive recurrent, and all the k-point motions are aperiodic; consequently Theorem 3.5 shows that, for any probability measure μ in $P(X)$, the sequence $\{\mu * \beta^n\}$ converges weakly in $P(X)$.*

Proof. The proof, which uses BLACKWELL's Renewal Theorem, may be found in DARLING and MUKHERJEA [7].

The state of knowledge about Examples 2.3A, B, C, and D is much less satisfactory; in our 1989 paper three different approaches were used to try to determine values of the parameters α and δ in Example 2.3C which allow the stochastic automaton to *survive*, but none gives a full solution.

In Example 2.3D, it may be conjectured that, excluding trivial (i.e. essentially deterministic) cases, $\{\beta^n\}$ either converges weakly in $P(M, \vartheta_0)$ to a Dirac measure on the zero matrix (i.e. the contact process dies out), or else the measure of every compact set in (M, ϑ_0) converges to zero (i.e. the number of living sites goes to infinity in the contact process). Even the simplest examples, oriented bond and site percolation, are not yet well understood from the point of view of convolution of measures on M.

References

[1] Bellman, R., Limit theorem for non-commutative operations I, Duke Math. J. 21 (1954), 491–500

[2] Berger, Marc A., Central limit theorem for products of random matrices, Trans. Amer. Math. Soc. 285 (1984), 777–803

[3] Bougerol, P., Tightness of products of random matrices and stability of linear stochastic systems, Ann. Prob. 15 (1987), 40–74

[4] Brown, D., On clans of nonnegative matrices, Proc. Amer. Math. Soc., 671–674

[5] Clark, W. E., Remarks on the kernel of a matrix semigroups, Czechoslovak Math. J. 15 (90), 305-309

[6] Darling, R. W. R., and A. Mukherjea, Stochastic flows on a countable set, J. of Theoretical Probability 1 (1988), 121–147

[7] Darling, R. W. R., and A. Mukherjea, Stochastic automata: discrete time voter models and contact processes, Preprint

[8] Durrett, R., "Lecture Notes in Particle Systems and Percolation", Wadsworth, Pacific Grove

[9] Durrett, R. and R. H. Schonmann, Stochastic growth models, In "Percolation Theory and the Ergodic Theory of Interacting Systems", H. Kesten, Ed., Springer-Verlag, New York

[10] Furstenberg, H., Non-commuting random products, Trans. Amer. Math. Soc. 108 (1963), 377–428

[11] Furstenberg, H. and H. Kesten, Products of random matrices, Ann. Math. Stat. 31 (1960), 457–469

[12] Grenander, U., Probabilities on algebraic structures, Wiley, Stockholm

[13] Griffeath, D., Cyclic random competition, Notices of the Amer. Math. Soc. 35 (1988), 1472–1480

[14] Guivarc'h, Y., Quelques propriétés asymptotiques des produits de matrices aléatoires, Lecture Notes in Math., Springer-Verlag New York etc., 774 (1980), 176–250

[15] Guivarc'h, Y. and A. Raugi, Products of random matrices: Convergence theorems, Contemporary Math. 50 (1986), Amer. Math. Soc., 31–54

[16] Harris, T. E., Contact interactions on a lattice, Ann. Prob. 2 (1974), 969–988

[17] Hognas, G., A note on products of random matrices, Stat. & Prob. Letters 5 (1987), 367–370

[18] Holley, R. and T. M. Liggett, Generalized potlatch and smoothing processes, Z. Wahrsch. verw. Gebiete 55 (1981), 165–195

[19] Kesten, H. and F. Spitzer, Convergence in distribution of products of random matrices, Z. Wahrscheinlichkeitstheorie verw. Gebiete 67 (1984), 363–386

[20] Liggett, T. M., "Interacting Particle Systems", Springer-Verlag, New York

[21] Mukherjea, A., Limit theorems: Stochastic matrices, ergodic Markov chains, and measures on semigroups, Probabilistic Analysis and Related topics, Vol. 2, Academic Press, 143–203

[22] —, Convergence in distribution of products of random matrices: a semigroup approach, Trans. Amer. Math. Soc. 303 (1987), 395–411

[23] Mukherhea, A. and N. Tserpes, Measures on Topological Semigroups: Convolution Products and Random Walks, Lecture Notes in Math., Springer-Verlag, New York, 547 (1975),

[24] Tutubalin, V. N., On limit theorems for the product of random matrices, Theor. Probab. Appl. 10 (1964), 15–27

[25] Wolfram, S., Theory and Applications of Cellular Automata, World Scientific, Singapore (1986)

Some trends and directions in the investigation of congruences on $S(X)$

K. D. Magill, Jr.

1. Introduction

The editors of this book were kind enough to ask me to contribute an article discussing trends and directions in the theory of semigroups of continuous selfmaps. At present, there are at least four broad areas where people are actively doing research on $S(X)$, the semigroup of all continuous selfmaps of a topological space X. These are (1) Homomorphisms from $S(X)$, into $S(Y)$, (2) Finitely generated subsemigroups of $S(X)$, (3) Green's relations and related topics for $S(X)$ and (4) Congruences on $S(X)$. A diacussion covering each of the four topics would be far too lengthy to be appropriate here. Consequently, I have decided to cover only one of these topics, namely congruences on $S(X)$.

2. Congeneric congruences

When X is discrete, $S(X)$ coincides with the full transformation semigroup on X. In this respect, A. I. Mal'cev [18] can be regarded as being the first one to make a contribution to our knowledge about congruences on $S(X)$. In 1952 he determined the complete lattice of congruences on $\mathcal{T}_X$, the full transformation semigroup on X, and this was a task of considerable difficulty. He showed that the lattice is generated by three types of congruences. There is a very readable account of his results in [6] so we will discuss them no further here.

In looking at the congruences on $S(X)$ in general, it is very natural to try to determine those congruences ρ on $S(X)$ with the property that $S(X)/\rho$ is isomorphic to $S(Y)$ for some generated space Y (i.e., Y is T_1 and $\{f^{-1}(y): f \in S(Y), y \in \text{Ran } f\}$ is a subbasis for the closed subsets of Y where Ran f is the range of f). We refer to such a congruence as a *congeneric congruence*. As we will soon see, there are not many. In what follows $\mathcal{C}$ will denote the collection of all components of a space X and the component of a point x will be denoted by C_x.

Definition 2.1. A topological space X is *C-admissible* if it satisfies the following conditions.

(2.1.1) X is completely regular and Hausdorff.

(2.1.2) The arc components of X coincide with the components of X.

(2.1.3) Suppose $\mathcal{A} \subseteq \mathcal{C}$, $\cup\mathcal{A}$ is open and $x \in \cup\mathcal{A}$. Then there exists $\mathcal{B} \subseteq \mathcal{A}$ such that $x \in \cup\mathcal{B}$ and $\cup\mathcal{B}$ is clopen.

(2.1.4) X contains a subset H such that $H \cap C$ is a singleton for each $C \in \mathcal{C}$ and for each open subset V of X, $\cup\{C_a : a \in V \cap H\}$ is also open.

Any completely regular locally pathwise connected Hausdorff space is C-admissible but there exist C-admissible spaces which are not even locally connected. We borrow an example from [16].

Example 2.2. Let $X = Q \times I$ where Q is the space of rational numbers and I is the closed unit interval. X is certainly completely regular and Hausdorff and one verifies that (2.1.3) holds. For the set H required by (2.1.4) choose any $a \in I$ and a set $H = \{(r, a) : r \in Q\}$. Thus X is C-admissible but not only is it not locally connected, its components aren't even open.

Next we define three congruences δ,ω,and γ on $S(X)$. Let

$$\delta = \{(f, f) : f \in S(X)\},$$
$$\omega = S(X) \times S(X),$$
$$\gamma = \{(f, g) : \text{ for any } C \in \mathcal{C} \text{ both } f[C] \text{ and } g[C] \text{ are contained in the same component of } X\}.$$

The congruences δ and ω are, in a sense, trivial and ω and γ will coincide precisely when X is connected. At any rate we are now in a position to state

Theorem 2.3. (Magill, Misra and Tewari [16]) *Let X be a C-admissible space. Then the only congeneric congruences on $S(X)$ are δ,ω and γ. Moreover, the quotient semigroup is isomorphic in these instances to $S(Y)$ where Y is homeomorphic to X, the one-point space and $X/\mathcal{C}$, respectively where $X/\mathcal{C}$ is the quotient space formed by identifying each component of X to a point.* □

3. The largest and the smallest proper congruences on $S(X)$

By a proper congruence on $S(X)$ we mean any congruence ρ which differs from both δ and ω. To my knowledge, the first one to study congruences on $S(X)$ after A. I. Mal'cev was E. G. Shutov [21] and he really was thinking about semigroups of continuous selfmaps whereas Mal'cev probably was not. Shutov discovered the largest proper congruence ρ on $S(I)$ where I is the closed unit interval

Theorem 3.1. (Shutov [20]) *For $f, g \in S(I)$, define $(f, g) \in \rho$ if whenever one of the functions is injective on a nondegenerate subinterval of I, then the two functions agree on that subinterval. The relation ρ is the largest proper congruence on $S(I)$.* □

SHUTOV's techniques can easily be modified to produce the result following the next definition.

Definition 3.1. A topological space X has the *internal extension property* if every continuous map from a closed subset of X into X can be extended to a continuous selfmap of X.

Theorem 3.3. (Magill [11]) *Let X be any compact N-dimensional subspace of Euclidean N-space with the internal extension property. For $f, g \in S(X)$, define $f, g \in \rho$ if, whenever one of the functions is injective on any copy H of the N-cell, then the two functions agree on H. The relation ρ is the largest proper congruence on $S(X)$.* □

The previous result shows that there are a large number of spaces whose semigroups have largest proper congruences. Nevertheless, there are also many spaces whose semigroups do not have this property. We first need to introduce some terminology. It so happens that there are spaces which arise naturally in various contexts in the consideration of semigroups of continuous selfmaps. These are the *local dendrites with finite branch number*. First of all, a dendrite is a Peano continuum which contains no simple closed curves and a local dendrite is a Peano continuum with the property that each point has a neighborhood which is a dendrite. Let X be a local dendrite, let $x \in X$ and let D be any dendrite which is a neighborhood of X. The number of components of $D \setminus \{x\}$ does not depend upon D. It is referred to as the rank of x in X and is denoted by $\mathrm{Rank}(x, X)$. The point x is said to be an *endpoint* if $\mathrm{Rank}(x, X) = 1$, a *local cutpoint* if $\mathrm{Rank}(x, X) > 1$ and a *branch point* if $\mathrm{Rank}(x, X) > 2$. The *branch number* of X, denoted $\mathrm{Br}(X)$ is defined to be the sum of the ranks of all the branch points of X. Evidently, $\mathrm{Br}(X)$ is finite if and only if X has only finitely many branch points and each of these has finite rank. If $\mathrm{Br}(X)$ is finite, then X is, by definition, a *local dendrite with finite branch number*. Finally, we wish to single out a very special type of dendrite. Let R denote the space of real numbers. Any space which is homeomorphic to

$$\cup_{n=1}^{N} \{(x, y) \in R^2 : y = x/n \quad, 0 \leq x \leq 1\}$$

for some positive integer N is referred to as a *star*.

Theorem 3.4. (Magill [13]) *Suppose X is a local dendrite with finite branch number which is not a star and has a unique branch point of greatest rank. Then $S(X)$ has no largest proper congruence.* □

Conjecture 3.5. Let X be a local dendrite with finite branch number. Then $S(X)$ has a largest proper congruence if and only if X is an arc.

Theorem 3.6. *$S(R)$ has no largest proper congruence where R is the space of real numbers.* □

Proof. Let

$$V = \{P \circ h : h \text{ is an increasing homeomorphism from } R \text{ onto } R \text{ and } P \text{ is an odd degree polynomial}\}.$$

It follows from Theorems 2.1 and 2.5 of [1] that $S(X)\backslash V$ is a prime ideal of $S(X)$. Next let W consist of all those elements in $S(X)$ which are not homeomorphisms from R onto R. In view of Theorem 3.5 of [5] W is also a prime ideal of $S(X)$. Then let

$$\sigma = (V \times V) \cup \big((S(X)\backslash V) \times \big(S(X)\backslash V)\big)$$

and

$$\rho = (W \times W) \cup \big((S(X)\backslash W) \times \big(S(X)\backslash W)\big) .$$

Both σ and ρ are distinct maximal congruences so we see that $S(R)$ cannot have a largest proper congruence.

Conjecture 3.7. $S(R^N)$ has no largest proper congruence for each Euclidean N-space R^N.

Although there are many spaces with no largest proper congruence, we will see that it is really quite rare for a space to have no smallest proper congruence. In what follows, X will be a set and $T(X)$ will be any semigroup of selfmaps of X which contains the identity map and all constant maps.

Definition 3.8. An equivalence relation ρ on X is a *T-equivalence* if $\big(f(x), f(y)\big) \in \rho$ for all $(x,y) \in \rho$ and $f \in T(X)$. The collection of all T-equivalences on X will be denoted by Teq(X).

The complete lattice of congruences on $T(X)$ will be denoted by $\mathrm{Con}\big(T(X)\big)$. For each $R \in \mathrm{Con}\big(T(X)\big)$, we let $\gamma(R) = \{(x,y) \in X \times X : (\langle x\rangle, \langle y\rangle) \in R\}$ where $\langle x\rangle$ and $\langle y\rangle$ denote the constant functions which map everything into x and y, respectively. For any set Y, we let $\Delta(Y) = \{(y,y) : y \in Y\}$. For each $\rho \in \mathrm{Teq}(X)$, let $C(\rho) = \{(\langle x\rangle, \langle y\rangle) : (x,y) \in \rho\} \cup \Delta\big(T(X)\big)$.

Theorem 3.9. (Hofmann and Magill [8]) *γ is a monotone map from* $\operatorname{Con}(T(X))$ *into* $\operatorname{Teq}(X)$ *and C is a monotone map from* $\operatorname{Teq}(X)$ *into* $\operatorname{Con}(T(X))$ *such that $\gamma \circ C$ is the identity map on* $\operatorname{Teq}(X)$. *Moreover, $\rho \subseteq \gamma(R)$ if and only if $C(\rho) \subseteq R$ for each $\rho \in \operatorname{Teq}(X)$ and $R \in \operatorname{Con}(T(X))$ so that (ρ, C) is a covariant Galois connection between* $\operatorname{Con}(T(X))$ *and* $\operatorname{Teq}(X)$. □

Let $\operatorname{ACon}(T(X))$ and $\operatorname{ATeq}(X)$ denote the collections of atoms of $\operatorname{Con}(T(X))$ and $\operatorname{Teq}(X)$, respectively. Although the map C is evidently injective, the map γ is generally far from being injective so we cannot hope for $C \circ \gamma$ to be the identity map on $\operatorname{Con}(T(X))$. However, we do have

Theorem 3.10. (Hofmann and Magill [8]) *γ maps* $\operatorname{ACon}(T(X))$ *into* $\operatorname{ATeq}(X)$ *and C maps* $\operatorname{ATeq}(X)$ *into* $\operatorname{ACon}(T(X))$. *Moreover, $\gamma \circ C$ is the identity map on* $\operatorname{ATeq}(X)$ *and $C \circ \gamma$ is the identity map on* $\operatorname{ATeq}(X)$ *so that γ actually maps* $\operatorname{ACon}(T(X))$ *bijectively onto* $\operatorname{ATeq}(X)$ *and, similarly, C maps* $\operatorname{ATeq}(X)$ *bijectively onto* $\operatorname{ACon}(T(X))$. □

This result is then used to prove

Theorem 3.11. (Hofmann and Magill [8]) *$T(X)$ has a smallest proper congruence if and only if X has a smallest proper T-equivalence. Furthermore, if A is the smallest proper congruence on $T(X)$ then $\gamma(A)$ is the smallest proper T-equivalence on X. Similarly, if α is the smallest proper T-equivalence on X, then $C(\alpha)$ is the smallest proper congruence on $T(X)$.* □

The latter result reduces the problem of finding the smallest proper congruence on $T(X)$ (if it exists) to that of finding the smallest T-equivalence on X. For each pair of distinct points $x, y \in X$, let

$$\eta(x, y) = \{(f(x), f(y)) : f \in T(X)\} \cup \{(f(y), f(x)) : f \in T(X)\}$$

and let $\eta_e(x, y)$ be the smallest equivalence relation on X containing $\eta(x, y)$. Next let

$$\sigma = \bigcup \{\eta_e(x, y) : \quad x, y \in X \quad \text{and} \quad x \neq y\}$$

and we have

Theorem 3.12. (Hofmann and Magill [8]) *X has a smallest proper T-equivalence if and only if $\sigma \neq \Delta(X)$. Moreover, when X does have a smallest proper T-equivalence, that equivalence is σ.* □

There is another equivalence relation on X which, for many of the semigroups $T(X)$, coincides with the T-equivalence σ. Let

$$\mu = \{(x,y) \in X \times X : \text{ for each pair of distinct points } a, b \in X, \text{ some } f \in T(X) \text{ maps } \{a,b\} \text{ onto } \{x,y\}\}$$

and let μ_e denote the smallest equivalence relation containing μ.

Theorem 3.13. (Hofmann and Magill [8]) *Either* $\mu_e = \Delta(X)$ *or* $\mu_e = \sigma$. □

So if one can show that $\mu_e \neq \Delta(X)$, one has verified not only that there is a smallest proper T-equivalence on X but that it is, in fact μ_e. The relation μ_e seems to be easier to deal with than the relation σ. The problem is that there are instances ([8] Example 2.21) where X does have a smallest proper T-equivalence but $\mu_e = \Delta(X)$. Nevertheless, these instances do seem to be rare.

The latter results can then be used to determine the smallest proper congruence on $S(X)$ for a large number of spaces X. Let

$$\sigma(\mathcal{G}) = \{(\langle x\rangle, \langle y\rangle) : x \text{ and } y \text{ belong to precisely the same open subsets of } X\}.$$

Theorem 3.14. (Hofmann and Magill [8]) *Suppose* X *is not* T_0. *Then* $\sigma(\mathcal{G})$ *is the smallest proper congruence on* $S(X)$. □

Let us recall that a space is said to be *Aleksandrov discrete* if it is T_0 and arbitrary intersections of open sets are open. For such a space X and each point $x \in X$, let G_x denote the intersections of all open sets containing x and define a relation $\mathcal{A}$ on X by

$$\mathcal{A} = \{(x,y) \in X \in X : G_x \cap G_y \neq \emptyset\} .$$

Then let $\mathcal{A}_e$ be the smallest equivalence relation on X containing A and we have

Theorem 3.15. (Hofmann and Magill [8]) *Let* X *be an Aleksandrov discrete space. Then* $S(X)$ *has a smallest proper conruence* σ. *If* X *is discrete, then*

$$\sigma = \{(\langle x\rangle, \langle y\rangle) : (x,y) \in X \times X\} \cup \Delta(S(X))$$

while

$$\sigma = \{(\langle x\rangle, \langle y\rangle) : (x,y) \in \mathcal{A}_e\} \cup \Delta(S(X))$$

if X *is not discrete.* □

It follows immediately from the last two results that the semigroup of every finite topological space has a smallest proper conruence. Next we recall that a space is said to be *totally separated* if for each pair of distinct points, there is a clopen set containing one of them but not the other. For any space X let

$$\mathcal{P} = \{(x,y) \in X \times X : \text{there is a path in } X \text{ from } x \text{ to } y\}.$$

Theorem 3.16. (Hofmann and Magill [8]) *Suppose X is either totally separated or is completely regular, Hausdorff and contains an arc. Then $S(X)$ has a smallest proper congruence σ. In the former case*

$$\sigma = \{(\langle x\rangle, \langle y\rangle) : (x,y) \in X \times X\} \cup \Delta(S(X))$$

while in the latter case

$$\sigma = \{(\langle x\rangle, \langle y\rangle) : (x,y) \in \mathcal{P}\} \cup \Delta(S(X)) . \qquad \square$$

The kernel $K(X)$ of $S(X)$ consists of all the constant functions and we denote by $\sigma(K(X))$ the Rees factor congruence which identifies all elements in $K(X)$ to a point.

Theorem 3.17. (Hofmann and Magill [8]) *Let X be any locally connected, compact N-dimensional subspace of Euclidean N-space whose components all have the internal extension property. Then the following statements are equivalent.*
(3.17.1) *$\sigma(K(X))$ is the smallest proper congruence on $S(X)$.*
(3.17.2) *$S(X)$ has a largest proper congruence.*
(3.17.3) *$S(X)$ has no prime ideals.*
(3.17.4) *X is pathwise connected.* $\square$

In [7] DE GROOT proved the existence of 2^c 1-dimensional, connected, locally connected subspaces of the Euclidean plane whose semigroups consist entirely of the constant functions together with the identity map. No two of the spaces are homeomorphic but their semigroups are all mutually isomorphic, being left zero semigroups with identities of equal cardinality. Let X be any one of these space. The congruences on $S(X)$ which differ from the universal congruences are precisely the relations of the form $\rho \cup \{(e,e)\}$ where e is the identity map and ρ is an equivalence relation on $S(X) \backslash \{e\}$. Evidently $S(X)$ has no smallest proper congruence. This fact also follows from two of our previous results. Let $\eta(x,y)$, $\eta_e(x,y)$ and σ be as defined in the discussion preceding Theorem 3.12. It is immediate that for each pair of distinct points $x, y \in X$, $\eta_e(x,y) = \eta(x,y) = \Delta(X) \cup \{(x,y),(y,x)\}$

and therefore $\sigma = \Delta(X)$. It then follows from Theorem 3.11 that X has no smallest proper T-equivalence and thus from Theorem 3.12 that $S(X)$ has no smallest proper conruence. $S(X)$, however, does have a largest proper congruence and that is $(S(X)\backslash\{e\}) \times (S(X)\backslash\{e\}) \cup \{(e,e)\}$.

4. Unifying congruences on *S*(*X*)

The definition of unifying congruence was motivated by the largest proper congruence on $S(I)$ introduced by Shutov in [21]. A family $\mathcal{A}$ of subsets of X is referred to as a *unifying family* if for each $A \in \mathcal{A}$ and $f \in S(X)$, if f is injective on A, then $f[A] \in \mathcal{A}$. For each unifying family $\mathcal{A}$ of subsets of X one can associate, in a natural way, a congruence $\sigma(\mathcal{A})$ on $S(X)$. Define $(f,g) \in \sigma(\mathcal{A})$ if anytime one of the functions is injective on some $A \in \mathcal{A}$, then the two functions agree on A. It is a routine matter to verify that $\sigma(\mathcal{A})$ is a congruence on $S(X)$ and we refer to such a congruence as a unifying congruence.

There is an abundance of unifying congruences on $S(X)$ for most spaces X. However, the closed unit interval I is an exception. In the following result, $\Delta\big(S(I)\big)$ is the diagonal of $S(I) \times S(I)$, α is the Rees factor congruence obtained by identifying all constant functions on I to a point, β is the Rees factor congruence obtained by identifying all those functions which are not injective on any nondegenerate subinterval to a point and $\mathcal{A}$ is the collection of all subarcs of I.

Theorem 4.1. (Magill, Misra and Tewari [15]) *Let $\sigma(\mathcal{B})$ be any unifying congruence on $S(I)$ where the unifying family $\mathcal{B}$ consists of closed subsets. Then $\sigma(\mathcal{B})$ coincides with either $\Delta\big(S(I)\big)$, α,β or $\sigma(\mathcal{A})$.* □

The function σ from the collection of unifying families to the collection of congruences is many-to-one. We do have some results which tell us when $\sigma(\mathcal{A}) = \sigma(\mathcal{B})$ for certain unifying families $\mathcal{A}$ and $\mathcal{B}$ but first we need some terminology. An open subset G of a space X is said to be *pliant* if there exists a map f from $c\ell\ G$ into X such that $f(x) = x$ for each $x \in \mathcal{F}r\ G$ and $f(a) \neq a$ for some $a \in G$ where $c\ell$ and $\mathcal{F}r$ denote closure and frontier, respectively. The space X is *malleable* if it has a basis of pliant sets. A unifying family $\mathcal{A}$ of subsets of X will be referred to as an *$\mathcal{R}$-unifying* family if it consists of malleable retracts of X.

The next result shows us how to determine when $\sigma(\mathcal{A}) = \sigma(\mathcal{B})$ when $\mathcal{A}$ and $\mathcal{B}$ are $\mathcal{R}$-unifying families.

Theorem 4.2. (Magill [10]) *Let $\mathcal{A}$ and $\mathcal{B}$ be two $\mathcal{R}$-unifying families. Then $\sigma(\mathcal{A}) \subseteq \sigma(\mathcal{B})$ if and only if each $B \in \mathcal{B}$ is the closure of the union of members of $\mathcal{A}$. Consequently, $\sigma(\mathcal{A}) = \sigma(\mathcal{B})$ if and only if each $A \in \mathcal{A}$ is the closure of the union of* *members of $\mathcal{B}$ and, similarly, each $B \in \mathcal{B}$ is the closure of the union of members of $\mathcal{A}$.* □

Example 4.3. Let

$$H = \{(x,y) \in I^2 : 1/4 \leq x \leq 3/4 \quad \text{and} \quad 0 \leq y \leq 1\}$$
$$V = \{(x,y) \in I^2 : 0 \leq x \leq 1/4 \quad \text{and} \quad y = 0\}$$
$$W = \{(x,y) \in I^2 : 3/4 \leq x \leq 1 \quad \text{and} \quad y = 0\} .$$

Then let $A = V \cup H$ and $B = V \cup H \cup W$. Finally, let $\mathcal{A}$ consist of all subspaces of I^2 which are homeomorphic to A and let $\mathcal{B}$ consist of all subspaces of I^2 which are homeomorphic to B. Both $\mathcal{A}$ and $\mathcal{B}$ are $\mathcal{R}$-unifying families which are completely distinct and yet $\sigma(\mathcal{A}) = \sigma(\mathcal{B})$ according to Theorem 4.2.

Example 4.4. Let

$$X_n = \cup\{(x,y) \in R^2 : y = x/j \ , \quad 0 \leq x \leq 1, \quad j = 1, \ldots n\}$$

and let $\mathcal{A}_n$ denote the collection of all subspaces of the Euclidean N-cell I^N which are homeomorphic to X_n. Each of these spaces is a dendrite (in fact, they are stars) and since dendrites are absolute retracts [4, p. 138], each family $\mathcal{A}_n$ is an $\mathcal{R}$-unifying family and it follows from Theorem 3.2 that $\{\sigma(\mathcal{A}_n)\}_{n=2}^{\infty}$ is an infinite collection of distinct unifying congruences on $S(I^N)$ so that this is another instance where the higher dimensional N-cells differ from I. Actually, it follows from Theorem 4.2 that $\{\sigma(\mathcal{A}_n)\}_{n=2}^{\infty}$ is an infinite chain.

At this point we know of only finitely many congruences on $S(I)$, i.e., those described in Theorem 4.1, and this leads to

Problem 4.5. Determine if $S(I)$ has infinitely many congruences.
Of course, if we could find infinitely many ideals in $S(I)$ we would have infinitely many congruences, namely the corresponding Rees factor congruences. Right now, we only know of two ideals of $S(I)$. The first is $K(I)$ the collection of all constant functions and the second is the family J of all functions which are not injective on any nondegenerate subinterval of I. We had observed previously that $K(I)$ is the smallest ideal of $S(I)$. At any rate, this is quite easy to verify. It is also not difficult to verify that J is the largest ideal of $S(I)$ which, in any event, follows immediately from Theorem 3.3 of [5]. Consequently, any other ideal of $S(I)$ must lie between $K(X)$ and J. The ideal J certainly contains many elements in addition to the constant functions. For example, every continuous nowhere differentiable selfmap of I belongs to J. This is an immediate consequence of the fact that if a continuous function f is injective on some nondegenerate subinterval A, then

f is strictly monotonic on A and by a well-known result in measure theory, is differentiable almost everywhere on A. To say that $K(X)$ and J are the only two ideals of $S(I)$ would be to say that J is a principal ideal and is generated by any continuous nowhere differentiable selfmap of I. It is difficult to believe that this would be true. Anyway, this leads to

Problem 4.6. Determine if $S(I)$ has infinitely many ideals.
Evidently, a positive answer to 4.6 implies a positive answer to 4.5 while a negative answer to 4.6 does nothing to settle 4.5. On the other hand, a negative answer to 4.5 implies a negative answer to 4.6 while a positive answer to 4.5 need not settle 4.6 (it depends upon what the congruences look like).

We next turn our attention to $\mathcal{R}$-unifying families consisting of local dendrites with finite branch number. The next result is similar in spirit to Theorem 4.2 but it is somewhat easier to apply.

Theorem 4.7. (Magill [10]) *Let X be any topological space and let $\mathcal{A}$ and $\mathcal{B}$ be $\mathcal{R}$-unifying families consisting of local dendrites with finite branch number. Then $\sigma(\mathcal{A}) \subseteq \sigma(\mathcal{B})$ if and only if each $B \in \mathcal{B}$ is the union of sets from $\mathcal{A}$. Consequently, $\sigma(\mathcal{A}) = \sigma(\mathcal{B})$ if and only if each $A \in \mathcal{A}$ is the union of sets from $\mathcal{B}$ and, similarly, each $B \in \mathcal{B}$ is the union of sets from $\mathcal{A}$.* □

We need to recall some terminology from [10]. For any collection $\mathcal{A}$ of subspaces of X, we denote by $\langle \mathcal{A} \rangle$ the collection of all subspaces of X which are homeomorphic to some $A \in \mathcal{A}$. If $\mathcal{A}$ is a subcollection of some $\mathcal{R}$-unifying family of compact subsets of X, then $\langle \mathcal{A} \rangle$ is also an $\mathcal{R}$-unifying family of subsets of X and if $\mathcal{A}$ is an $\mathcal{R}$-unifying family, then $\langle \mathcal{A} \rangle = \mathcal{A}$. If $\mathcal{A}$ is a unifying family of compact subsets but not an $\mathcal{R}$-unifying family, it may happen that $\langle \mathcal{A} \rangle$ properly contains $\mathcal{A}$. A subfamily $\mathcal{A}_1$ of an $\mathcal{R}$-unifying family $\mathcal{A}$ is said to be a set of generators for $\mathcal{A}$ if $\langle \mathcal{A}_1 \rangle = \mathcal{A}$. If $\mathcal{A}$ contains a finite set of generators, then we say that $\mathcal{A}$ is *finitely generated.* Finally, a nonempty collection of nonempty subsets $\mathcal{A}$ of a topological space X is said to be *independent* if no $A \in \mathcal{A}$ is the union of copies of sets from $\mathcal{A} \backslash \{A\}$.

Theorem 4.8. (Magill [10]) *Let $\mathcal{A}$ be an $\mathcal{R}$-unifying family of compact subsets of X with a finite set of generators $\mathcal{A}_1$. Then $\mathcal{A}_1$ contains an independent subcollection $\mathcal{A}_2$ such that $\sigma(\mathcal{A}) = \sigma(\langle \mathcal{A}_2 \rangle)$.* □

We observed previously that σ is in general a many-to-one map. However, the next two results show that it is one-to-one on a certain subcollection of unifying families.

Theorem 4.9. (Magill [10]) *Let $\mathcal{A}$ and $\mathcal{B}$ be two independent finite subcollections of some $\mathcal{R}$-unifying family of local dendrites with finite branch numbers. Then the following statements are equivalent.*

(4.9.1) $\sigma(\langle\mathcal{A}\rangle) = \sigma(\langle\mathcal{B}\rangle)$.

(4.9.2) *There is a one-to-one correspondence between the sets in $\mathcal{A}$ and those in $\mathcal{B}$ such that corresponding sets are homeomorphic.*

(4.9.3) $\langle\mathcal{A}\rangle = \langle\mathcal{B}\rangle$. □

The next result is an immediate consequence of Theorems 4.8 and 4.9.

Corollary 4.10. *Let $\mathcal{A}$ and $\mathcal{B}$ be finite subsets of some $\mathcal{R}$-unifying family of local dendrites with finite branch numbers. Then $\sigma(\langle\mathcal{A}\rangle) = \sigma(\langle\mathcal{B}\rangle)$ if and only if $\langle\mathcal{A}\rangle = \langle\mathcal{B}\rangle$.* □

5. The partially ordered family of continuum congruences

If a unifying family $\mathcal{A}$ consists of subcontinua of X then $\sigma(\mathcal{A})$ is referred to as a *continuum congruence* and the partially ordered family of all continuum congruences on $S(X)$ will be denoted by $\mathrm{Con}_C(S(X))$. A collection $\mathcal{K}_c$ of subcontinua of X is referred to as a *characteristic* collection if no two distinct continua in $\mathcal{K}_c$ are homeomorphic and each subcontinuum of X is homeomorphic to a subcontinuum in $\mathcal{K}_c$. Finally, denote by $\mathrm{Ind}(\mathcal{K}_c)$ the family of all independent subcollections of $\mathcal{K}_c$. If X happens to be a local dendrite with finite branch number we can partially order $\mathrm{Ind}(\mathcal{K}_c)$ by defining $\mathcal{A} \leq \mathcal{B}$ if each $B \in \mathcal{B}$ is the union of copies of sets from $\mathcal{A}$. It is immediate that this relation is reflexive and transitive and it follows from Theorem 4.9 that it is also antisymmetric.

Theorem 5.1. (Magill [10]) *Let X be a local dendrite with finite branch number. Then $\mathrm{Con}_C(S(X))$ is order isomorphic to $\mathrm{Ind}(\mathcal{K}_c)$.* □

One then uses this result to determine precisely when $\mathrm{Con}_C(S(X))$ is a lattice. In any event, it is finite in view of Theorem 2.1 of [12].

Theorem 5.2. (Magill [10]) *Let X be a local dendrite with finite branch number. Then $\mathrm{Con}_C(S(X))$ is always a lower semilattice and it is a lattice if and only if X is the union of copies of K for every subcontinuum K of X.* □

Example 5.3. We borrow this example from [10]. Let

$$X = \{(x, y) \in R^2 : x^2 + y^2 = 1\} \cup \{(0, y) \in R^2 : 1 \leq y \leq 2\}.$$

A characteristic family $\mathcal{K}_c$ of subcontinua of X is obtained by choosing a point, an arc, a triod, a simple closed curve and X itself. These subcontinua are represented by the following figures.

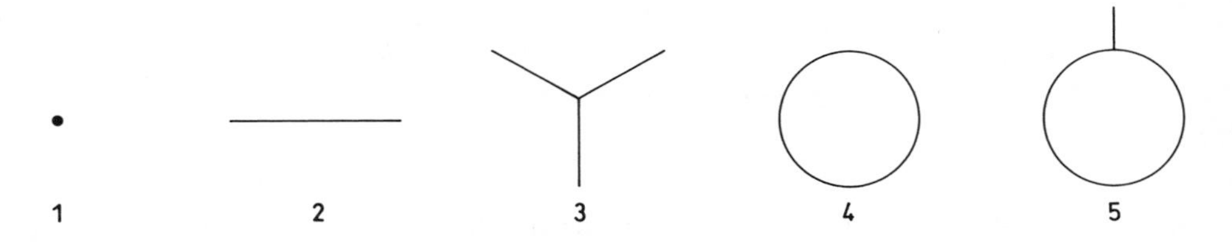

We denote these subcontinua by their corresponding integers.
Thus $\mathcal{K}_c = \{1, 2, 3, 4, 5\}$ and one sees that

$$\operatorname{Ind}(\mathcal{K}_c) = \{\{1\}, \{2\}, \{3\}, \{4\}, \{5\}, \{3, 4\}, \{4, 5\}\} .$$

It is evident that X is not the union of copies of 4 so according to Theorem 5.2$\operatorname{Con}_C\big(S(X)\big)$ is not a lattice. By Theorem 5.1 it is isomorphic to $\operatorname{Ind}(\mathcal{K}_c)$ so its diagram is as follows.

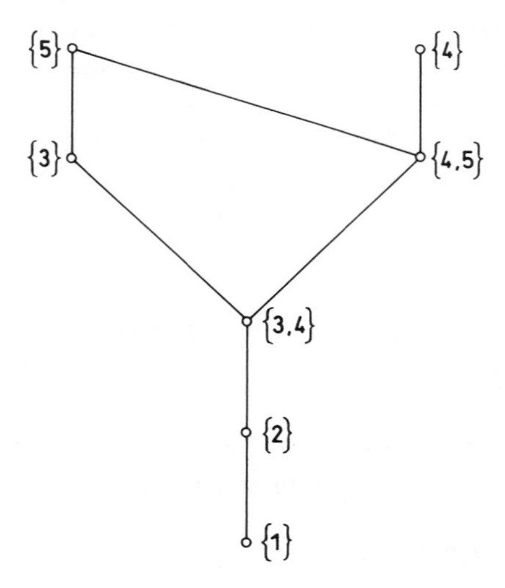

Example 5.4. In this and in some of the subsequent discussions, we will be dealing with subspaces of the Euclidean plane. We think it will be apparent from the figures just what these spaces are, so in the interest of conserving space we will not give an analytic description of them. Let X be the subspace of the plane which is represented by the following figure.

Up to homeomorphism it has nine different subcontinua and they are represented by

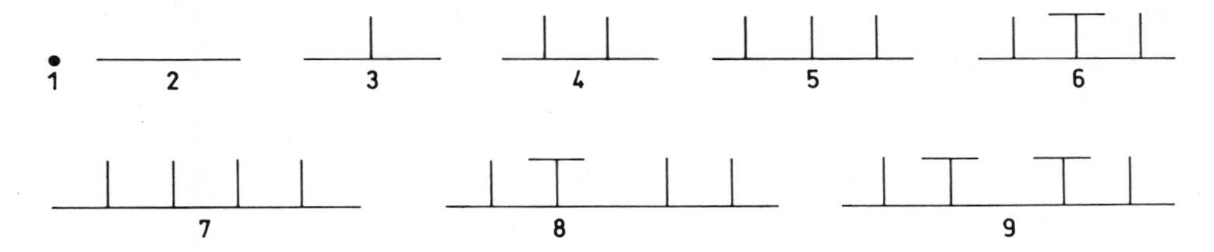

and one sees that in this case

$$\mathrm{Ind}(\mathcal{K}_c) = \{\{1\}, \{2\}, \{3\}, \{4\}, \{5\}, \{6\}, \{7\}, \{8\}, \{9\}, \{6, 7\}\}.$$

Since X is the union of copies of each of its subcontinua, we are assured by Theorem 5.2 that $\mathrm{Con}_C\big(S(X)\big)$ is a lattice. By inspecting $\mathrm{Ind}(\mathcal{K}_c)$, we see that its lattice is represented by the following diagram.

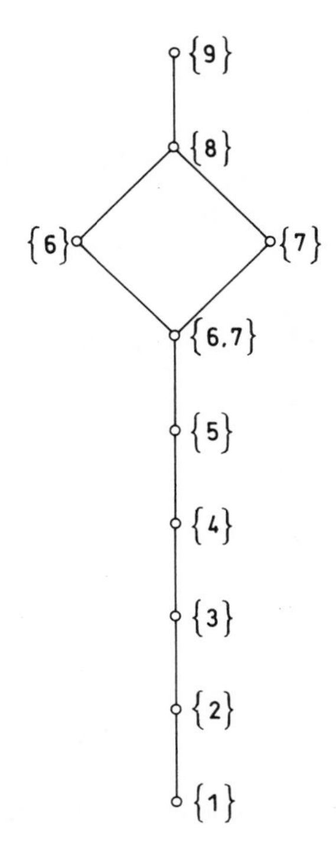

In the previous example, $\mathrm{Con}_C\big(S(X)\big)$ was almost a chain. Our next result characterizes those spaces within the Peano continua for which $\mathrm{Con}_C\big(S(X)\big)$ is a finite chain. But first we need to define a few things. $\mathcal{D}_R\big(S(X)\big)$ denotes the collection of regular $\mathcal{D}$-classes of $S(X)$. We define a $\mathcal{J}$-class to be regular it if contains at least one regular element and we let $\mathcal{J}_R\big(S(X)\big)$ denote the partially orderd family of all regular $\mathcal{J}$-classes of $S(X)$. That is, $J_f \leq J_g$ if the principal ideal generated by f is contained in that generated by g. A *bed of nails* is any space which is homeomorphic to

$$B_n = \{(k,y) : k = 1,\dots,n,\ 0 \leq y \leq 1\} \cup \{(x,0) : 0 \leq x \leq n+1\}$$

and a *triple* T is any space which is homeomorphic to

$$\begin{aligned} T_3 =&\{(x,0) : -2 \leq x \leq 2\} \cup \{(-1,y) : -1 \leq y \leq 0\} \cup \\ &\{(1,y) : -1 \leq y \leq 0\} \cup \{(0,y) : 0 \leq y \leq 1\} \cup \\ &\{(x,1) : -1 \leq x \leq 1\} \end{aligned}$$

Recall that a *star* was defined preceding Theorem 3.4.

Theorem 5.5. (Magill [12]) *The following statements are equivalent for a Peano continuum X.*

(5.5.1) $\mathrm{Con}_C\big(S(X)\big)$ *is a finite chain.*

(5.5.2) $\mathcal{J}_R\big(S(X)\big)$ *is a finite chain.*

(5.5.3) $\mathrm{Con}_C\big(S(X)\big)$ *and* $\mathcal{J}_R\big(S(X)\big)$ *are finite and have the same number of elements.*

(5.5.4) $\mathrm{Con}_C\big(S(X)\big)$ *and* $\mathcal{D}_R\big(S(X)\big)$ *are finite and have the same number of elements.*

(5.5.5) *X is either a point, a simple closed curve, a triple T, a star or a bed of nails.* □

The equivalence of (5.5.2) and (5.5.5) was first established by Subbiah and myself for plane Peano continua in [17].

If $\mathcal{A}$ is a unifying family which consists of all the copies of one particular subcontinuum of X, then $\sigma(\mathcal{A})$ is referred to as a *principal continuum congruence* and the partially ordered family of all principal continuum congruences on $S(X)$ will be denoted by $\mathrm{Con}_{PC}\big(S(X)\big)$.

A local subsemigroup of a semigroup S is any subsemigroup of the form vSv where v is an idempotent of S. Denote the collection of all local subsemigroups of S by Loc(S) and define an equivalence relation ε on Loc(S) by $(vSv, wSw) \in \varepsilon$ if each of the subsemigroups can be embedded in the other. Let ELoc(S) denote the collection of equivalence classes and partially order ELoc(S) by $\varepsilon(vSv) \leq \varepsilon(wSw)$ (where the latter represent equivalence classes) if vSv can be embedded in wSw.

In the statement of the following theorem, a number of the spaces under consideration will contain subspaces which are stars. Such a subspace will be represented by any one of the following figures.

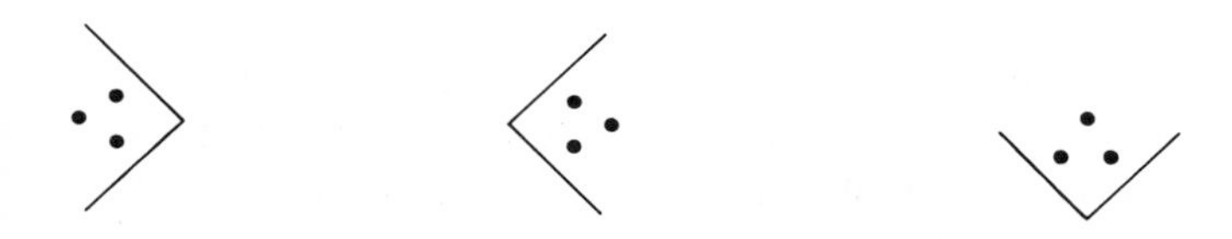

Theorem 5.6. (Magill and Misra [14]) *The following statements are equivalent for any Peano continuum X.*

(5.6.1) $\mathrm{Con}_{PC}(S(X))$ *and* $\mathcal{J}_R(S(X))$ *are isomorphic finite lattices.*

(5.6.2) $\mathrm{Con}_{PC}(S(X))$ *and* $\varepsilon\mathrm{Loc}(S(X))$ *are isomorphic finite lattices.*

(5.6.3) *X is either a simple closed curve or is homeomorphic to a subcontinuum of one of the spaces represented by the following figures.* □

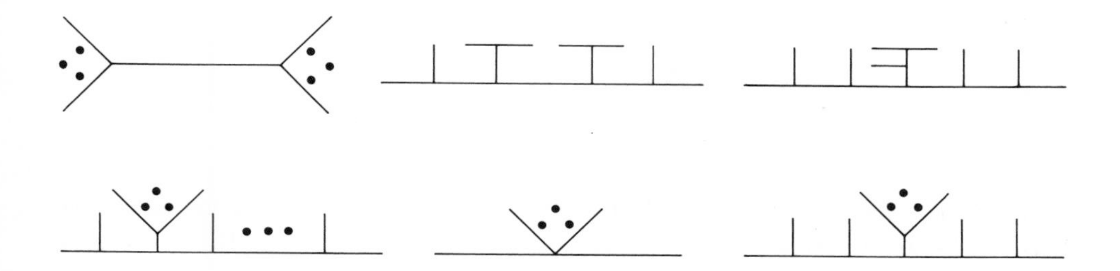

The first figure in the second row may have any finite number of "upright spikes".

Corollary 5.7. (Magill and Misra [14]) *Let X be any nondegenerate Peano continuum with no cut points. Then the following statements are equivalent.*

(5.7.1) $\mathrm{Con}_{PC}(S(X))$ *and* $\mathcal{J}_R(S(X))$ *are isomorphic finite lattices.*

(5.7.2) $\mathrm{Con}_{PC}(S(X))$ *and* $\mathrm{ELoc}(S(X))$ *are isomorphic finite lattices.*

(5.7.3) *X is a simple closed curve.* □

6. Congruences and the relation *V*

The relation V was introduced by SCHEIN in [19], and that paper was subsequently translated into English [20]. It is defined for any semigroup S as follows:

$$V = \{(a,b) \in S \times S : a \text{ and } b \text{ are mutually inverse to one another}\}.$$

In [9] KOCH and MADISON investigated those congruences on regular semigroups which are under $\mathcal{L}$ and posed the problem of characterizing the congruences which commute with V. This problem has been solved for a great many $S(X)$. For any $f \in S(X)$ let

$$\pi(f) = \{f^{-1}(y) : y \in \text{Ran } f\}$$

where Ran f denotes the range of f and then let

$$\begin{aligned}
\mathcal{E} &= \{(x,y) \in X \times X : x \text{and} y \text{belongto} \\
&\quad \text{thesamecomponentof} X\} \\
\beta_1 &= \{(f,g) : \pi(f) = \pi(g) \text{andRan } f = \text{Ran } g = \\
&\quad \{a,b\} \text{with} a \neq b \text{and} (a,b) \in \mathcal{E}\} \\
\beta_2 &= \{(f,g) : \pi(f) = \pi(g), \text{Ran } f = \text{Ran } g \\
&\quad \text{and} |\text{Ran } f| = 2\} \\
\alpha_1 &= \{(\langle x\rangle, \langle y\rangle) : (x,y) \in E\} \cup \Delta\big(S(X)\big) \\
\alpha_2 &= \{(\langle x\rangle, \langle y\rangle) : (x,y) \in X \times X\} \cup \Delta\big(S(X)\big) \\
\alpha_3 &= \alpha_1 \cup \beta_1, \ \ \alpha_4 = \alpha_2 \cup \beta_1, \ \ \alpha_5 = \alpha_2 \cup \beta_2.
\end{aligned}$$

One shows without too much effort that each α_i, $1 \le i \le 5$ is a congruence on $S(X)$. Moreover, these five congruences are all distinct if and only if X is not connected. If X is connected, they all coincide.

Theorem 6.1. (Barit, Magill and Subbiah [3]) *Let X be a nondiscrete, locally connected, normal, Hausdorff space whose components are all pathwise connected and let σ be any congruence on $S(X)$. Then the following statements are equivalent.*

(6.1.1) $\sigma \subseteq \mathcal{L}$.

(6.2.2) $\sigma \subseteq \mathcal{D}$.

(6.1.3) $V \circ \sigma = \sigma \circ V$.

(6.1.4) *Either* $\sigma = \Delta\big(S(X)\big)$ *or* $\sigma = \alpha_1$ *for some* i *where* $1 \le i \le 5$. □

The equivalence of (6.1.3) and (6.1.4) was originally established in [2]. As we pointed out in [3], one shows quite easily that the only congruence on $S(X)$ under

$\mathcal{R}$ is $\Delta(S(X))$. Next we let $\mathcal{K}$ denote the collection of all compact subspaces K of X with the property that each component of K is *nondegenerate, open* (*in* K) and *pathwise connected.* A unifying family consisting of members of $\mathcal{K}$ will be referred to as a $\mathcal{K}$-*unifying family* and the corresponding congruence will be referred to as a $\mathcal{K}$-*unifying congruence.* Let X be any completely regular Hausdorff space which contains an arc and let $\sigma(\mathcal{B})$ be a $\mathcal{K}$-unifying congruence. Since $\sigma(\mathcal{B}) \neq \Delta(S(X))$, it follows from one of our previous observations that $\sigma(\mathcal{B}) \not\subseteq \mathcal{R}$. We also have

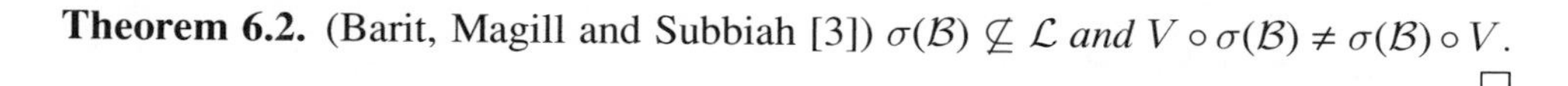

Theorem 6.2. (Barit, Magill and Subbiah [3]) $\sigma(\mathcal{B}) \not\subseteq \mathcal{L}$ *and* $V \circ \sigma(\mathcal{B}) \neq \sigma(\mathcal{B}) \circ V$. □

It would seem that $\mathcal{K}$-unifying congruences have little to do with Green's relations or the V relation. Suppose, however, we let , $\rho = R \times R$ where R denotes the collection of all nonconstant regular elements of $S(X)$ and let $\sigma(\mathcal{A})$ denote the $\mathcal{K}$-unifying congruence on $S(X)$ where $\mathcal{A}$ consists of all subarcs of X. Then we have

Theorem 6.3. (Barit, Magill and Subbiah [3]) *Let X be any nondegenerate pathwise connected normal space and let $\sigma(\mathcal{B})$ be any $\mathcal{K}$-unifying congruence on $S(X)$. Then the following statements are equivalent.*

(6.3.1) $$\sigma(\mathcal{B}) \cap \rho \subseteq \mathcal{R}.$$

(6.3.2) $$\big(V \circ \sigma(\mathcal{B})\big) \cap \rho = V \cap \rho.$$

(6.3.3) $$(\sigma(\mathcal{B}) \circ V) \cap \rho = V \cap \rho.$$

(6.3.4) $$\sigma(\mathcal{B}) = \sigma(\mathcal{A}).$$ □

Problem 6.4. Determine the complete lattice of congruences on $S(X)$ (different classes of spaces are likely to require different techniques). We know some congruences on $S(X)$ which are not unifying congruences. Are they abundant or not? At this point, we just don't know.

References

[1] Baird, B. B. and K.D. Magill, Jr., Green's $\mathcal{R}$-, $\mathcal{D}$-, and $\mathcal{H}$-relations for generalized polynomials, to be submitted

[2] Barit, W., K. D. Magill, Jr. and S. Subbiah, The congruences on $S(X)$ which commute with V, Semigroup Forum, 36 (1987), 351–364

[3] —, Green's relations, the V-relation, and congruences on $S(X)$, Semigroup Forum 37 (1988), 313–324

[4] Borsuk, K., "Theory of retracts", Polska Akademia Nauk Monografie Matematyczne, Polish Scientific Publishers, Warszawa, 1967

[5] Cezus, F. A., K. D. Magill, Jr. and S. Subbiah, Maximal ideals of semigroups of endomorphisms, Bull. Austral. Math. Soc. 12 (1975), 211–225

[6] Clifford, A. H. and G. B. Preston, "Algebraic theory of semigroups", Vol. II, Math. Surv., No. 7, Amer. Math. Soc., Providence, Rhode Island, 1967

[7] de Groot, J., Groups represented by homeomorphism groups, I, Math. Annalen 138 (1959), 80–102

[8] Hofmann, K. H. and K. D. Magill, Jr., The smallest proper congruence on $S(X)$, Glasgow Math. J. 30 (1988), 301–313

[9] Koch, R. J. and B. L. Madison, Congruences under $\mathcal{L}$ on regular semigroups, Simon Stevin 57 (1983), 273–283

[10] Magill, Jr., K. D., Congruences on semigroups of continuous selfmaps, Semigroup Forum 29 (1984), 159–182

[11] —, The largest proper congruence on $S(X)$, Internat. J. Math. & Math. Sci. 7 (1984), 663–666

[12] —, Semigroups for which the continuum congruences form finite chains, Semigroup Forum 30 (1984), 221–230

[13] —, On a family of ideals of $S(X)$, Semigroup Forum 35(1987), 321–339

[14] Magill, Jr., K. D. and P. R. Misra, Principal continuum congruences, regular $\mathcal{J}$-classes and local subsemigroups, to appear

[15] Magill, Jr., K. D. , P. R. Misra and U. B. Tewari, Unifying congruences on $S(I)$, Semigroup Forum 21 (1980), 363–371

[16] —, Epimorphisms from $S(X)$ onto $S(Y)$, Can. J. Math. 38 (1986), 538–551

[17] Magill, Jr., K. D. and S. Subbiah, Regular $\mathcal{J}$-classes and semigroups of continua, Semigroup Forum 22 (1981), 159–179

[18] Mal'cev, A. I., Symmetric groupoids, Mat. Sbornik 31 (1952), 136–151

[19] Schein, B. M., Theory of semigroups and its applications, I, Izdat. Saratov Univ., Saratov, (1965), 286–324

[20] —, On the theory of inverse semigroups and generalized groups, Amer. Math. Soc. Transl. 113 (1979), 89–122

[21] Shutov, E. G., Homomorphisms of certain semigroups of continuous functions, Sibirsk Mat. Z. 4 (1963), 693–701

List of contributors

John W. Baker, Department of Pure Mathematics, University of Sheffield, Sheffield, S102TN, England

Christian Berg, Matematiske Institut, Københavns Universitet, Universitetsparken 5, DK-2100 København, Denmark

R. W. R. Darling, Department of Mathematics, University of South Florida, Tampa, Fl 33620-5700, USA

Herbert Heyer, Mathematisches Institut, Universität Tübingen, Auf der Morgenstelle 10, D-7400 Tübingen 1, West Germany

Joachim Hilgert, Mathematisches Institut, Universität Erlangen, Bismarckstr. 1 1/2, D-8520 Erlangen, West Germany

Neil Hindman, Department of Mathematics, Howard University, Washington, DC 20059, USA

Karl H. Hofmann, Technische Hochschule Darmstadt, Fachbereich Mathematik, Schlossgartenstr. 7, D-6100 Darmstadt, West Germany

Ivan Kupka, Département de Mathématiques, Université de Paris VI, 4, place Jussieu, F-75230 Paris, Cedex 05, France

Anthony To-Ming Lau, Department of Mathematics, University of Alberta, Edmonton, T6G2G1, Alberta, Canada

Jimmie D. Lawson, Department of Mathematics, Louisiana State University, Baton Rouge, LA 70803, USA

Kenneth D. Magill, Jr., Department of Mathematics, SUNY Buffalo, 106 Diefenbach Hall, Buffalo, NY 14214, USA

Michael W. Mislove, Department of Mathematics, Tulane University, New Orleans, LA 70118, USA

Arunava Mukherjea, Department of Mathematics, University of South Florida, Tampa, FL, 33620-5700, USA

John S. Pym, Department of Pure Mathematics, University of Sheffield, Sheffield, S102TN, England

Lex E. Renner, Department of Mathematics, University of Western Ontario, Middlesex College, London N6A5B7, Ontario, Canada

Wolfgang A. F. Ruppert, Institut für Mathematik und Angewandte Statistik, Universität für Bodenkultur, Gregor-Mendel-Str. 33, A-1180 Wien, Austria

Jean-Pierre Troallic, Département de Mathématiques, Faculté des Sciences et des Techniques, Université de Rouen, F-76130 Mont-Saint-Aignan, France